Conjugated Objects

Conjugated Objects

Developments, Synthesis, and Applications

edited by

Atsushi Nagai
Koji Takagi

PAN STANFORD PUBLISHING

Published by

Pan Stanford Publishing Pte. Ltd.
Penthouse Level, Suntec Tower 3
8 Temasek Boulevard
Singapore 038988

Email: editorial@panstanford.com
Web: www.panstanford.com

British Library Cataloguing-in-Publication Data
A catalogue record for this book is available from the British Library.

Conjugated Objects: Developments, Synthesis, and Applications

ISBN 978-981-4774-03-1 (Hardcover)
ISBN 978-1-315-22962-1 (eBook)

Printed in the USA

Contents

Preface

Even if I say that it's definitely used, a π-conjugated system is a system which isn't exaggerated in chemistry. π-Conjugated objects, constructed from a conjugated frame, contain connected π-orbitals with delocalized electrons in compounds with alternating single and multiple bonds, which, in general, may lower the overall energy of the molecule and increase stability. Lone pairs, radicals, or carbenium ions may be part of the objects.

Conjugation is the overlap of one π-orbital with another across an intervening σ-bond. Conjugated objects have a region of overlapping π-orbitals bridging the interjacent single bonds. They allow delocalization of π-electrons across all the adjacent aligned π-orbitals. The π-electrons do not belong to a single bond or atom but rather to a group of atoms. Examples of the largest conjugated objects are graphene, graphite, conducting polymers, and carbon nanotubes.

Conjugation is possible by means of alternating single and double bonds. As long as each contiguous atom in a chain has available π-orbitals, the system can be considered conjugated. For instance, furan is a five-membered ring with two alternating double bonds and an oxygen in position. Oxygen has two lone pairs, one of which occupies a π-orbital on that position, thereby maintaining the conjugation of the five-membered ring. The presence of a nitrogen in the ring or group α to the ring like a carbonyl group (C=O), an imine group (C=N), a vinyl group (C=C), or an anion will also suffice as a source of π-orbitals to maintain conjugation. There are also other ways of conjugation. Homoconjugation is an overlap of two π-systems separated by a nonconjugating group, such as CH_2. For example, the molecule CH_2=CH-CH_2-CH=CH_2 (1,4-pentadiene) is homoconjugated because the two C=C double bonds, which are a π-system because each double bond contains one π-bond, are separated by one CH_2 group.

Next, the introduction of well-known π-conjugated objects shows two examples, conjugated cyclic compounds, like aromatic and nonaromatic compounds, and conjugated systems in pigments, like pthalocyanine and prophyrin. First, cyclic compounds can be partly or completely conjugated. Annulenes, completely conjugated monocyclic hydrocarbons, may be aromatic, nonaromatic, or antiaromatic. Conjugated, planar, cyclic compounds that follow Hückel's rule are aromatic and exhibit an unusual stability. The classic example benzene has a system of 6 π-electrons, which form the benzene ring along the planar σ-ring with its 12 electrons. In nonaromatic compounds, not all compounds with alternating double and single bonds are aromatic. Cyclco-octatetraene, for instance, possesses alternating single and double bonds. The molecule typically adopts a tub conformation. Because the π-orbitals of the molecule do not align themselves well in this nonplanar molecule, the electrons are not as easily shared between the carbon atoms. The molecule can be still considered conjugated, but it is neither aromatic nor antiaromatic in order that it is not planar. Second, conjugated objects have unique and interesting properties that give rise to strong colors. Many pigments make use of conjugation electron systems, such as the long conjugated hydrocarbon chain in β-carotene, resulting in a strong orange color. When an electron in the system absorbs a photon of light of the right wavelength, it can be promoted to a higher energy level. Most of these electronic transitions are from one conjugated π-system molecular orbital (MO) with an even kind of symmetry to another conjugated π-system MO with an odd kind of symmetry; π to π*, but electrons from other states can also be promoted to a π-system MO (π to π* as happens in charge transfer complexes). Often a HOMO to LUMO transition is made by an electron if it is allowed by the selection rules for electromagnetic transitions. Conjugated objects of fewer than eight conjugated double bonds absorb only in the ultraviolet region and are colorless to the human eye. With every double bond added, the system absorbs photons of longer wavelength and lower energy, and the compound ranges from yellow to red color. Compounds that are blue or green typically do not rely on conjugated double bonds alone. This absorption of light in the ultraviolet to visible spectrum can be quantified using ultraviolet-visible spectroscopy and forms the basis for the entire field of photochemistry. Conjugated objects that are widely used for synthetic pigments and dyes are diazo and azo compounds and phthalocyanine. A general compound is given as an example here for the reader.

This book, *Conjugated Objects: Developments, Synthesis, and Applications,* has a selection of subjects that offer a plain compilation, as a textbook for graduate and undergraduate students, of basic learning and current trends in π-conjugated systems for successful development and application by all writers (young scientists) working in broad research areas, such as new design of unique π-conjugation, catalysts, self-assembly, charge transfer complexes, liquid crystalline molecules, supramolecules, and nanostructures, by using π-conjugated small and/or macro-objects, organically or electrochemically.

Atsushi Nagai
The University of Texas Southwestern Medical Center, USA
Institute for Molecular Science, Department of Materials Molecular Science, Japan

Chapter 1

Unique Electronic and Stereochemical Properties of Salen Complexes

Takuya Kurahashi
Institute for Molecular Science, National Institutes of Natural Sciences, Myodaiji, Okazaki, Aichi 444-8787, Japan
kurahasi@ims.ac.jp

Salen complexes, which bear diimine and diphenolate around metal, have been widely utilized as a basic skeleton for catalysts and materials. This chapter briefly describes syntheses and applications of salen complexes. The main focus of this chapter is to describe unique redox and stereochemistry of salen complexes, which are of fundamental importance for excellent functionalities.

1.1 Introduction

1.1.1 Syntheses of Salen Complexes

A salen ligand is readily obtained in a multigram scale by the condensation of a salicylaldehyde and a diamine, typically in

Conjugated Objects: Developments, Synthesis, and Applications
Edited by Atsushi Nagai and Koji Takagi

ISBN 978-981-4774-03-1 (Hardcover), 978-1-315-22962-1 (eBook)
www.panstanford.com

refluxing EtOH (Fig. 1.1).[1] Nonsymmetrical salen ligands could also be selectively prepared.[2] Resulting salen ligands are usually pure enough without a tedious purification procedure. Metal insertion reactions could be carried out under mild conditions, which is particularly important in preparing nonsymmetrical salen complexes without isomerization. The complexation with metal produces a robust structure that is resistant to the hydrolysis. The ease of synthesis is one of the great benefits in using the salen system.

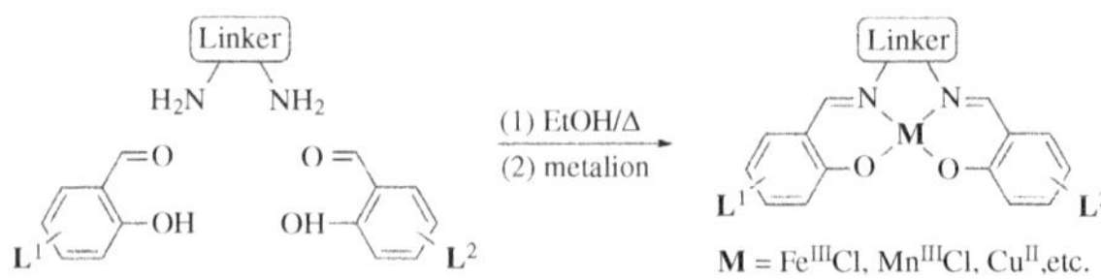

Figure 1.1 Preparation of salen complexes.

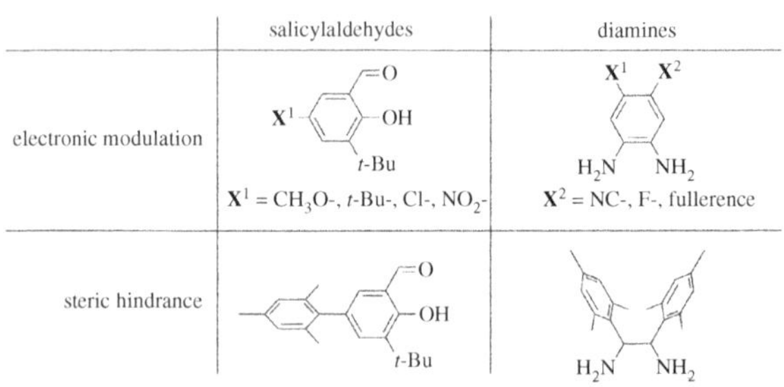

Figure 1.2 Selected salicylaldehydes and diamines that have been utilized for the synthesis of salen complexes.

Another important advantage is that metal complexes having various electronic and stereochemical properties are created simply changing salicylaldehydes and diamines. Figure 1.2 shows selected salicylaldehydes and diamines which have been used to prepare salen complexes. Electronic properties of salen complexes are greatly altered incorporating electron-donating or electron-withdrawing CH_3O-, *t*-Bu-, Cl-, and NO_2- substituents to the salicylaldehyde.[3] You can also use diamines to control electronic properties of salen complexes. Substituted *o*-phenylenediamines are typical examples. In addition to simple substituents such as F- and NC-,[4] C_{60} fullerene was successfully attached to the *o*-phenylenediamine for the synthesis of salen complexes.[5] Likewise, stereochemical properties are altered incorporating sterically demanding groups to the salicylaldehyde[6] or diamine.[7] Combining these salicylaldehydes and diamines in a modular manner, it is relatively easy to construct a wide variety of salen complexes to search for desirable functions. The salen system was even used as a compound library for one- and two-photon live cell imaging.[4]

1.1.2 Salen Complex as an Asymmetric Catalyst

When chiral components are introduced to salicylaldehydes and/or diamines, one could obtain well-defined chiral environments around metal in salen complexes. By use of chiral environments, a number of salen and related complexes have been successfully utilized as asymmetric catalysts. Those who are interested in this subject can refer to previous excellent reviews.[8–11]

Figure 1.3 shows a good example to see how the chiral environment is created using a salen complex.[12] This complex has characteristic axial chirality in the salicylaldehyde moiety in addition to the chiral ethylene diamine moiety. As you can see from the X-ray crystal structure, the aromatic rings attached to the salen complex generate a distinct chiral space around manganese ion.

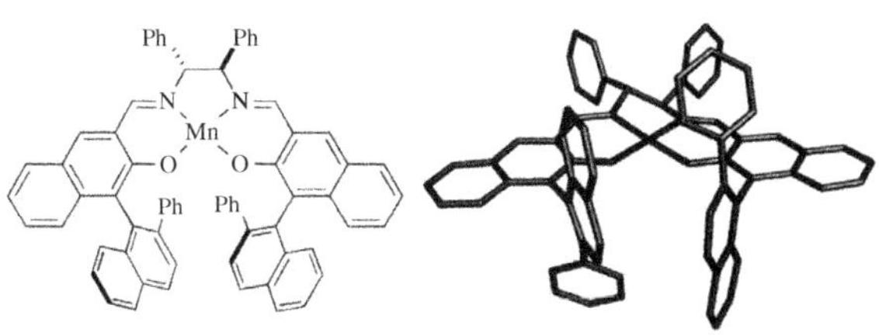

Figure 1.3 Unique chiral environment created on the salen platform.

A chiral salen ligand shown in Fig. 1.4 possesses a quite simple structure with a *trans*-cylcohexane-1,2-diamine moiety as the only chiral unit. But this simple salen ligand is now known as one of the best chiral ligands for asymmetric catalysis. This ligand is utilized as an excellent catalyst for asymmetric epoxidation of olefins with manganese ion[13] and hydrolytic kinetic resolution of epoxides with a cobalt ion.[14] The metal complexes with this simple salen ligand adopt an almost planar conformation, which triggered a great deal of controversy concerning the origin of high enantioselectivity.[11] This topic is discussed in more detail in the Section 1.2.1.

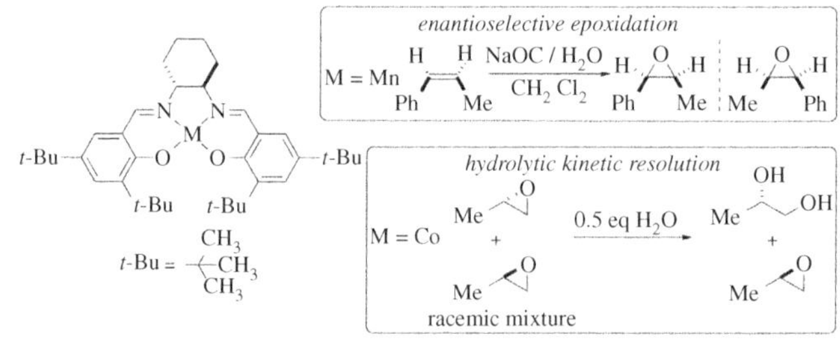

Figure 1.4 Asymmetric catalysis by the chiral salen complex.

1.1.3 Role of a Redox-Active Salen Ligand in Catalysis

Salen complexes have electron-rich phenolates, which could be reversibly converted to phenoxyl radicals under oxidizing conditions. The redox cycle between phenolate and phenoxyl radical can function as an additional redox site beside the metal ion. Redox noninnocence of salen ligands is quite an important property you should keep in mind in using salen complexes.

The best example to show the importance of the redox noninnocent salen ligand is alcohol oxidations by Cu^{II}(salen), which has been intensively studied as a model of galactose oxidase, a copper-containing metalloenzyme that catalyzes the aerobic oxidation of alcohols to aldehydes. It was demonstrated that under the catalytic conditions in the presence of air, the Cu^{II}(salen) complex generates a salen ligand radical with Cu^{II} ion, Cu^{II}(salen$^{+\bullet}$), which functions as a true catalytic species for the oxidation of benzyl alcohol to benzaldehyde (Fig. 1.5).[15] It was proposed that a phenoxyl radical might abstract a hydrogen atom from benzyl alcohol to generate a phenol, which then provides a hydrogen atom to O_2, regenerating a phenoxyl radical.

It is now clarified that salen ligand radicals are readily generated with other metal ions. However, the redox and spectroscopic properties are significantly different depending on the metal ion. The details are described in Sections 1.2.2–1.2.4.

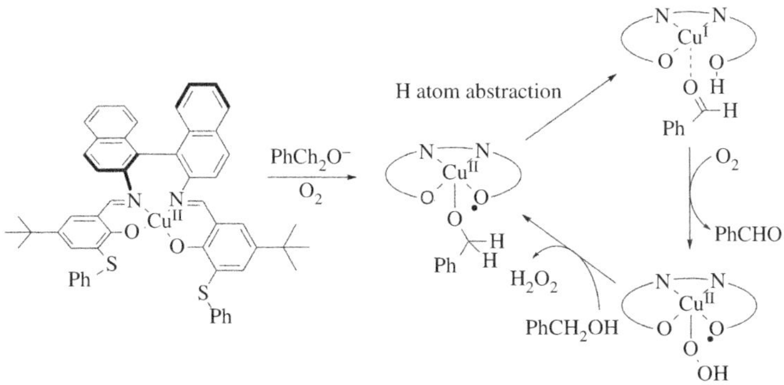

Figure 1.5 Oxidation of benzyl alcohol via hydrogen atom abstraction by a phenoxyl radical as a key step.

1.1.4 Building Block for Multimetallic Systems

Salen systems can be readily extended to multimetallic systems ranging from well-defined dinuclear complexes to polymeric materials and metal-organic frameworks. One of the methods of choice is to incorporate two salicylaldehyde functionalities in one unit. As shown in Fig. 1.6a, when a chiral binaphtholdialdehyde is utilized, a dinuclear salen complex with unique stereochemical properties is obtained.[16] Thanks to the two metal centers arranged in the appropriate geometry, the resulting dinuclear cobalt complex shows exceptional levels of enantioselectivity and activity in the epoxide polymerization.

Another way to prepare multimetallic systems is to incorporate additional metal-binding site to the salicylaldehyde unit. As shown in Fig. 1.6b, a salen ligand bearing the carboxylic acid functionality yields a multimetallic Cu^{II}(salen) complex via the coordination of the carboxylate to Cu^{II} ion upon the reaction with $Cu(OAc)_2{\cdot}6H_2O$.[17] It was reported that the resulting complex shows an unusual solvent-induced transformation between a triangular macrocycle and a helical coordination polymer.

Construction of salen-based multimetallic systems is an expanding research field. Recent excellent reviews are available to get an overview of this exciting topic.[18–21]

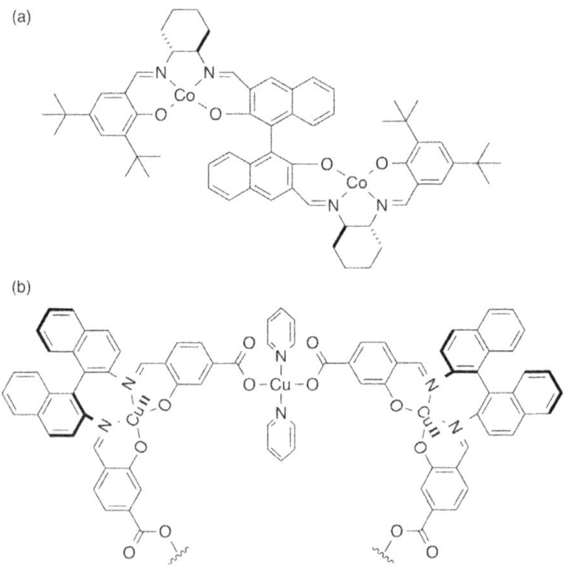

Figure 1.6 Examples of multimetallic salen complexes.

1.2 Electronic and Stereochemical Properties Revealed by Experimental Studies

Most of salen complexes having interesting functions are paramagnetic species with unpaired electrons. Precise characterization of these paramagnetic species requires spectroscopic techniques that differ from the characterization of organic compounds.[22] This section describes our own studies on paramagnetic salen complexes. Previously, density functional theory (DFT) calculations were extensively utilized to obtain mechanistic insight for catalysis by salen complexes. But our experimental studies have clarified previously unknown electronic and stereochemical properties of salen complexes, which are of fundamental importance for future developments.

1.2.1 Role of Axial Ligands in Stereochemical Properties

Our studies start with the question why the salen catalyst having a very simple structure shown in Fig. 1.4 exhibits high enantioselectivity. It was believed that enantioselective reactions require elaborated chiral environments around metal ion as seen for active sites in metalloenzymes. However, the salen catalyst in Fig. 1.4 apparently lacks such a chiral environment. This question prompted us to study stereochemical properties of salen complexes in detail.

We first focused on the oxidation state of the metal ion,[23] which is usually manganese(III) under ambient conditions. A stabilized manganese(IV) complex was synthesized using electron-donating N_3 as external axial ligands. As seen from X-ray crystal structures in Fig. 1.7, the Mn^{III}(salen)(N_3)(CH_3OH) complex adopts a planar conformation, which is also the case for other Mn^{III}(salen)(X) complexes.[24] But it is quite interesting to note that the Mn^{IV}(salen)(N_3)$_2$ complex, one-electron oxidized product from Mn^{III}(salen)(N_3)(CH_3OH), shows a stepped conformation with one of two salicylidene rings pointing upward and the other pointing downward. The characteristic stepped conformation is created by the bias from the *trans*-cyclohexane-1,2-diamine moiety as the only chiral unit in this catalyst.

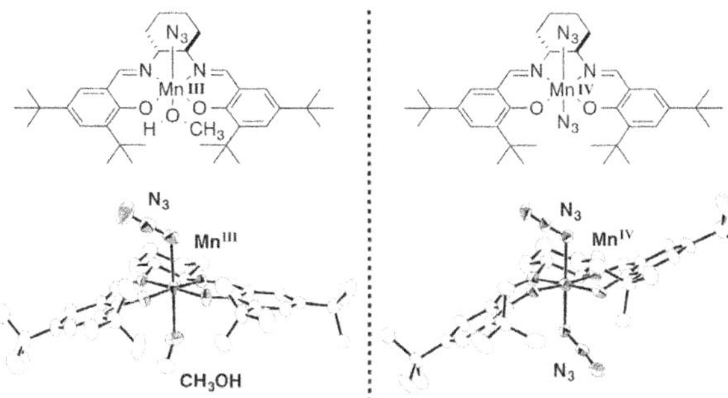

Figure 1.7 X-ray crystal structures of Mn^{III}(salen)(N_3)(CH_3OH) and Mn^{IV}(salen)(N_3)$_2$ complexes.

The observation from X-ray crystal structures alone is not enough to draw any conclusion on conformational properties, because crystal packing often alters shapes of flexible molecules. To study conformational changes in solution, we employed circular dichroism (CD) spectroscopy, which is a useful spectroscopic technique for chiral molecules.[25]

Figure 1.8 shows CD spectra of Mn^{III}(salen)(N_3)(CH_3OH) and Mn^{IV}(salen)(N_3)$_2$ complexes. For the comparison of CD spectra, one should pay attention to the intensity of UV-Vis absorptions. The comparison of CD bands in the region from 400–600 nm is not valid, because the absorption spectra in this region are substantially different between $Mn^{III}(salen)(N_3)(CH_3OH)$ and $Mn^{IV}(salen)(N_3)_2$. In contrast, the absorptions in the 200–300 nm region, which are assigned as arising from the π–π* transition of the salen ligand, are almost the same. Then, we can reliably compare CD bands in this region to obtain conformational difference. As you can see, the $Mn^{IV}(salen)(N_3)_2$ complex shows a much stronger CD band at 275 nm than the $Mn^{III}(salen)(N_3)(CH_3OH)$ complex. This indicates that in solution, the $Mn^{IV}(salen)(N_3)_2$ complex adopts a more chirally distorted conformation than the $Mn^{III}(salen)(N_3)(CH_3OH)$ complex, which is consistent with the observation from X-ray crystal structures. This study clarified a rather unexpected result that stereochemical properties of a salen complex are affected by the oxidation state of central metal ion.[23]

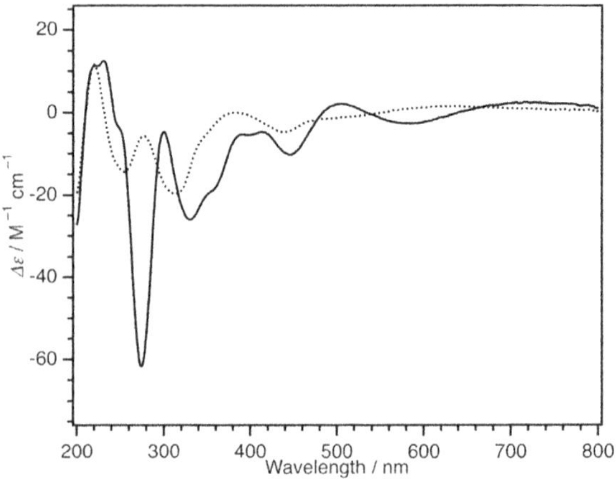

Figure 1.8 Circular dichroism spectra of (*R*,*R*)-$Mn^{III}(salen)(N_3)(CH_3OH)$ (dotted line) and (*R*,*R*)-$Mn^{IV}(salen)(N_3)_2$ (solid line).

We then investigated the effect of external axial ligands and synthesized a series of manganese(IV) salen complexes bearing CF_3CH_2O, Cl, and NO_3, in addition to N_3.[26] X-ray crystal structures of $Mn^{IV}(salen)(CF_3CH_2O)_2$ and $Mn^{IV}(salen)(Cl)_2$ were successfully obtained. Notably the $Mn^{IV}(salen)(Cl)_2$ complex shows an almost planar conformation, while the $Mn^{IV}(salen)(CF_3CH_2O)_2$ complex adopts a stepped conformation, which is also observed for $Mn^{IV}(salen)(N_3)_2$, as shown in Fig. 1.7. The solution conformations of these manganese(IV) complexes were investigated with CD spectroscopy, which shows that the intensity of CD in the region 200–300 nm is considerably increased in the order $Cl < NO_3 << N_3 < CF_3CH_2O$. This is a clear indication that $Mn^{IV}(salen)(CF_3CH_2O)_2$ and $Mn^{IV}(salen)(N_3)_2$ adopt a conformation in which the salen ligand is more chirally distorted in solution. The N_3 and CF_3CH_2O ligands are tightly bound to metal ion, while the Cl and NO_3 ligands are weakly bound to metal ion. This experimental observation clarified another surprising finding that the external axial ligands also play a critical role in determining the conformation of a salen complex.[26]

These experimental findings lead to the proposal that tight binding of the external axial ligands from both sides along the axial direction is a key factor to amplify the chirality of *trans*-cyclohexane-1,2-diamine to the entire salen complex molecule (Fig. 1.9). Tightly bound external axial ligands may serve to enhance the energy difference among possible conformers by confining the movement of the Mn ion along the axial direction. In the case of tightly bound N_3 and CF_3CH_2O, the stability of the stepped conformation might be significantly enhanced, leading to an increased distribution of this conformer relative to the others. But in the case of weakly bound Cl and NO_3, the energy difference among conformers remains very small, resulting in a mixture of multiple conformers in solution. In this regard, the manganese(IV) ion is definitively favorable compared to the manganese(III) ion, because the manganese(IV) complex can accommodate two anionic axial ligands due to the 2+ charge of a salen complex. In addition, the manganese(III) ion has a *d* electron in the d_{z^2} orbital, which disturbs tight binding of the external axial ligands.

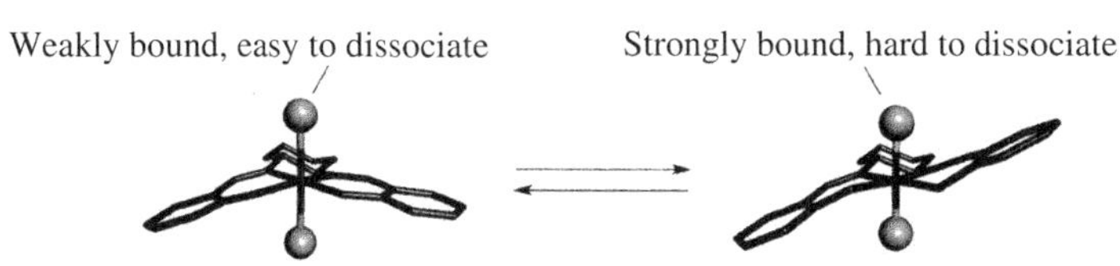

Figure 1.9 Proposed mechanism for the formation of a chirally distorted stepped conformation.

An important lesson from the present experimental studies is that the structure of salen ligands alone may not be enough to determine the stereochemistry of salen complexes. Additionally, you need to consider the oxidation state of metal ion and the external axial ligands for designing new asymmetric catalysts. Interestingly, we also found that stereochemical properties of salen complexes are completely different between manganese(IV) and titanium(IV).[28]

A next topic of interest is how the chirally distorted stepped conformation is related to the enantioselectivity in epoxidation reactions. It was strongly expected that the stepped conformation should lead to much higher enantioselectivity than the planar conformation. To test this hypothesis, nonreactive external axial ligands such as CF_3CH_2O and N_3 on manganese(IV) are replaced with a reactive oxidant.[29,30] Iodosylarenes were employed as an oxidant. As expected, the reaction of $Mn^{IV}(salen)(Cl)_2$ with iodosylarenes yielded iodosylarene adducts of a manganese(IV) salen complex. The X-ray crystal structure of the iodosylmesitylene adduct was successfully obtained, as shown in Fig. 1.10. Upon the reaction with $Mn^{IV}(salen)(Cl)_2$, the neutral iodosylmesitylene, 1,3,5-$(CH_3)_3C_6H_2IO$, is converted to the anionic 1,3,5-$(CH_3)_3C_6H_2IClO^-$ ligand by binding the Cl^-counteranion on the I atom. The anionic 1,3,5-$(CH_3)_3C_6H_2IClO^-$ ligands tightly bind to the manganese(IV) ion, which induces a chirally distorted stepped conformation.

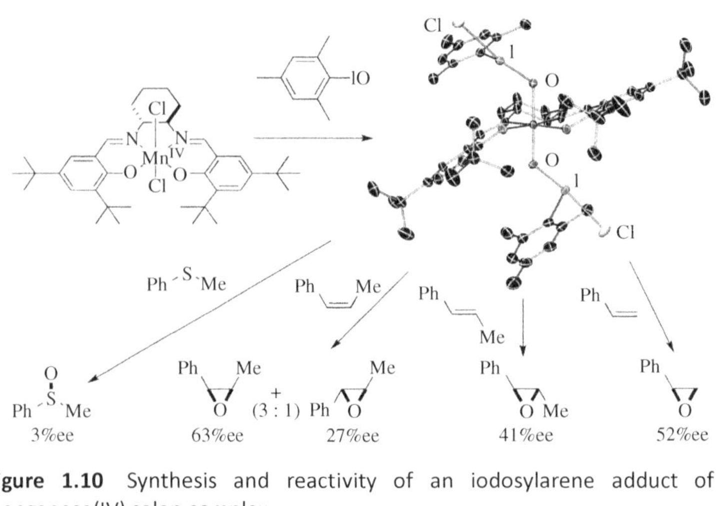

Figure 1.10 Synthesis and reactivity of an iodosylarene adduct of a manganese(IV) salen complex.

Stoichiometric reactions of the iodosylmesitylene adduct with thioanisole and styrenes were carried out. Enantioselectivity of the single-turnover reaction with thioanisole is negligibly small (3%ee), which is consistent with low enantioselectivity in the catalytic reactions of sulfides. In contrast, the single-turnover reactions with styrenes yield moderate enantioselectivity. The reaction with *cis*-β-methyl styrene gives a mixture of *cis*-β-methyl styrene oxide (63%ee) and *trans*-β-methyl styrene oxide (27%ee).

In the case of the catalytic reactions using Mn^{III}(salen)(Cl) as a catalyst and NaOCl as an oxidant, *cis*-β-methyl styrene oxide is obtained in high enantioselectivity (92%ee) and the isomeric *trans*-β-methyl styrene oxide is not detected.[13] Embarrassingly, the single-turnover reaction by the iodosylmesitylene adduct with the chirally distorted stepped conformation gave a poor result as compared to the catalytic reaction by the original Mn^{III}(salen)(Cl) with a planar conformation. This indicates that the chirally distorted stepped conformation is not the only factor to determine the enantioselectivity. Further studies are needed to clarify the whole picture for the asymmetric catalysis by this simple catalyst.

1.2.2 Characterization of Salen Ligand Radicals

Although salen complexes have a long history, it is only very recently that salen ligand radicals are fully accepted as well-characterized species. Characterization of salen ligand radicals is not an easy task. One of the difficulties is that even X-ray crystal structures do not necessarily yield fruitful information to distinguish between radical and nonradical. According to the previous reports, a salen ligand radical with nickel(II) shows a ring deformation to a quinoid structure in one of the phenolate rings, which is indeed clear evidence for a radical character for this moiety.[31] But another salen ligand radical with nickel(II), where the structure of the salen ligand is only slightly different, does not show such a ring deformation at all.[32]

Why are the structures of salen ligand radicals different? The structural difference of salen ligand radicals comes from the difference of electronic structures. The salen ligand radical with a deformed phenolate ring is a class II mixed-valence compound with a localized radical on one of two phenolate rings (Fig. 1.11, top).

In contrast, the salen ligand radical showing no ring deformation belongs to a class III mixed-valence compound with a delocalized radical over two phenolate rings (Fig. 1.11, bottom). Because of the delocalization of a radical, it is inherently difficult to detect a structural deformation arising from the formation of a salen ligand radical. These two classes of salen ligand radicals also show significantly different spectroscopic properties, which complicate the characterization of salen ligand radicals.

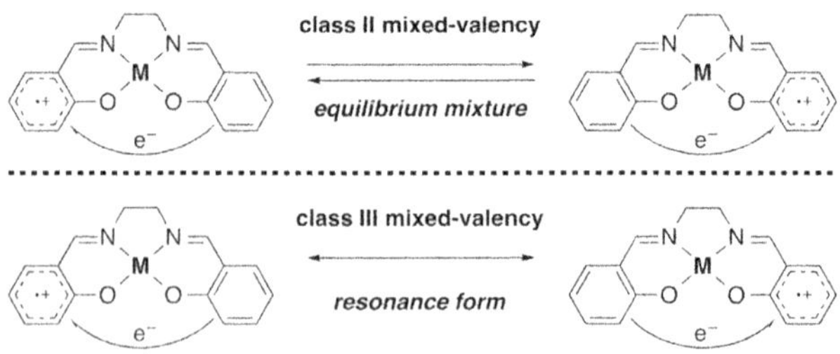

Figure 1.11 Localized versus delocalized salen ligand radicals.

In the beginning of the studies to clarify functional roles of salen ligand radicals, it is very important to develop spectroscopic techniques to precisely characterize salen ligand radicals. One of the excellent probes for salen ligand radicals is ^{1}H nuclear magnetic resonance (NMR) spectroscopy. To use ^{1}H NMR spectroscopy for studies of paramagnetic molecules, you always need to be very careful for the assignment of signals. For this purpose, selectively deuterated salen ligands were prepared, as shown in Fig. 1.12, and ^{2}H NMR spectra were measured for metal complexes with these deuterated ligands.[26,27,33]

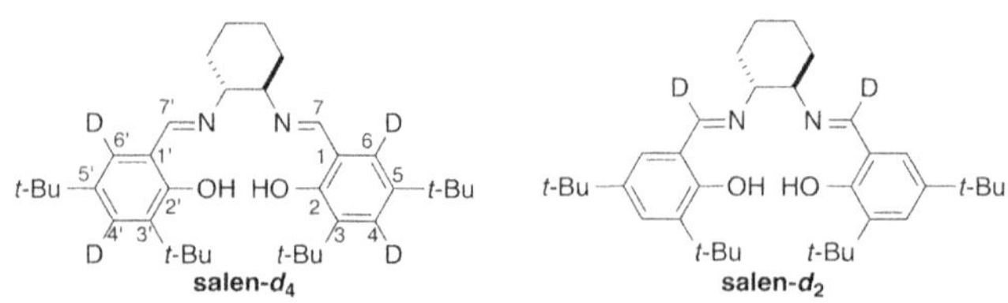

Figure 1.12 Selectively deuterated salen ligands. In salen-d_4, ^{2}H atoms are selectively incorporated into the phenolate rings (80% D) and the *tert*-butyl groups (7% D). In salen-d_2, the 7/7′ position of the azomethine group is selectively deuterated (99.5% D).

Figure 1.13 shows ^{1}H and ^{2}H NMR spectra of the salen ligand and the salen complex with iron(III) as typical examples. The ^{2}H NMR spectra of salen-d_4 show broad signals at 7.30/7.00 and 1.39/1.22 (Fig. 1.13a, middle), which arise from phenolate and *tert*-butyl groups, respectively. The salen-d_2 ligand shows only one ^{2}H NMR signal at 8.31 (Fig. 1.13a, bottom), which is assigned as arising from the azomethine group. In the case of the diamagnetic salen ligand, these assignments are readily obtained without ^{2}H NMR spectra of selectively deuterated samples.

But the ^{1}H NMR spectrum of the paramagnetic iron(III) complex is utterly different (Fig. 1.13b, top).[34] The ^{1}H NMR signals are distributed over a 150 ppm range, and it is impossible to assign these signals from their positions like the assignment of ^{1}H NMR signals of organic molecules.

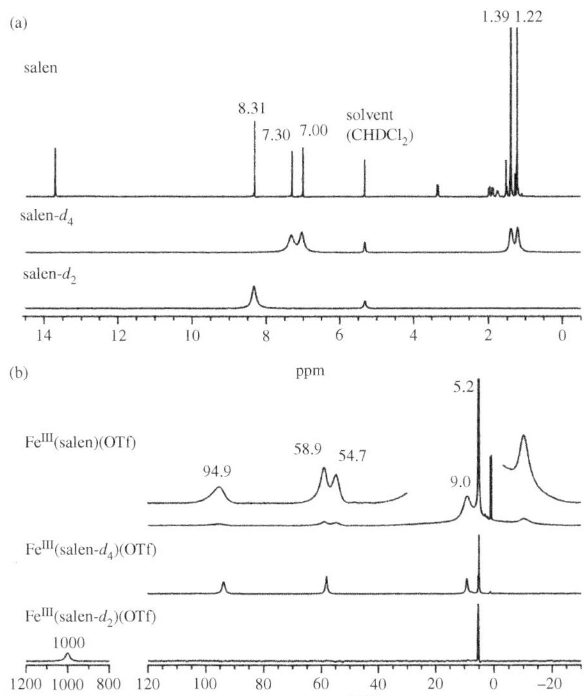

Figure 1.13 (a) ^{1}H NMR spectrum of salen (top), ^{2}H NMR spectra of salen-d_4 (middle), and salen-d_2 (bottom). (b) ^{1}H NMR spectrum of Fe^{III}(salen)(OTf) (top), ^{2}H NMR spectra of Fe^{III}(salen-d_4)(OTf) (middle), and Fe^{III}(salen-d_2)(OTf) (bottom).

The ^{2}H NMR spectrum of Fe^{III}(salen-d_4)(OTf) (Fig. 1.13b, middle) unambiguously shows that the signals at 94.9 and 58.9 ppm come from the phenolate protons. The signals at 9.0 and 5.2 ppm are readily assigned as arising from the *tert*-butyl groups, because paramagnetic shifts of the *tert*-butyl protons should be smaller than the phenolate protons that are directly attached to the π–system. The ^{2}H NMR signal at 1000 ppm is assigned as arising from the azomethine proton (Fig. 1.13b, bottom), which cannot be detected with ^{1}H NMR. In addition to the assignment of the signals, ^{2}H NMR has another advantage in studying paramagnetic molecules. ^{2}H NMR signals of paramagnetic molecules are generally sharper than the corresponding ^{1}H NMR signals (compare the ^{1}H and ^{2}H NMR signals in Fig. 1.13b). Detection of ^{1}H NMR signals from paramagnetic molecules often suffers from severe broadening especially at lower temperature.

^{1}H and ^{2}H NMR spectroscopy was utilized for the characterization of the salen ligand radical with manganese(III). The most important point here is to distinguish between the salen ligand radical with manganese(III) and the manganese(IV) salen complex (Fig. 1.14). These two species are valence tautomers that could be converted with each other via intramolecular electron transfer.

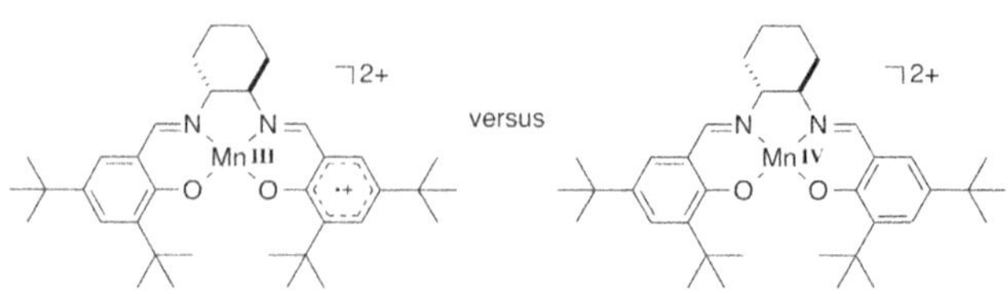

Figure 1.14 Valence tautomers; manganese(III) salen ligand radical versus manganese(IV).

Figure 1.15 shows ^{2}H NMR spectra of Mn(salen-d_4)(X)$_2$,[27] where X is an external axial ligand that coordinates to manganese from the direction perpendicular to the salen ligand (see Fig. 1.7, for the X-ray crystal structure of Mn(salen)(N_3)$_2$). When X is CF_3CH_2O, N_3, NO_3, and Cl, the ^{2}H NMR signals from the *tert*-butyl groups appear in a region from 2.6 to 4.8 ppm. But when X is a weakly coordinating SbF_6, the ^{2}H NMR signals from the *tert*-butyl groups are shifted to –3.6 and –22.0 ppm. The significant paramagnetic shift of the *tert*-butyl groups clearly indicates the presence of an unpaired electron on the phenolate rings in close vicinity to the *tert*-butyl groups. This is strong evidence for the ligand radical character in Mn(salen-d_4)(SbF_6)$_2$.

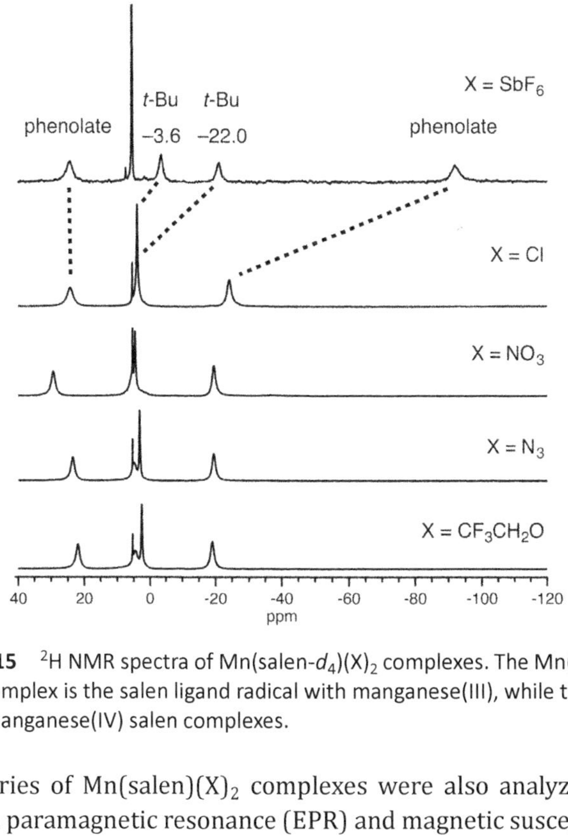

Figure 1.15 ^{2}H NMR spectra of Mn(salen-d_4)(X)$_2$ complexes. The Mn(salen-d_4)(SbF_6)$_2$ complex is the salen ligand radical with manganese(III), while the others are the manganese(IV) salen complexes.

A series of Mn(salen)(X)$_2$ complexes were also analyzed with electron paramagnetic resonance (EPR) and magnetic susceptibility measurements, which are frequently utilized to characterize paramagnetic molecules. But EPR and magnetic susceptibility data are essentially the same for the salen ligand radical with manganese(III) and the manganese(IV) salen complex.[27]

Absorption spectroscopy is also an important spectroscopic technique in characterizing paramagnetic molecules. In the case of salen ligand radicals, we can expect absorptions around 400 nm as arising from the π–π* transitions of the phenoxyl radical moiety.[35] But it was difficult to reliably assign the π–π* transitions of the phenoxyl radical among other overlapping transitions such as ligand-to-metal charge transfer (LMCT) that might appear around this region.

It is also expected that salen ligand radicals show absorptions that arise from ligand-to-ligand charge transfer (LLCT) transitions from the phenolate to the phenoxyl radical. The LLCT transitions usually appear in the near-infrared (NIR) region apart from other transitions. Once the LLCT transitions are precisely assigned, the LLCT transitions could be utilized as a reliable marker for salen ligand radicals without interference from other transitions.

Our strategy to assign the LLCT transitions in salen ligand radicals is shown in Fig. 1.16. The redox potential of the phenolate moiety is altered changing the substituent of the phenolate. The redox potential of the phenolate is increased in the order CH_3O < *t*-Bu < Cl. Accordingly, the excitation energy required for photoinduced intramolecular electron transfer is increased. Thus, the LLCT transitions are expected to shift to higher energy in the order CH_3O < *t*-Bu < Cl.

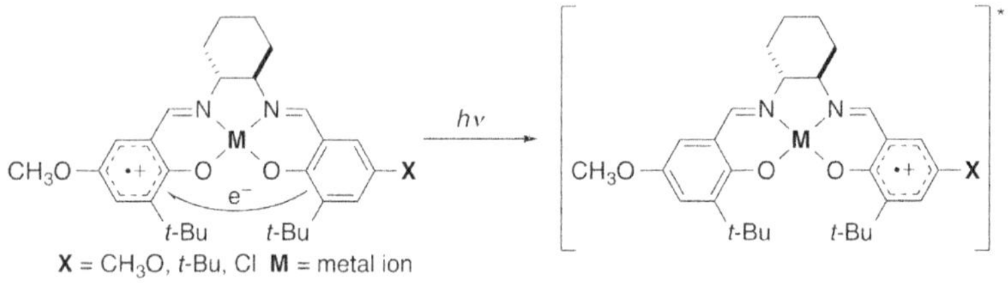

Figure 1.16 Precise assignment of the LLCT transitions in salen ligand radicals.

Figure 1.17 shows UV-Vis-NIR absorptions of salen ligand radicals with manganese(III). The symmetrical salen ligand radical (X = CH_3O in Fig. 1.16) shows a very weak absorption at 1270 nm (Fig. 1.17a). Changing the substituent of the phenolate shifts the broad absorption to 1015 and 890 nm (X = *t*-Bu and Cl, respectively) (Figs. 1.17b and 1.17c). Such absorption shifts are exactly consistent with the assignment as the LLCT transitions. The NIR absorption observed here is thus another strong evidence for the formation of salen ligand radicals. It was demonstrated that ^{1}H and ^{2}H NMR and NIR absorption spectroscopy is effectively utilized to characterize salen ligand radicals.[27] Spectroscopic techniques described herein would be applied for other systems containing a ligand radical.

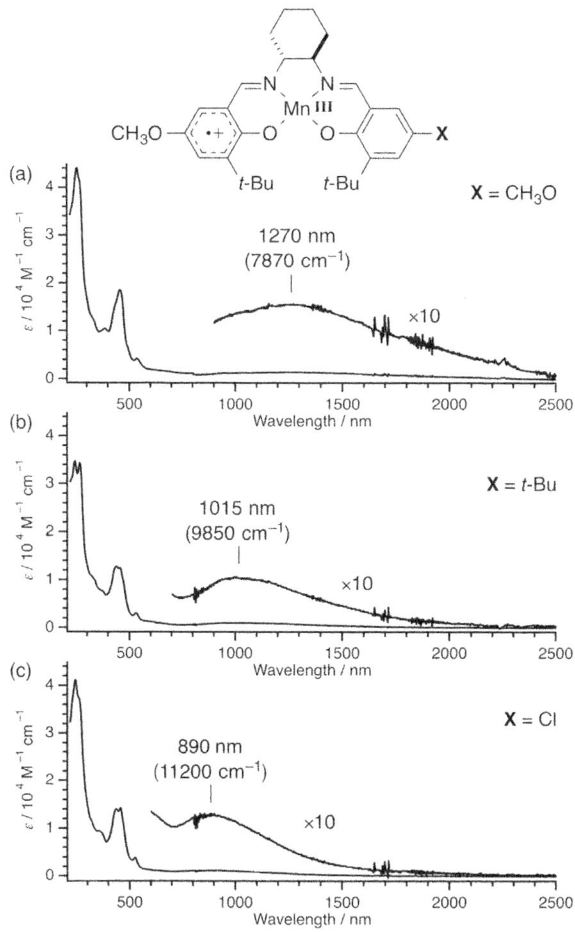

Figure 1.17 UV-Vis-NIR absorptions of salen ligand radicals with manganese(III). (a) X = CH_3O, (b) X = *t*-Bu, and (c) X = Cl.

1.2.3 Localized versus Delocalized Salen Ligand Radicals

One of the important findings in the experimental studies of salen ligand radicals is that the electron transfer from the phenolate to the phenoxyl radical is drastically modulated by the metal ion that bridges two redox centers. Modulation of the electron transfer leads to the difference of the electronic structure of salen ligand radicals, that is, localized versus delocalized salen ligand radicals, as shown in Fig. 1.11.

The salen ligand radical with manganese(III) is classified as the class II mixed-valence compound because absorption spectral shifts shown in Fig. 1.17 are fully consistent with the localized radical. In addition, the observed bandwidth at half-height is in accord with the theoretical value calculated by the Marcus–Hush equation.[27] In the case of the class II salen ligand radical, the phenolate as the electron donor and the phenoxyl radical as the electron acceptor have moderate electronic interaction and electron transfer is occurring between two redox centers.

In sharp contrast, the salen ligand radicals with nickel(II), palladium(II), and platinum(II) were reported to have a character of class III mixed-valence compounds.[36] In class III mixed-valence compounds, two redox centers (phenolate and phenoxyl radical in this case) have such strong electronic interaction that delocalization of the unpaired electron occurs as a consequence of disappearance of the activation barrier for electron transfer. The assignment as class III mixed-valence compounds was previously obtained from the relatively narrow NIR absorptions.

The NIR absorption of the salen ligand radical with nickel(II) was further analyzed by our method using nonsymmetrical salen complexes shown in Fig. 1.16. As shown in Fig. 1.18a, the symmetrical salen ligand radical with nickel(II) shows two absorptions at 1950 and 2140 nm. The broad NIR absorption at 1950 nm shifts to 1580 and 1400 nm upon changing one of the CH_3O substituents to *t*-Bu and Cl (Figs. 1.18b and 1.18c), consistent with the LLCT transitions arising from the class II mixed valency. The origin of the 1950 nm absorption for the Ni^{II}–radical is exactly the same as the 1230 nm absorption for the Mn^{III}–radical in Fig. 1.17, although the intensity is significantly different.

In contrast, the relatively narrow NIR absorption at 2140 nm (Fig. 1.18a) is characteristic only for the symmetrical salen ligand radical with nickel(II). Such a narrow NIR absorption is not observed for manganese(III) radicals (Fig. 1.17). It is interesting to note that the narrow absorption at 2140 nm for the symmetrical complex (Fig. 1.18a) completely disappears for nonsymmetrical complexes (Figs. 1.18b and 1.18c), in which the radical should be localized on one of the phenolate rings. It is thus strongly indicated that the narrow NIR absorption at 2140 nm arises from the class III mixed valency.

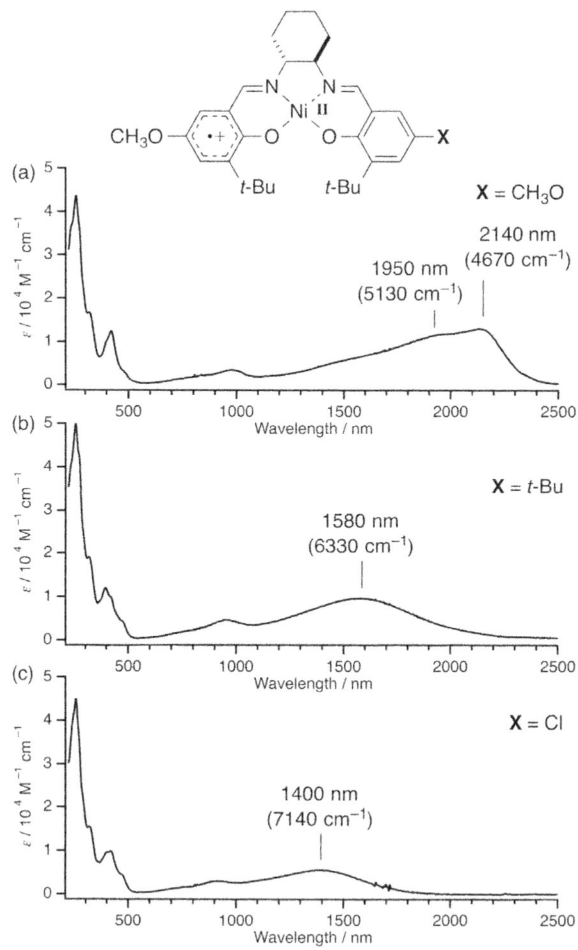

Figure 1.18 UV-Vis-NIR absorptions of salen ligand radicals with nickel(II). (a) X = CH_3O, (b) X = *t*-Bu, and (c) X = Cl.

This study demonstrates a fascinating possibility that electron transfer could be drastically modulated exchanging the metal ion that bridges the two redox centers. The exact factors to modulate electron transfer in salen ligand radicals still remain to be investigated. But the control of electron transfer and the resulting NIR absorption in minimal systems like the present case is quite interesting in creating new functional materials.

1.2.4 Stable Salen Ligand Radical with Cobalt

Phenoxyl radicals including salen ligand radicals were believed to be unstable active species.[35] Indeed, the salen ligand radicals with copper(II),[37] nickel(II),[38] manganese(III)[39], and iron(III)[7] reported previously were prepared under inert atmosphere or at low temperature. However, rather surprisingly, our spectroscopic studies have shown that the Co(salen)(CF_3SO_3) complex, which is a stable isolable solid under ambient conditions, has a character of a salen ligand radical.[33]

The Co(salen)(CF_3SO_3) complex is prepared by aerobic oxidation under biphasic CH_2Cl_2-CF_3SO_3H/H_2O conditions. The Co(salen)(L) complex (L = CH_3CO_2, Cl, etc.), which is well known as an excellent catalyst for hydrolytic kinetic resolution of racemic epoxides,[14] is conventionally described as a cobalt(III) salen complex. Recent theoretical calculations raised a new proposal that the Co(salen)(H_2O)(SbF_6) complex contains appreciable contribution from a Co^{II}(salen$^{+\bullet}$) electronic structure,[40] while other theoretical calculations for Co(salen)(Cl) indicated a triplet Co^{III}(salen) electronic structure.[41]

To evaluate rather confusing theoretical results, we carried out an experimental study and obtained clear evidence for a salen ligand radical. The ^1H NMR signal from the CH_3O substituent incorporated to the phenolate ring shows a considerable paramagnetic shift, which cannot be explained by the Co^{III}(salen) (CF_3SO_3) electronic structure. Furthermore, the ^1H NMR signal from the CH_3O substituent shows a larger paramagnetic shift exchanging the substituent on the other phenolate ring from CH_3O to *t*-Bu and Cl (Fig. 1.19). This observation is clearly indicative of a salen ligand radical character for Co(salen) (CF_3SO_3).

Oxidizing ability of a stable salen ligand radical with cobalt is low, as estimated from the redox potential. The salen ligand radical with cobalt may not function as the one-electron oxidant unlike the other salen ligand radicals.

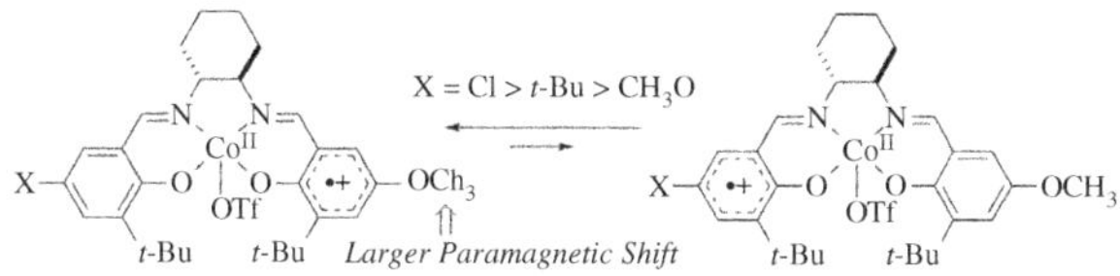

Figure 1.19 Evidence for a salen ligand radical with cobalt.

1.3 Conclusions

Excellent catalytic activity of manganese and cobalt salen complexes prompted us to investigate their electronic and stereochemical properties. One of new findings is that the conformation of salen complexes is altered by unexpected two factors. One is the electronic properties of the central metal such as the oxidation state. The other is the external axial ligands that coordinate to the central metal from the direction perpendicular to the salen ligand. This finding is important in designing catalysts and materials with well-defined stereochemistry.

Another important finding is that the ligand moiety of a salen complex is readily oxidized to yield a stable salen ligand radical. Salen ligand radical complexes show distinct spectroscopic properties depending on the electronic structure. The most characteristic spectroscopic feature of a salen ligand radical complex is a variety of NIR absorptions, which arise from localization versus delocalization of a radical on two phenolate rings connected via metal ion. The present study has shown that spectroscopic properties could be controlled changing the metal ion and redox properties of the phenolate rings. This finding is a very promising result to create new light-absorbing material using a very simple system like salen.

Acknowledgments

All of the studies described above were carried out in collaboration with Prof. Hiroshi Fujii (Nara Women's University). Studies on iodosylarene adducts were carried out by Dr. Chunlan Wang as her PhD work. The author acknowledges financial support from JSPS KAKENHI (grant numbers 23550086 and 15K05462) and the Naito Foundation.

References

1. Larrow, J. F., Jacobsen, E. N., Gao, Y., Hong, Y., Nie, X., Zepp, C. M. *J. Org. Chem.*, **59**, 1939–1942 (1994).
2. Kleij, A. W. *Eur. J. Inorg. Chem.*, 193–205 (2009).
3. Palucki, M., Finney, N. S., Pospisil, P. J., Güler, M. L., Ishida, T., Jacobsen, E. N. *J. Am. Chem. Soc.*, **120**, 948–954 (1998).
4. Xie, D., Jing, J., Cai, Y. B., Tang, J., Chen, J. J., Zhang, J. L. *Chem. Sci.*, **5**, 2318–2327 (2014).
5. Lebedeva, M. A., Chamberlain, T. W., Davies, E. S., Mancel, D., Thomas, B. E., Suyetin, M., Bichoutskaia, E., Schröder, M., Khlobystov, A. N. *Chem. Eur. J.*, **19**, 11999–12008 (2013).
6. Hutson, G. E., Türkmen, Y. E., Rawal, V. H. *J. Am. Chem. Soc.*, **135**, 4988–4991 (2013).
7. Kurahashi, T., Kobayashi, Y., Nagatomo, S., Tosha, T., Kitagawa, T., Fujii, H. *Inorg. Chem.*, **44**, 8156–8166 (2005).
8. Matsunaga, S., Shibasaki, M. *Chem. Commun.*, 1044–1057 (2014).
9. Matsumoto, K., Saito, B., Katsuki, T. *Chem. Commun.*, 3619–3627 (2007).
10. Baleizão, C., Garcia, H. *Chem. Rev.*, **106**, 3987–4043 (2006).
11. McGarrigle, E. M., Gilheany, D. G. *Chem. Rev.*, **105**, 1563–1602 (2005).
12. Hashihayata, T., Punniyamurthy, T., Irie, R., Katsuki, T., Akita, M., Moro-oka, Y. *Tetrahedron*, **55**, 14599–14610 (1999).
13. Jacobsen, E. N., Zhang, W., Muci, A. R., Ecker, J. R., Deng, L. *J. Am. Chem. Soc.*, **113**, 7063–7064 (1991).
14. Tokunaga, M., Larrow, J. F., Kakiuchi, F., Jacobsen, E. N. *Science*, **277**, 936–938 (1997).
15. Wang, Y., DuBois, J. L., Hedman, B., Hodgson, K. O., Stack, T. D. P. *Science*, **279**, 537–540 (1998).
16. Hirahata, W., Thomas, R. M., Lobkovsky, E. B., Coates, G. W. *J. Am. Chem. Soc.*, **130**, 17658–17659 (2008).
17. Heo, J., Jeon, Y. M., Mirkin, C. A. *J. Am. Chem. Soc.*, **129**, 7712–7713 (2007).
18. Clarke, R. M., Storr, T. *Dalton Trans.*, **43**, 9380–9391 (2014).
19. Whiteoak, C. J., Salassa, G., Kleij, A. W. *Chem. Soc. Rev.*, **41**, 622–631 (2012).
20. Glaser, T. *Chem. Commun.*, **47**, 116–130 (2011).

21. Akine, S., Nabeshima, T. *Dalton Trans.*, 10395–10408 (2009).
22. Lawrence Que, Jr. *Physical Methods in Bioinorganic Chemistry*, University Science Books: Sausalito (2000).
23. Kurahashi, T., Fujii, H. *Inorg. Chem.*, **47**, 7556–7567 (2008).
24. Pospisil, P. J., Carsten, D. H., Jacobsen, E. N. *Chem. Eur. J.*, **2**, 974–980 (1996).
25. Rodger, A., Nordén, B. *Circular Dichroism & Linear Dichroism*, Oxford University Press: New York (1997).
26. Kurahashi, T., Hada, M., Fujii, H. *J. Am. Chem. Soc.*, **131**, 12394–12405 (2009).
27. Kurahashi, T., Fujii, H. *J. Am. Chem. Soc.*, **133**, 8307–8316 (2011).
28. Kurahashi, T., Hada, M., Fujii, H. *Inorg. Chem.*, **53**, 1070–1079 (2014).
29. Wang, C., Kurahashi, T., Inomata, K., Hada, M., Fujii, H. *Inorg. Chem.*, **52**, 9557–9566 (2013).
30. Wang, C., Kurahashi, T., Fujii, H. *Angew. Chem. Int. Ed.*, **51**, 7809–7811 (2012).
31. Orio, M., Jarjayes, O., Kanso, H., Philouze, C., Neese, F., Thomas, F. *Angew. Chem. Int. Ed.*, **49**, 4989–4992 (2010).
32. Storr, T., Wasinger, E. C., Pratt, R. C., Stack, T. D. P. *Angew. Chem. Int. Ed.*, **46**, 5198–5201 (2007).
33. Kurahashi, T., Fujii, H. *Inorg. Chem.*, **52**, 3908–3919 (2013).
34. Kurahashi, T., Fujii, H. *Bull. Chem. Soc. Jpn.*, **85**, 940–947 (2012).
35. Chaudhuri, P., Wieghardt, K. *Prog. Inorg. Chem.*, **50**, 151–216 (2001).
36. Shimazaki, Y., Stack, T. D. P., Storr, T. *Inorg. Chem.*, **48**, 8383–8392 (2009).
37. Storr, T., Verma, P., Pratt, R. C., Wasinger, E. C., Shimazaki, Y., Stack, T. D. P. *J. Am. Chem. Soc.*, **130**, 15448–15459 (2008).
38. Shimazaki, Y., Yajima, T., Tani, F., Karasawa, S., Fukui, K., Naruta, Y., Yamauchi, O. *J. Am. Chem. Soc.*, **129**, 2559–2568 (2007).
39. Kurahashi, T., Kikuchi, A., Tosha, T., Shiro, Y., Kitagawa, T., Fujii, H. *Inorg. Chem.*, **47**, 1674–1686 (2008).
40. Kochem, A., Kanso, H., Baptiste, B., Arora, H., Philouze, C., Jarjayes, O., Vezin, H., Luneau, D., Orio, M., Thomas, F. *Inorg. Chem.*, **51**, 10557–10571 (2012).
41. Kemper, S., Hrobárik, P., Kaupp, M., Schlörer, N. E. *J. Am. Chem. Soc.*, **131**, 4172–4173 (2009).

Biography

The author received his BS (with Prof. Yoshihiko Ito and Prof. Michinori Suginome), MS (with Prof. Hisanobu Ogoshi and Prof. Tadashi Mizutani), and PhD (with Prof. Junichi Yoshida) from Kyoto University, Japan, where he studied molecular recognition and regioselective functionalization of carbohydrates by synthetic host molecules. He then moved to Nagoya University, Japan, for postdoctoral work with Prof. Minoru Isobe, where he studied metalloenzymes isolated from the diapause eggs of the silkworm to reveal the molecular mechanism of the termination of insect diapause. He joined the Institute for Molecular Science as assistant professor, where he started research on bioinorganic chemistry field with Prof. Hiroshi Fujii. His research has focused on unveiling the structure–function relationships of molecular systems having unpaired electrons, such as reactive species generated from metalloenzyme models and molecular catalysts.

Chapter 2

Fused Polycyclic Aromatic Compounds: [*n*]Acenes, [*n*]Helicenes, and Their Heterocyclic Analogues

Koji Nakano
Department of Organic and Polymer Materials Chemistry,
Tokyo University of Agriculture and Technology, 2-24-16 Naka-cho,
Koganei, Tokyo 184-8588, Japan
k_nakano@cc.tuat.ac.jp

Polycyclic aromatic compounds, such as [*n*]acenes, [*n*]helicenes, and their heterocyclic analogues, have been widely studied because of their attractive function such as semiconducting, emission, and chiroptical properties. To create new functional materials based on these classes of π-conjugated compounds, the development of efficient synthetic method is indispensable. This chapter will briefly survey [*n*]acenes, [*n*]helicenes, and their heterocyclic analogues mainly from the viewpoint of synthetic methods.

Conjugated Objects: Developments, Synthesis, and Applications
Edited by Atsushi Nagai and Koji Takagi

ISBN 978-981-4774-03-1 (Hardcover), 978-1-315-22962-1 (eBook)
www.panstanford.com

2.1 Introduction

Fused π-conjugated compounds have attracted much attention because of their optical and electronic properties, which allow them to be applicable to organic field-effect transistors (OFETs),[1] organic light-emitting diodes (OLEDs),[2] and organic photovoltaics (OPVs).[3] Their properties depend on many factors such as the number of fused rings, a connectivity of aromatic rings, the incorporated elements, and so on. Therefore, a numerous number of fused π-conjugated compounds have been synthesized and characterized to study structure–property relationship and to develop high-performance functional materials.

The basic component of fused π-conjugated compounds is the benzene ring. The fusion of two benzene rings results in naphthalene where two carbon atoms and one C–C bond are shared (*ortho*-fused) (Fig. 2.1). The fusion of three benzene rings gives two isomers, depending on their connectivity. One is anthracene, which consists of three *ortho*-fused benzene rings in a linear arrangement. Another is phenanthrene, in which naphthalene is fused with a benzene ring at the C1–C2 bond to form a bent structure. Further annelation of benzene ring(s) to anthracene in a linear manner gives the so-called [*n*]acenes such as tetracene and pentacene. A zigzag annelation of benzene ring(s) to phenanthrene affords [*n*]phenacenes such as chrysene and picene. These [*n*]acenes and [*n*]phenacenes possess planar structures and demonstrate promising properties as organic semiconductors for OFETs. On the other hand, an angular annelation of benzene ring(s) to phenanthrene gives [*n*]helicenes with nonplanar helical structures. [*n*]Helicenes composed of six or more benzene rings exhibit stable helical chirality because of an intramolecular steric repulsion, which endows them with unique chiroptical properties.

Heteroaromatic rings are also important modules for fused π-conjugated compounds since the introduction of heteroatom(s) into a π-conjugated framework is an effective way to control the electronic structure of them. In particular, thiophene-fused compounds have been well-studied in terms of synthesis and application as organic functional materials. For example, a series of thiophene-fused pentacyclic aromatic compounds were synthesized and found to demonstrate high semiconducting properties.

Representative examples are shown in Fig. 2.2. In these compounds, benzene and/or thiophene rings are annelated in a linear fashion. Accordingly, these compounds are called as thienoacenes. Anthra[2,3-*b*:6,7-*b*']dithiophene (**1**) is apparently isoelectronic with pentacene, the corresponding pentacyclic hydrocarbon acene. However, compounds **2** and **3** should be considered not as [*n*]acenes but as [*n*]phenacenes in the sense of electronic structure. The sulfur atom in a thiophene ring donates two electrons to the π-system like an ethenylene unit in [*n*]phenacenes. Thus, compounds **2** and **3** are considered to be isoelectronic with picene, the corresponding pentacyclic [*n*]phenacene. Nevertheless, these compounds are described as heteroacenes for the sake of expedience in this chapter.

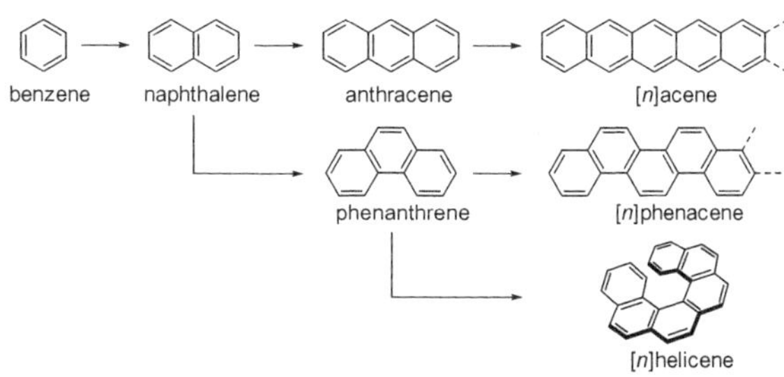

Figure 2.1 Structures of [*n*]acenes, [*n*]phenacenes, and [*n*]helicenes.

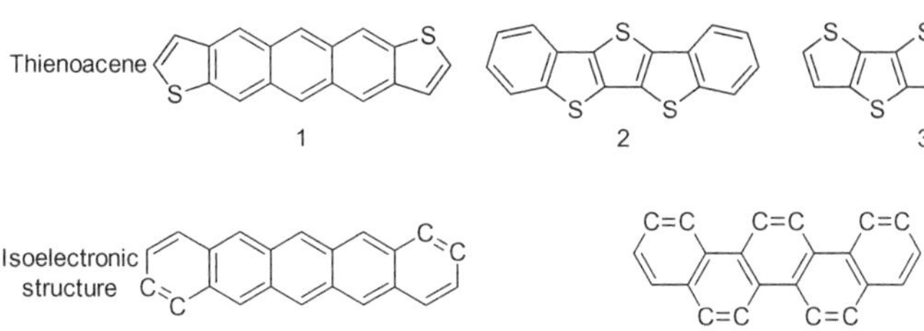

Figure 2.2 Molecular structures of pentacyclic thienoacenes.

This chapter will briefly survey [*n*]acenes, [*n*]helicenes, and their heterocyclic analogues mainly from the viewpoint of synthetic methods. These compounds have been known over 100 years and been synthesized via a variety of strategies. Accordingly, there have been some excellent review articles and books which cover these compounds comprehensively. On the other hand, this chapter will describe the selected methods which have been historically significant or recently developed (mainly in this century) for the readers who become interested in fused polycyclic aromatic compounds. In addition to the synthetic methods, their properties and applications as organic functional materials are also described briefly. This chapter also covers heteroacenes and heterohelicenes containing silole or phosphole ring(s), which are not aromatic rings, since they are known to exhibit excellent optical/optoelectronic properties.

2.2 [*n*]Acenes

2.2.1 Introduction

Numerous studies on acenes have been conducted for understanding their electronic structures, stability, aromaticity as well as for developing their application to organic electronics. With an increase in the number of fused benzene rings, the highest occupied molecular orbital–lowest unoccupied molecular orbital (HOMO–LUMO) energy gaps decrease.[4] In addition, the aromaticity per ring and stability are also decrease with the number of fused benzene rings. This trend can be understood qualitatively by the aromatic sextet proposed by Clar.[5] According to the Clar's theory, [*n*]acenes possess only one aromatic sextet ring (Fig. 2.3). Thus, the aromaticity is shared among the rings, resulting in lower aromaticity per ring and lower stability for higher [*n*]acenes. In fact, pentacene has been known to be very sensitive to oxygen and light to give the endoperoxide **4** or the dimer **5** (Scheme 2.1). The lower stability as well as low solubility of hexacene, heptacene, and higher [*n*]acenes has prevented them from being isolated and well-characterized until recently. Accordingly, there have been a limited number of reports on them, and the synthesis of higher acenes is still a challenging goal.[6] Although a huge number of synthetic methods have been developed for lower [*n*]acenes ($n \leq 3$), this chapter will focus on higher [*n*]acenes ($n \geq 4$).

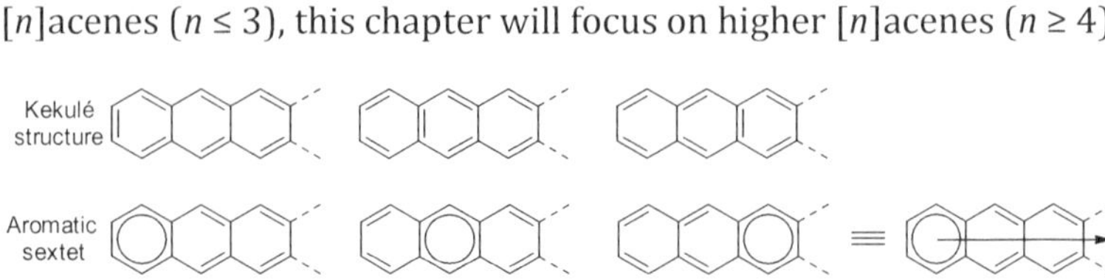

Figure 2.3 Clar's aromatic sextet.

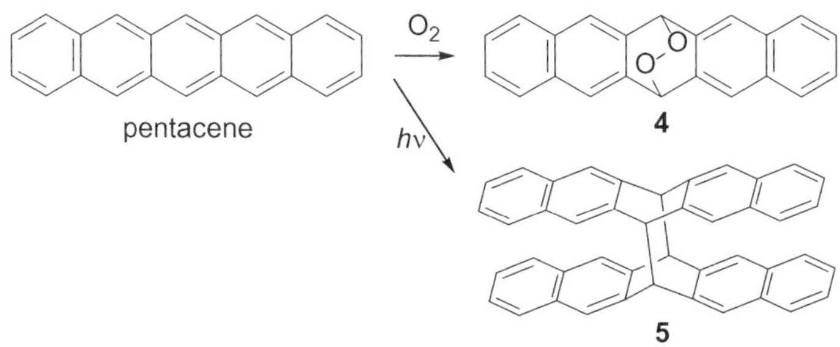

Scheme 2.1 Reactions of pentacene under ambient conditions.

2.2.2 Hydrocarbon [*n*]Acenes

Benzene, naphthalene, and anthracene can be extracted from coal. On the other hand, higher [*n*]acenes can be obtained through multistep synthesis. For examples, pentacene can be synthesized via aldol condensation between two equivalents of *o*-phthalaldehyde and one equivalent of 1,4-cyclohexanedione and the subsequent reduction of the obtained pentacene-6,13-dione **6** (Scheme 2.2).[7] Other synthetic strategies for [*n*]acenes are summarized in some excellent review articles.[6c]

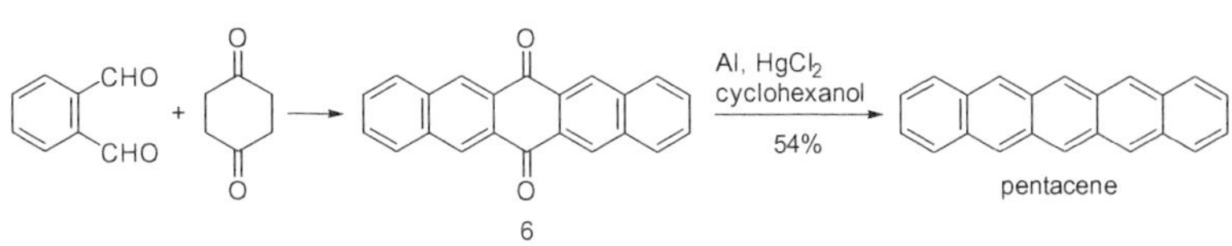

Scheme 2.2 Representative synthetic route to pentacene.

Introduction of substituents is efficient strategy to control electronic structure, stability, solubility, molecular arrangement in the solid state, and so on. Accordingly, a variety of substituted [*n*]acenes have been synthesized. In 2001, Anthony and coworkers developed a series of alkynylpentacenes **7**, which can be synthesized by the reaction of pentacene-6,13-dione **6** with an alkynylmetal reagent and the following aromatization (Fig. 2.4a).[8] The obtained dialkynylpentacene **7** shows higher stability and solubility than the parent pentacene and can be applied as solution-processable semiconducting materials (vide infra). They also succeeded in synthesizing dialkynylheptacenes **8** in 2005, which are stable enough for characterization (Fig. 2.4b).[9] This is the pioneering work for synthesizing higher [*n*]acenes.

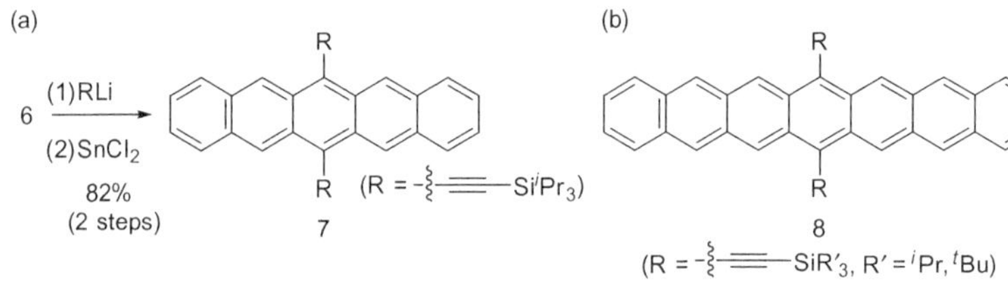

Figure 2.4 (a) Synthesis of dialkynylpentacene and (b) molecular structure of dialkynylheptacene.

Miller and coworkers also succeeded in the synthesis of heptacenes in 2009.[10] They used the Diels–Alder reaction for ring construction to synthesize heptacene **9** (Scheme 2.3). The steric effect of *ortho*-dialkylphenyl group provides photooxidative resistance on the acene core. In addition, the thioether substituents were found to be more effective for enhancing photooxidative resistance than silylalkynyl groups which are used in compounds **7** and **8**. They also achieved the synthesis of nonacene **10** by using the same synthetic strategy in 2010.[11] This is the first example for a persistent nonacene derivative.

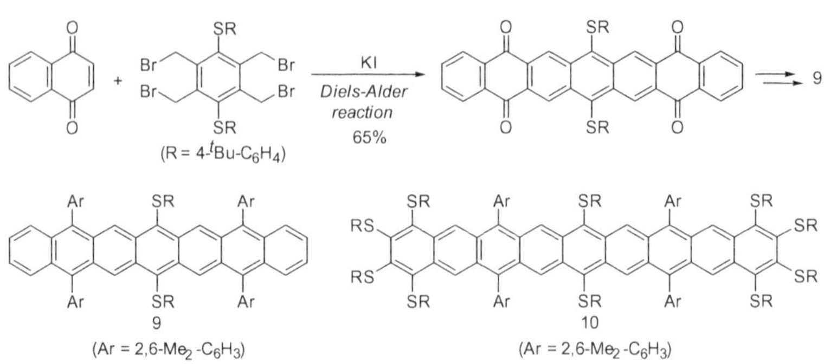

Scheme 2.3 Synthesis of substituted heptacene and nonacene.

Recent advances in transition-metal-mediated ring-construction reactions enable the efficient bottom-up synthesis of substituted acenes. In 2000, Takahashi et al. reported a novel homologation method as a pioneering work. Starting from phthalates **11**, reduction, bromination, and the reaction with alkynyllithium afford diynes **12** (Scheme 2.4).[12] Treatment of diynes **12** with Cp_2ZrBu_2 provides zirconacyclopentadienes **13**, and the subsequent reaction with acetylenedicarboxylate gives hydroanthracene derivatives **14**. Finally, aromatization is achieved by oxidation with DDQ to afford anthracene derivatives **15**. The resulting anthracene derivatives **15** possess a phthalate moiety and can be transformed to pentacene derivatives **16** via the same reaction sequence. Accordingly, this homologation method can add a naphthalene moiety per one cycle. A variety of substituted naphthacenes and pentacenes were synthesized by using this homologation method.[13]

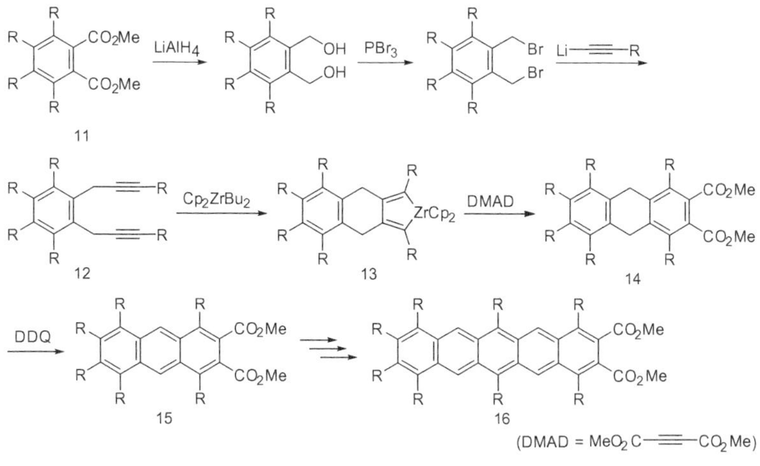

Scheme 2.4 Homologation strategy for substituted [*n*]acenes.

Catalytic version of homologation for [*n*]acene synthesis has also been investigated. The typical strategy is the transition-metal-catalyzed coupling of one equivalent of aromatic substrates with two equivalents of internal alkynes to provide the next higher [*n*]acenes (Scheme 2.5a). A range of efforts have been devoted in terms of catalytic conditions and the starting aromatic compounds, while most of them were applied to the synthesis of substituted naphthalenes and anthracenes. In this context, Miura and coworkers developed rhodium-catalyzed oxidative coupling of arylboron reagents with alkynes and applied it to the synthesis of naphthacene derivatives **17** in 2011 (Scheme 2.5b).[14] This method involves C–H bond cleavage which is attractive in the view of atom economy and efficiency. By changing the combination of the two starting materials, a variety of substituted naphthacenes **17** were synthesized in moderate yields. Very recently, Cramer and Pham reported the rhodium-catalyzed aromatic homologation which involves double C–H bond cleavage.[15] They successfully synthesized naphthacene derivative **18** by using this reaction (Scheme 2.5c).

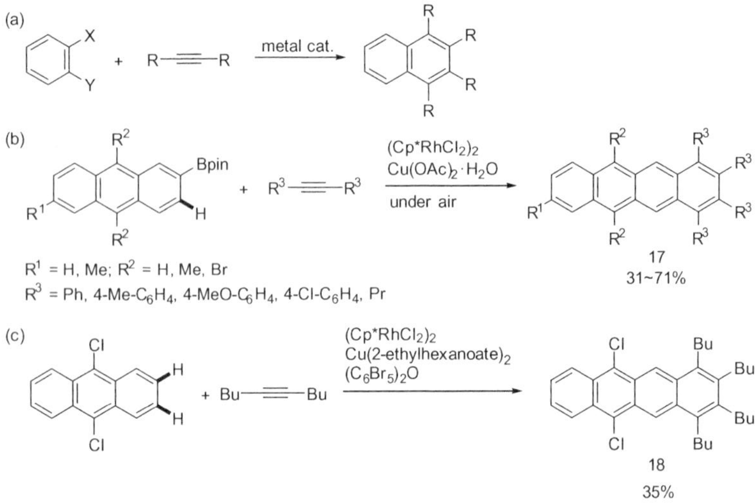

Scheme 2.5 Transition-metal-catalyzed homologation method for [*n*]acene synthesis.

2.2.3 Heteroacenes

One of the major classes of heteroacenes is the one with heteroaromatic rings fused at the terminal(s) in a linear fashion. For example, anthradithiophenes (ADTs) **19**, which are isoelectronic with pentacene, as shown in Fig. 2.2, were reported by Katz and coworkers in 1998 (Fig. 2.5).[16] These compounds can be synthesized with the corresponding thiophene-2,3-dicarbaldehydes as a starting material through the same reaction sequence described in Scheme 2.2. However, regioselectivity of this reaction is poor, resulting in an inseparable mixture of *syn-/anti-*isomers. Dialkynyl derivatives

20[17] and their furan analogues anthradifurans (ADFs) **21**[18] were reported by the Anthony group and the Chow group, respectively, while they are also a mixture of *syn-/anti*-isomers (Fig. 2.5).

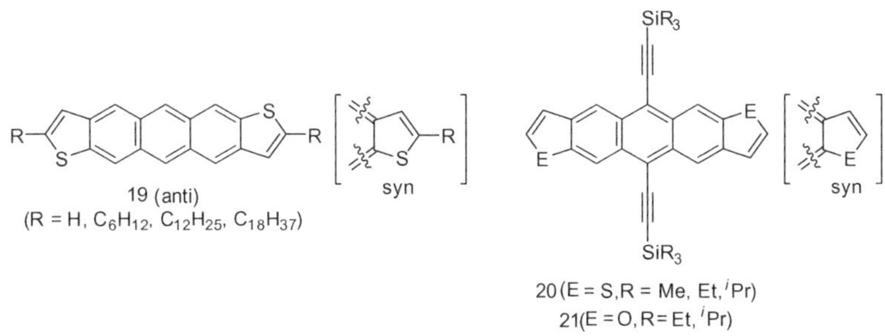

Figure 2.5 Molecular structures of anthradithiophenes (ADTs) and anthradifurans (ADFs).

Recently, isomerically pure *syn*- and *anti*-ADTs were successfully synthesized. In 2012, Jurchescu, Tykwinski, and coworkers reported the synthesis of *syn*-ADTs **22** which involves the regioselective aldol condensation as a key reaction (Scheme 2.6a).[19] *anti*-ADTs **23** were synthesized with use of well-designed substrates by the Geerts group[20] and the Takimiya group[21] independently (Schemes 2.6b and 2.6c). Takimiya and coworkers also investigated the synthesis of isomerically pure ADF,[21] naphtodithiophenes (NDTs)[22] and naphthodifurans (NDFs),[23] which are thiophene- and furan analogues of naphthacene, by using the cyclization reaction from dialkynylanthracene and dialkynylnaphthalenes as precursors.

Another class of heteroacenes is the one with a benzene-heteroaromatic ring alternating structure. Typical compound is benzo[1,2-*b*:4,5-*b*']bis[1]benzothiophene (BBBT). This compound is called thienoacene since benzene and thiophene rings are fused in a linear fashion. However, it is isoelectronic not with pentacene, the corresponding hydrocarbon acene, but with dibenzo[*a*,*h*]anthracene (Fig. 2.6). The conventional synthetic strategy utilized the Bradsher reaction with bis([1]benzothiophene-3-yl)methane.[24a] In 2007, Takimiya and coworkers reported the efficient method to synthesize BBBT using Bergman cycloaromatization (Scheme 2.7a).[24b] More π-extended compound, dibenzo[*b*,*b*']thieno[2,3-*f*:5,4-*f*']bis[1] benzothiophene (DBTBBT), was tried to be synthesized by intramolecular ring-closing condensation of (methysulfinyl)biaryl, while the obtained material was a mixture of the regioisomers (Scheme 2.7b).[25] Recently, dinaphtho[2,3-*d*:2',3'-*d'*]benzo[1,2-*b*:4,5-*b'*]dithiophenes **24**, which are also π-extended compounds with a BBBT core, were successfully synthesized and characterized by Takeya group via intramolecular cyclization with Newman–Kwart rearrangement (Scheme 2.7c).[26]

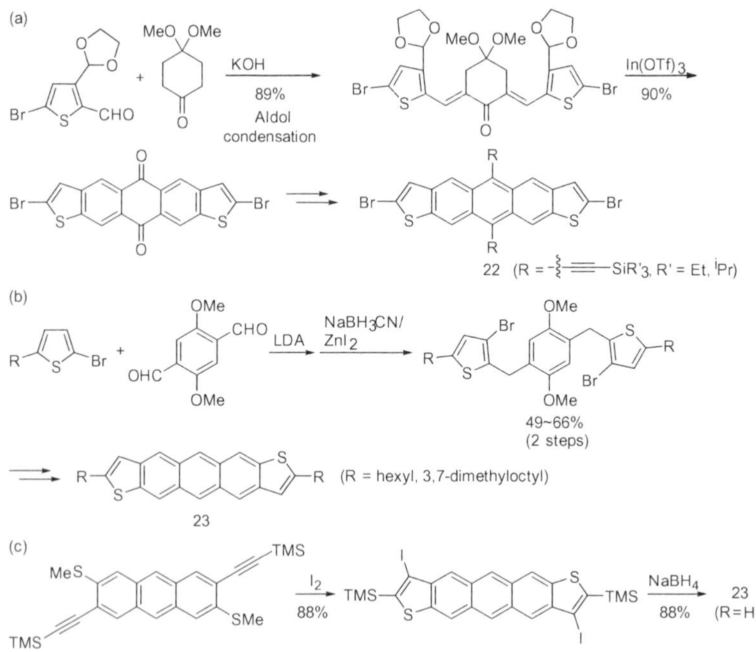

Scheme 2.6 Synthesis of isomerically pure ADTs.

Figure 2.6 Molecular structures of BBBT and dibenzo[*a*,*h*]anthracene.

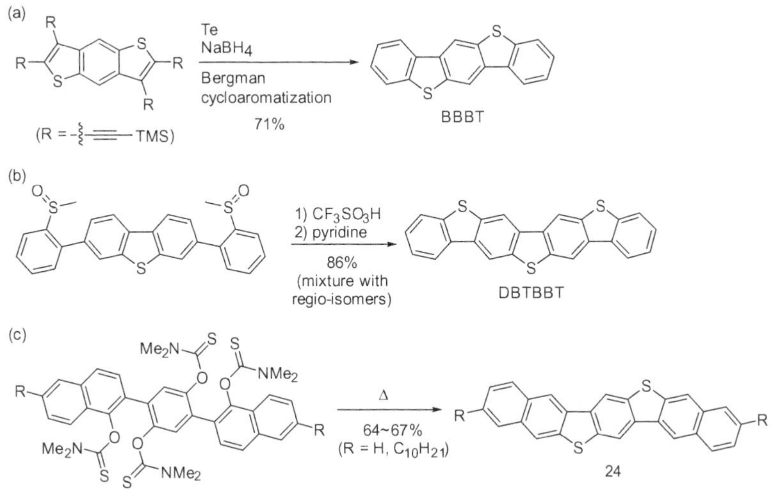

Scheme 2.7 Synthesis of thienoacenes with a BBBT skeleton.

In addition to thiophene-fused heteroacenes, pyrrole-fused ones have been also developed. The representative is indolo[3,2-*b*] carbazole which is a nitrogen analogue of BBBT. Fischer indole synthesis and Cadogan cyclization have been applied to synthesize indolo[3,2-*b*]carbazole derivatives and their higher homologue **25**, while these methods potentially gave isomers with a bent structure (Scheme 2.8).[27]

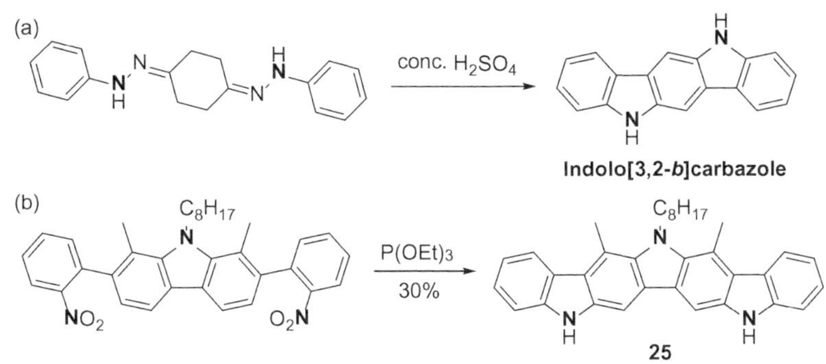

Scheme 2.8 Synthesis of pyrrole-fused heteroacenes.

The third class of heteroacenes is the one where two or more heteroaromatic rings are fused directly. The most-intensely studied skeleton in this class is [1]benzothieno[3,2-*b*]benzothiophene (BTBT) with a thieno[3,2-*b*]thiophene substructure. Since Takimiya and coworkers reported the high performance of BTBT derivatives as organic semiconductors (vide infra),[28] a series of BTBT derivatives and more π-extended homologues were synthesized and characterized thoroughly. The conventional synthetic approach for the BTBT skeleton is double cyclization from benzylidene dichloride, stilbene-2,2′-disulfonate, or 2,2′-dibromostilbene. Recently, Takimiya and coworkers developed new synthetic routes to the BTBT skeleton from commercially available *o*-chlorobenzaldehyde, which are easily applicable to multigram-scale synthesis (Scheme 2.9, routes a and b).[29] Transition-metal-catalyzed ring construction approaches were also developed. For example, the Takimiya group reported the efficient procedure which includes the cyclization of 2-[(trimethylsilyl)ethynyl]thioanisole with PhSCl, bromination, and the final ring construction via palladium-catalyzed intramolecular C–H arylation (Scheme 2.9, route c).[30] Miura and coworkers developed another procedure by using the palladium-catalyzed intramolecular decarboxylative C–C bond-forming reaction (Scheme 2.9, route d).[31]

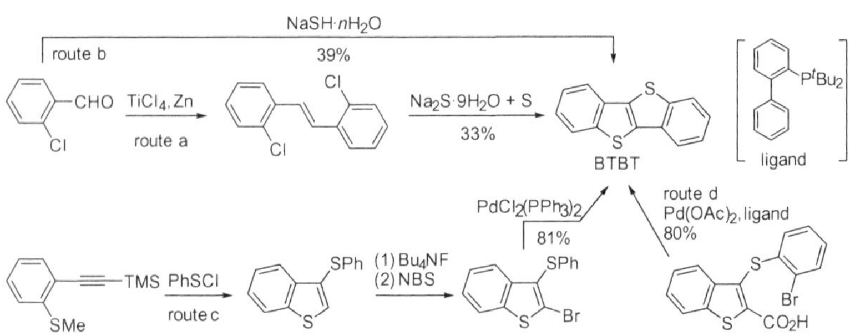

Scheme 2.9 Synthesis routes to BTBT.

More π-extended thienoacenes with the BTBT substructure were also synthesized. Takimiya and coworkers reported the synthesis of dinaphtho[2,3-*b*:2′,3′-*f*]thieno[3,2-*b*]thiophenes (DNTT)[32] and dianthra[2,3-*b*:2′,3′-*f*]thieno[3,2-*b*]thiophene (DATT),[33] which are benzo- and naphtho-annelated homologues of BTBT, by using iodine-mediated cyclization of the corresponding bis[(methylthio)aryl]ethenes (Scheme 2.10). In addition, thienoacenes with two thieno[3,2-*b*]thiophene substructures have been developed by using the similar synthetic strategies described above.[34]

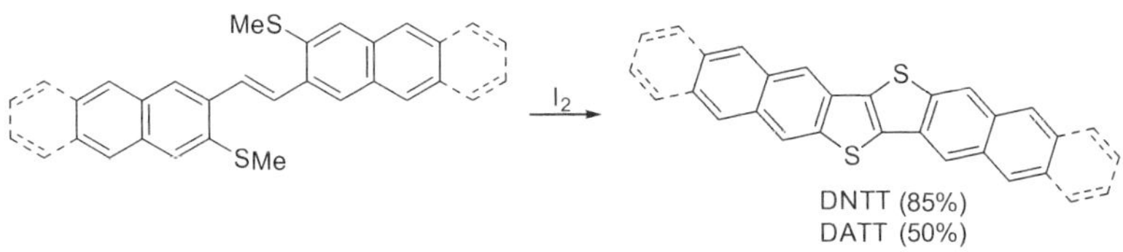

Scheme 2.10 Synthesis of DNTT and DATT.

As shown in Fig. 2.2, the fused oligothiophene skeletons with three or more thiophene rings have also been incorporated into thienoacenes. Two major strategies to construct the dithieno[3,2-*b*;2',3'-*d*]thiophene (DTT) core are the bridging with a sulfur atom at 3- and 3'-positions of a 2,2'-bithiophene unit (route a) and the C–C bond formation at 2- and 2'-positions of a di(thiophen-2-yl)sulfane unit (route b) (Scheme 2.11a). By using these synthetic strategies, a variety of thienoacenes have been synthesized.[35] Yamaguchi and coworkers developed unique method to construct the DTT core by using the intramolecular triple cyclization of bis(*o*-haloaryl) diynes and the following dechalcogenation (Scheme 2.11b).[36] They successfully synthesized dibenzo-fused DTT **26** and higher thinoacenes with up to eight thiophene rings.

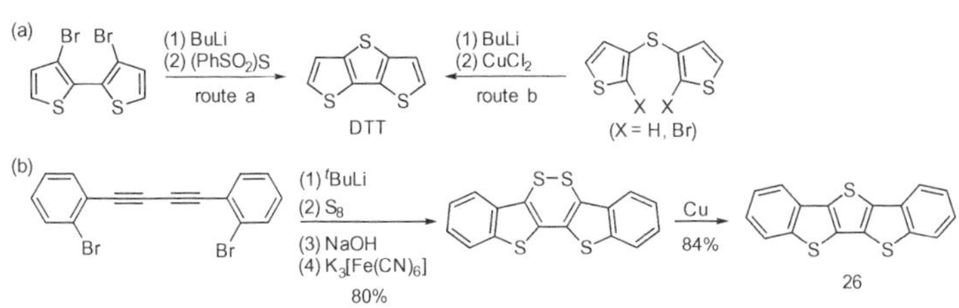

Scheme 2.11 Synthesis of thienoacenes with a dithieno[3,2-*b*;2',3'-*d*] thiophene (DTT) core.

There have been some literatures which have investigated the synthesis of indolo[3,2-*b*]indole (II) derivatives, which are nitrogen analogues of BTBTs. Indolo[3,2-*b*]indole derivatives were conventionally synthesized using reductive cyclization of dinitrobenzils (Scheme 2.12a),[37] Fischer indole synthesis with phenylhydrazines and indolin-3-ones,[38] and so on, while these methods generally result in low yield and/or require multistep synthesis. In this context, more convenient method to II and dinaphtho[2,3-*b*:2′,3′-*f*]pyrrolo[3,2-*b*]pyrrole (DNPP) has been reported recently that utilizes reductive cyclization of 6,12-dibenzo[*b*,*f*][1,5]diazocines (Scheme 2.12b).[39] Transition-metal-catalyzed ring construction via intramolecular C–H/C–H coupling reaction was also applied to the synthesis of indolo[3,2-*b*] indole derivative **27**, as reported by Kanai, Kuninobu, and coworkers (Scheme 2.12c).[40]

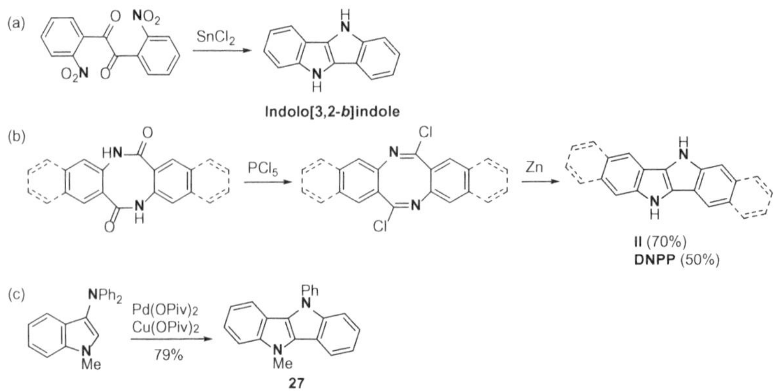

Scheme 2.12 Synthetic approach to indolo[3,2-*b*]indole derivatives.

Silole- or phosphole-containing compounds have been known to exhibit efficient luminescence properties and/or electron-transporting properties. Thus, incorporation of these rings into heteroacene structures has been investigated.[41] Heteroacenes **28** and **29** with a benzene-silole alternating structure were first reported by the Murakami group[42] and the Xu group[43] independently by using iridium-catalyzed [2+2+2] cycloaddition between diynes and alkynes (Scheme 2.13a) and the conventional dibenzosilole construction method (Scheme 2.13b, route a), respectively. Shimizu, Hiyama, and coworkers applied their efficient method for dibenzosilole synthesis, which includes palladium-catalyzed intramolecular coupling of 2-(arylsilyl)aryltriflates, to the synthesis of silole-fused heteroacene **29** (Scheme 2.13b, route b).[44] Compound **28** was reported to show a high fluorescence quantum yield of >90%. Heteroacene **30** with a silolo[3,2-*b*]silole substructure, which is a silicon analogue of BTBT, was first synthesized by Barton and coworkers through a rearrangement of 5,6-disiladibenzo[*c,g*]cyclooctyne derivative.[45] Tamao, Yamaguchi, and coworkers developed an efficient method to construct the structure via intramolecular reductive double cyclization from bis(2-silylphenyl)ethynes (Scheme 2.13c) and synthesized silolo[3,2-*b*]silole-fused compounds **30** and **31**.[46] The synthesis of phosphole analogues were also achieved.[47] For example, compounds **32** and **33** were synthesized via radical cyclization, as reported by Liu's group and Kawashima's group independently (Scheme 2.14).[47a,b] A diastereomeric mixture of *trans*- and *cis*-**33**, which can be separated by silica-gel column chromatography, was obtained under the condition with BEt_3/O_2. Yamaguchi and coworkers expanded their double cyclization strategy described above to the synthesis of benzophospholo[3,2-*b*]benzophospholes **34** (Scheme 2.14c).[47c] This reaction also gave a diastereomeric mixture, and the diastereomers can be separated by chromatography. The resulting compounds were found to show a significantly high fluorescence quantum yield of 99%. Heteroacenes with a phosphole-thiohene fused structure have been developed by Baumgartner and coworkers.[48] The versatile reactivity of the trivalent phosphorus atom allows easy derivatization to afford a series of phospholo[3,2-*b*:4,5-*b*']bis[1]benzothiophene derivatives **35** (Scheme 2.14d). These derivatives exhibit moderate fluorescence quantum yields (<72%). The oxygen analogue of **35** was also synthesized and applied to OLEDs.[49]

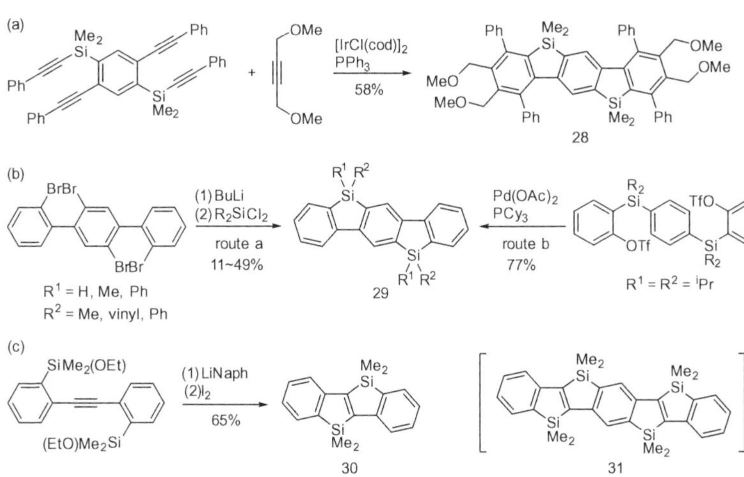

Scheme 2.13 Synthesis of silole-fused heteroacenes.

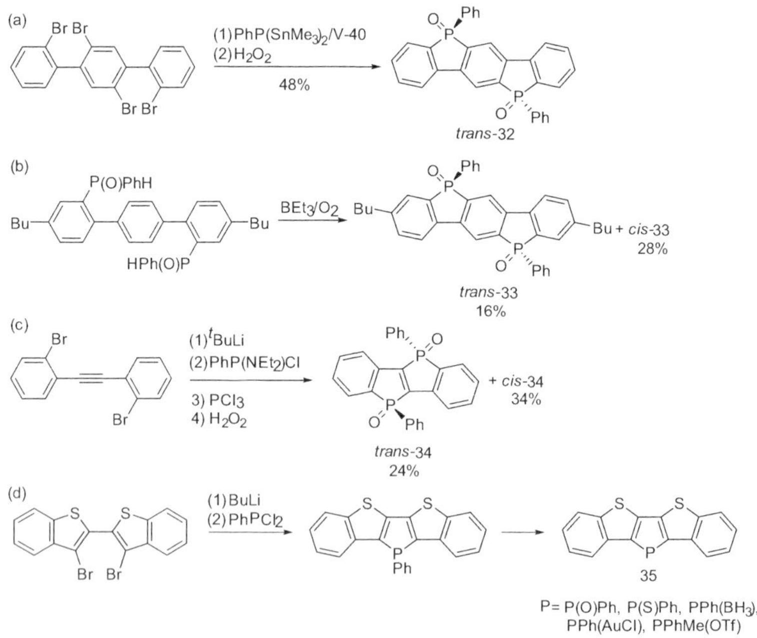

Scheme 2.14 Synthesis of phosphole-fused heteroacenes.

2.2.4 Semiconducting Properties of [*n*]Acenes and Heteroacenes

[*n*]Acenes and heteroacenes have been considered to be most promising semiconducting materials for OFETs.[1g,50] Transistors act as an on/off switch and are indispensable for the integrated circuits. OFETs typically consist of an organic semiconducting layer, a gate insulator, a source electrode, a drain electrode, and a gate electrode (Fig. 2.7a). When no bias is applied at the gate electrode, the current between the source and drain electrodes is near zero since there is no conductive channel in the organic semiconducting layer (off state). On the other hand, once a negative or positive bias is applied to the gate electrode, positive or negative charge is injected to organic semiconducting layer, forming a conducting channel between the source and drain electrodes. In this state, current flow (I) can be allowed by applying a voltage between the source and

drain electrodes (on state). The important properties of OFETs are mobility (μ_{FET}) and *I*on/*I*off ratio. Since the current is in proportion to mobility, high mobility is required for application. Depending on the application, a typical target mobility is ~0.5 cm^2/Vs, which enables OFETs to be competitive with amorphous Si. The *I*on/*I*off ratio is the ratio of the current in the on (maximum gate voltage) and off states and should be greater than 10^5 for application.

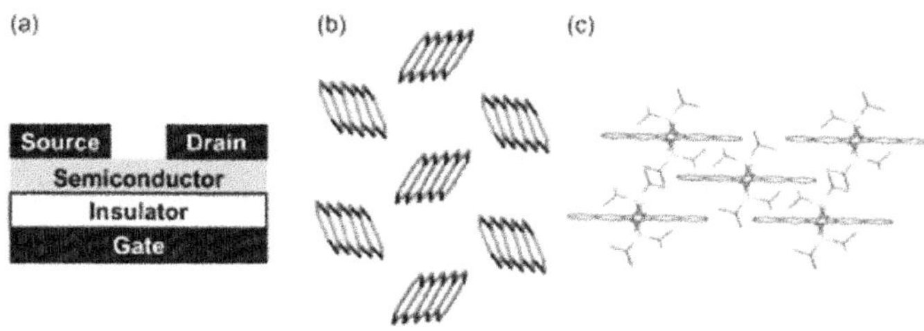

Figure 2.7 (a) Top-contact-type FET device and (b) molecular packing of pentacene and (c) dialkynylpentacene 7 (R = -CCSi^{i}Pr$_3$).

Pentacene has been known to demonstrate a high mobility of ~3 cm^2/Vs and is considered as a benchmark organic semiconducting material.[51] In the solid state, pentacene forms a herringbone packing structure (Fig. 2.7b). Such a molecular arrangement is favorable for charge transport because it allows two-dimensional electronic interaction. Acene-based organic semiconductors with high mobility tend to form this packing structure. Another promising packing structure for high mobility is a bricklayer one which allows two-dimensional face-to-face electronic interaction between π-conjugated planes (Fig. 2.7c). Typical materials with this packing structure are dialkynylpentacene **7**.[8]

Some heteroacenes have been reported to demonstrate mobility comparable to or higher than that of pentacene. Takimiya and coworkers reported high mobility of 1.0–2.0 cm^2/Vs with BTBT derivative **36a** in 2006.[28] After this report, device fabrication methods have been optimized, and a very high mobility of 43 cm^2/Vs (average: 25 cm^2/Vs) was reported with BTBT **36b** by using an off-center spin-coating method in 2013.[52] OFET characteristics of the selected molecules are summarized in Table 2.1.

Table 2.1 FET characteristics of selected compounds.

Compound	Mobility (cm^2/Vs)	On/off	Threshold voltage: V_{th} (V)
7[53]	0.4	10^6	–
19 (R = Hex)[16]	0.15 ± 0.02	–	–
20 (R = Et)[17]	1.0	10^7	–
21 (R = Et)[18]	0.11	8×10^4	18
23 (R = H)[21]	0.20–0.30	2.8×10^6	-3.3
37[21]	0.50–0.60	6.5×10^5	0.5
24[26]	16 (avg: 12.1)	10^6	−10 to 0
38[27a]	0.07–0.12	10^6–10^7	–
36a[28]	1.0–2.0	$>10^7$	–
DNTT[32]	2.1–2.9	10^7	−11 ± 2.0
DATT[33]	2.3–3.0	5×10^5	16.5
39[54]	5.4 - 8.8	10^5	-26 to −27
DNPP[39a]	0.014	4.1×10^5	-22.5

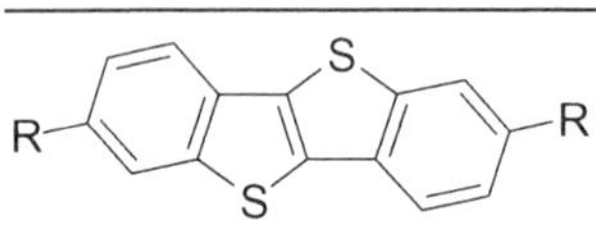

36a: R = Ph
36b: R = C_8H_{17}

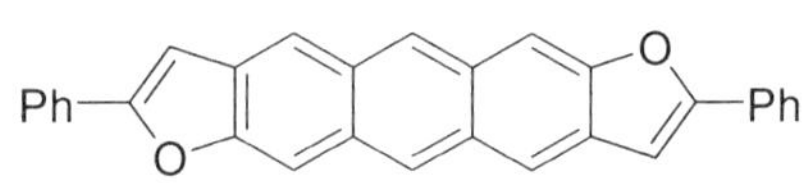

37

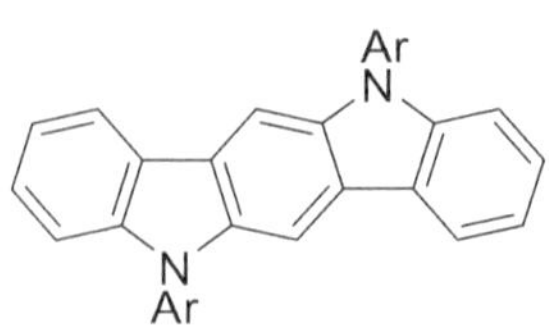

38
(Ar = 4-Oct-C_6H_4)

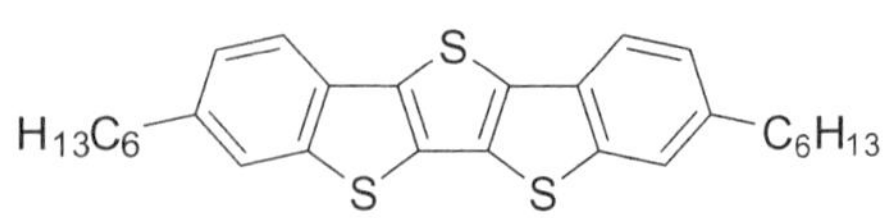

39

2.3 [*n*]Helicenes

2.3.1 Introduction

Helicenes have been known for more than 100 years since the first synthesis in 1903.[55] Helicenes composed of six or more benzene rings exhibit stable helical chirality because of an intramolecular steric repulsion. The *P* (plus) isomer possesses a right-handed helix, and the *M* (minus) isomer does a left-handed helix. The stable helical structure of helicenes provides unique chiroptical properties such as optical rotatory power, circular dichroism, and circularly polarized luminescence, compared to usual organic chiral molecules. After 50 years later from the first synthesis of helicene, Newman et al. reported the preparation of enantiomerically pure [6]helicene for the first time.[56] They synthesized racemic [6]helicene by using Friedel–Crafts acylation and separated the enantiomers through charge transfer complexation with enantiomerically pure 2-(2,4,5,7-tetranitro-9-fluorenylideneaminooxy)propionic acid (TAPA). It is not too much to say that this result opened up the greater interest in the chiroptical properties of helicenes as well as efficient synthetic methods including asymmetric versions.

2.3.2 Hydrocarbon [*n*]Helicenes

Conventional and common approach to helicenes has been photocyclization of stilbene skeleton. In 1967, the Scholz group[57] and the Martin group[58] independently reported the synthesis of [4]helicene and [7]helicene, respectively, by using oxidative photocyclization method. After that, the Martin group successfully synthesized higher helicenes (up to [14]helicene). This reaction requires high dilution condition to prevent an intermolecular [2+2] dimerization, which is a disadvantage for large-scale synthesis. However, because of easy access to substrates (e.g. by Wittig reaction) and its simplicity, this photocyclization approach is still the most common and powerful method for helicene synthesis. Very recently, the synthesis of [16]helicene, which is the longest carbohelicene ever reported, was achieved via sextuple photocyclization by Fujita, Murase, and Mori (Scheme 2.15).[59] They found that the [2]+[1]+[1]+[2] sequence is efficient for higher helicene synthesis and proposed that such a sequence tends to form spontaneous helical folding under photoirradiation. This should be a novel guideline for synthesizing higher helicenes.

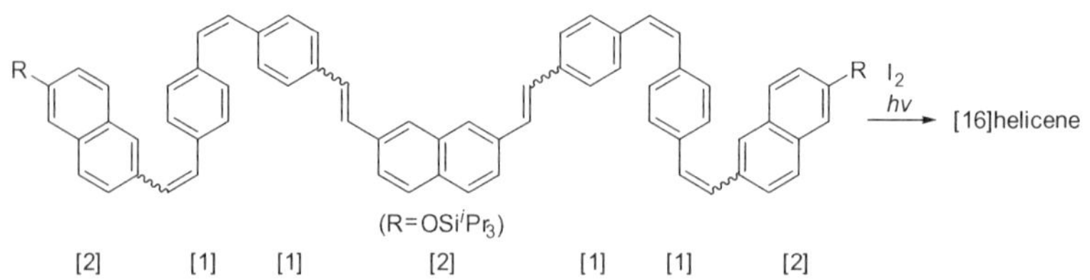

Scheme 2.15 Synthesis of [16]helicene.

As described above, Newman et al. reported the preparation of enantiomerically pure [6]helicene through charge transfer complexation with enantiomerically pure TAPA. After that, some asymmetric syntheses of helicenes were reported with chiral auxiliaries (Scheme 2.16).[60] In addition, the absolute asymmetric synthesis of [6]helicene was achieved by using circularly polarized light, as reported by Kagan and coworkers, while the enantiomeric excess was very low (<0.20%).[61]

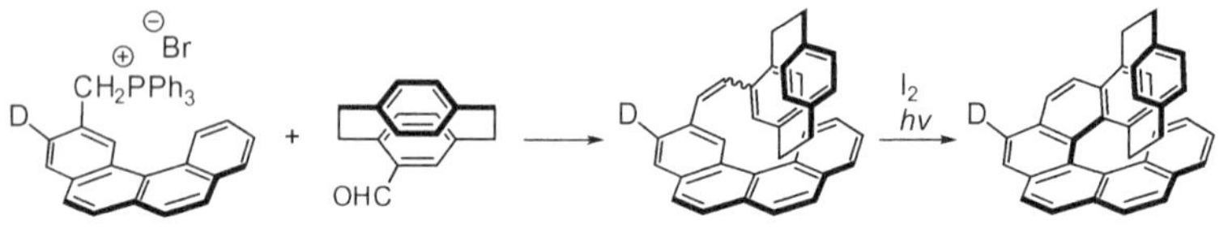

Scheme 2.16 Asymmetric synthesis of [6]helicene with chiral auxiliary.

Another conventional and common method is based on Diels–Alder cycloaddition, which was first achieved by Weidlich in 1938.[62] This method can be applied to large-scale synthesis. In addition, the cycloaddition is tolerant of a variety of substituents on dienes and dienophiles. Accordingly, this method is useful for synthesizing multisubstituted helicenes. In the 1990s and 2000s, Katz and coworkers made a great contribution to this Diels–Alder method and the whole of helicene chemistry. They prepared a number of helicenes **40** by the reaction of divinylarenes with 1,4-benzoquinone (Scheme 2.17a).[63] The quinone moiety can be reduced to hydroquinone to afford fully aromatic helicenes. They also investigated some helicene aggregates which showed enhanced chiroptical properties. Asymmetric synthesis with the Diels–Alder approach was also achieved by using enantiopure benzoquinone **41** (Scheme 2.17b).[64]

(a)

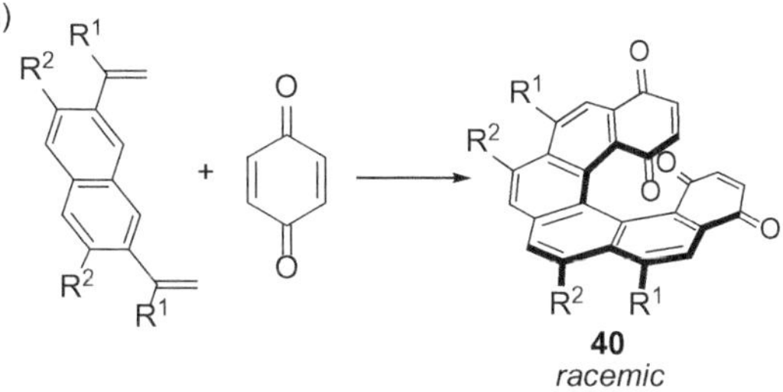

(b)

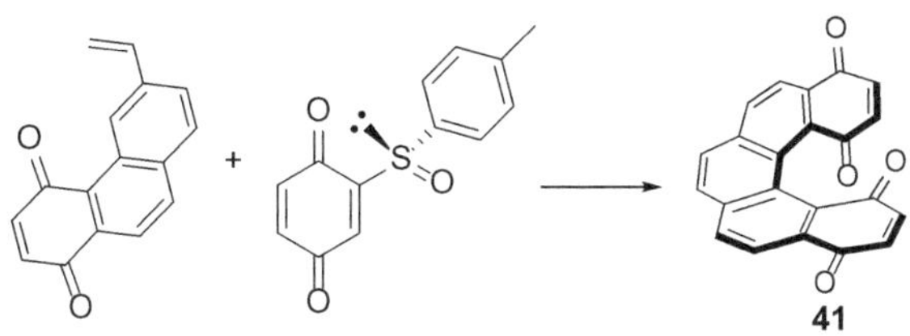

41
80% ee

Scheme 2.17 Synthesis of helicenes via Diels–Alder reaction.

As in the Newman's work for the first [6]helicene synthesis, Friedel–Crafts-type intramolecular cyclization has historically been applied to helicene synthesis.[56b] As a new approach with Friedel–Crafts-type intramolecular cyclization, Ichikawa and coworkers reported the helicene synthesis from 1,1-difluoro-1-alkenes with two tethered aryl groups in 2008.[65a,b] The domino Friedel–Crafts-type intramolecular cyclization in the presence of magic acid ($FSO_3H \cdot SbF_5$) and the following dehydrogenation afforded the target helicenes **42** in good yields (Scheme 2.18a). An efficient synthesis of diamino[6]helicene **43** was also achieved by Gaucher and coworkers by using Friedel–Crafts-type intramolecular cyclization from the dinitrile precursor (Scheme 2.18b).[65c]

In this century, transition-metal-catalyzed ring construction has attracted much attention in helicene synthesis. In particular, a metal-catalyzed [2+2+2] cycloaddition method has been intensely studied because of its modular approach, wide substrate scope, and applicability to the enantioselective synthesis. In 1998, Stará, Starý, and coworkers reported an innovative synthetic method for helicene-like molecule **44** by using the cobalt-catalyzed intramolecular [2+2+2] cycloaddition of triynes (Scheme 2.19a).[66] Subsequently, they developed the synthesis of fully aromatic carbohelicene **45** via the intramolecular [2+2+2] cycloaddition and the following dehydrogenation or acetic acid elimination.[67] This strategy can be applied to the synthesis of higher helicenes. Indeed, they successfully synthesized [11]helicene in 2009.[68] An asymmetric version was also achieved by using a chiral ligand, while the enantioselectivities were moderate.

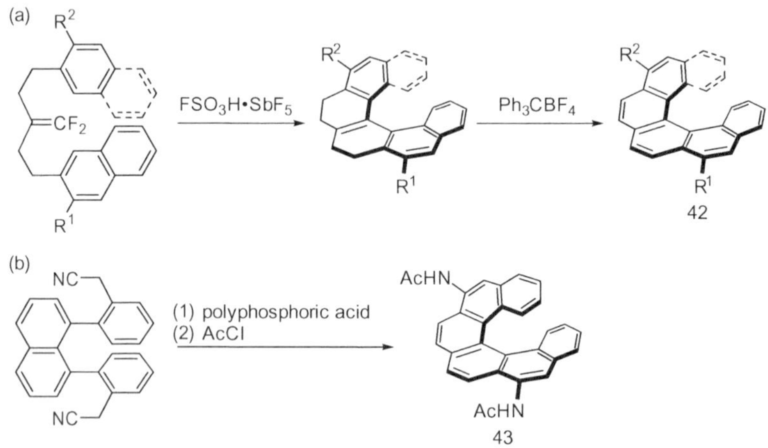

Scheme 2.18 Synthesis of helicenes via Friedel–Crafts-type reactions.

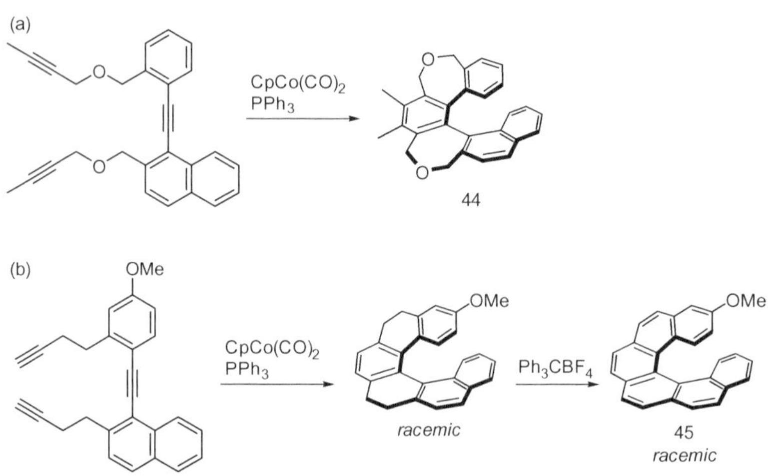

Scheme 2.19 Synthesis of helicenes via metal-catalyzed intramolecular [2+2+2] cycloaddition.

Pérez, Guitián, and coworkers developed another [2+2+2] cycloaddition approach for the synthesis of [5]helicenes in which the palladium-catalyzed cyclotrimerization of arynes is applied. The reaction with 1-(trimethylsilyl)naphthalen-2-yl trifluoromethanesulfonate gave a mixture of regioisomers including [5]helicene **46** (Scheme 2.20a).[69] On the other hand, [7]helicene **47** was obtained selectively despite the possibility of regioisomer formation.[70] This compound also possesses a [5]helicene moiety and is called double helicene. They also developed co-trimerization of arynes and alkynes for the synthesis of [5]helicenes.

(a)

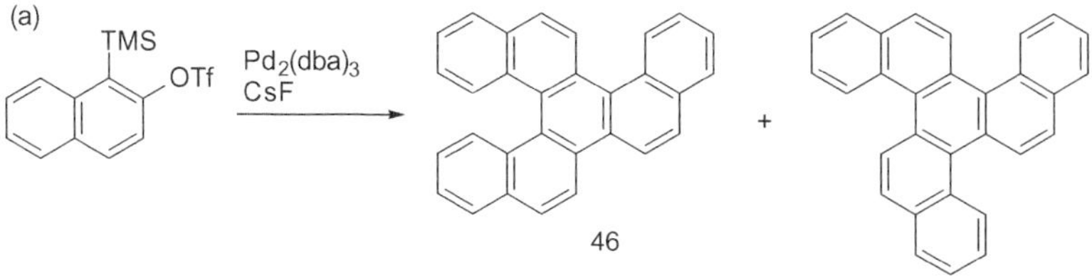

46

(b)

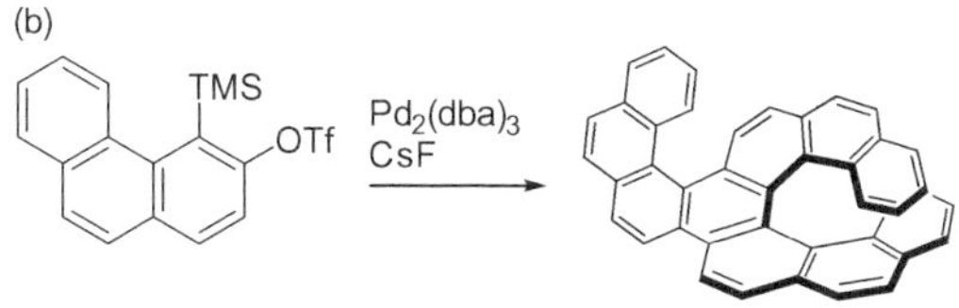

47

Scheme 2.20 Synthesis of helicenes via metal-catalyzed [2+2+2] cycloaddition of arynes.

Ring-closing olefin metathesis has been recognized as a powerful tool to construct cyclic compounds. In 2006, Collins et al. reported the efficient preparation of [5]-, [6]-, and [7]helicenes from the corresponding divinylbiaryl precursors by using a ruthenium-based second-generation Grubbs catalyst (Scheme 2.21a).[71] They further applied this method to asymmetric synthesis. [7]Helicene with high enantiopurity (up to 80% ee) was obtained from racemic 3,3'-divinyl-4,4'-biphenanthryl by kinetic resolution (Scheme 2.21b).[72] During the optimization of reaction conditions, they found that an olefin additive and solvent have a great impact on enantiomer selectivity.

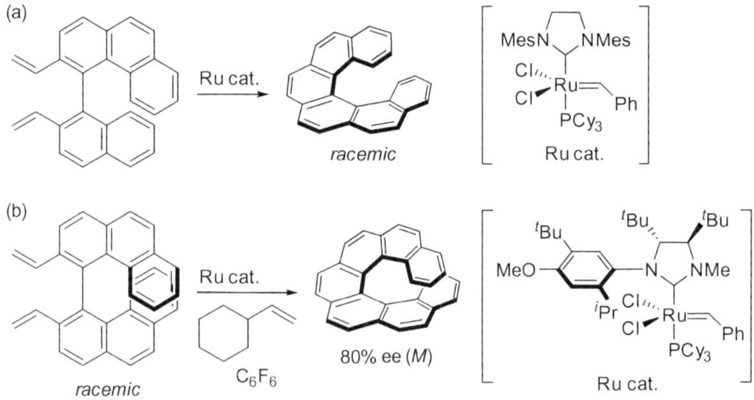

Scheme 2.21 Synthesis of helicenes via ring-closing olefin metathesis.

Transition-metal-catalyzed ring construction with C–H bond cleavage was also applicable to helicene synthesis. Kamikawa et al. developed palladium-catalyzed double C–H arylation of *Z,Z*-bis(bromostilbene)-type precursors to produce [5]- and [6]helicenes **48** (Scheme 2.22).[73] The reaction was found to give better yield with electron-deficient substrates.

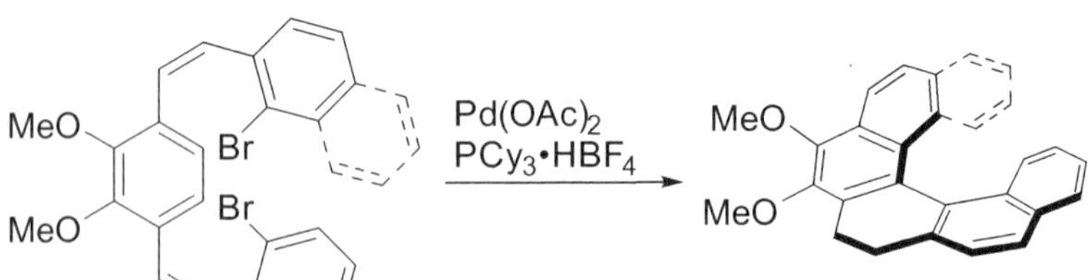

48

Scheme 2.22 Synthesis of helicenes via C–H bond cleavage.

2.3.3 Heterohelicenes

Introduction of heteroatom(s) into a helicene framework is useful for controlling properties and reactivity.[74] Thus, a variety of heterohelicenes have been synthesized. Basically, the synthetic methods for carbohelicenes described above can be applied to the synthesis of heterohelicenes. As for carbohelicene synthesis, photocyclization of stilbene skeleton has been widely used (Schemes 2.23a and 2.23b). Thiahelicenes fused with thiophene ring(s), such as **49**,[75] have been intensely studied among heterohelicenes. Recently, Hasobe and coworkers reported the synthesis of the tetrathia[9]helicene **50** and found a significant enhancement of the absolute fluorescence quantum yield (Φ = 0.27) by oxidation of the parent tetrathia[9]helicene (Scheme 2.23c).[76] Synthesis of other heterohelicenes, such as azahelicene **51** containing pyridine ring(s),[77] has also been investigated (Scheme 2.23b). In addition to photocyclization, the Diels–Alder approach has been applied to heterohelicene synthesis, as reported by Katz and coworkers (Scheme 2.23d).[78]

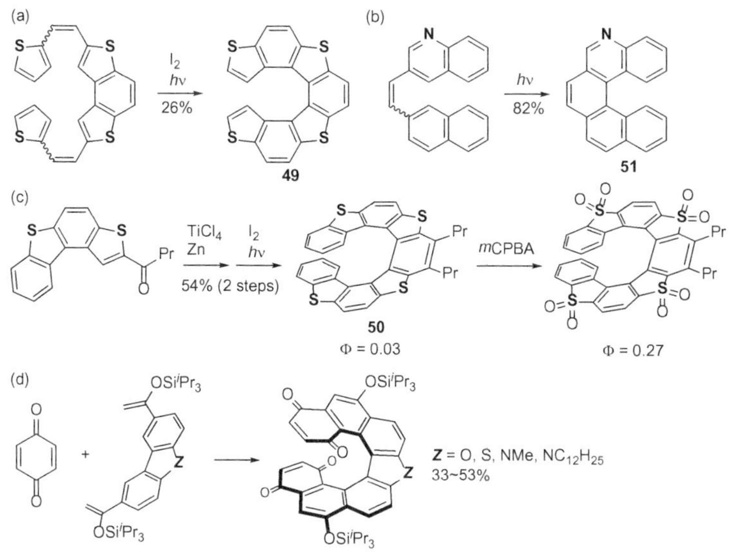

Scheme 2.23 Synthesis of heterohelicenes via photocyclization and Diels–Alder cycloaddition.

As shown in Scheme 2.11a, a thiophene ring can be constructed by the reaction of dilithiated biaryls with bis(phenylsulfonyl)sulfide. By using this strategy, Rajca and coworkers successfully synthesized thiahelicene **52** in which the helical π-conjugated framework is constructed only by thiohpene rings (Scheme 2.24).[79] Asymmetric synthesis of (*M*)-**52** was achieved by using (−)-sparteine in lithiation, while the enantiomeric excess is moderate (<47% ee). Higher thia[11]helicene was also synthesized.[80]

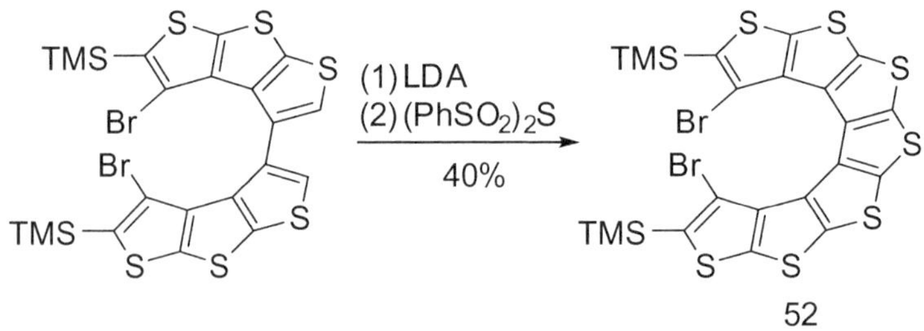

Scheme 2.24 Synthesis of thiahelicene which consists of only thiophene rings.

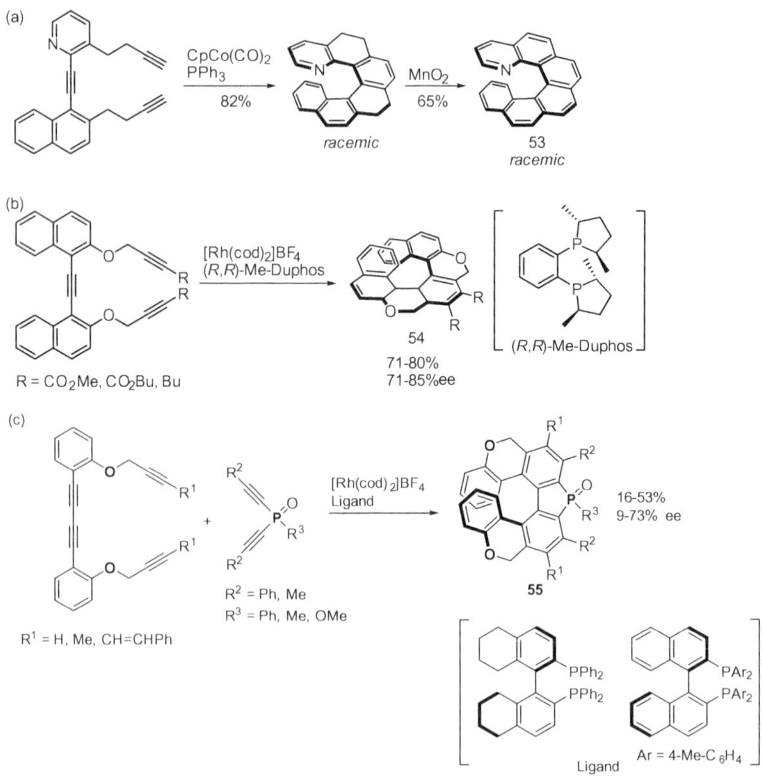

Scheme 2.25 Synthesis of heterohelicenes by [2+2+2] cycloaddition.

The metal-catalyzed [2+2+2] cycloaddition method have been applied to heterohelicene synthesis. In 2008, Stará, Starý, and coworkers utilized this reaction for synthesizing aza[6]helicene

53 (Scheme 2.25a).[81] Tanaka and coworkers intensely studied the application of metal-catalyzed [2+2+2] cycloaddition method to asymmetric synthesis of helicene-like molecules. In 2007, they reported the rhodium-catalyzed intramolecular enantioselective [2+2+2] cycloaddition of triynes to produce pyran-fused helicene-like compounds **54** (Scheme 2.25b).[82] Relatively high enantioselectivity was achieved by using (*R,R*)-Me-Duphos as a ligand. Subsequently, they expanded this method to prepare [9]helicene-like molecules via double intramolecular [2+2+2] cycloaddition of tetraynes and dialkynylketones or dialkynylphosphines, resulting in cyclopentadienone- or phosphole-fused (**55**) compounds (Scheme 2.25c).[83]

2.4 Synthesis of Heteroacenes and Heterohelicenes via Ring Construction by Carbon–Heteroatom Bond Formation

2.4.1 Introduction

In the last two decades, transition-metal-catalyzed carbon–heteroatom bond formation reactions have been significantly developed. The representative achievement is the Buchwald–Hartwig coupling reaction reported in 1995 in which a C–N bond formation proceeds via a palladium-catalyzed cross-coupling reaction of an aryl halide with an amine.[84] This reaction is a very powerful tool to make a C–N bond and has been applied to the synthesis of a wide range of compounds including natural products, biologically active compounds, and organic electronic/optoelectronic materials.

In 2003, Nakano, Nozaki, Hiyama, and coworkers reported the application of the Buchwald–Hartwig C–N coupling reaction to the carbazole synthesis (Scheme 2.26a).[85] The carbazole skeleton is contained in a variety of alkaloids from plants and organic (opt) electronic materials. Thus, it has been highly desired to develop an efficient synthetic method for carbazoles. In their method, primary amines and 2,2'-di(pseudo)halobiphenyls are coupled, forming two C–N bonds on the nitrogen atom. Accordingly, this method is called double *N*-arylation. A variety of di(pseudo)halobiphenyls and amines with electron-donating or electron-withdrawing substituent(s) can be employed to give multisubstituted carbazoles in moderate to high yields (Scheme 2.26). Following their report, this method has been applied to the synthesis of carbazole-containing natural products and organic functional materials.[86]

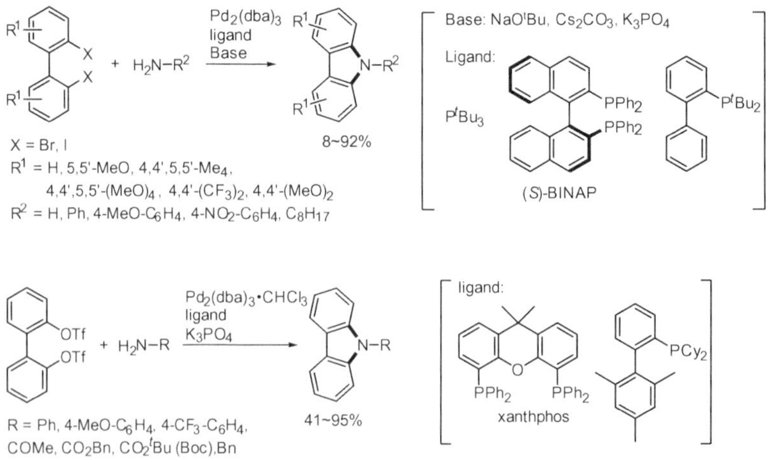

Scheme 2.26 Double *N*-arylation for carbazole synthesis.

The following section will summarize the synthesis of heteroacenes and heterohelicenes through ring construction with carbon–heteroatom bond formation, mainly focusing on the author's work. In addition to the synthesis with a C–N bond formation reaction described above, other synthetic examples via formation of a carbon–heteroatom bond, such as C–O, C–Si, and C–P bonds, are also described.

2.4.2 Heteroacene Synthesis via Ring Construction Including Carbon–Heteroatom Bond Formation

Nakano and coworkers expanded their double *N*-arylation strategy to the synthesis of pyrrole-fused heteroacenes.[87] They designed the synthetic route from terphenyls **56** where twice double *N*-arylation proceeds to construct two pyrrole rings (Scheme 2.27a). Through optimization of reaction conditions, they found that concentration of terphenyls and the suitable combination of a base and an additive are important for predominant formation of the desired indolo[3,2-*b*] carbazoles **57**. By introducing a substituent on a terphenyl moiety, dissymmetrically substituted indolo[3,2-*b*]carbazoles **57b** and **57c** were synthesized for the first time. A cyano-group was found to induce a red shift in the UV-Vis absorption spectrum compared to the parent indolo[3,2-*b*]carbazole **57a** and alkyl-substituted **57b**. Pyrrolo[3,2-*b*:4,5-*b*']dicarbazoles **58**, which possess more-π-extended system than indolo[3,2-*b*]carbazoles, were also prepared via intramolecular *N*-arylation in the presence of a copper catalyst, as reported by Leclerc and coworkers.[88]

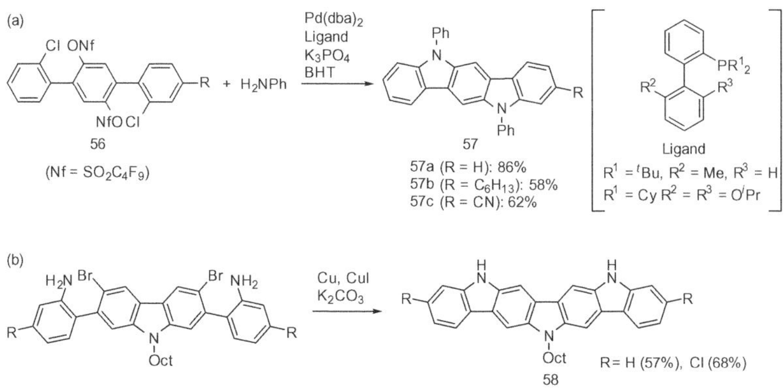

Scheme 2.27 Synthesis of indolo[3,2-*b*]carbazole derivatives.

The double *N*-arylation strategy was further applied to other pyrrole-containing heteroacenes. For example, Nakano and coworkers reported the synthesis of benzofuro[3,2-*b*]carbazoles **59** in 2008 (Scheme 2.28).[89] The reaction proceeds well to give the target compounds in high yields. UV-Vis spectra of them showed blue-shifted absorption bands compared to that of indolo[3,2-*b*]carbazole **58a**. In addition, cyclic voltammetry analysis of **59a** indicates lower HOMO energy level of **59a** than that of indolo[3,2-*b*] carbazole **58a**. These data indicate less efficient π-conjugation of a benzofuro[3,2-*b*] carbazole core compared to an indolo[3,2-*b*]carbazole core. Single-crystal X-ray crystallography revealed that two molecules form a cofacial and antiparallel stacking pair, probably because of the steric effect of the unsymmetrically placed phenyl group. The cofacial pairs further stack in a cofacial manner to form a one-dimensional stacking column.

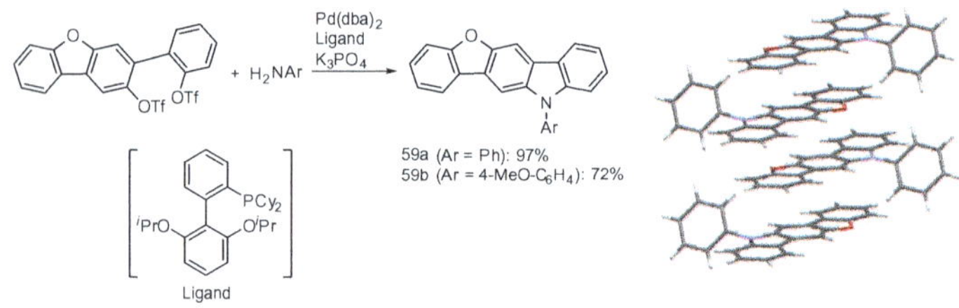

Scheme 2.28 Synthesis of benzofuro[3,2-*b*]carbazoles.

Very recently, Nakano and coworker also reported the synthesis of benzofuro[3,2-*b*]indoles (BFIs, **60**) and indolo[3,2-*b*]indoles (IIs, **61**), which are oxygen/nitrogen and nitrogen analogues of BTBTs, by using palladium-catalyzed double *N*-arylation (Scheme 2.29).[90] This is the first example of benzofuro[3,2-*b*]carbazole synthesis via double *N*-arylation, while the synthesis of indolo[3,2-*b*]indoles was achieved a year prior to their report.[91] Electron-donating substituents on anilines were tolerant to give the corresponding BFIs in moderate yields. The slightly electron-withdrawing fluoro-substituent is also applicable, while more electron-withdrawing trifluoromethyl and acyl groups are not suitable for the reaction. The same reaction conditions can be applied to the II synthesis to afford the target IIs **61** in moderate yields. Pyrrole is known to exhibit higher aromatic character than furan. Accordingly, more effective π-conjugation over the entire 18-π-electron system of the II skeleton is achieved, resulting in the red-shifted absorption spectrum.

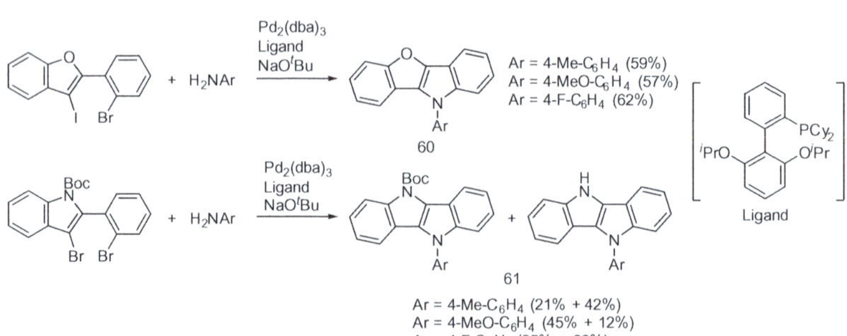

Scheme 2.29 Synthesis of benzofuro- and indolo[3,2-*b*]indoles.

In addition to a C–N bond, transition metal catalysis can make a carbon–heteroatom bond, such as C–O, C–S, C–Si, and C–P bonds. Catalytic reactions for these carbon–heteroatom bonds have also been significantly studied, and been applied to construct dibenzoheterole skeletons which are heteroatom analogues of carbazole. The typical strategy to dibenzoheteroles is intramolecular carbon–heteroatom bond formation from the corresponding biphenyls (Scheme 2.30a). Indeed, Nakano and coworkers reported the synthesis of dibenzo[*d*,*d*']benzo[1,2-*b*:4,5-*b*']difurans (DBBDF) **62** via palladium-catalyzed intramolecular C–O bond formation (Scheme 2.30b).[87] The synthesis of DBBDF was already investigated via cyclodehydrogenation of 1,4-diphenoxybenzene with $Pd(OAc)_2$ in AcOH before their report, while by-products such as a regioisomer were also obtained. In contrast, in their method, C–O bond formation occurred at the carbon atom bearing the chloride-leaving group, leading to the selective synthesis. Similar to the synthesis of indolo[3,2-*b*]carbazoles, concentration of terphenyls and the combination of K_3PO_4 and BHT are important for high yields. Single-crystal X-ray crystallography revealed that unsubstituted DBBDF forms a herringbone packing structure like pentacene, which is favorable for charge transport. Indeed, DBBDF derivatives can work as organic semiconductors.

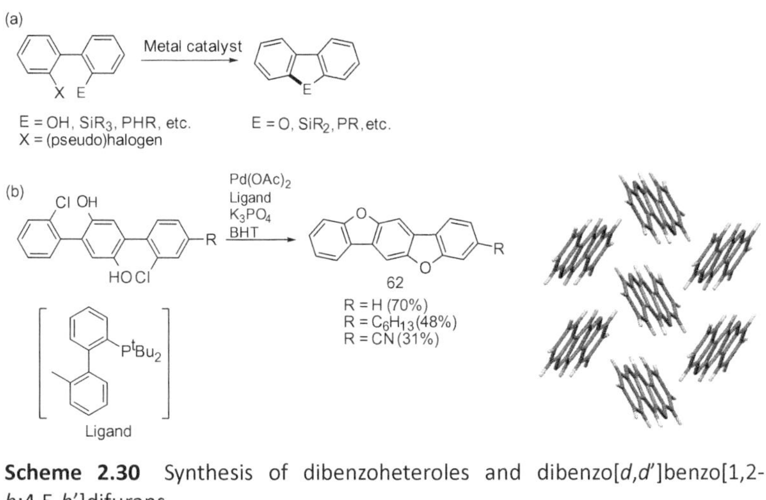

Scheme 2.30 Synthesis of dibenzoheteroles and dibenzo[*d*,*d*']benzo[1,2-*b*:4,5-*b*']difurans.

Kuninobu, Takai, and coworkers revealed that dibenzosiloles **63** were obtained through the reaction of 2-silylbiphenyls in the presence of rhodium catalyst (Scheme 2.31a).[92] Noteworthy is that the regioselectivity is perfect, and the possible sterically unfavorable regioisomer was not obtained. By utilizing this advantage, they successfully synthesized ladder-type dibenzo[*d*,*d*']benzo[1,2-*b*:4,5-*b*']disilole **64**. They also developed dibenzophosphole-oxide synthesis via a C–H bond functionalization and an intramolecular C–P bond formation. Biphenyl-2-ylhydrophosphine oxides were found to be transformed to dibenzophosphole oxides **65** in high yield (Scheme 2.31b).[93] They demonstrated the first synthesis of ladder-type dibenzo[*d*,*d*']benzo[1,2-*b*:4,5-*b*']diphosphole oxide **66**.

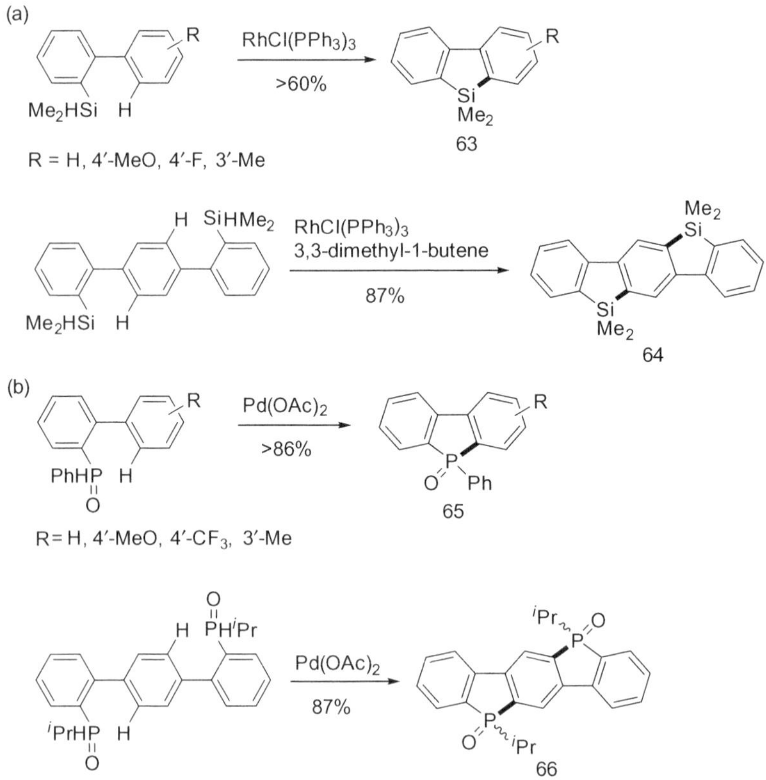

Scheme 2.31 Synthesis of silole- or phosphole-fused π-conjugated compounds.

2.4.3 Heterohelicene Synthesis via Ring Construction Including Carbon–Heteroatom Bond Formation

As shown in the previous section, the palladium-catalyzed carbon–heteroatom bond formation is a powerful method to construct heterole-fused π-conjugated compounds. Nakano, Nozaki, and coworkers applied their method to helicene synthesis. Starting from 4,4'-biphenanthryl-3,3'-diol, transformation to bisnonaflate and the following double *N*-arylation with aniline gave aza[7]helicene **67** in high yield (Scheme 2.32a).[94] Absolute configuration of the biphenanthryl moiety is maintained during the ring construction. Thus, a single enantiomer of **67** can be readily prepared from enantiomerically pure bisnonaflate by a stereospecific double *N*-arylation reaction. Oxa[7]helicene **68** was also synthesized via transformation of 4,4'-biphenanthryl-3,3'-diol to monononaflate and the subsequent intramolecular *O*-arylation (Scheme 2.32b). Oxa[7] helicene **68** was found to racemize under the reaction condition, indicating a lower racemization barrier of oza[7]helicene than that of aza[7]helicene **67**. Accordingly, a stereospecific conversion to enantioenriched oxa[7]helicene required a shorter reaction time. Introduction of a heteroatom allows regioselective functionalization: the regioselective bromination proceeds to give dibromo- and tetrabromohetero[7]helicenes, which are useful platforms for multisubstituted hetero[7]helicenes (Scheme 2.32c).

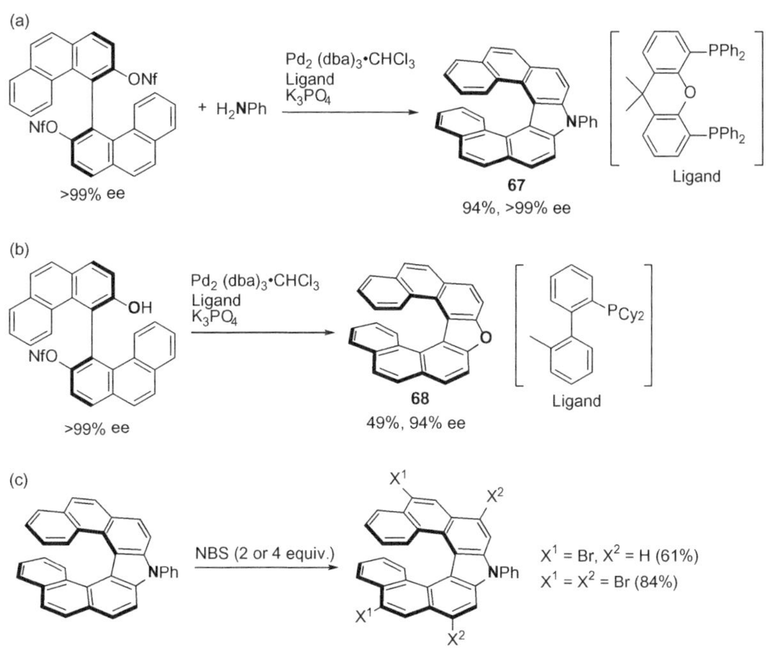

Scheme 2.32 Synthesis of aza- and oxa[7]helicenes.

Nakano and coworkers also prepared phospha[7]helicene **69** via stepwise C–P bond formation by using palladium catalysts (Scheme 2.33).[95] Single-crystal X-ray crystallography revealed that phospha[7]helicene **69** forms a columnar arrangement along the helical axis. They proposed that the phosphole-oxide moiety induces a dipole moment and the dipole moment vector perpendicular to the helical axis allows such an arrangement.

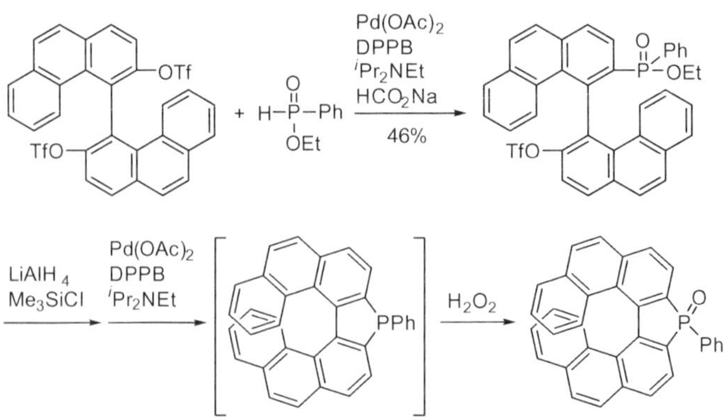

Scheme 2.33 Synthesis of phospha[7]helicene.

2.5 Conclusions

In this chapter, [*n*]acenes, [*n*]helicenes, and their heterocyclic analogues are briefly reviewed mainly from the viewpoint of synthetic methods. In addition to the conventional synthetic approaches, transition-metal-catalyzed reactions have been found to be a powerful tool and have become an indispensable method of synthesizing these classes of π-conjugated compounds. I hope new molecular designs along with further progress in synthetic methods will lead to novel functional materials based on these π-conjugated compounds in the future.

References

1. (a) Anthony, J. E. *Chem. Rev.*, **106**, 5028 (2006); (b) Mei, J., Diao, Y., Appleton, A. L., Fang, L., Bao, Z. *J. Am. Chem. Soc.*, **135**, 6724 (2013); (c)

Wang, C., Dong, H., Hu, W., Liu, Y., Zhu, D. *Chem. Rev.*, **112**, 2208 (2012); (d) Takimiya, K., Shinamura, S., Osaka, I., Miyazaki, E. *Adv. Mater.*, **23**, 4347 (2011); (e) Allard, S., Forster, M., Souharce, B., Thiem, H., Scherf, U. *Angew. Chem. Int. Ed.*, **47**, 4070 (2008); (f) Murphy, A. R., Fréchet, J. M. J. *Chem. Rev.*, **107**, 1066 (2007); (g) Bao, Z., Locklin, J. *Organic Field-Effect Transistors*, CRC Press: Boca Raton (2007).

2. (a) Müllen, K., Scherf, U. *Organic Light-Emitting Devices: Synthesis, Properties and Application*, Wiley-VCH Verlag GmbH & Co. KGaA: Weinheim (2006); (b) Kafafi, Z. *Organic Electroluminescence*, CRC Press: Boca Raton (2005); (c) Kraft, A., Grimsdale, A. C., Holmes, A. B. *Angew. Chem. Int. Ed.*, **37**, 402 (1998).
3. (a) Scharber, M. C., Mühlbacher, D., Koppe, M., Denk, P., Waldauf, C., Heeger, A. J., Brabec, C. J. *Adv. Mater.*, **18**, 789 (2006); (b) Sun, S.-S., Sariciftci, N. S. *Organic Photovoltaics: Mechanism, Materials, and Devices*, CRC Press: Boca Raton (2005).
4. Ruiz-Morales, Y. *J. Phys. Chem. A*, **106**, 11283 (2002).
5. (a) Clar, E. *The Aromatic Sestet*, John Wiley & Sons Ltd: New York (1972); (b) Solà, M. *Front. Chem.*, **1** (2013).
6. (a) Ye, Q., Chi, C. *Chem. Mater.*, **26**, 4046 (2014); (b) Zade, S. S., Bendikov, M. *Angew. Chem. Int. Ed.*, **49**, 4012 (2010); (c) Bendikov, M., Wudl, F., Perepichka, D. F. *Chem. Rev.*, **104**, 4891 (2004).
7. (a) Goodings, E. P., Mitchard, D. A., Owen, G. *J. Chem. Soc. Perkin Trans.*, **1**, 1310 (1972); (b) Pramanik, C., Miller, G. P. *Molecules*, **17**, 4625 (2012).
8. Anthony, J. E., Brooks, J. S., Eaton, D. L., Parkin, S. R. *J. Am. Chem. Soc.*, **123**, 9482 (2001).
9. Payne, M. M., Parkin, S. R., Anthony, J. E. *J. Am. Chem. Soc.*, **127**, 8028 (2005).
10. Kaur, I., Stein, N. N., Kopreski, R. P., Miller, G. P. *J. Am. Chem. Soc.*, **131**, 3424 (2009).
11. Kaur, I., Jazdzyk, M., Stein, N. N., Prusevich, P., Miller, G. P. *J. Am. Chem. Soc.*, **132**, 1261 (2010).
12. Takahashi, T., Kitamura, M., Shen, B. J., Nakajima, K. *J. Am. Chem. Soc.*, **122**, 12876 (2000).
13. (a) Takahashi, T., Li, S., Huang, W. Y., Kong, F. Z., Nakajima, K., Shen, B. J., Ohe, T., Kanno, K. *J. Org. Chem.*, **71**, 7967 (2006); (b) Takahashi, T., Kashima, K., Li, S., Nakajima, K., Kanno, K. I. *J. Am. Chem. Soc.*, **129**,

15752 (2007); (c) Li, S., Li, Z., Nakajima, K., Kanno, K., Takahashi, T. *Chem. Asian J.*, **4**, 294 (2009); (d) Takahashi, T., Li, Y., Hu, J., Kong, F., Nakajima, K., Zhou, L., Kanno, K. *Tetrahedron Lett.*, **48**, 6726 (2007).

14. Fukutani, T., Hirano, K., Satoh, T., Miura, M. *J. Org. Chem.*, **76**, 2867 (2011).
15. Pham, M. V., Cramer, N. *Angew. Chem. Int. Ed.*, **53**, 3484 (2014).
16. Laquindanum, J. G., Katz, H. E., Lovinger, A. J. *J. Am. Chem. Soc.*, **120**, 664 (1998).
17. Payne, M. M., Parkin, S. R., Anthony, J. E., Kuo, C. C., Jackson, T. N. *J. Am. Chem. Soc.*, **127**, 4986 (2005).
18. Watanabe, M., Su, W. T., Chang, Y. J., Chao, T. H., Wen, Y. S., Chow, T. J. *Chem. Asian J.*, **8**, 60 (2013).
19. Lehnherr, D., Waterloo, A. R., Goetz, K. P., Payne, M. M., Hampel, F., Anthony, J. E. Jurchescu, O. D., Tykwinski, R. R. *Org. Lett.*, **14**, 3660 (2012).
20. Tylleman, B., Vande Velde, C. M. L., Balandier, J. Y., Stas, S., Sergeyev, S., Geerts, Y. H. *Org. Lett.*, **13**, 5208 (2011).
21. Nakano, M., Niimi, K., Miyazaki, E., Osaka, I., Takimiya, K. *J. Org. Chem.*, **77**, 8099 (2012).
22. Shinamura, S., Osaka, I., Miyazaki, E., Nakao, A., Yamagishi, M., Takeya, J., Takimiya, K. *J. Am. Chem. Soc.*, **133**, 5024 (2011).
23. Nakano, M., Mori, H., Shinamura, S., Takimiya, K. *Chem. Mater.*, **24**, 190 (2012).
24. (a) Ahmed, M., Ashby, J., Ayad, M., Meth-Cohn, O. *J. Chem. Soc. Perkin Trans. 1,* 1099 (1973); (b) Ebata, H., Miyazaki, E., Yamamoto, T., Takimiya, K. *Org. Lett.*, **9**, 4499 (2007).
25. Sirringhaus, H., Friend, R. H., Wang, C. S., Leuningerb, J., Müllen, K. *J. Mater. Chem.*, **9**, 2095 (1999).
26. Mitsui, C., Okamoto, T., Yamagishi, M., Tsurumi, J., Yoshimoto, K., Nakahara, K., Soeda, J., Hirose, Y., Sato, H., Yamano, A., Uemura, T., Takeya, J. *Adv. Mater.*, **26**, 4546 (2014).
27. (a) Wu, Y. L., Li, Y. N., Gardner, S., Ong, B. S. *J. Am. Chem. Soc.*, **127**, 614 (2005); (b) Yudina, L. N., Bergman, J. *Tetrahedron*, **59**, 1265 (2003); (c) Grotta, H. M., Riggle, C. J., Bearse, A. E. *J. Org. Chem.*, **26**, 1509 (1961); (d) Bouchard, J., Wakim, S., Leclerc, M., *J. Org. Chem.*, **69**, 5705 (2004).
28. Takimiya, K., Ebata, H., Sakamoto, K., Izawa, T., Otsubo, T., Kunugi, Y. *J. Am. Chem. Soc.*, **128**, 12604 (2006).

29. (a) Saito, M., Osaka, I., Miyazaki, E., Takimiya, K., Kuwabara, H., Ikeda, M. *Tetrahedron Lett.*, **52**, 285 (2011); (b) Saito, M., Yamamoto, T., Osaka, I., Miyazaki, E., Takimiya, K., Kuwabara, H., Ikeda, M. *Tetrahedron Lett.*, **51**, 5277 (2010).

30. Mori, T., Nishimura, T., Yamamoto, T., Doi, I., Miyazaki, E., Osaka, I., Takimiya, K. *J. Am. Chem. Soc.*, **135**, 13900 (2013).

31. Minami, S., Hirano, K., Satoh, T., Miura, M. *Tetrahedron Lett.*, **55**, 4175 (2014).

32. Yamamoto, T., Takimiya, K. *J. Am. Chem. Soc.*, **129**, 2224 (2007).

33. Niimi, K., Shinamura, S., Osaka, I., Miyazaki, E., Takimiya, K. *J. Am. Chem. Soc.*, **133**, 8732 (2011).

34. (a) Yang, Y. S., Yasuda, T., Adachi, C. *Bull. Chem. Soc. Jpn.*, **85**, 1186 (2012); (b) Yamamoto, T., Nishimura, T., Mori, T., Miyazaki, E., Osaka, I., Takimiya, K. *Org. Lett.*, **14**, 4914 (2012).

35. Zhang, X. N., Côté, A. P., Matzger, A. J. *J. Am. Chem. Soc.*, **127**, 10502 (2005).

36. (a) Okamoto, T., Kudoh, K., Wakamiya, A., Yamaguchi, S. *Org. Lett.*, **7**, 5301 (2005); (b) Okamoto, T., Kudoh, K., Wakamiya, A., Yamaguchi, S. *Chem. Eur. J.*, **13**, 548 (2007).

37. (a) Murray, M. M., Kaszynski, P., Kaisaki, D. A., Chang, W., Dougherty, D. A. *J. Am. Chem. Soc.*, **116**, 8152 (1994); (b) Ruggli, P., Zaeslin, H. *Helv. Chim. Acta*, **18**, 845 (1935).

38. Grinyov, A. N., Ryabova, S. Y. *Khim. Geterotsikl.*, 199 (1982).

39. (a) Qiu, L., Yu, C., Zhao, N., Chen, W., Guo, Y., Wan, X., Yang, R., Liu, Y. *Chem. Commun.*, **48**, 12225 (2012); (b) Qiu, L., Wang, X., Zhao, N., Xu, S., An, Z., Zhuang, X., Lan, Z., Wen, L., Wan, X. *J. Org. Chem.*, **79**, 11339 (2014).

40. Saito, K., Chikkade, P. K., Kanai, M., Kuninobu, Y. *Chem. Eur. J.*, **21**, 8365 (2015).

41. (a) Stolar, M., Baumgartner, T. *Chem. Asian J.*, **9**, 1212 (2014); (b) Fukazawa, A., Yamaguchi, S. *Chem. Asian J.*, **4**, 1386 (2009).

42. Matsuda, T., Kadowaki, S., Goya, T., Murakami, M. *Org. Lett.*, **9**, 133 (2007).

43. Li, L. C., Xiang, J. F., Xu, C. H. *Org. Lett.*, **9**, 4877 (2007).

44. (a) Shimizu, M., Mochida, K., Katoh, M., Hiyama, T. *Sci. China Chem.*, **54**, 1937 (2011); (b) Shimizu, M., Mochida, K., Hiyama, T. *Angew. Chem. Int. Ed.*, **47**, 9760 (2008).

45. Serby, M., Ijadi-Maghsoodi, S., Barton, T. J., presented in part at the 33rd Symposium on Organosilicon Chemistry, Saginaw, MI (2000).

46. Yamaguchi, S., Xu, C. H., Tamao, K. *J. Am. Chem. Soc.*, **125**, 13662 (2003).

47. (a) Hanifi, D., Pun, A., Liu, Y. *Chem. Asian J.*, **7**, 2615 (2012); (b) Furukawa, S., Haga, S., Kobayashi, J., Kawashima, T. *Org. Lett.*, **16**, 3228 (2014); (c) Fukazawa, A., Hara, M., Okamoto, T., Son, E. C., Xu, C. H., Tamao, K., Yamaguchi, S. *Org. Lett.*, **10**, 913 (2008).

48. Dienes, Y., Eggenstein, M., Kárpáti, T., Sutherland, T. C., Nyulászi, L., Baumgartner, T. *Chem. Eur. J.*, **14**, 9878 (2008).

49. Chen, H., Delaunay, W., Li, J., Wang, Z. Y., Bouit, P. A., Tondelier, D., Geffroy, B., Mathey, F., Duan, Z., Réau, R., Hissler, M. *Org. Lett.*, **15**, 330 (2013).

50. (a) Pron, A., Gawrys, P., Zagorska, M., Djurado, D., Demadrille, R. *Chem. Soc. Rev.*, **39**, 2577 (2010); (b) *Organic Electronics II: More Materials and Applications*, WILEY-VCH Verlag GmbH & Co. KGaA: Weinheim (2012).

51. (a) Kelly, T. W., Boardman, L. D., Dunbar, T. D., Muyres, D. V., Pellerite, M. J., Smith, T. P. *J. Phys. Chem. B*, **107**, 5877 (2003); (b) Lin, Y. Y., Gundlach, D. J., Nelson, S. F., Jackson, T. N. *IEEE Electron Device Lett.*, **18**, 606 (1997); (c) Klauk, H., Halik, M., Zschieschang, U., Schmid, G., Radlik, W., Weber, W. *J. Appl. Phys.*, **92**, 5259 (2002).

52. Yuan, Y. B., Giri, G., Ayzner, A. L., Zoombelt, A. P., Mannsfeld, S. C. B., Chen, J. H., Nordlund, D., Toney, M. F., Huang, J. S., Bao, Z. N. *Nat. Commun.*, 5 (2014).

53. Sheraw, C. D., Jackson, T. N., Eaton, D. L., Anthony, J. E. *Adv. Mater.*, **15**, 2009 (2003).

54. Miyata, Y., Yoshikawa, E., Minari, T., Tsukagoshi, K., Yamaguchi, S. *J. Mater. Chem.*, **22**, 7715 (2012).

55. Meisenheimer, J., Witte, K. *Chem. Ber.*, **36**, 4153 (1903).

56. (a) Newman, M. S., Lutz, W. B., Lednicer, D. *J. Am. Chem. Soc.*, **77**, 3420 (1955); (b) Newman, M. S., Lednicer, D. *J. Am. Chem. Soc.*, **78**, 4765 (1956).

57. Scholz, M., Mühlstädt, M., Dietz, F. *Tetrahedron Lett.*, **8**, 665 (1967).

58. Flammang-Barbieux, M., Nasielski, J., Martin, R. H. *Tetrahedron Lett.*, **8**, 743 (1967).

59. Mori, K., Murase, T., Fujita, M. *Angew. Chem. Int. Ed.*, **54**, 6847 (2015).

60. Tribout, J., Martin, R. H., Doyle, M., Wynberg, H. *Tetrahedron Lett.*, **13**, 2839 (1972).

61. Moradpour, A., Nicoud, J. F., Balavoine, G., Kagan, H., Tsoucaris, G. *J. Am. Chem. Soc.*, **93**, 2353 (1971).

62. Weidlich, H. A. *Chem. Ber.*, **71**, 1203 (1938).

63. (a) Willmore, N. D., Liu, L. B., Katz, T. J. *Angew. Chem. Int. Ed.*, **31**, 1093 (1992); (b) Katz, T. J., Liu, L. B., Willmore, N. D., Fox, J. M., Rheingold, A. L., Shi, S. H., Nuckolls, C., Rickman, B. H. *J. Am. Chem. Soc.*, **119**, 10054 (1997).

64. (a) Carreño, M. C., Hernández-Sánchez, R., Mahugo, J., Urbano, A. *J. Org. Chem.*, **64**, 1387 (1999); (b) Carreño, M. C., García-Cerrada, S., Urbano, A. *Chem. Commun.*, 1412 (2002).

65. (a) Fuchibe, K., Jyono, H., Fujiwara, M., Kudo, T., Yokota, M., Ichikawa, J. *Chem. Eur. J.*, **17**, 12175 (2011); (b) Ichikawa, J., Yokota, M., Kudo, T., Umezaki, S. *Angew. Chem. Int. Ed.*, **47**, 4870 (2008); (c) Pieters, G., Gaucher, A., Prim, D., Marrot, *J. Chem. Commun.*, 4827 (2009).

66. (a) Stará, I. G., Starý, I., Kollárovič, A., Teplý, F., Šaman, D., Tichý, M. *J. Org. Chem.*, **63**, 4046 (1998); (b) Stará, I. G., Starý, I., Kollárovič, A., Teplý, F., Vyskočil, S., Šaman, D. *Tetrahedron Lett.*, **40**, 1993 (1999).

67. (a) Teplý, F., Stará, I. G., Starý, I., Kollárovič, A., Šaman, D., Vyskočil, S., Fiedler, P. *J. Org. Chem.*, **68**, 5193 (2003); (b) Teplý, F., Stará, I. G., Starý, I., Kollárovič, A., Luštinec, D., Krausová, Z., Šaman, D., Fiedler, P. *Eur. J. Org. Chem.*, 4244 (2007).

68. Sehnal, P., Stará, I. G., Šaman, D., Tichý, M., Míšek, J., Cvačka, J., Rulíšek, L., Chocholoušová, J., Vacek, J., Goryl, G., Szymonski, M., Císařová, I., Starý, I. *Proc. Natl. Acad. Sci. U.S.A.*, **106**, 13169 (2009).

69. (a) Pēna, D., Pérez, D., Guitián, E., Castedo, L. *Org. Lett.*, **1**, 1555 (1999); (b) Pēna, D., Cobas, A., Pérez, D., Guitián, E., Castedo, L. *Org. Lett.*, **2**, 1629 (2000).

70. Pēna, D., Cobas, A., Pérez, D., Guitián, E., Castedo, L. *Org. Lett.*, **5**, 1863 (2003).

71. Collins, S. K., Grandbois, A., Vachon, M. P., Côté, J. *Angew. Chem. Int. Ed.*, **45**, 2923 (2006).

72. Grandbois, A., Collins, S. K. *Chem. Eur. J.*, **14**, 9323 (2008).

73. Kamikawa, K., Takemoto, I., Takemoto, S., Matsuzaka, H. *J. Org. Chem.*, **72**, 7406 (2007).

74. Shen, Y., Chen, C. F. *Chem. Rev.*, **112**, 1463 (2012).

75. (a) Dopper, J. H., Oudman, D., Wynberg, H. *J. Am. Chem. Soc.*, **95**, 3692 (1973); (b) Groen, M. B., Schadenberg, H., Wynberg, H. *J. Org. Chem.*, **36**, 2797 (1971).

76. Yamamoto, Y., Sakai, H., Yuasa, J., Araki, Y., Wada, T., Sakanoue, T., Takenobu, T., Kawai, T., Hasobe, T. *J. Phys. Chem. C*, **120**, 7421 (2016).

77. Bazzini, C., Brovelli, S., Caronna, T., Gambarotti, C., Giannone, M., Macchi, P., Meinardi, F., Mele, A., Panzeri, W., Recupero, F., Sironi, A., Tubino, R. *Eur. J. Org. Chem.*, 1247 (2005).

78. (a) Phillips, K. E. S., Katz, T. J., Jockusch, S., Lovinger, A. J., Turro, N. J. *J. Am. Chem. Soc.*, **123**, 11899 (2001); (b) Dreher, S. D., Weix, D. J., Katz, T. J. *J. Org. Chem.*, **64**, 3671 (1999).

79. (a) Rajca, A., Wang, H., Pink, M., Rajca, S. *Angew. Chem. Int. Ed.*, **39**, 4481 (2000); (b) Rajca, A., Miyasaka, M., Pink, M., Wang, H., Rajca, S. *J. Am. Chem. Soc.*, **126**, 15211 (2004).

80. Miyasaka, M., Rajca, A., Pink, M., Rajca, S. *J. Am. Chem. Soc.*, **127**, 13806 (2005).

81. Míšek, J., Teplý, F., Stará, I. G., Tichý, M., Šaman, D., Císařová, I., Vojtíšek, P., Starý, I. *Angew. Chem. Int. Ed.*, **47**, 3188 (2008).

82. Tanaka, K., Kamisawa, A., Suda, T., Noguchi, K., Hirano, M. *J. Am. Chem. Soc.*, **129**, 12078 (2007).

83. (a) Fukawa, N., Osaka, T., Noguchi, K., Tanaka, K. *Org. Lett.*, **12**, 1324 (2010); (b) Sawada, Y., Furumi, S., Takai, A., Takeuchi, M., Noguchi, K., Tanaka, K. *J. Am. Chem. Soc.*, **134**, 4080 (2012).

84. (a) Guram, A. S., Rennels, R. A., Buchwald, S. L. *Angew. Chem. Int. Ed.*, **34**, 1348 (1995); (b) Louie, J., Hartwig, J. F. *Tetrahedron Lett.*, **36**, 3609 (1995).

85. (a) Kuwahara, A., Nakano, K., Nozaki, K. *J. Org. Chem.*, **70**, 413 (2005); (b) Nozaki, K., Takahashi, K., Nakano, K., Hiyama, T., Tang, H. Z., Fujiki, M., Yamaguchi, S., Tamao, K. *Angew. Chem. Int. Ed.*, **42**, 2051 (2003).

86. (a) Hung, T. Q., Thang, N. N., Hoang, D. H., Dang, T. T., Ayub, K., Villinger, A., Lochbrunner, S., Flechsig, G.-U., Langer, P. *Eur. J. Org. Chem.*, **2015**, 1007 (2015); (b) Evenson, S. J., Pappenfus, T. M., Delgado, M. C. R., Radke-Wohlers, K. R., Navarrete, J. T. L., Rasmussen, S. C. *Phys. Chem. Chem. Phys.*, **14**, 6101 (2012); (c) Ueno, A., Kitawaki, T., Chida, N. *Org. Lett.*, **10**, 1999 (2008).

87. Kawaguchi, K., Nakano, K., Nozaki, K. *J. Org. Chem.*, **72**, 5119 (2007).

88. Wakim, S., Bouchard, J., Blouin, N., Michaud, A., Leclerc, M. *Org. Lett.* **6**, 3413 (2004).

89. Kawaguchi, K., Nakano, K., Nozaki, K. *Org. Lett.*, **10**, 1199 (2008).

90. Truong, M. A., Nakano, K. *J. Org. Chem.*, **80**, 11566 (2015).

91. Hung, T. Q., Hancker, S., Villinger, A., Lochbrunner, S., Dang, T. T., Friedrich, A., Breitsprecher, W., Langer, P. *Org. Biomol. Chem.*, **13**, 583 (2015).

92. Ureshino, T., Yoshida, T., Kuninobu, Y., Takai, K. *J. Am. Chem. Soc.*, **132**, 14324 (2010).

93. Kuninobu, Y., Yoshida, T., Takai, K. *J. Org. Chem.*, **76**, 7370 (2011).

94. Nakano, K., Hidehira, Y., Takahashi, K., Hiyama, T., Nozaki, K. *Angew. Chem. Int. Ed.*, **44**, 7136 (2005).

95. Nakano, K., Oyama, H., Nishimura, Y., Nakasako, S., Nozaki, K. *Angew. Chem. Int. Ed.*, **51**, 695 (2012).

Biography

Dr. Koji Nakano received his PhD from Kyoto University, Japan, under the guidance of Prof. Tamejiro Hiyama in 2005. During his PhD study,

he developed novel metal catalysts for asymmetric copolymerization of epoxides with carbon dioxide and acquired organic synthetic chemistry, organimetallic chemistry, and polymer chemistry. He started his academic career at the University of Tokyo (Prof. Nozaki's group), Japan, in 2002 as research associate and assistant professor. He developed a variety of catalysts for ring-opening polymerization of cyclic monomers. In addition, he expanded his research field to the synthesis and application of π-conjugated compounds, such as heteroacenes and heterohelicenes. In 2012, he moved to Tokyo University of A&T as senior assistant professor and was promoted to associate professor in 2015. His current research interest is the development of new functional materials based on π-conjugated compounds and polymers.

Chapter 3

Poly(*ortho*-Phenylene)s and Their Derivatives: Synthesis, Conformation, and Physical Properties

Koichiro Mikami
Polymer Chemistry Laboratory, Sagami Chemical Research Institute (SCRI), Hayakawa, Ayase, Kanagawa 2743-1, Japan
koichiro.mikami@sagami.or.jp

Polyphenylenes are one of the simplest π-conjugated polymers, and *para*- and *meta*-type polyphenylenes and a variety of complex carbon-based architectures are reported. In contrast to these, research of *ortho*-substituted poly-/oligophenylenes had lagged behind *para*-/*meta*-type polyphenylenes and these complex architectures. In this chapter, recent development of the synthetic methodology for poly(*ortho*-phenylene)s and their derivatives, the resultant poly/oligomer's conformation, and unique properties are reviewed.

Conjugated Objects: Developments, Synthesis, and Applications
Edited by Atsushi Nagai and Koji Takagi

ISBN 978-981-4774-03-1 (Hardcover), 978-1-315-22962-1 (eBook)
www.panstanford.com

3.1 Introduction

A fundamental understanding of substitution position effects, as well as chemical structure and functional groups, on target molecules/ materials is a key research topic in chemistry and materials science. This is particularly important because the influence observed at the molecular level can propagate into the bulk beyond the nano- and mesoscales, which has an impact on material properties and functionality. In this chapter, I focus on the recent development of the elusive *ortho-substituted oligo-/polyphenylenes* molecules (Fig. 3.1) and of their related derivatives, poly(quinoxaline-2,3-diyl)s.

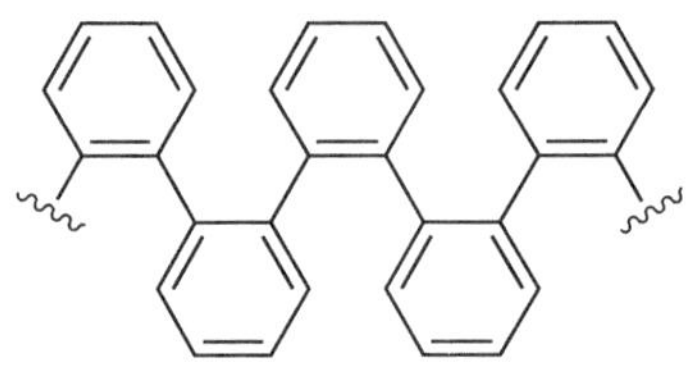

Figure 3.1 Chemical structure of an *ortho*-substituted polyphenylene.

π-Conjugated nanocarbons, such as polyphenylenes, carbon nanorings, fullerenes, graphene, and carbon nanotubes that are obtained by condensation/fusion of aromatic rings, have been considered as promising candidates for organic semiconducting materials and are enabling applications as materials for thin-film solar cells, transistors, and light-emitting diodes. Polyphenylenes constructed by connecting many benzene units through single bonds are one of the simplest π-conjugated polymers. By arranging the positions of any aryl substitutions and varying the reaction type, a variety of complex carbon-based architectures are accessed, for example, branched-type compounds, dendrimers and dendrons, sheet-type compounds, and graphene nanoribbons, which have been extensively studied for its synthetic strategy, functions, properties, and applications.[1] On the contrary, research of *ortho*-substituted poly/oligophenylenes where benzene rings are simply connected at the *ortho*-positions, had lagged behind these complex architectures and *para-/meta*-type polyphenylenes until quite recently, due to the lack of an effective synthetic methodology for oligo/poly(*ortho*-phenylene)s. This is despite recent and significant developments in C–C bond-forming reactions, for example, using transition metal catalyzed cross-coupling reactions. In this chapter, the synthetic methodology will first be reviewed, followed by discussion of the resulting poly/oligomer's conformation and physical properties.

3.2 Synthesis of Oligo-/Poly(*ortho*-Phenylene)s

3.2.1 Stepwise Synthesis through Transition Metal–Mediated Reactions

Stepwise methodologies with transition metal mediated coupling reactions have been developed for the synthesis of oligo(*ortho*-phenylene)s. In the early stages of research, Ibuki et al. demonstrated a copper-mediated Ullmann coupling reaction between iodobiphenyl and diidobiphenyl derivatives, in which the yield of the resulting material was insufficient due to formation of undesired by-products.[2,3] In 1998, Simpkins et al. reported an iterative synthesis of oligo(*ortho*-phenylene)s by a Suzuki–Miyaura cross coupling reaction between **1** and 1-bromo-2-iodobenzene followed by bolyration of the end group of the resulting compounds (Fig. 3.2).[4] In this fashion, the synthesis could be extended up to 9-mer. Single-crystal X-ray analysis of the 6-mer compound revealed that the oligomer adopted a helical geometry with 3 monomer units per turn, with the benzene units stacked with a face-to-face orientation along the helical axis (see also section 3). Manabe et al. have developed an alternative Suzuki–Miyaura cross-coupling reaction in which the activation of a phenolic hydroxyl group was combined with anhydrous triflic acid.[5] Recently, Hartley et al. modified Manabe's approach and achieved an efficient synthesis of oligo(*ortho*-phenylene)s up to 12-mer through a Suzuki–Miyaura cross-coupling reaction using **2**, followed by triflation of the phenolic hydroxyl group at the end group of the resulting oligomers.[6] This approach can be applied to the synthesis of a variety of oligo(*ortho*-phenylene)s with different side chains and end groups (see also Section 3.3).

Not only oligo(*ortho*-phenylene)s but also oligo(*ortho*-arylene)s (oligo(naphthalene-2,3-diyl) and oligo(quinoline-2,3-diyl)) have been synthesized by an iterative stepwise approach. Ito et al. developed a synthetic strategy for oligo(naphthalene-2,3-diyl)s via Negishi-type cross-coupling reactions between 2-naphthylzinc bromide and 2-bromonaphthalene derivatives.[7] Single-crystal X-ray analysis of the resulting 4-mer revealed that the oligomer adopted a helical geometry similar to oligo(*ortho*-phenylene)s. Oligo(quinoline-2,3-diyl)s obtained through Suzuki–Miyaura cross-coupling reactions have also been reported.[8] In 2013, Suginome et al. developed an anthranilamide-masked strategy[9], which is associated with two types of transition metal–catalyzed reactions, Ru-catalyzed *ortho*-directed silylation followed by iodination and a Suzuki–Miyaura-type cross-coupling reaction. Through this strategy, they prepared oligo(*ortho*-phenylene) with three repeat units and oligo(naphthalene-2,3-diyl)s with four repeat units. These approaches would be also effective toward synthesis of various oligo(*ortho*-arylene)s.

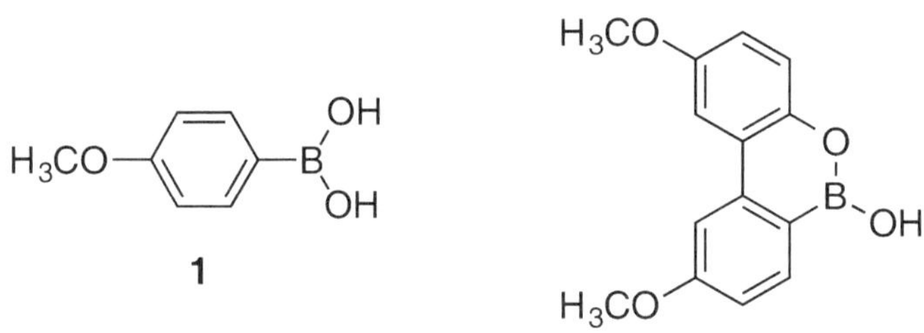

Figure 3.2 Chemical structures of **1** and **2**.

3.2.2 Polymerization by Homocoupling Reactions

The synthesis of oligo(*ortho*-phenylene) through a Yamamoto cross-coupling reaction of biphenyl derivatives (**3**) having bromides at both end groups has been reported (Fig. 3.3A).[10] Although the production up to 18-mer is detected by MALDI-TOF mass spectrometry, 6mer was only isolated by HPLC. Fukushima and Aida et al. developed a synthetic approach providing oligo(*ortho*-phenylene)s with high molecular weight (Fig. 3.3B).[11] This approach is associated with the homocoupling reaction of oligomers synthesized in a stepwise manner: lithiation of oligo(*ortho*-phenylene)s (8-mer) having bromines at the chain end via a halogen–metal exchange reaction using *t*-BuLi, followed by a oxidative homocoupling reaction using CuCN and duroquinone. Through purification with recycling HPLC,

16-, 24-, 32-, 40-, and 48-mer products were systematically obtained. The key finding for obtaining longer oligomers lay in the suppression of intramolecular reactions and undesired side reactions by use of oligomer homocoupling.

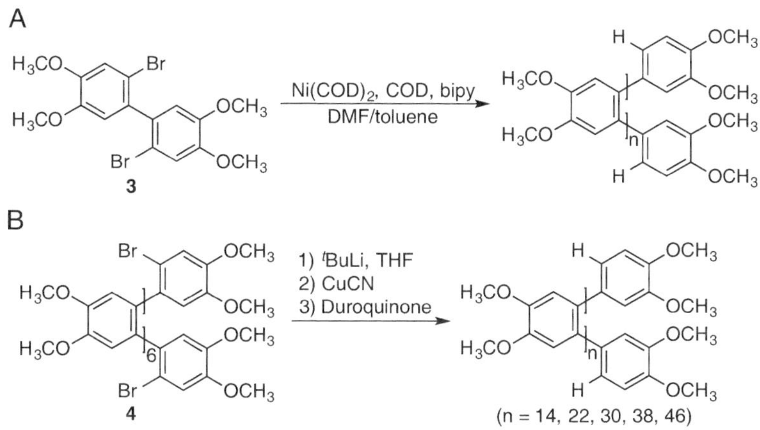

Figure 3.3 Polymerization to afford (A) poly- and (B) oligo(*ortho*-phenylene)s through homocoupling reactions.

3.2.3 Polymer Reactions

Ito and Nozaki et al. have reported a formal aryne(synthon) polymerization strategy based on [2.2.1]oxabicyclic alkene polymerization followed by aromatization of the resultant polymers under strongly acidic conditions (Fig. 3.4).[12] The polymerization proceeds via a coordination/insertion mechanism when using a complex prepared with π-allyl palladium chloride dimer and silver hexafluoroantimonate. A variety of monomers can be polymerized using this strategy to yield polymers containing various aromatic groups in the polymer main chain. Furthermore, this approach enables copolymerization between formal aryne and carbon monoxide[13] or formal aryne and olefin,[14] which eventually provides an aromatic polyketone/ketal species and polyolefins consisting of aromatic structures in the polymer main chain. These moieties are difficult to synthesize through conventional polymerization approaches. While not a direct polymerization route for yielding poly(*ortho*-pehnylene)s, this method would be a strong strategy for preparing a variety of poly(*ortho*-arylene)s and related polymers.

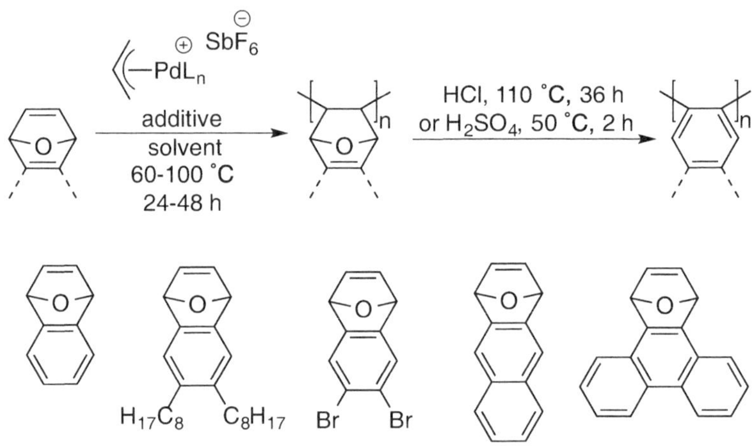

Figure 3.4 Formal aryne polymerization.

Clancy and Dichtel et al. reported the synthesis of polyphenylenes containing *ortho*-substituted arylene units in the main chain via benzannulation through [2+2+2] cycloaddition of poly(phenylene ethynylene).[15] Although poly(*ortho*-phenylene)s had not been prepared in this report, it would be a quite interesting future approach to give *ortho*-polyarylenes via a polymerization reaction.

3.2.4 Direct Polymerization of Aromatic Components

The strategies as mentioned above are not straightforward approaches to poly(*ortho*-phenylene)s. Polymerization via cross-coupling reactions and oxidative coupling reactions between aromatic components would be obviously one of the most promising approaches to poly(*ortho*-phenylene)s. However, although various approaches including oxidative polymerization of toluene and chlorobenzeneas well as catechol electropolymerization were attempted for the direct synthesis of poly(*ortho*-phenylene)s, these approaches did not provide the desired polymers without defects such as undesired intramolecular reaction and cyclic trimerization products.[16–26]

On the other hand, aryne is a good candidate monomer. Aryne, which results from the abstraction of two neighboring hydrogen atoms of an aromatic ring, is one of the oldest reactive intermediates. Since its discovery, the reactivity and unique structural features of aryne have been extensively investigated theoretically and experimentally, and therefore, aryne has been used in key reactions for both total synthesis and nanomaterial functionalization.[27–34] Although the highly distorted triple bond was considered as a promising chemical bond for addition and insertion/coordination polymerization, polymerization of aryne was not achieved for a long time. The possibility of aryne as a monomer for the polymerization synthesis for poly(*ortho*-arylene)s was first revealed by Ihara and coworkers.[35] They found copolymerization between an in situ generated aryne and pyridine took place to afford alternating copolymers (Fig. 3.5). However, homopolymerization of aryne did not proceed even by treatment of various initiators, including radical initiators (AIBN and V-70), transition metal complexes ($MoCl_5$, WCl_6 and $Ni(acac)_2$), and nucleophiles (Grignard reagents, amines, and phosphines).

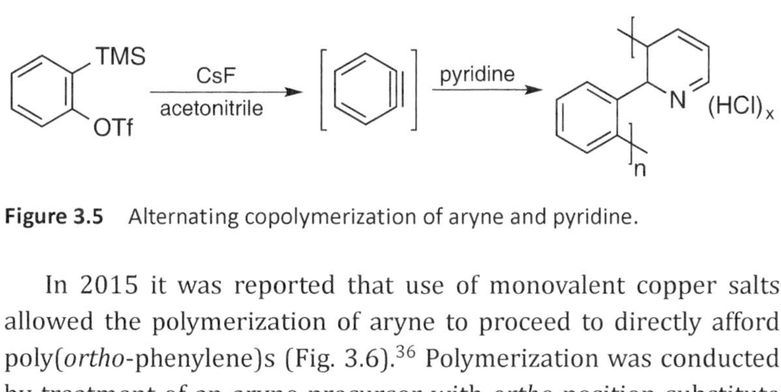

Figure 3.5 Alternating copolymerization of aryne and pyridine.

In 2015 it was reported that use of monovalent copper salts allowed the polymerization of aryne to proceed to directly afford poly(*ortho*-phenylene)s (Fig. 3.6).[36] Polymerization was conducted by treatment of an aryne precursor with *ortho*-position substitute trimethylsilyl/triflate groups and CsF/18-crown-6 ether as a fluoride source in the presence of monovalent copper salts. Other metal pieces, such as Au, Pd, and KCN, did not smoothly polymerize the monomer. By changing the copper/monomer precursor feed ratio, the molecular weight of the resultant polymers was varied, demonstrating that polymerization proceeded in a chain-growth rather than a step-growth polymerization mechanism. Theoretical exploration using density functional theory (DFT) calculations (B3LYP/6-31+G(d) level of theory) indicated that one of the plausible reaction pathways is a coordination/insertion mechanism, where the aryne coordinates to the CuCN metal center, inserts into the Cu–CN bond, and is followed by coordination/insertion reactions; this repeats iteratively. Although the monomer scope is still limited, this approach is a promising reaction for affording various poly(*ortho*-arylene)s.

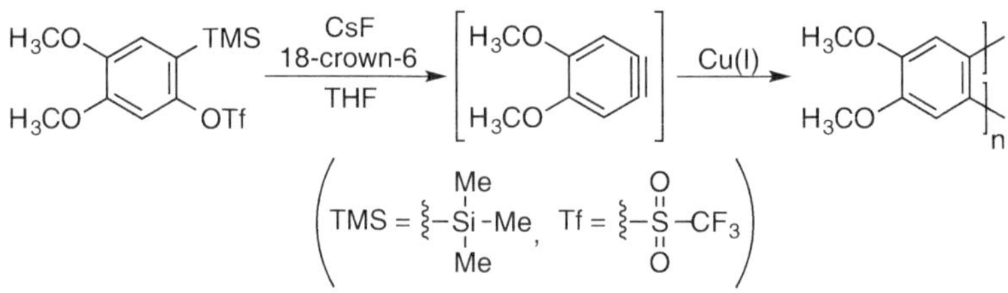

Figure 3.6 Aryne polymerization.

3.3 Conformation of Poly(*ortho*-Phenylene)s

3.3.1 Open and Closed Helix Models

Two limiting plausible conformations are considered for the ground state of oligo/poly(*ortho*-phenylene)s: closed and open helical geometries, as shown in Fig. 3.7.[6] A variety of oligo(*ortho*-phenylene)s with different side chains have been synthesized. The conformational behaviors have been extensively studied using 1D and 2D nuclear magnetic resonance (NMR), UV-Vis, circular dichroism (CD), fluorescence spectroscopy, and cyclic voltammetry, as well as computational approaches using DFT. These oligomers tended to adopt a closed helical geometry, with the slow interconversion among different conformers on the NMR timescale being one of the characteristic aspects of *ortho*-substituted poly/oligophenylenes. The stability of closed helical structures highly depends on steric effects at the end groups[37] and aromatic stacking interaction among monomer units.[38,39]

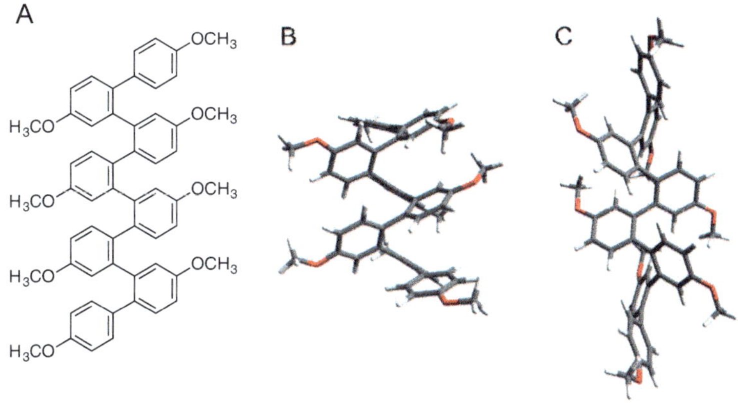

Figure 3.7 (A) Chemical structure of oligo(*ortho*-phenylene) and its two plausible major conformations obtained by DFT: (B) closed and (C) open helical geometries.

3.3.2 Oligo(*ortho*-Phenylene)s with Disubstituted Side Chains on Monomer Units

Oligo(*ortho*-phenylene)s with dimethoxy groups in each monomer units have been intensely studied by Fukushima and Aida et al (Fig. 3.3B).[11,37] The oligomer structure as revealed by single-crystal X-ray analysis was a perfect 3_1-helical conformation with a highly condensed π-cloud along its helical pitch. The interplanar distance between the cofacially oriented phenylene units is 3.232–3.285 Å, which is shorter than the intersheet distance of graphite (3.35 Å). It should be noted that the oligo(*ortho*-phenylene) (8mer) with two nitro group at both ends shows chiral symmetry breaking during crystallization, which means that right- or left-handed helices separately crystallize during recrystallization of the racemic mixtures.

Notably, the helical structure responds to redox stimulus and shows spring-like motion during one-electron oxidation.[11] This unique behavior is also elucidated by single-crystal X-ray analysis of the neutral form and the one-electron-oxidized form obtained by oxidation with tris(4-bromophenyl) ammoniumyl hexachloroantimonate). The interplanar distances between proximal phenylene units of the oxidized form (3.192–3.252 Å) are

0.02–0.07 Å shorter than those of the neutral form. That the decrease of the interplanar distance by one-electron oxidation is not driven by crystal packing but radical cation delocalization over the stacked π-array is supported by DFT (Fig. 3.8). Furthermore, a long-lasting chiroptical memory effect is also observed by one electron oxidation. The half-life at –10°C ($t_{1/2}$ = 44 h) of the oxidized species is 450 times longer the neutral species. The racemization profile followed first-order kinetics with a rate constant at –10°C of 2.18 × 10^{-6} s^{-1}. The activation energy for the helical inversion of the oxidized form is 96.6 kJ/mol, which is 8.3 kJ/mol higher than the neutral form. In the neutral form, the helical inversion activation energy is 88.3 kJ/mol and $t_{1/2}$is 352s, which is also amenable to first-order kinetics with a rate constant of 9.84 × 10^{-4} s^{-1}.

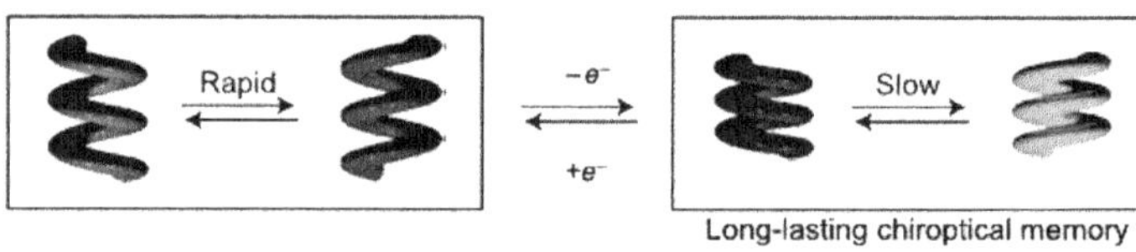

Figure 3.8 Redox-responsible spring-like motion and long-lasting chiroptical memory of oligo(*ortho*-phenylene) with nitro groups at both end groups.

The oligomer terminal groups and the solvent nature significantly affects oligomer helical folding. With terminal groups such as CH_3, CH_2OH, Br, COOBn, and NO_2, folding yields a perfect helical conformation in acetonitrile (Fig. 3.9).[37] Although the ^1H NMR spectra in $CDCl_3$, toluene-d_8, DMSO-d_6, and DMF-d_7 are quite complicated, spectra in acetonitrile-d_3 are exceptionally simple and shows C_2 symmetry, supporting that the oligomer adopts 3_1-helical geometry. The CH_2 protons of the benzyl ester group are observed to be diastereotopic, which supports perfect helical folding of the oligomers. The authors concluded that steric factors rather than end group polarity would dominate the helical folding process in solvent. The helical inversion barrier of the Br-terminated oligomer at 298 K is $t_{1/2}$ = 5.5 h. Furthermore, variable-temperature NMR observation of the diastereomeric signal coalescence temperature for the terminal benzyl ester groups revealed a helical inversion rate constant of k_c = 244 s^{-1}. Analysis using the Erying equation yielded a free-energy change of activation for the helical inversion as $\Delta G^\ddagger$ = 62.5 kJ/mol.

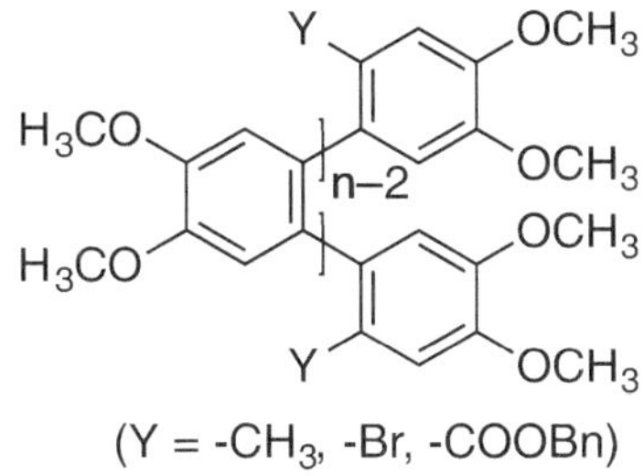

Figure 3.9 Chemical structures of poly(*ortho*-phenylene)s with CH_3, Br, or COOBn groups at both end groups.

3.3.3 Oligo(*ortho*-Phenylene)s with and without Monosubstituted Side Chains on Monomer Units

Hartley et al. performed systematic conformational studies of mainly monosubstituted oligo(*ortho*-phenylene)s.[6,38–43] The single-crystal X-ray exploration revealed that most of the oligomers adopted a perfectly helical geometry, with partial misfolding at the end units being observed in the case of some oligomers.[6] It was also pointed out that unusual behaviors occurred which are not observed in *para*- and *meta*-derivatives :(i) significant bathochromic shifts in UV-Vis spectra, suggesting that conjugation extends much further along the *ortho*-phenylene backbone; (ii) significant hypsochromic shifts in the fluorescence spectra; and (iii) largely length-independent oxidation potentials (and HOMO energies). The unusual hypsochromic shift is rationalized using an excited-state time-dependent DFT (TD-DFT) exploration.[42] These theoretical calculations indicated that in the excited state, compression of the spring-like compounds occurs, and for the longer oligomer, this compression is localized to one end of the oligomer (Fig. 3.10). On the other hand, these observations appear to partially contradict to those of the oligomers with dimethoxy side chains (as shown above), which show a lack of chain-length dependency in the UV-Vis spectra and are devoid of extended π-conjugation. These differences would be associated with stability of the helical conformation in solution.

The role of arene–arene interactions on each monomer unit are revealed experimentally and theoretically[38,39], with the end groups having a remarkable influence on helical folding.[37] As a result, simple rules are proposed: (i) a specific sequence of biaryl torsional angles is prohibited because of steric strain, and (ii) conformer stability is otherwise a simple function of the number of aromatic stacking interactions possible for the given geometry. These suggestions indicate that oligomer helical folding can be arranged by controlling π-repulsion among aromatic components per three monomer units. These rules were elucidated through conformational studies of several series of *ortho*-phenylene oligomers having various side chains on each repeating unit; these groups were nearly electron withdrawing in nature (Fig. 3.11).[39] These side chains improved helical folding, and especially strong electron-withdrawing groups (OTf and CN) resulted in an undetectable misfolding in solution, reflecting the enhancement of intramolecular arene–arene stacking interactions. Especially, triflate-substituted oligomers adopts a perfect helical folding even in $CDCl_3$, in sharp contrast to the dimethoxy counterpart, which only properly folds in polar solvents(acetonitrile). The relationship between helical structure stability and arene–arene interactions through side chain are rationalized by DFT at a PCM/B97-D/TZV-(2d,2p) level of theory and Hammett substituent constants.

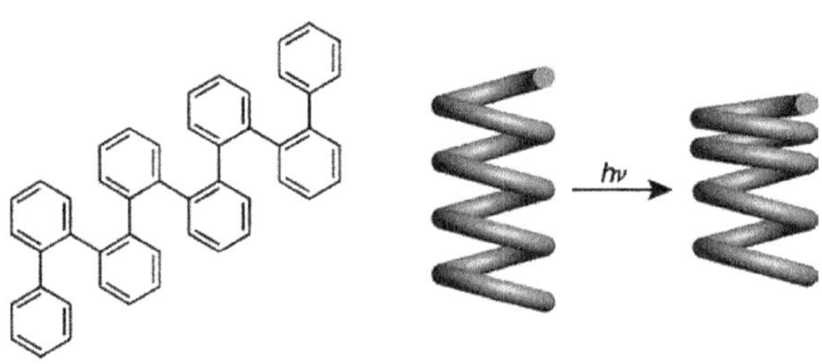

Figure 3.10 Excited-state structure of an unsubstituted oligo(*ortho*-phenylene) as predicted by TD-DFT.

By attaching dimethylamino and nitro groups at both ends, push and pull (donor and acceptor)-type oligo(*ortho*-phenylene)s without side chains can be prepared (Fig. 3.12).[43] The major conformer is a highly symmetric helical geometry (ca. 50%) on the NMR timescale. Although a charge transfer band is observed in the 3- and 4-mer species, this band is not observed in longer oligomers, and the CT interaction between the push and pull substituents attenuates quickly compared to the other usual π-conjugated version. On trifluoroacetic acid (TFA) protonation of the dimethyamino groups, the observed CT bands disappear, and the resulting spectra show a systematic bathochromic shift with increasing *n* values (mer length). Interestingly, protonation results in favoring of helical geometry (ca. 80%). Furthermore, on protonation, the effective conjugation length is significantly extended as compared to unsubstituted *ortho*-phenylene oligomers.

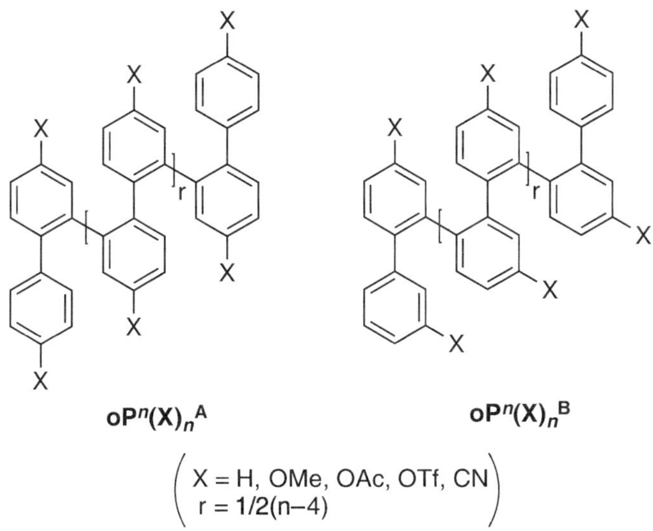

Figure 3.11 Chemical structures of oligo(*ortho*-phenylene)s with a different substituent nature and orientation.

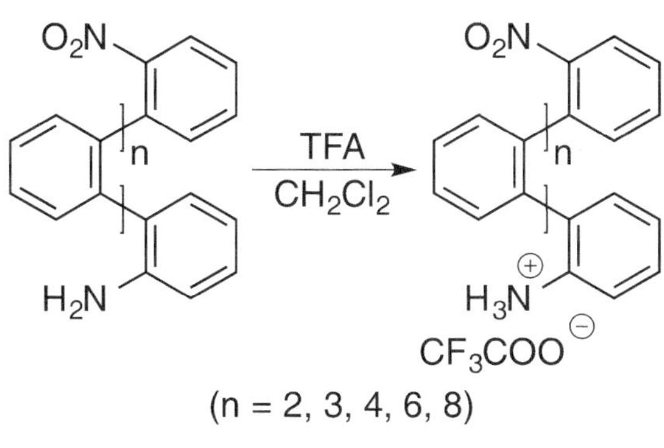

Figure 3.12 Chemical structure of a donor–acceptor-type oligo(*ortho*-phenylene) and protonation reaction with TFA.

3.4 Physical Properties and Applications of Poly(*ortho*-Phenylene)s

3.4.1 As a Surface Modifier for π-Conjugated Disks

As mentioned above, oligo/poly(*ortho*-phenylene)s adopt a helical geometry, and therefore, as expected, these oligomers and polymers show totally different physical properties from *para*- and *meta*-type polyphenylenes. Kajitani, Fukushima, and Aida et al. reported a possible surface modifier for π-conjugated molecules (Fig. 3.13).[44] As also mentioned in other chapters, π-conjugated molecules are associated with applications such as organic electronics, including thin-film solar cells, field-effect transistors, and light-emitting diodes. On the other hand, it is also quite important not only to develop novel π-conjugated molecules but also to control and arrange the stacking orientation of these π-conjugated components over large areas in order to fully realize the potential of the molecules. This remains a very challenging issue. In 2013, it was found that oligo(*ortho*-phenylene) s have brilliant possibilities for these fields. Researchers took note of the intrinsically geometrical aspects of oligo(*ortho*-phenylene)s, which adopt an approximately cylindrical helical geometry when projected along their longer skeletal axis; the resulting structure looks like triphenylene. This molecular feature allowed researchers to suppose that these molecules might be miscible with disk-shaped liquid crystal materials. Consequently, it was found that a polymeric oligo(*ortho*-phenylene) octamer and its monomeric unit serve as first surface modifiers for homeotropic columns of a variety of discotic liquid crystals, such as triphenylene, phthalocyanine and hexabenzocoronene derivatives, up to a macroscopic size range. It is assumed that the reason these discotic liquid crystal molecules stack homeotropically is a result of interactions of the glass substrate surface functional groups with oligomer end group or linking groups, which are anisotropically and regularly absorbed onto glass surfaces (Fig. 3.13). In addition, the oligo(*ortho*-phenylene) tetramer does not work as a surface modifier, indicating that a cylindrical helical shape is also crucial for realizing this unique function.

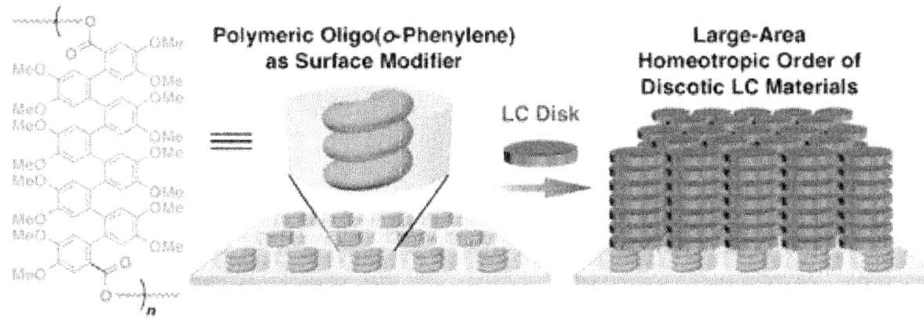

Figure 3.13 Schematic illustration of oligo(*ortho*-phenylene)s acting as a surface modifier.

3.5 Poly(Quinoxaline-2,3-Diyl)s: "Cousin" of Poly(*ortho*-Phenylene)s

Poly(quinoxaline-2,3-diyl)s (Fig. 3.14) are a kind of *ortho*-substituted polyarylenes, and are considered as structural "cousins" of poly(*ortho*-phenylene)s. The synthetic methodology, in contrast to the case of poly/oligo(*ortho*-phenylene)s, was already reported by Ito et al. in 1990,[45] with the conformation and physical/functional properties having been extensively studied. In this last section, I review the synthesis, conformation, and physical properties of this poly(*ortho*-phenylene) cousin.

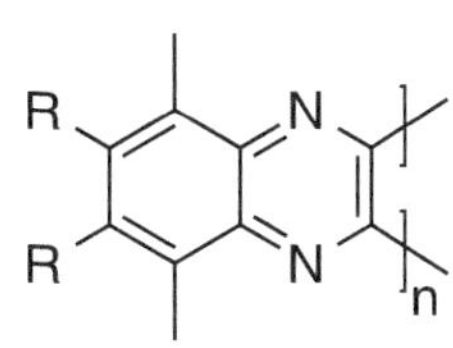

Figure 3.14 A representative chemical structure of poly(quinoxaline-2,3-diyl).

3.5.1 Synthesis of Poly(Quinoxaline-2,3-Diyl)s

Poly(quinoxaline-2,3-diyl)s are synthesized through Pd-catalyzed aromatization polymerization reactions (Fig. 3.15).[45] These reactions are living polymerizations in nature, and block copolymers of poly(quinoxaline-2,3-diyl) and poly(isocyanide)s have been synthesized.[46] A Ni(II) salt is also useful for facile synthesis.[47,48]

These polymerization systems are one of the most established living polymerization systems, similar to anionic polymerization.

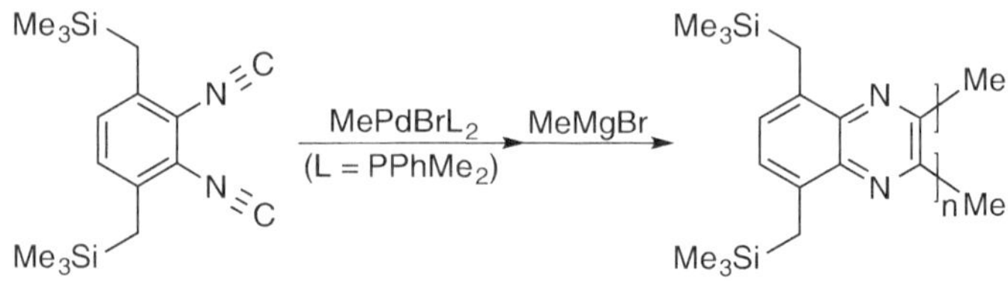

Figure 3.15 A general synthetic approach for poly(quinoxaline-2,3-diyl) through aromatization.

3.5.2 Conformation of Poly(Quinoxaline-2,3-Diyl)s

In contrast to poly(*ortho*-phenylene)s which adopt a closed helical geometry, poly(quinoxaline-2,3-diyl) adopts an open form (Figs. 3.7C, 3.16);[49,50] this is associated with steric hindrance of the two bulky substituents on the monomeric unit aromatic components. One-handed polymers are often selectively induced by asymmetric polymerization with a chiral initiator;[49,51–53] this may also be controlled with a chiral side chain.[54] The helical bias of these polymers is amenable to the "sergeant and soldiers principle"[55,56] and the "majority rule" effects[57]. Interestingly, the conformation,

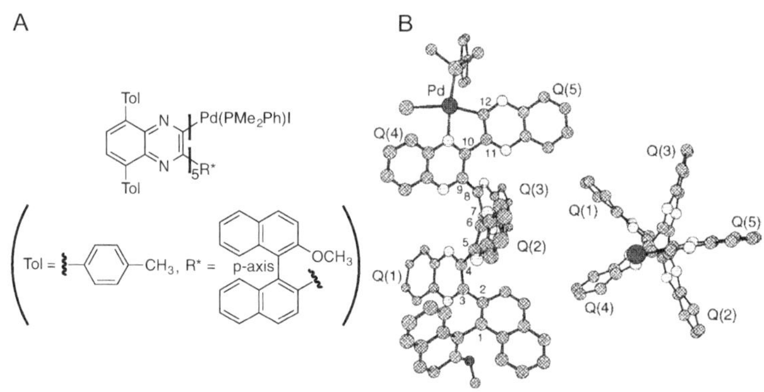

Figure 3.16 (A) Chemical structure of oligo(qunoxaline-2,3-diyl) and (B) its helical structure obtained by single-crystal X-ray diffraction (binaphthyl, tolyl, phosphine, and iodo groups are omitted. Right: side view; left: top view).

despite its quite rigid helical structure, can be changed by a variety of stimuli, including solvent nature[58,59] and pressure (Fig. 3.17).[60] For example, the poly(quinoxaline-2,3-diyl) with (*S*)-2-butoxymethyl side chains in 1,2-dichloroethane solution shows helical inversion from right- to left-handed, depending on pressure.[60]

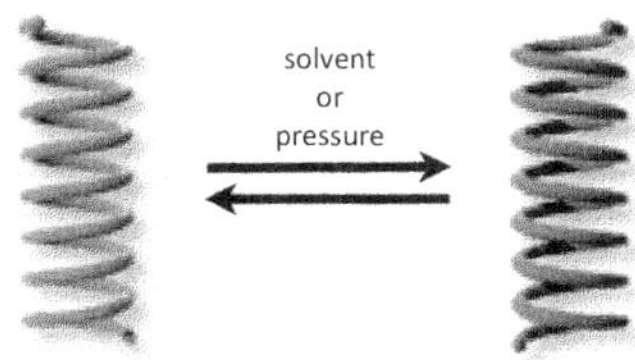

Figure 3.17 Solvent- and pressure-dependent helical inversion of poly(qunoxaline-2,3-diyl).

3.5.3 Circular Polarized Luminescence and Reflection

Blue circular polarized luminescence (CPL) is observed from solutions of poly(quinoxaline-2,3-diyl) with (*S*)-2-methyl butoxy side chains.[61] Solvent-dependent helical inversion [chloroform (*M*-helix) and 1,1,1-trichloroethane (*P*-helix)] induces an inversion of the CPL one-handedness. Furthermore, Suginome et al. reported that random copolymers of poly(quinoxaline-2,3-diyl)s with chiral side chains, achiral side chains, and chloromethyl side chains at the terminus show interesting CPL reflection as films (Fig. 3.18).[62] After annealing films in $CHCl_3$ vapor, selective reflection of one-handed CPL is observed in visible region. The chiral side-chain ratio enabled control of the selective reflection mean wavelength, with a decreasing ratio giving a red-shifted reflection. Cholesteric liquid-crystal formation is a driving force of the reflection.[62,63] When the polymer film was annealed under 1,2-DCE vapor, the one-handed CPL reflection inversed. It was also proposed that incrementing the distance between the polymer chains by increasing the chlorooctyl side-chain length made the cholesteric pitch elongate, resulting in a red shift of the structural color. Moreover, Langmuir–Blodgett films containing anisotropically oriented poly(quinoxaline-2,3-diyl)s can also be prepared[64], with further applications expected.

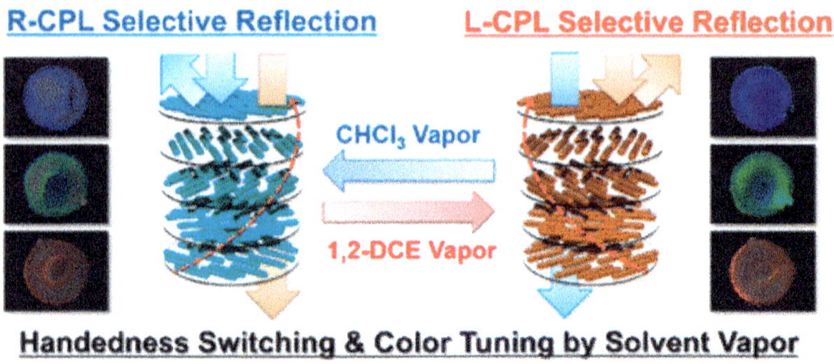

Figure 3.18 Schematic illustration of CPL reflection from a poly(quinoxaline-2,3-diyl) film and photograph of its films.

3.5.4 Chiral Catalysts for Asymmetric Reactions

One of sophisticated functions in biopolymers is chiral catalytic ability. While it has been attempted to use artificial helical polymers for catalysis applications, observed enantioselectivities are moderate. Suginome and coworkers have suggested some criteria to be fulfilled in order to improve selectivity:[65]

- Very high purity for a one-handed screw sense
- A highly stable helical structure, even in solution, with no racemization or denaturation
- Side chains, which may be modified and onto which catalytically active sites can be introduced without affecting the helical structure

Poly(qunoxaline-2,3-diyl)s are rigid rod helical polymers with a quite long, persistent helical length, and can be synthesized through "living" polymerization approaches, providing controlled molecular weights, polydispersity, and well-defined end groups. These advantages fulfill the above criteria. In fact, palladium-catalyzed reactions using various types of polyquinoxaline with phosphino ligand groups show various asymmetric reactions involving styrene hydrosilylation[66], Suzuki–Miyaura cross coupling reactions[67], silaborative C–C bond cleavage of *meso*-methylenecyclopropanes[68], and the arylative ring-opening reaction of 1,4-epoxy-1,4-dihydronaphthalene (Fig. 3.19).[69] These reactions give the desired products with high yield and enantioselectivity. It should be noted that the absolute chirality of the products are inversed when the chiral polymers undergo helical inversion with a suitable solvent.

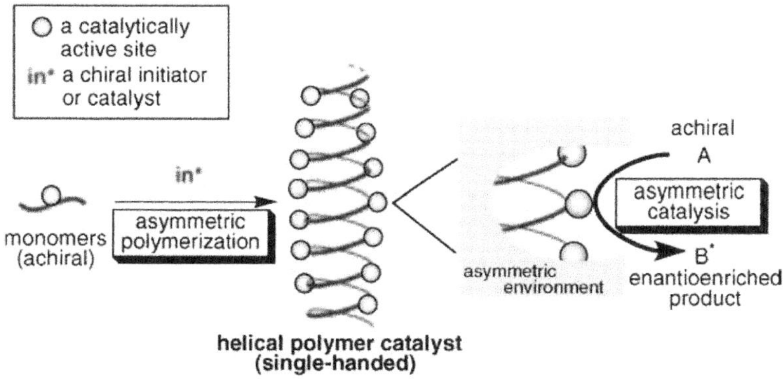

Figure 3.19 Schematic illustration of asymmetric reactions using poly(quinoxaline-2,3-diyl) with a chiral catalyst nature.

3.6 Conclusions and Outlook

In this chapter, I reviewed the synthesis, conformational behaviors, and function/properties of oligo/poly(*ortho*-phenylene)s and its related polymer, poly(quinoxaline-2,3-diyl)s. Efficient synthetic approaches to these poly(*ortho*-arylene)s have been developed, and the unusual functionalities unlike conventional π-conjugated polymers have been gradually revealed. On the other hand, one future research topic is to investigate the major contribution to the totally different conformational behaviors and functionality observed between oligo/poly(*ortho*-phenylene)s and poly(quinoxaline-2,3-diyl)s. The history of poly(*ortho*-arylene)s is still shallow compared to other π-conjugated polymers, and further development of robust synthetic methodologies and the foundation of fundamental function/property relationships should be the primary task.

References

1. Berresheim, A. J., Müller, M., Müllen, K. *Chem. Rev.*, **99**, 1747–1785 (1999).

2. Ibuki, E., Ozasa, S., Murai, K. *Bull. Chem. Soc. Jpn*, **48**, 1868–1874 (1975).
3. Ozasa, S., Fujioka, Y., Fujiwara, M., Ibuki, E. *Chem. Pharm. Bull.*, **28**, 3210–3222 (1980).
4. Blake, A. J., Cooe, P. A., Doyle, K. J., Gair, S., Simpkins, N. S. *Tetrahedron Lett.*, **39**, 9093–9096 (1998).
5. Ishikawa, S., Manabe, K. *Chem. Lett.*, **35**, 164–165 (2006).
6. He, J., Crase, J. L., Wadumethringe, S. H, Thakur, K., Dai, L., Zou, S., Rathore, R., Hartley, C. S. *J. Am. Chem. Soc.*, **132**, 13848–13857 (2010).
7. Motomura, T., Nakamura, H., Suginome, M. Murakami, M., Ito, Y. *Bull. Chem. Soc. Jpn.*, **78**, 142–146 (2005).
8. Suginome, M., Noguchi, H., Murakami, M. *Chem. Lett.*, **36**, 1036–1037 (2007).
9. Koyanagi, M., Eichenauer, N., Ihara, H., Yamamoto, T., Suginome, M. *Chem. Lett.*, **42**, 541–543 (2013).
10. Ormsby, J. L., Black, T. D., Hilton, C. L., Bharat, King, B. T. *Tetrahedron*, **64**, 11370–11378 (2008).
11. Ohta, E., Sato, H., Ando, S., Kosaka, A., Fukushima, T., Ushiyama, H., Yamashita, K., Aida, T. *Nat. Chem.*, **3**, 68–73 (2011).
12. Ito, S., Takahashi, K., Nozaki, K. *J. Am. Chem. Soc.*, **136**, 7574–7550 (2014).
13. Ito, S., Wang, W., Nozaki, K. *Macromolecules*, **48**, 1959–1962 (2015).
14. Ito, S., Wang, W., Nozaki, K. *Polym. J.*, **47**, 474–480 (2015).
15. Arslan, H., Saathoff, J. D., Bunck, D. N., Clancy, P., Dichtel, W. R. *Angew. Chem. Int. Ed.*, **51**, 12051–12054 (2012).
16. Wittig, G., Lehmann, G. *Chem. Ber.*, **90**, 875–892 (1957).
17. Winkler, H. J. S., Wittig, G. *J. Org. Chem.*, 28, 1733–1740 (1963).
18. Wittig, G., Klar, G. *Liebigs Ann. Chem.*, **704**, 91–108 (1967).
19. Kovacic, P., Uchic. J. T., Hsu, L.-C. *J. Polym. Sci. Part A: Polym. Chem.*, **5**, 945–964 (1967).
20. Kovacic, P., Ramsey, J. S. *J. Polym. Sci. Part A: Polym. Chem.*, **7**, 111–125 (1967).
21. Hsing, C.-F., Jones, M. B., Kovacic, P. *J. Polym. Sci. Part A: Polym. Chem.*, **19**, 973–984 (1981).
22. Tour, J. M. *Adv. Mater.*, **6**, 190–198 (1994).
23. Xu, J., Liu, H., Pu, S., Li, F., Luo, M. *Macromolecules*, **39**, 5611–5616 (2006).
24. Dong, B., Zheng, L., Xu. J., Liu, H., Pu, S. *Polymer*, **48**, 5548–5555 (2007).

25. Ma, M., Liu, H., Xu, J., Li, Y., Wan, Y. *J. Phys. Chem. C*, **111**, 6889–6896 (2007).

26. Vaisin, E., Williams, V. E. *Macromolecules*, **41**, 2994–2997 (2008).

27. Hoffmann, R. W. *Dehydrobenzene and Cycloalkynes*, Academic Press: New York (1967).

28. Nakamura, Y., Takano, N., Nishimura, T., Yashima, E., Sato, M., Kudo, T., Nishimura, J. *Org. Lett.*, **3**, 1193–1196 (2001).

29. Wenk, H. H., Winkler, M., Sander, M. *Angew. Chem. Int. Ed.*, **5**, 502–528 (2003).

30. Pellissier, H., Santelli, M. *Tetrahedron*, **59**, 701–730 (2003).

31. Dyke, A. M., Hester, A. J. Lloyd-Jones, G. C. *Synthesis*, **24**, 4093–4112 (2003).

32. Grampe, C. M., Carreira, E. M. *Angew. Chem. Int. Ed.*, **51**, 3766–3778 (2012).

33. Tadross, P. M., Stoltz. B. M. *Chem. Rev.*, **112**, 3550–3577 (2012).

34. Wu, C., Shi, F. A. *Asian J. Org. Chem.*, **2**, 116–125 (2013).

35. Ihara, E., Kurokawa, A., Koda, T., Muraki, T., Itoh, T., Inoue, K. *Macromolecules*, **38**, 2167–2172 (2005).

36. Mizukoshi, Y., Mikami, K., Uchiyama, M. *J. Am. Chem. Soc.*, **137**, 74–77 (2015).

37. Ando, S., Ohta, E., Kosaka, A., Hashizume, D., Koshino, H., Fukushima, T., Aida, T. *J. Am. Chem. Soc.*, **134**, 11084–11087 (2012).

38. Mathew, S. M., Engle, J. T., Ziegler, C. J., Hartley, C. S. *J. Am. Chem. Soc.*, **135**, 6714–6722 (2013).

39. Mathew, S., Crandall, L. A., Ziegler, C. J., Hartley, C. S. *J. Am. Chem. Soc.*, **136**, 16666–16675 (2014).

40. Hartley, C. S., He, J. *J. Org. Chem.*, **75**, 8627–8636 (2010).

41. Mathew, S. M., Hartley, C. S. *Macromolecules*, **44**, 8425–8432 (2011).

42. Hartley, C. S. *J. Org. Chem.*, **76**, 9188–9191 (2011).

43. He, J., Mathew, S. M., Cornett, S. D., Grundy, S. C., Hartley, C. S. *Org. Biomol. Chem.*, **10**, 3398–3405 (2012).

44. Kajitani, T., Suna, Y., Kosaka, A., Osawa, T., Fujikawa, S., Takata, M., Fukushima, T., Aida, T. *J. Am. Chem. Soc.*, **135**, 14564–14567 (2013).

45. Ito, Y., Ihara, E., Murakami, M. *J. Am. Chem. Soc.*, **112**, 6446–6447 (1990).

46. Yamada, T., Suginome, M. *Macromolecules*, **43**, 3999–4002 (2010).

47. Ito, Y., Ihara, E., Murakami, M. *Polym. J.*, **24**, 297–299 (1992).

48. Nagata, Y., Ke, Y.-Z., Suginome, M. *Chem. Lett.*, **44**, 53–55 (2015).
49. Ito, Y., Ihara, E., Murakami, M. *Angew. Chem. Int. Ed. Engl.*, **31**, 1509–1510 (1992).
50. Ito, Y., Ihara, E., Murakami, M., Sisido, M. *Macromolecules*, **25**, 6810–6813 (1992).
51. Yamada, T., Noguchi, H., Nagata, Y., Suginome, M. *J. Polym. Sci. Part A: Polym. Chem.*, **48**, 898–904 (2010).
52. Ito, Y., Miyake, T., Hatano, S., Shima, R., Ohara, T., Suginome, M. *J. Am. Chem. Soc.*, **120**, 11880–11893 (1998).
53. Ito, Y., Miyake, T., Ohara, T., Suginome, M. *Macromolecules*, **31**, 1697–1699 (1998).
54. Nagata, Y., Nishikawa, T., Suginome, M. *Chem. Comm.*, **48**, 11193–11195 (2012).
55. Nagata, Y., Nishikawa, T., Suginome, M. *J. Am. Chem. Soc.*, **137**, 4070–4073 (2015).
56. Nagata, Y., Yamada, T., Adachi, T., Akai, Y., Yamamoto, T. *J. Am. Chem. Soc.*, **135**, 10104–10113 (2013).
57. Ke, Y.-Z., Nagata, Y., Yamada, T., Suginome, M. *Angew. Chem. Int. Ed.*, **54**, 9333–9337 (2015).
58. Yamada, T., Nagata, Y., Suginome, M. *Chem. Commun.*, **46**, 4914–4916 (2010).
59. Nagata, Y., Nishikawa, T., Suginome, M. *J. Am. Chem. Soc.*, **136**, 15901–15904 (2014).
60. Nagata, Y., Takeda, R., Suginome, M. *Chem. Commun.*, **51**, 11182–11185 (2015).
61. Nagata, Y., Nishikawa, T., Suginome, M. *Chem. Commun.*, **50**, 9951–9953 (2014).
62. Nagata, Y., Takagi, K., Suginome, M. *J. Am. Chem. Soc.*, **136**, 9895–9861 (2014).
63. Ito, Y., Ihara, E., Uesaka, T., Murakami, M. *Macromolecules*, **25**, 6711–6713 (1992).
64. Ito, Y., Miyake, T., Suginome, M., Katakura, R., Mitsuishi, M., Miyashita, T. *Polym. J.*, **42**, 406–410 (2010).
65. Yamamoto, T., Suginome, M. *Angew. Chem. Int. Ed.*, **48**, 539–542 (2009).
66. Yamamoto, T., Yamada, T., Nagata, Y., Suginome, M. *J. Am. Chem. Soc.*, **132**, 7899–7901 (2010).

67. Akai, Y., Yamamoto, T., Nagata, Y., Ohmura, T., Suginome, M. *J. Am. Chem. Soc.*, **134**, 11092–11095 (2012).

68. Yamamoto, T., Akai, Y., Suginome, M. *Angew. Chem. Int. Ed.*, **53**, 12785–12788 (2014).

69. Akai, Y., Konnert, L., Yamamoto, T., Suginome, M. *Chem. Commun.*, **51**, 7211–7214 (2015).

Biography

Koichiro Mikami was born in Kanagawa in 1981. He received his BS (2006), MS (2008), and PhD (2011) in applied chemistry from Kanagawa University, Japan, under the direction of Prof. Tsutomu Yokozawa. After receiving his PhD, he worked as a research associate with the Wooley Group at Texas A&M University in 2011, where he developed glucose-based polycarbonate polymerization via an organocatalytic approach. In 2012–2015, he joined the Uchiyama Group; as a postdoctoral fellow in 2012 he worked at the Advanced Elements Chemistry Laboratory at RIKEN; in 2013–2014, he received a JSPS postdoctoral fellowship and worked in the University of Tokyo. In 2015, he was awarded a special postdoctoral researcher fellowship for 3 years at RIKEN and worked in the Uchiyama Elements Chemistry Laboratory for 4 months. With the Uchiyama Group, he studied the development of novel reactions and functional molecules, analyzed theoretical reaction pathways, and predicted theoretical spectra using DFT calculations. From April to July in 2015, he also concurrently worked as a researcher at the Yokozawa Group of Kanagawa University. Starting August 2015, Dr. Mikami joined the faculty of the Sagami Research Chemical Institute, where he is presently a research scientist. His research interests include the development of novel polymerization reactions, synthesis of polymers with unique functionality and properties for engineering and bioapplication fields, and theoretical analysis with DFT calculations.

Chapter 4

Electrochemistry of Conjugated Polymers: Synthesis, Properties, and Reactions

Shinsuke Inagi
Department of Chemical Science and Engineering, School of Materials and Chemical Technology, Tokyo Institute of Technology, 4259 Nagatsuta-cho, Midori-ku, Yokohama 226-8502, Japan
inagi@echem.titech.ac.jp

π-Conjugated polymers are electrochemically (redox) active and therefore widely studied in application to organic electronics. This chapter consists of the fundamentals of electrochemistry of organic compounds and materials, dealing with the electrosynthesis of π-conjugated polymers by electro-oxidative or reductive polymerization. The concept of electrochemical functionalization of π-conjugated polymers is also described briefly.

4.1 Introduction

Electrochemical synthesis of organic compounds has been developed as organic electrosynthesis so far to produce a variety

Conjugated Objects: Developments, Synthesis, and Applications
Edited by Atsushi Nagai and Koji Takagi

ISBN 978-981-4774-03-1 (Hardcover), 978-1-315-22962-1 (eBook)
www.panstanford.com

of functional compounds via electrochemically generated species as a key intermediate.[1,2] In an electrolytic cell equipped with an anode, a cathode, and a reference electrode containing a supporting electrolyte/solvent (organic solvent, water, ionic liquid), potentials applied on both the anode and the cathode can oxidize and reduce organic substrates, and subsequent reactions take place (Fig. 4.1a). It is easy to set an electrode potential by using a potentiostat, a power supply with controlling electrolytic parameters. Constant current electrolysis using two-electrode system consisting of an anode and a cathode is also useful for organic electrosynthesis.

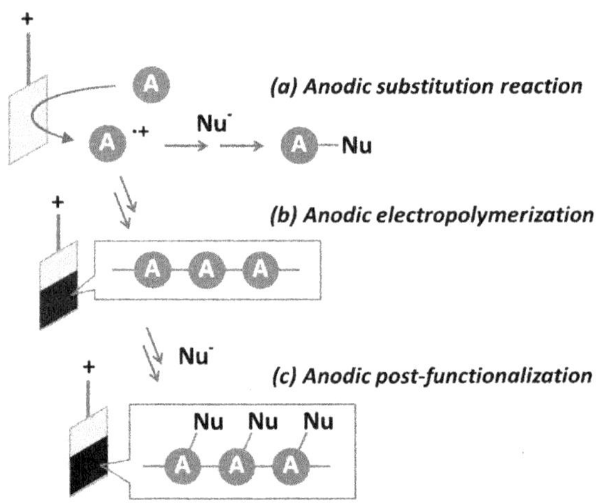

Figure 4.1 Schematic illustration of anodic reactions of an organic molecule (A).

It is possible to expand the methodology to polymer synthesis. Electrochemical polymerization utilizes electrogenerated species of a substrate for polymerization as a monomer or an initiator. In the former case, electrogenerated species of an aromatic monomer couples with a polycondensation process to give a π-conjugated polymer, in which the p-orbital of the carbon atoms on the aromatics overlaps throughout the polymer main chain. Such π-conjugated polymers are intrinsically conductive and thus called conducting polymers (Fig. 4.1b).[3,4] Electrosynthetic methods are also available for activation of π-conjugated polymers for further functionalization by using ionic species in the polymer chains. For example, the anodic electrolysis of π-conjugated polymers can undergo nucleophilic reactions (Fig. 4.1c). These electrosynthetic methods are described in the following sections.

4.2 Electrochemical Polymerization

Electrochemical polymerization of aromatic monomers is a powerful tool to give π-conjugated polymers. Generally they are semiconducting materials but chemical or electrochemical doping (see 4.2.4. electrochemical doping) imparts electrical conductivity to them. The conducting polymers are obtained on the surface of working electrode as films because the electron transfer of monomer and its coupling reaction proceeds at the near surface of the electrode, and then an insoluble polymeric product deposits on it. Although chemical electron transfer of monomers in solution is also available to produce the corresponding π-conjugated polymers, the obtained insoluble polymer powder has difficulty in processing for application. The easy film formation property during electrochemical polymerization is useful for application in electrochemical devices such as sensors and displays.[5,6]

4.2.1 Electro-Oxidative Polymerization of Aromatic Monomers

Electron-rich aromatic and heteroaromatic monomers, as listed in Fig. 4.2, can be easily oxidized on an anode surface to form their radical cation states. The radical cations couple to form a carbon–carbon bond, and the following deprotonation leads to afford a neutral dimer (Fig. 4.3). The generated conjugated dimer has a lower oxidation potential than that of monomer, so further oxidation of the dimer leads to the oligomerization and polymerization. The highly conjugated polymers are no longer soluble in an electrolytic medium to be deposited on the anode surface as a film. The deposited polymer is oxidatively doped during the application of potential for polymerization, thus the film formed is not passive but still conductive to carry out continuous electrochemical reactions on the electrode. In this procedure, the coupling position of the monomer can be predicted by spin density distribution on the aromatic ring.[7] The introduction of an electroauxiliary such as a silyl group can realize dominant coupling at the specific position.[8]

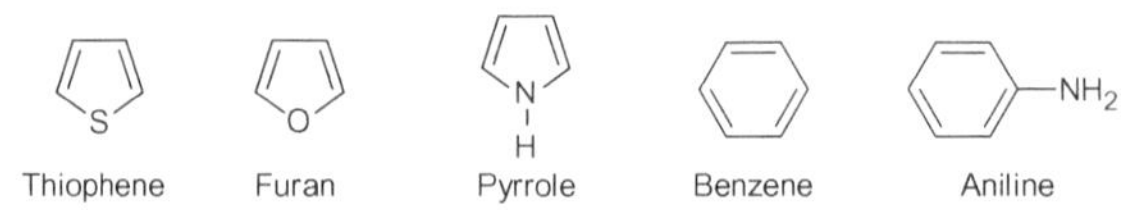

Figure 4.2 Aromatic and heteroaromatic monomers for anodic electropolymerization.

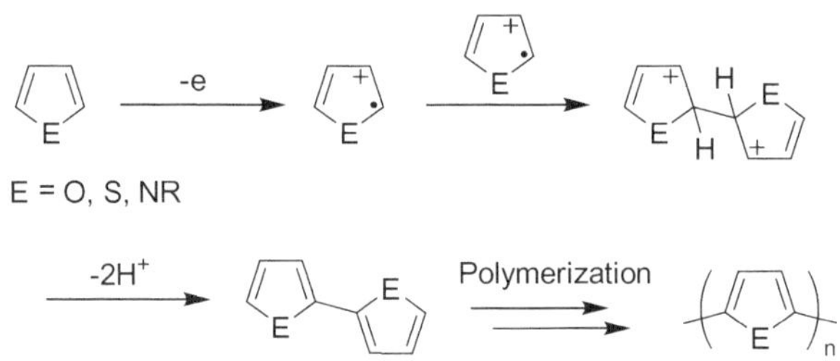

Figure 4.3 Anodic reaction of heteroaromatics, providing their coupled dimers and subsequent polymerization.

4.2.2 Method of Electrochemical Polymerization

For electrochemical polymerization, there are several methods such as a potential sweep method, a potentiostatic (constant-potential) method, and a galvanostatic (constant-current) method. One should choose a suitable method depending on monomers and conditions used. Cyclic voltammetry (CV) measurement of a monomer in a three-electrode setup gives information on oxidation potential for electrochemical polymerization. To prevent overoxidation of a product polymer, the application of higher potential to the working electrode for a long period should be avoided since the product polymer with extended π-conjugation is easily oxidized compared to the corresponding monomer. The potential sweep method and the potentiostatic method are rather milder polymerization conditions than the galvanostatic method.

In the potential sweep method, polymerization is carried out with a CV analyzer. A working electrode, a counterelectrode, and a reference electrode are put together in an electrolytic cell containing an electrolyte and a monomer. The repeating potential sweeps in a potential range determined from the advanced CV measurement produce product polymer films on the working electrode and this electron transfer process can be monitored by CV, as shown in Fig. 4.4. Figure 4.4 shows typical voltammograms of pyrrole during electropolymerization. The first scan represents the oxidation current of the pyrrole monomer and then the polymerization proceeds at the surface of the working electrode, resulting in showing an irreversible voltammogram. In the course of deposition of the product polypyrrole during repeating scans, anodic and cathodic currents derived from redox of polyryrrole at a lower potential range gradually increase.

In a potentiostatic method, the three-electrode system is used and the potential applied to the working electrode is controlled by a potentiostat with monitoring charge passed. In a galvanostatic method, a two-electrode system consisting of a working electrode and a counterelectrode is used and constant current is passed between them by a DC power source. This process cannot control the potential on the working electrode, which may cause several problems such as overoxidation of product polymers.

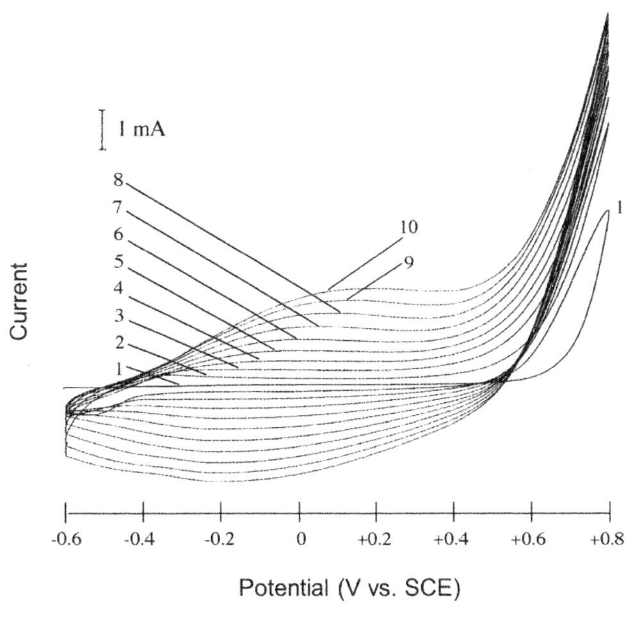

Figure 4.4 Typical cyclic voltammograms of pyrrole during repetitive potential scans (10 cycles) in water containing sodium perchlorate (0.1 M) for providing polypyrrole films on a working electrode.

4.2.3 Conditions for Electrochemical Polymerization

Concentration of monomer is one important factor. If the polymerization does not proceed at an electrode surface, the use of much monomer may be effective to promote the film formation of π-conjugated polymers. The choice of material for the working electrode is also important in terms of not only overpotential but also compatibility to π-conjugated polymer films.

In addition to conventional electrolytic solutions such as supporting salts/solvents, room-temperature ionic liquids are known to be suitable for polymerization media.[9] There are several advantages in electropolymerization in ionic liquids: (i) electropolymerization proceeds rapidly, (ii) compatibility of a product film with an electrode surface is good, and (iii) product polymers show dense and smooth surfaces, high electric conductivity, high electrochemical capacitance, good reversibility of redox cycles.

Recently the boron trifluoride–ether complex has been attracting attention as a new polymerization medium.[10] Boron trifluoride acts as a Lewis acid to form a complex with an aromatic monomer, which results in reducing the oxidation potential of the monomer. Therefore, this method is effective for polymerization of aromatic monomers having a relatively high oxidation potential. The reduction of polymerization potential can avoid overoxidation of product polymers.

4.2.4 Electrochemical Doping

If a polymer film on a working electrode is obtained by electropolymerization, it should be subjected to an electrolytic solution without monomer (monomer-free electrolyte) in order to study the redox behavior of the π-conjugated polymers. Typically a pair of broad redox responses is observed in the voltammogram. The injection or removal of electrons to or from π-conjugated polymers results in the formation of polarons and bipolarons in the repeating structure (commonly known as doping), producing considerable variations in the physical properties of the polymer itself and imparting features such as drastic color changes and electrical conductivity. The charges thus generated along the polymer must be compensated for by the addition of neighboring ions (or dopants) and the insertion and release of such dopants may induce volume changes in the conjugated polymer. The doping amount can be estimated from the amount of charge generated in the conducting polymers.

4.2.5 Electroreductive Polymerization of Aromatic Monomers

Aromatic dihalides are representative monomers for electroreductive polymerization. Cathodic reduction of an aromatic dihalide generates a radical anion, followed by polymerization accompanying elimination of the halide (Fig. 4.5). The bond formation takes place at the specific position of the monomer, where the halogen atom is substituted.

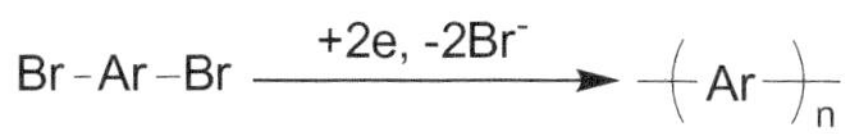

Figure 4.5 Scheme of cathodic electropolymerization of aromatic dihalide.

On electroreductive polymerization of 1,4-dibromopyridine, addition of a catalytic amount of a nickel(II) complex is effective and the complex works as a mediator. Electrogenerated nickel(0) promotes dehalogenation of the dibromopyridine, followed by its polymerization.[11] Electro-oxidative polymerization of pyridine is generally difficult because of its high oxidation potential; however, this reductive polymerization enables to produce polypyridines.

Poly(*p*-phenylenevinylene) (PPV) is also synthesized by electroreductive polymerization. As shown in Fig. 4.6, the electrochemical reduction of $\alpha,\alpha,\alpha',\alpha'$-tetrabromo-*p*-xylene generates a quinodimethane intermediate and its subsequent polymerization gives a poly(xylylene). This is further cathodically reduced to be poly(*p*-phenylenevinylene).[12]

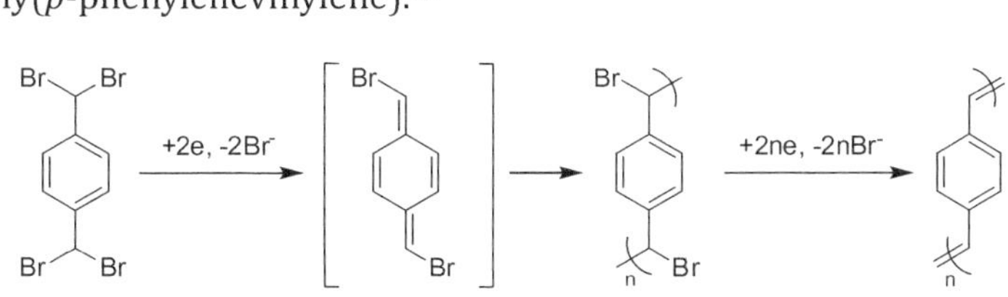

Figure. 4.6 Electroreductive synthesis of PPV by multiple reduction of xylene derivatives.

4.2.6 Copolymer Synthesis by Electrochemical Methods

Polymers consisting of two or more different monomer units, so-called copolymers, are of great importance because they possess various functionalities derived from each monomer component. For example, the combination of monomers with different electrical properties enables the tuning of bandgaps of π-conjugated polymers. There have been a lot of reports on electrochemical copolymerization, usually achieved by simply mixing different monomers in an electrolyte.[13,14] However, monomers for copolymerization should possess close oxidation potentials each other. The obtained polymer films show intermediate optical properties of each homopolymer, which can be tuned by the feeding ratios of the monomers. To produce alternating copolymers, one approach is to prepare a monomer composed of two or more units beforehand, followed by electrochemical polymerization. 2,5-Dithienyl-3,4-disubstituted thiophene derivatives are candidates for monomers for alternating copolymers. As shown in Fig. 4.7, electrochemical polymerization of such terthiophene derivatives having 3-fluoro-4-hexylthiophene unit proceeds to give an alternating structure of bithiophene and the substituted thiophene.[15]

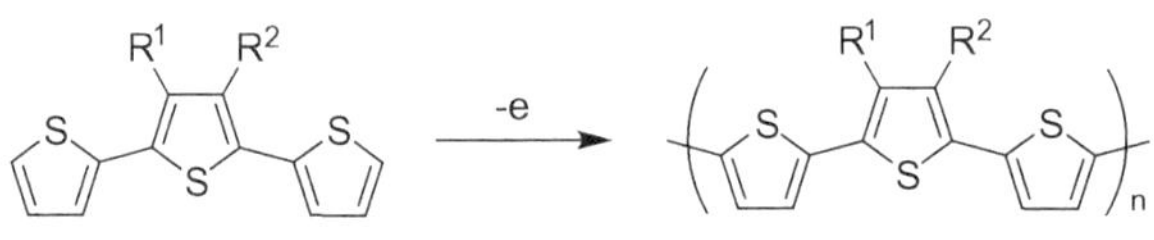

R^1 = F, R^2 = C_6H_{13}

Figure 4.7 Electropolymerization of fluorinated terthiophene derivatives.[15]

4.2.7 Application of Conducting Polymers

Physical properties of π-conjugated polymers are drastically changed by a doping process. Table 4.1 summarizes the application of π-conjugated polymers. The undoped (neutral) state of π-conjugated polymers is generally semiconducting, and it can be applied to materials for transistors, photovoltaic cells, and electroluminescence devices. On the other hand, electrical conductivity appears in the doped state of π-conjugated polymers, which is useful as a conducting material, antistatics, and capacitor material. The doping/dedoping process of π-conjugated polymers is also useful for electrochromic devices, sensors, and actuator applications.

Table 4.1 Applications of π-conjugated polymer films

Property	Application
Conductivity (doped state)	Conducting material, antistatics, capacitor material
Doping/de-doping process	Electrochromic material, sensor, actuator, memory
Semiconductivity (undoped state)	Transistor, photovoltaic cell, electroluminescent material

4.3 Electrochemical Postfunctionalization of π-Conjugated Polymers

4.3.1 Anodic Substitution Reaction of Conducting Polymers

In the previous section, the contribution of electrochemistry to π-conjugated polymer synthesis was described. Electrochemistry is very powerful to oxidize or reduce small molecules to generate reactive species or catalyst for polymerization. However, to address desired polymeric materials via electrochemistry, another approach is possible: to use a polymer reaction process in which polymers are transformed into other structures. In this method, various polymers with different functionalities can be produced from a precursor polymer. Moreover, polymer reactions also enable the introduction of many functional groups, which may inhibit direct polymerization of monomers possessing such functional groups. From these points of view, this kind of polymer reaction (post-functionalization) is recognized as a strong methodology in polymer synthesis.

In the field of polymer chemistry, polymer reaction by electrochemical techniques has not been actively investigated because many polymers work as an insulator. However, conducting polymers can involve electron transfer at electrode surface and thus electrochemical postfunctionalization is potentially realized.

π-Conjugated polymers can easily be oxidized to form doped state, possessing polarons or bipolarons in the main chain. The generated cation species are generally stabilized in the conjugated backbones with nonnucleophilic dopants but are reactive with nucleophiles to undergo nucleophilic substitution reactions on the π-conjugated polymers. Such electrochemical polymer reactions enable fine-tuning of the degree of reaction manipulated by charge passed in the solid state (simple work-up process). Therefore, the electrochemical polymer reaction has attracted much attention as a novel way of postfunctionalization of π-conjugated polymers.[16]

Pickup and coworkers reported pioneering works on electrochemical polymer reactions. They focused on the degradation mechanism of a π-conjugated polymer, poly(3-methylthiophene) (P3MT), by nucleophilic reaction of water.[17] In this case, the resulting polymer undergoes isomerization and becomes its nonconducting form. On the other hand, the substitution reaction with chloride ions would give chlorinated P3MT having good conductivity (Fig. 4.8). P3MT prepared on an electrode by electropolymerization was used as a substrate for the anodic substitution in the presence of tetraethylammonium chloride (Et_4NCl) as a supporting electrolyte and a chlorine source.[18,19] The CV of P3MT showed a large oxidation current in comparison to the CV measured with tetraethylammonium perchlorate (Et_4NClO_4), a nonnucleophilic salt. Additionally, the oxidation process was irreversible. These results indicate that following chemical reactions occurred as well as anodic doping. X-ray analysis evidenced the existence of C–Cl covalent bonds in the P3MT film, where the nucleophilic substitution reaction of Cl^- proceeded at the 4-position of the repeating thiophene rings (Fig. 4.8). The CV of the obtained chlorinated P3MT film in Et_4NClO_4/CH_3CN showed reversible redox waves derived from doping and de-doping similar to those observed in unmodified P3MT, while the oxidation potential of chlorinated P3MT slightly anodically shifted. This was thought to be the electron-withdrawing effect of the chlorine substituent. They also reported bromination and alkoxylation based on the same mechanism; however, iodination did not proceed due to its poor nucleophilicity. In addition, they reported the electrochemical polymer reaction using polypyrrole as a substrate. This postfunctionalization is a powerful methodology in terms of realizing not only the facile tuning of the electron density of π-conjugated polymers but also the synthesis of methoxy-substituted polythiophenes, which cannot be obtained by electropolymerization of a methoxythiophene monomer.

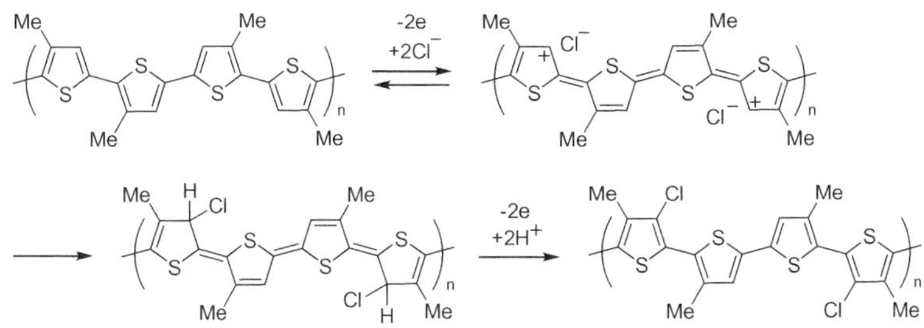

Figure 4.8 Electrochemical doping of P3MT and following nucleophilic substitution by chloride ions.

Bélanger and coworkers reported electrochemical chlorination of poly(3-(4-fluorophenyl)thiophene) in a similar manner to that of P3MT.[20] They also demonstrated the complete recovery of the original starting material following the application of a negative potential (−2 V vs. Ag/Ag^+), due to one-electron reduction and H atom transfer from the electrolyte.

Iyoda and coworkers reported electrochemical substitution of pyridine into the 4-position of poly(3-hexylthiophene) (Fig. 4.9).[21] An electropolymerized poly(3-hexylthiophene) film on a Pt plate was subjected to anodic oxidation in the presence of pyridine derivatives (0.1 M). By constant potential or potential sweep electrolysis, pyridinium moieties were successfully introduced and confirmed by Fourier transform infrared (FTIR) and CV measurement. The amount of introduction pyridinium moieties was achieved up to 60%.

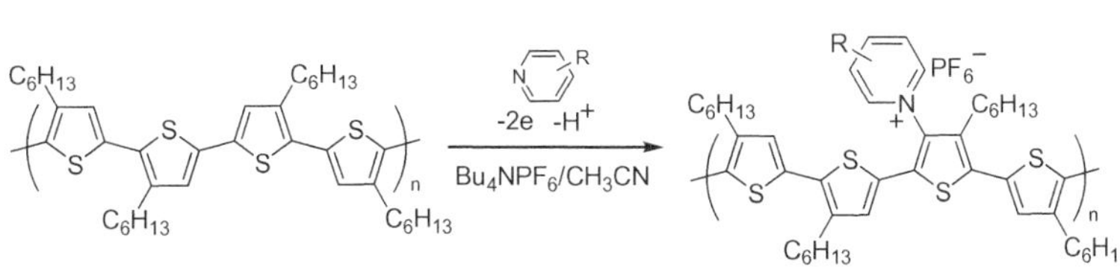

Figure 4.9 Electrochemical introduction of pyridinium moieties into poly(3-hexylthiophene).

Fabre and coworkers contributed to this field with anodic cyanation of poly(1,4-dimethoxybenzene).[22,23] The anodic oxidation of electropolymerized poly(1,4-dimethoxybenzene) at 1.2 V versus Ag/Ag^+ in the presence of tetraethylammonium cyanide (Et_4NCN) involved the aromatic nucleophilic substitution of hydrogen on the dimethoxy benzene unit rather than the replacement of aromatic methoxy group by CN (Fig. 4.10). The cyanation resulted in the linkage of the ca. 1.4 CN group per dimethoxybenzene unit as evidenced by elemental analysis and X-ray photoelectron spectroscopy (XPS) analysis.

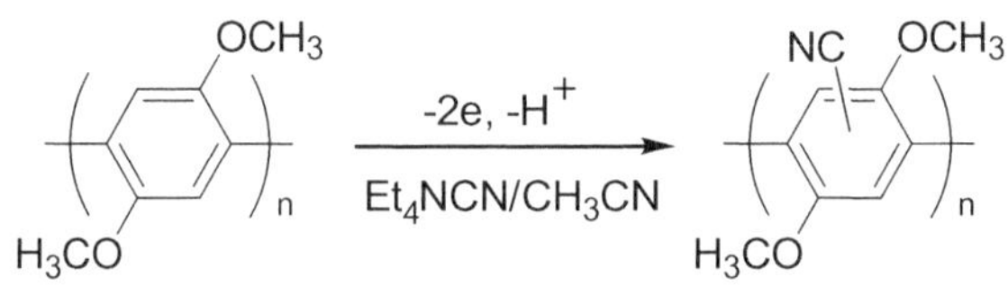

Figure 4.10 Anodic cyanation of poly(1,4-dimethoxybenzene).

In the research mentioned above, characterization of postfunctionalized polymers was conducted in the solid state by IR, UV-Vis measurements, and X-ray analyses. However, other analytic methods in solution such as nuclear magnetic resonance (NMR), which is powerful for providing information on chemical structures, are not available. Inagi, Fuchigami, and coworkers recently developed the electrochemical polymer reactions with soluble conducting polymers.[16]

An alternating copolymer composed of thiophene and 9,9-dioctylfluorene was prepared by the Suzuki–Miyaura polycondensation reaction (Fig. 4.11). This poly(thiophene-alt-fluorene) (PTF) possesses a reactive site for oxidative halogenation (thiophene unit) and a solubilizing moiety (fluorene with long alkyl chains). Thus the chemical structure of PTF was determined by NMR in chloroform-*d*, and the number-average molecular weight (M_n) and the polydispersity index (M_w/M_n) of PTF were estimated as 8700 and 2.1, respectively, by gel permeation chromatography (GPC). The cast-coated film from the solution of PTF on a Pt plate was used as an anode and oxidized by constant current technique in an electrolyte containing 0.1 M Et_4NCl in CH_3CN (poor solvent for PTF).[24] After a passage of 30 F/mol of electricity, the film fixed on the anode was washed with methanol to remove supporting salts and dissolved in chloroform-*d* for ^{1}H NMR analysis. Degree of chlorination was estimated by ^{1}H NMR. M_n and M_w/M_n of the product polymer was 9000 and 2.3, respectively. These results strongly suggest that the electrochemical chlorination of the polymer film proceeded selectively at the thiophene ring without side reactions like chain dissociation or crosslinking.

It was also revealed that the amount of chlorine introduced could be controlled by changing the charge passed. By passing a current from 5 F/mol, 10 F/mol, and 20 F/mol for the PTF film, the degree of chlorination per thiophene unit, *m*, was determined as 0.66, 1.02, and 1.46, respectively.[25] The onset potentials of each film became more positive, indicating that the chlorinated PTF gained tolerance toward oxidation. As described here, the electrochemical polymer reaction is a good methodology to modify conducting polymers in a well-controlled manner, benefited by the facile manipulation of electrolytic conditions. Moreover, the halogenation of poly(fluorene-*alt*-bithiophene)[26] and poly(3-hexylthiophene)[27] was successfully accomplished.

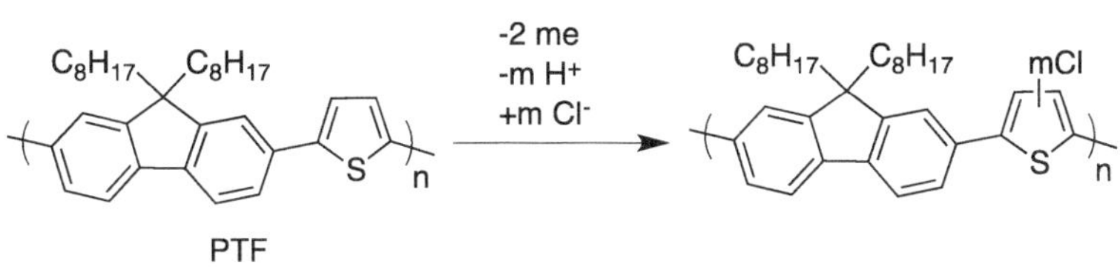

Figure 4.11 Electrochemical chlorination of a fluorene-thiophene-based copolymer (PTF).

The fluorination of organic compounds is one of the great challenges in organic synthesis toward application to pharmaceuticals, agrichemicals, and organic semiconductors. Anodic fluorination of organic molecules is one of the most successful processes, in which the electrochemical oxidation and subsequent nucleophilic fluorination are involved under mild conditions.[28] Based on this chemistry, the electrochemical fluorination of a conducting polymer film was carried out (Fig. 4.12). A copolymer composed of 9,9-dioctylfluorene and 9,9-diarylsulfanylfluorene (M_n = 6400) was synthesized by coupling polymerization. Anodic fluorodesulfurization of the polymer proceeded almost completely by a passage of 24 F/mol of current in an electrolytic cell containing neat Et_4NF-5HF (ionic liquid), evidenced by ^1H and ^{19}F NMR and elemental analysis.[29,30] Remarkably, the electrolysis of a nonconjugated fluorine-based polymer did not afford the fluorinated product at all, indicating that the anodic doping of conjugated polyfluorene promoted the fluorine ion penetration into the polymer film as a dopant, and the consequent substitution reaction proceeded.

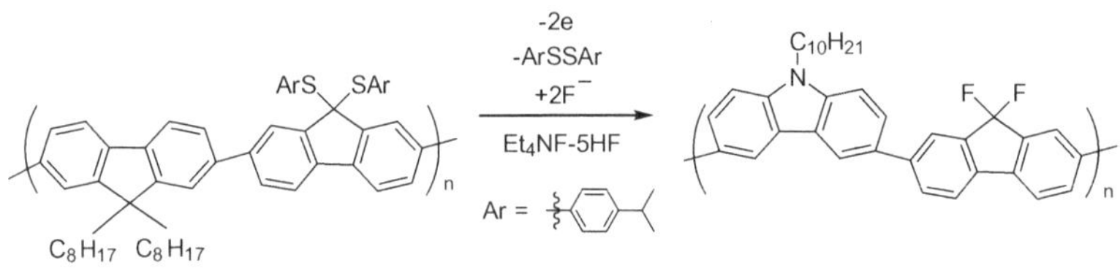

Figure 4.12 Electrochemical fluorination of a fluorene-based π-conjugated polymer.

A more powerful donor–acceptor-type alternating copolymer consisting of 9-alkylcarbazole and 9,9-difluorofluorene was prepared by the electrochemical method similarly.[31] This was accomplished by electrolyzing a precursor polymer containing arylsulfanylfluorene units in Et_4NF-5HF. Unfortunately, the conversion which was achieved under these conditions was only approximately 50% (Fig. 4.13); the oxidation of the alkylcarbazole units competed with that of the arylsulfanyl groups, which prevented effective fluorodesulfurization during electrolysis.

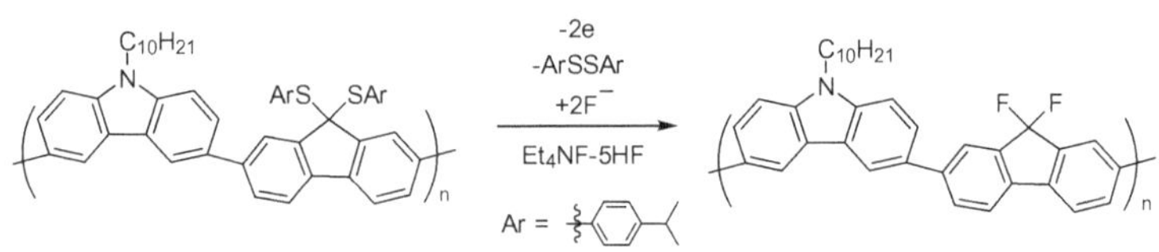

Figure 4.13 Anodic fluorodesulfurization of a polyfluorene derivative containing an alkylcarbazole moiety.

4.3.2 Cathodic Reaction and Paired Reactions

Such electrochemical polymer reactions can also be performed on a cathode initiated by cathodic doping. Poly(9,9-dioctylfluorene-*alt*-9-fluorenone) (M_n = 4100) was prepared as a substrate, where the ketone group is possibly transformed into a methylene moiety by reductive hydrogenation. Electrolysis was carried out using tetraethylammonium-*p*-toluenesulfonate (Et_4NOTs) as a supporting electrolyte and 2-propanol as a solvent and a proton source (Fig. 4.14).[32] A zinc cathode was effective for the reduction of ketone because of its high overpotential for reduction of proton. A passage of 16 F/mol of charge under a constant current condition resulted in the quantitative formation of the hydrogenated product polymer. The degree of reaction could be controlled by electricity passed and the lowest unoccupied molecular orbital (LUMO) level was also tunable depending on the progress of reaction.

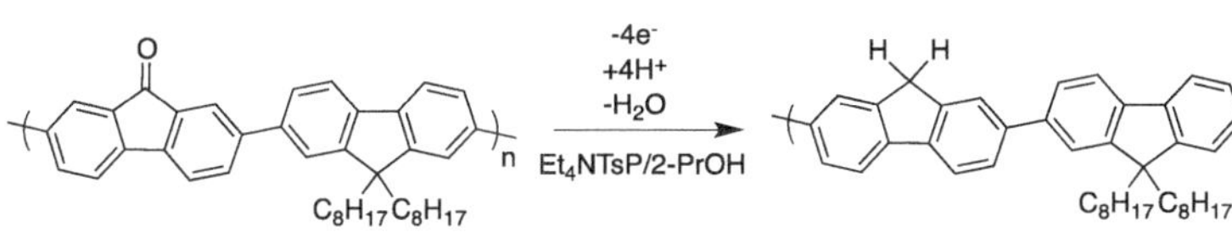

Figure 4.14 Electrochemical reduction of a ketone moiety in a fluorene-based π-conjugated polymer.

The electrochemical polymer reactions described above have been carried out only on an anode or a cathode with sacrificial reactions at a counterelectrode. To improve current efficiency in the electrolytic system, useful reactions should be carried out simultaneously at both electrodes, which is generally known as *paired electrolysis* in the field of organic electrochemistry.[33] Combining two independent anodic and cathodic polymer reactions into a single electrochemical cell is of great interest for doubled current efficiency; in addition, product polymers still attached on both electrodes are easily separated and purified. Such a parallel polymer reaction is impossible in the general solution process of the polymer reaction, and it could potentially be an economically and environmentally friendly process. As a proof-of-concept, Inagi and coworkers have performed parallel electrochemical polymer reactions from one precursor polymer, poly(9-fluorenol-*alt*-9,9-dioctylfluorene) (P1) (Fig 4.15). For anodic oxidation, two-electron oxidation on Pt afforded the 9-fluorenone-containing polymer, P2. On the other electrode, two-electron reduction on Zn provided the fluorine-containing polymer, P3. As electrons were used in a closed system, the current efficiency was almost doubled compared to the independent reactions. The obtained polymers exhibited a totally different emission property in the solid state, namely nonemissive for P2, yellow emission for P3, and blue emission for P1.[34]

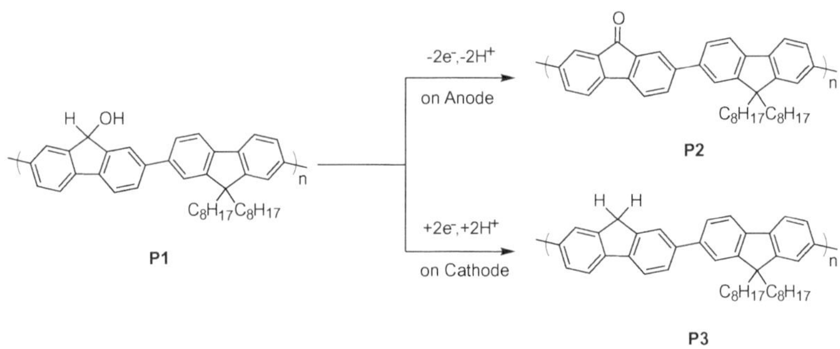

Figure 4.15 Paired electrochemical reactions of a fluorene-based π-conjugated polymer.

4.3.3 The CRS Method

Polyaniline is one of the most known conducting polymers due to its multiple doping behaviors driven by oxidation and protonation. An emeraldine base of polyaniline is an oxidized state having the reactivity with nucleophiles. Han and coworkers developed the concurrent reduction and substitution (CRS) method, in which an emeraldine base of polyaniline can be reduced to be a leucoemeraldine base state along with the incorporation of various nucleophiles such as dialkylamines.[35,36] When an emeraldine base of polyaniline prepared by electropolymerization was immersed into a solution of tetrabutylammonium fluoride in methanol, fluorine was introduced to the polyaniline main chain. Since only one fluorine could be introduced to one quinoidal structure, the repetitive treatment of the oxidation/CRS cycles provided the quantitative fluorination of the polyaniline (one fluorine per repeating aniline) (Fig. 4.16).[37] It was also found that the conductivity of the fluorinated polyaniline was higher than that of the precursor. Another approach to obtain the fluorinated polyaniline, that is, electropolymerization of 2-fluoroaniline, did not give a high-molecular-weight polymer. Therefore, this method has a good advantage to prepare functionalized polyanilines.

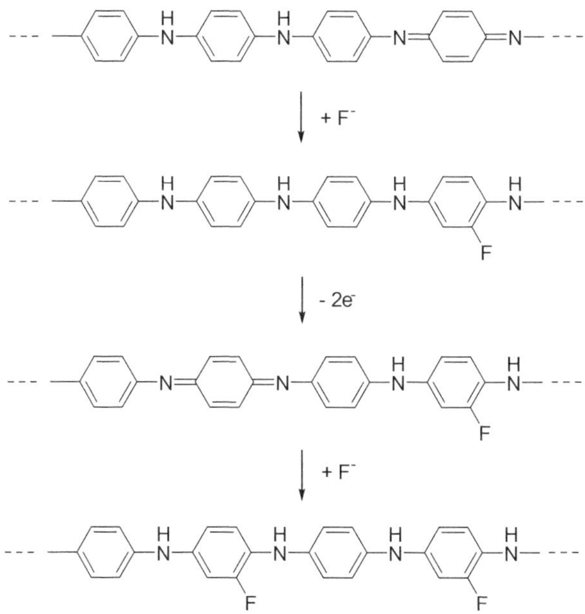

Figure 4.16 Concurrent reduction and substitution (CRS) method for fluorination of polyanilines.

4.4 Conclusions

In this chapter, electrochemical synthesis of π-conjugated polymers was described. The electropolymerization of aromatic monomers is a very strong methodology to prepare π-conjugated polymer films on electrodes. The obtained polymer films involve electron transfer easily to exhibit interesting electrochemical features arising from the conjugated polymer backbones. A new strategy of further functionalization of such π-conjugated polymers was successfully demonstrated through electrochemical doping and subsequent chemical reactions based on the concept of organic electrosynthesis. Recently, the solid-state electrochemical transformation of π-conjugated polymers can be achieved by using a heterogeneous condition of an electrode potential on a bipolar electrode, which is driven in a low concentration of electrolyte.[38–40] The gradient electrochromism,[41–43] composition gradient,[34,44] gradient surface modification[45], and π-conjugated polymer fiber formation[46] have been demonstrated by means of bipolar electrochemistry.

References

1. Hammerich, O., Speiser, B. *Organic Electrochemistry*, 5th ed., CRC Press: Boca Raton (2015).
2. Fuchigami, T., Atobe, M., Inagi, S. *Fundamentals and Applications of Organic Electrochemistry: Synthesis, Materials, Devices*, John Wiley & Sons: Hoboken (2014).
3. Skotheim, T. A., Reynolds, J. R. *Handbook of Conducting Polymers*, 3rd ed., CRC Press: Boca Raton (2007).
4. Inzelt, G. *Conducting Polymers*, Springer, Heidelberg (2008).
5. Heinze, J., Frontana-Uribe, B. A., Ludwigs, S. *Chem. Rev.*, **110**, 4724–4771 (2010).
6. Beaujuge, P. M., Reynolds, J. R. *Chem. Rev.*, **110**, 268–320 (2010).
7. Ando, S., Ueda, M. *Synth. Met.*, **129**, 207–213 (2002).
8. Lemaire, M., Büchner, W., Garreau, R., Hoa, K. A., Guyard, A., Roncali, J. *J. Electroanal. Chem.*, **281**, 293–298 (1990).
9. Sekiguchi, K., Atobe, M., Fuchigami, T. *Electrochem. Commun.*, **4**, 881–885 (2002).
10. Shi, G., Jin, S., Xue, G., Li, C. *Prog. Polym. Sci.*, **30**, 783–811 (2005).
11. Saito, N., Kanbara, T., Nakamura, Y., Yamamoto, T., Kubota, K. *Macromolecules*, **27**, 756–761 (1994).
12. Utley, J. H. P., Gruber, J. *J. Mater. Chem.*, **12**, 1613–1624 (2002).
13. Smith, E. L., Glidle, A., Mortimer, R. J., Ryder, K. S. *Phys. Chem. Chem. Phys.*, **9**, 6098–6105 (2007).
14. Nie, G., Qu, L., Xu, J., Zhang, S. *Electrochim. Acta*, **53**, 8351–8358 (2008).
15. Gohier, F., Frère, P., Roncali, J. *J. Org. Chem.*, **78**, 1497–1503 (2013).
16. Inagi, S., Fuchigami, T. *Macromol. Rapid Commun.*, **35**, 854–867 (2014)
17. Qi, Z., Pickup, P. G. *J. Chem. Soc., Chem. Commun.*, 1675–1676 (1992).
18. Qi, Z., Pickup, P. G. *Anal. Chem.*, **65**, 696–703 (1993).
19. Qi, Z., Rees, N., Pickup, P. G. *Chem. Mater.*, **8**, 701–707 (1996).
20. Soudan, P., Lucas, P., Breau, L., Bélanger, D. *Langmuir*, **16**, 4362–4366 (2000).
21. Li, Y., Kamata, K., Asaoka, S., Yamaguchi, T., Iyoda, T. *Org. Biomol. Chem.*, **1**, 1779–1784 (2003).

22. Fabre, B., Simonet, J. *J. Electroanal. Chem.*, **416**, 187–189 (1996).
23. Fabre, B., Kanoufi, F., Simonet, J. *J. Electroanal. Chem.*, **434**, 225–234 (1997).
24. Inagi, S., Hayashi, S., Hosaka, K., Fuchigami, T. *Macromolecules*, **42**, 3881–3883 (2009).
25. Inagi, S., Hosaka, K., Hayashi, S., Fuchigami, T. *J. Electrochem. Soc.*, **157**, E88–E91 (2010).
26. Hayashi, S., Inagi, S., Hosaka, K., Fuchigami, T. *Synth. Met.*, **159**, 1792–1795 (2009).
27. Hayashi, S., Inagi, S., Fuchigami, T. *Polym. Chem.*, **2**, 1632–1637 (2011).
28. Fuchigami, T., Inagi, S. *Chem. Commun.*, **47**, 10211–10223 (2011).
29. Inagi, S., Hayashi, S., Fuchigami, T. *Chem. Commun.*, 1718–1720 (2009).
30. Hayashi, S., Inagi, S., Fuchigami, T. *Macromolecules*, **42**, 3755–3760 (2009).
31. Hayashi, S., Inagi, S., Fuchigami, T. *Electrochemistry*, **78**, 114–117 (2010).
32. Inagi, S., Koseki, K., Hayashi, S., Fuchigami, T. *Langmuir*, **26**, 18631–18633 (2010).
33. Baizer, M. M., in *Organic Electrochemistry*, 3rd ed., eds., Lund, H., Baizer, M. M., Marcel Dekker: New York, Chapter 35 (1991).
34. Inagi, S., Nagai, H., Tomita, I., Fuchigami, T. *Angew. Chem. Int. Ed.*, **52**, 6616–6619 (2013).
35. Han, C. C., Jeng, R. C. *Chem. Commun.*, 553–554 (1997).
36. Han, C. C., Hong, S. P., Yang, K. F., Bai, M. Y., Lu, C. H., Huang, C. S. *Macromolecules*, **34**, 587–591 (2001).
37. Han, C. C., Chen, H. Y. *Macromolecules*, **40**, 8969–8973 (2007).
38. Loget, G., Zigah, D., Bouffler, L., Sojic, N. Kuhn, A. *Acc. Chem. Res.*, **46**, 2513–2523 (2013).
39. Fosdick, S. E., Knust, K. N., Scida, K. Crooks. R. M. *Angew. Chem. Int. Ed.*, **52**, 10438–10456 (2013).
40. Inagi, S. *Polym. J.*, **48**, 39–44 (2016).
41. Inagi, S., Ishiguro, Y., Atobe, M., Fuchigami, T. *Angew. Chem. Int. Ed.*, **49**, 10136–10139 (2010).
42. Ishiguro, Y., Inagi, S., Fuchigami, T. *Langmuir*, **27**, 7158–7162 (2011).
43. Inagi, S., Ishiguro, Y., Shida, N., Fuchigami, T. *J. Electrochem. Soc.*, **159**, G146–G150 (2012).

44. Ishiguro, Y., Inagi, S., Fuchigami, T. *J. Am. Chem. Soc.*, **134**, 4034–4036 (2012).
45. Shida, N., Ishiguro, Y., Atobe, M., Fuchigami, T., Inagi, S. *ACS Macro. Lett.*, **1**, 656–659 (2012).
46. Koizumi, Y., Shida, N., Ohira, M., Nishiyama, H., Tomita, I., Inagi, S. *Nat. Commun.*, **7**, 10404 (2016).

Biography

Shinsuke Inagi received his PhD from Kyoto University, Japan, in 2007 under the direction of Prof. Yoshiki Chujo, where he developed

the novel synthetic methods of sulfur-containing conjugated polymers. After postdoctoral research (Research Fellowship for Young Scientists of the Japan Society for the Promotion of Scientists, JSPS) at Kyoto University, he joined the group of Prof. Toshio Fuchigami as an assistant professor at Tokyo Institute of Technology in 2007. Then he started work on the electrochemistry of conjugated polymers. His achievements related to the field are the development of electrochemical polymer reactions and the bipolar electrochemistry of polymeric materials. He was promoted to lecturer in 2011 and associate professor in 2015. He received the Young Researcher Award of the Electrochemical Society of Japan (Sano Award) in 2013 and the Commendation for Science and Technology by the Minister of Education, Culture, Sports, Science and Technology (MEXT) in 2016. His current research interests include the electrochemical synthesis of polymeric materials.

Chapter 5

Chain-Growth Polymerization for the Synthesis of π-Conjugated Polymers

Yoshihiro Ohta and Tsutomu Yokozawa
Department of Materials and Life Chemistry, Kanagawa University, Rokkakubashi, Kanagawa-ku, Yokohama 221-8686, Japan
yokozt01@kanagawa-u.ac.jp

In this chapter, various types of chain-growth polymerization to synthesize well-defined π-conjugated polymers are reviewed. In the former session, mechanism, initiators, ligands of catalyst, and monomers of catalyst transfer Kumada–Tamao coupling polymerization are summarized. In the latter session, catalyst transfer Suzuki–Miyaura coupling polymerization, Stille coupling polymerization, and other methods including not via a metal catalyst transfer are described.

5.1 Introduction

π-Conjugated polymers are important class of materials owing to their potential applications in thin-film transistors (TFTs),

Conjugated Objects: Developments, Synthesis, and Applications
Edited by Atsushi Nagai and Koji Takagi

ISBN 978-981-4774-03-1 (Hardcover), 978-1-315-22962-1 (eBook)
www.panstanford.com

organic light-emitting diodes (OLEDs), and photovoltaic cells.[1–4] A lot of π-conjugated polymers have been conventionally prepared via step-growth polymerization, which affords polymers with uncontrolled molecular weight and polymer end groups, and broad molecular weight distribution.[5] This often leads to batch-to-batch differences and alters material properties, which is undesirable for reproducibility. In addition, synthesis of well-defined block copolymers containing π-conjugated polymers by step-growth polymerization is difficult. Therefore, the development of synthetic methods for well-defined π-conjugated polymers was required.

In 2004, Yokozawa and coworkers found that Kumada–Tamao coupling polymerization of 2-bromo-5-chloromagnesio-3-hexylthiophene **1a**, generated by reaction of 2-bromo-3-hexyl-5-iodothiophene with an exact amount of isopropylmagnesium chloride, with Ni(dppp)Cl_2 (dppp = 1,3-bis(diphenylphosphino) propane) in THF at room temperature proceeded in a chain-growth polymerization manner (Fig. 5.1). The M_n values of polymers controlled by the feed ratio of the monomer to the Ni catalyst and M_w/M_n values are low.[6] In addition, when the polymerization of **1a** was quenched with hydrochloric acid, the M_w/M_n ratios were around 1.1 up to M_n of 28,700.[7] After a detail study, they found that above polymerization involves intramolecular catalyst transfer process, leading to a chain-growth polymerization mechanism.[8] McCullough and coworkers also reported chain-growth polymerization of a zinc monomer[9] from 2-bromo-3-hexylthiophene and **1a** from the corresponding dibromothiophene in the presence of Ni(dppp) Cl_2 (Fig. 5.1).[10] After these reports, many kinds of π-conjugated polymers and architectures, including block copolymers, star polymers, and graft copolymers, have been synthesized via chain-growth polymerization.[11–17]

In this chapter, various types of chain-growth polymerization for the synthesis of well-defined π-conjugated polymers are described. The former part of this chapter is focused on catalyst transfer Kumada–Tamao coupling polymerization: the mechanism, initiators, ligands of catalysts, and monomers are summarized. The latter part is focused on recent progress in catalyst transfer Suzuki–Miyaura coupling polymerization and Stille coupling polymerization and other methods not via metal catalyst transfer.

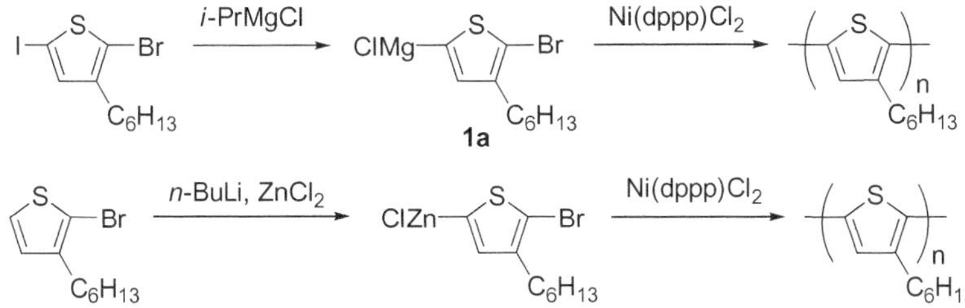

Figure 5.1 Synthesis of poly(3-hexylthiophene) by polymerization of **1a** and zinc monomer with Ni(dppp)Cl_2.

5.2 Kumada–Tamao Coupling Polymerization

5.2.1 Mechanism

Polymerization of **1a** was investigated in detail, and four important points were revealed:

- One polymer end group is Br and the other is H.
- The propagating end group is a polymer-Ni-Br complex.
- One Ni molecule forms one polymer chain.
- The initiator is a dimer of **1a** formed in situ.

On the basis of the above results, the mechanism of catalyst transfer condensation polymerization has been proposed (Fig. 5.2, Path 1).[8] Transmetallation of Ni(dppp)Cl_2 and 2 equivalents of **1a** proceeds, and then reductive elimination occurs to generate a Ni(0) complex. This Ni(0) complex does not diffuse to the reaction mixture but moves along intramolecular π-face, followed by oxidative addition of the intramolecular C–Br bond. Transmetallation of the produced Ni(II) complex and another **1a** occurs, followed by reductive elimination and intramolecular transfer of the Ni(0) catalyst to the next C–Br bond. Propagation continues in this manner and the Ni(0) catalyst transfers to the polymer end group.

In the case of Path 1 in Fig. 5.2, polymerization proceeds in only one direction. Kiriy and coworkers investigated a walking process of the Ni catalyst by use of Br-C_6H_4-Ni(dppe)-Br (dppe = 1,3-bis(diphenylphosphino)ethane) as an initiator (Fig. 5.3).[18] This polymerization afforded not only polythiophene with a

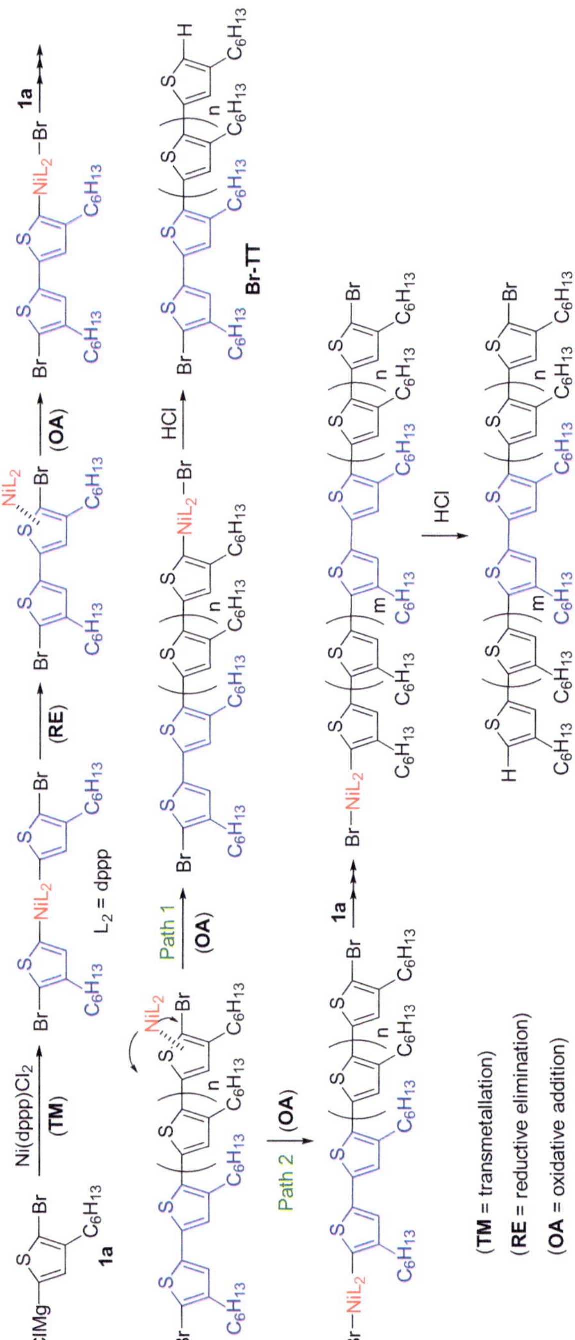

Figure 5.2 Mechanism of Kumada–Tamao catalyst transfer condensation polymerization of **1a** with $Ni(dppp)Cl_2$.

p-bromophenyl end group (Fig. 5.3, Path 1) but also polythiophene with the phenyl ring inside the main chain (Fig. 5.3, Path 2). Furthermore, the content of the product with the internal phenyl ring increased with an increasing degree of polymerization. On the basis of results of control experiments, Ni(0) catalyst did not diffuse the reaction mixture in this condition. These results mean that Ni(0) catalyst can "walk" along the polythiophene main chain to the opposite end and that the propagation reaction occurs on both sides of the polymer ends. In other words, polymerization proceeds in two directions.

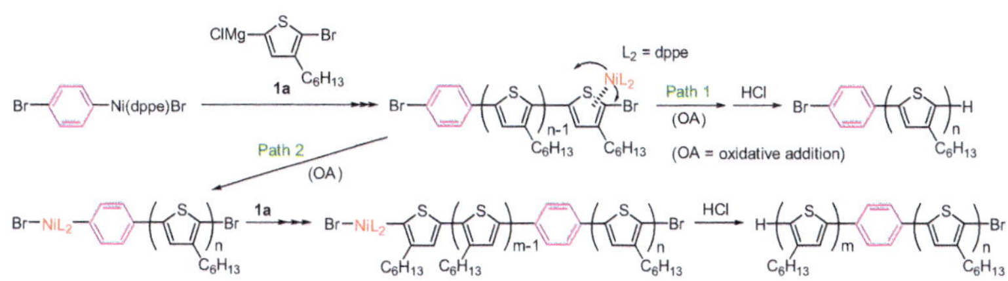

Figure 5.3 Unidirectional and bidirectional propagation of polymerization of **1a** with Br-C_6H_4-Ni(dppe)-Br.

Koeckelberghs and coworkers showed that bidirectional growth also occurs in the case of polymerization of **1a** with Ni(dppp)Cl_2 (Fig. 5.2, Path 2).[19] If unidirectional growth occurs, a tail-to-tail bithiophene unit with bromine exists (designated as Br-TT) (Fig. 5.2, Path 1). On the other hand, if bidirectional growth occurs, head-to-tail bithiophene units with bromine exist (designated as Br-HT) (Fig. 5.2, Path 2). They measured ^{1}H nuclear magnetic resonance (NMR) spectrum of P3HT obtained by polymerization with Ni(dppp)Cl_2 and calculated the ratio of Br-TT/(Br-HT + Br-TT). Consequently, Br-HT was always observed and the ratio of Br-TT/(Br-HT + Br-TT) decreased with an increasing degree of polymerization. These results indicate that the propagation reaction proceeds in two directions.

5.2.2 Initiators

When catalyst transfer condensation polymerization of **1a** with Ni(dppp)Cl_2 was conducted, the dimer of **1a**, generated by transmetallation of Ni(dppp)Cl_2 and 2 equivalents of **1a** in situ, acts as a chain initiator. On the other hand, external Ni initiators for catalyst transfer Kumada–Tamao coupling polymerization were reported. Kiriy and coworkers used $PhNi(PPh_3)_2Br$ as an initiator of polymerization of **1a** for synthesis of Ph-terminated P3HT, but polymerization was less controlled in comparison with polymerization with $Ni(dppp)Cl_2$.[20,21] Therefore, ArNi(dppp)X (X = Cl, Br) as a more effective initiator to synthesize Ar-terminated P3HT with low polydispersity was developed by Bronstein and Luscombe[22] and Senkovskyy et al.[23], respectively. In the case of Bronstein and Luscombe's method, ArNi(dppp)X was prepared by reaction of Ar-X with $Ni(PPh_3)_4$ to obtain Ni(II) complex, followed by ligand exchange reaction of the ArNi(II)X complex with dppp (Fig. 5.4). This method enables us to introduce protected functional group in Ar moiety.[24] In the case of Senkovskyy's method, Et_2Ni(2,2′-bipyridine) was utilized instead of $Ni(PPh_3)_4$ (Fig. 5.5).

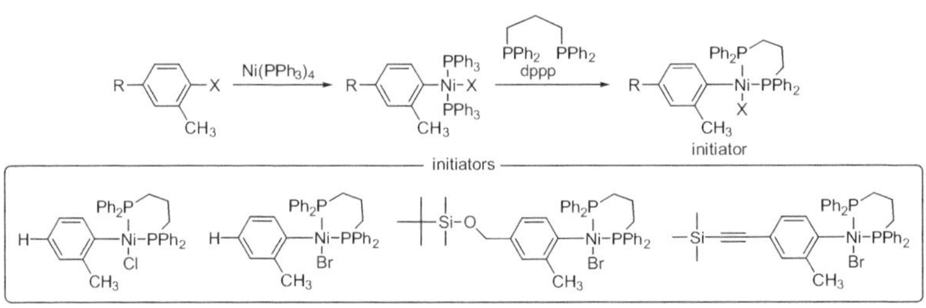

Figure 5.4 Synthesis of Ni initiators by using $Ni(PPh_3)_4$ for catalyst transfer condensation polymerization.

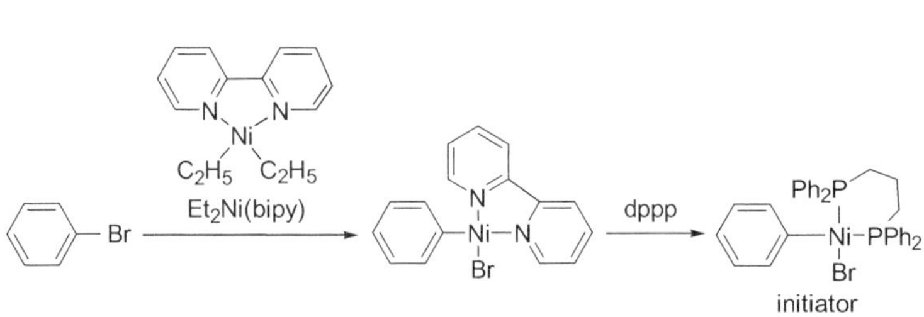

Figure 5.5 Synthesis of Ni initiator by using Et_2Ni(bipy) for catalyst transfer condensation polymerization.

Sommer and coworkers found that polymerization of **1a** with Ni complex **2** containing a benzothiadiazole unit proceeds in only one direction (Fig. 5.6). When transmetallation of **2** and **1a** occurs, followed by reductive elimination, produced Ni(0) catalyst is not able to transfer to the opposite end because this catalyst can't walk over the benzothiadiazole unit, leading to unidirectional growth.[25] These initiator methods enable us to synthesize pure AB diblock copolymers.

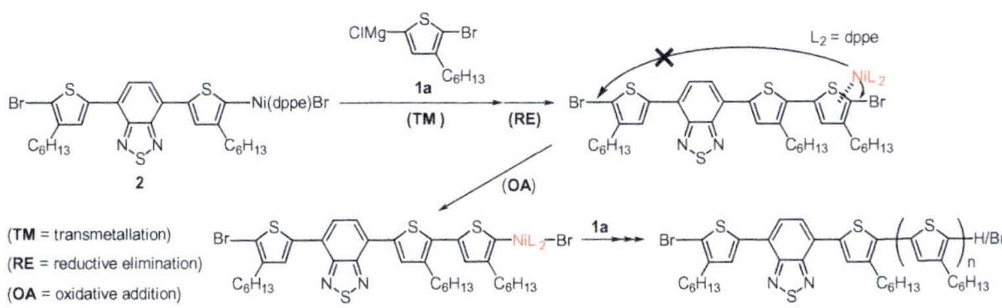

Figure 5.6 Unidirectional growth polymerization of **1a** with initiator **2** containing the benzothiadiazole unit.

5.2.3 Ligands of Catalysts

Molecular weight and molecular weight distribution of polymers obtained by catalyst transfer Kumada–Tamao coupling polymerization depend on the choice of the phosphine ligands of the Ni catalyst.[26,27] In the case of polymerization of **1a** with Ni(dppe)Cl_2, the molecular weight of the obtained polymer was slightly lower than that in the case of polymerization with Ni(dppp)Cl_2 and the molecular weight distribution was slightly broad. When polymerization with Ni(PPh_3)$_2$$Cl_2$, Ni(dppb)$Cl_2$ (dppb = 1,4-bis(diphenylphosphino) butane), and Ni(dppf)Cl_2 (dppf = 1,1′-bis-(diphenylphosphino) ferrocene) was conducted, lower-molecular-weight polymers with a broad molecular weight distribution were produced. These results indicate that Ni(dppp)Cl_2 is more suitable for catalyst transfer condensation polymerization of **1a** than these catalysts.

Mori and coworkers found that polymerization of 2-bromo-3-hexylthiophene with chloro magnesium 2,2,6,6-tetramethylpiperidide lithium chloride salt (TMPMgCl·LiCl) in the presence of [CpNiCl(SIPr)] (Cp = cyclopentadienyl, SIPr = 1,3-bis(2,6-diisopropylphenyl)imidazolidin-2-ylidene) gave high-molecular-weight P3HT (M_n= 233500), though the molecular weight was not controlled by the feed ratio of monomer to Ni catalyst, and the molecular weight distribution was broad (M_w/M_n = 3.49).[28] In contrast, Geng and coworkers reported a more effective catalyst, Ni(IPr)(acac)$_2$ (IPr = bis(2,6-diisopropylphenyl)imidazolin-2-ylidene, acac = acetylacetonate), which afforded high-molecular-weight P3HT ($M_n \leq 353000$, $M_w/M_n \leq 1.50$) in a controlled manner.[29] When polymerization with 0.15 mol% of Ni(IPr)(acac)$_2$ was carried out, the relationship of $\ln([M]_0/[M])$ versus time and of M_n versus conversion of **1a** were linear. This polymerization behavior might be attributed to strong coordination interaction between IPrNi(0) catalyst and the polymer backbone and to the formation of suitable oxidative addition complexes between the IPrNi(0) catalyst and a C–Br bond at the polymer chain end, leading to more efficient intramolecular catalyst transfer.

Mori and coworkers found that a Ni catalyst **3** bearing *N*-heterocyclic carbine (NHC) as a ligand was effective for the polymerization of chlorothiophene monomer **4**, which was generated in situ by the reaction of 2-chloro-3-hexylthiophene with TMPMgCl·LiCl.[30] Ni(dppe)Cl$_2$ was not effective for this polymerization. On the other hand, polymerization with **3** gave P3HT with controlled molecular weight and narrow molecular weight distribution (Fig. 5.7).

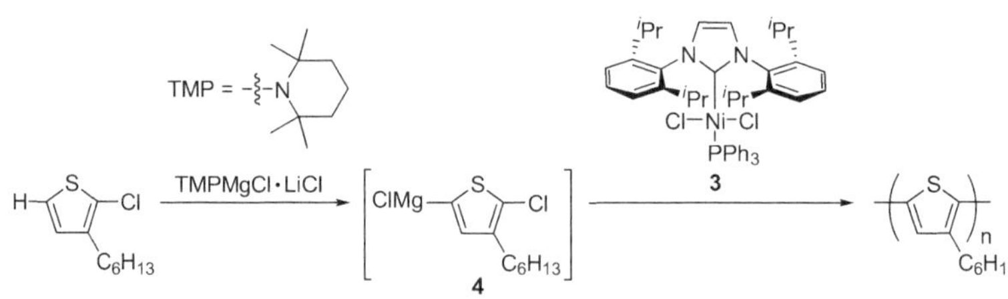

Figure 5.7 Chain-growth condensation polymerization of 2-chloro-3-substituted thiophenes with **3**.

5.2.4 Monomers

Various types of monomers for catalyst transfer Kumada–Tamao coupling polymerization have been reported. Thiophene monomer **1b** with a butyl side chain was polymerized in the presence of Ni(dppp)Cl$_2$ to give polythiophene with a relatively narrow molecular weight distribution, but the solubility of this polymer was lower than that of P3HT, resulting in the production of a polymer with an M_n of less than 5500 (Fig. 5.8).[31] The polymerization of thiophene monomer **1c** bearing an alkoxymethyl side chain also proceeded in a chain-growth polymerization manner. In this case, the suitable Ni catalyst was Ni(dppe)Cl_2 rather than Ni(dppp)Cl_2.[32] Catala and coworkers reported polymerization of thiophene monomer **1d** with ester moiety on the side chain. The molecular weight of poly-**1d** was increased with a decreasing amount of catalyst, and molecular weight distribution was relatively narrow (M_w/M_n = 1.25–1.5), indicating that this polymerization proceeds in a chain-growth polymerization mechanism.[33]

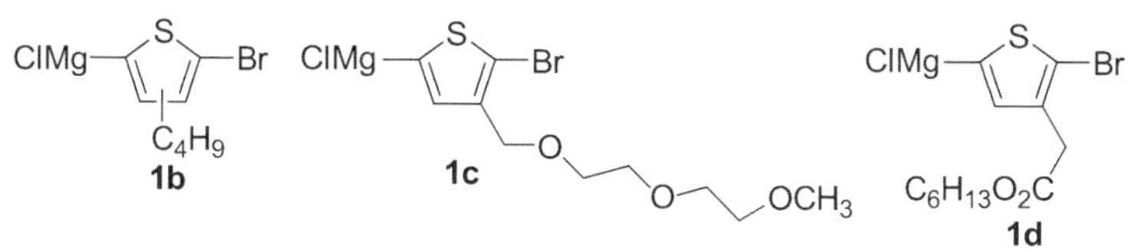

Figure 5.8 Thiophene monomers for Kumada–Tamao coupling polymerization.

In contrast with polymerization of **1a**, polymerization of reversed thiophene monomer **5** with Ni(dppp)Cl_2 did not proceed because the formation of a dimer, generated by transmetallation of Ni(dppp)Cl_2 and 2 equivalents of **5**, was prevented by steric hindrance of the side chain of **5** (Fig. 5.9).[34] On the other hand, Geng and Catala independently found that polymerization of **5** with Ni(dppp)Cl_2 was promoted by addition of LiCl (Fig. 5.9).[35,36] This polymerization showed a chain-growth polymerization character, though polymerization afforded polythiophene with a broad molecular weight distribution owing to slow initiation reaction. The addition of LiCl increases the reactivity of Grignard monomers and modifies Ni center by a halogen exchange reaction.

To reveal whether catalyst transfer Kumada–Tamao coupling polymerization is applicable to not only polythiophenes but also other π-conjugated polymers, Yokozawa and coworkers carried out the polymerization of a Grignard-type pyrrole monomer **6** in the presence of a Ni catalyst (Fig. 5.10).[37] In this polymerization, Ni(dppe)Cl_2 was more suitable than Ni(dppp)Cl_2 but polymerization afforded not only a polymer with narrow molecular weight distribution but also an oligomeric by-product. When polymerization with

$Ni(dppe)Cl_2$ in the presence of dppe as an additive was conducted at 0°C, production of an oligomeric by-product was suppressed, and the obtained polymer possessed a narrower molecular weight distribution. The relationships of M_n versus conversion and of M_n versus feed ratio were linear, indicating catalyst transfer chain-growth polymerization. As another heterocycle monomer, Seferos and coworkers reported polymerization of selenophene monomer **7** with $Ni(dppp)Cl_2$ to afford poly(3-hexylselenophene) with M_w/M_n of 1.3, though yield of the polymer was low.[38] They also carried out block copolymerization of selenophene monomers and thiophene monomers, yielding block copolymers with narrow molecular weight distribution in a moderate yield.[38,39]

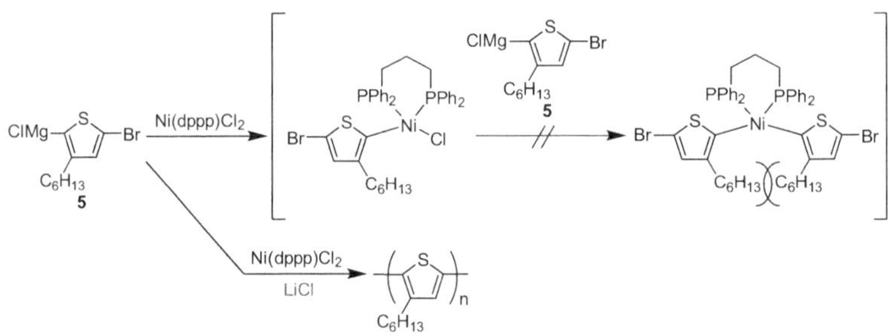

Figure 5.9 Catalyst transfer condensation polymerization of **5** with Ni(dppp) Cl_2 in the absence or presence of LiCl.

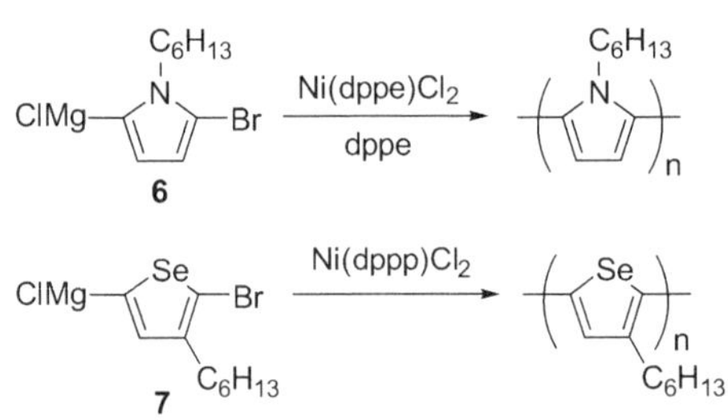

Figure 5.10 Synthesis of poly(*N*-hexylpyrrole) and poly(3-hexylselenophene) by catalyst transfer condensation polymerization.

To investigate whether catalyst transfer polymerization is applicable to monomer without a heteroatom in the aromatic ring, Yokozawa and coworkers conducted polymerization of phenylene monomer **8** with a Ni calatyst (Fig. 5.11). In this polymerization, addition of LiCl is important to obtain high-molecular-weight poly-**8** with narrow molecular weight distribution. When polymerization with Ni(dppe)Cl_2 in the presence of LiCl, the molecular weight of poly-**8** increased linearly in proportion to the conversion and feed ratio of monomer to Ni catalyst, and molecular weight distribution was narrow.[40] Poly(*m*-phenylene), which is no conjugated polymer, was also synthesized by the polymerization of **9** with Ni(dppp)Cl_2 and LiCl (Fig. 5.11).[41]

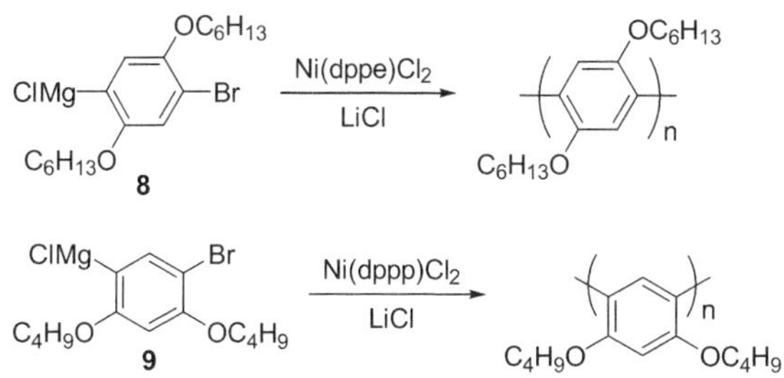

Figure 5.11 Synthesis of poly(*p*-phenylene) and poly(*m*-phenylene) by catalyst transfer condensation polymerization.

Polymerization of monomers with a longer molecular length, such as carbazole, fluorene, and cyclopentadithiophene, was studied. Polymerization of the carbazole monomer was conducted by McCullough and coworkers. The magnesium halogen exchange of 2,7-dibromo-*N*-octylcarbazole with *i*-PrMgCl·LiCl was very slow and almost no conversion was observed. On the other hand, lithium tributylmagnesate was effective for the magnesium halogen exchange. The polymerization of **10a** with Ni(dppp)Cl_2 afforded polycarbazole with M_n of 2600 and M_w/M_n of 1.23 (Fig. 5.12).[42] Ge et al. found that polymerization of **10b** instead of **10a** gave polycaarbazole with higher molecular weight, retaining narrow molecular weight distribution (M_n = 17400, M_w/M_n = 1.27).[43] McCullough and coworkers also reported synthesis of polyfluorene by Kumada–Tamao coupling polymerization of fluorine monomer **11a** with Ni(dppp)Cl_2 (Fig. 5.12).[44] The relationship of conversion versus time plots showed a linear increase initially and then leveled off with time, implying that polymerization proceeded in a chain-growth manner during the initial period but amount of chain termination was increased with time. When polymerization with

1.0 mol% of $Ni(dppp)Cl_2$ was conducted, polyfluorenes with a molecular weight of about 10,000 and with M_w/M_n values <1.2 were obtained. However, it would be difficult to control the molecular weight by the feed ratio of monomer to catalyst. In addition, when end-capping the polymer with methyl-MgBr and phenyl-MgBr was carried out, end-capping percentages were 20%–50%. In contrast, Geng and coworkers found more effective catalyst, $Ni(acac)_2$/dppp, for polyfluorenes (Fig. 5.12). When polymerization of **11a** with varying the amount of $Ni(acac)_2$/dppp from 11–1.1 mol% was carried out at 0°C, the M_n values increased in proportion to the converted **11a**/catalyst ratio, retaining narrow molecular weight distribution ($M_w/M_n \leq 1.24$). Furthermore, in monomer addition experiment, the gel permeation chromatography (GPC) profile of the product was clearly shifted toward the higher-molecular-weight region, while retaining a narrow molecular weight distribution. Polymerization of **11b** and **11c** with $Ni(acac)_2$/dppp also gave well-defined polymers.[45] Koeckelberghs and coworkers carried out synthesis of poly(cyclopentadithiophene) by the polymerization of **12** with $Ni(dppp)Cl_2$ (Fig. 5.12).[46] The M_n versus conversion and $\ln([M]_0/[M])$ versus time plots showed a clear linear relation, although the M_w/M_n values were about 2.0.

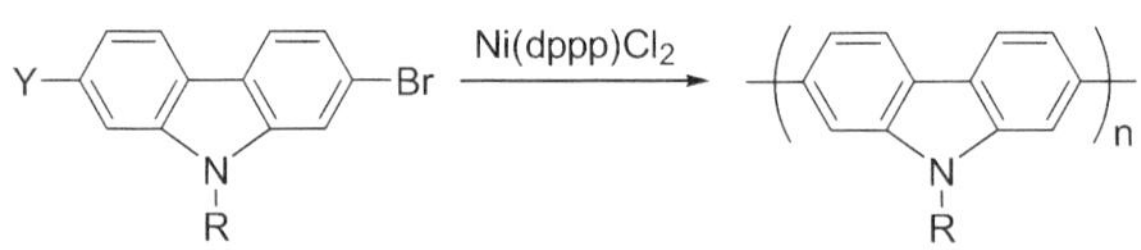

10

10a: R = C_8H_{17}, Y = MgLi, **10b**: R = $CH(C_8H_{17})_2$, Y = MgCl

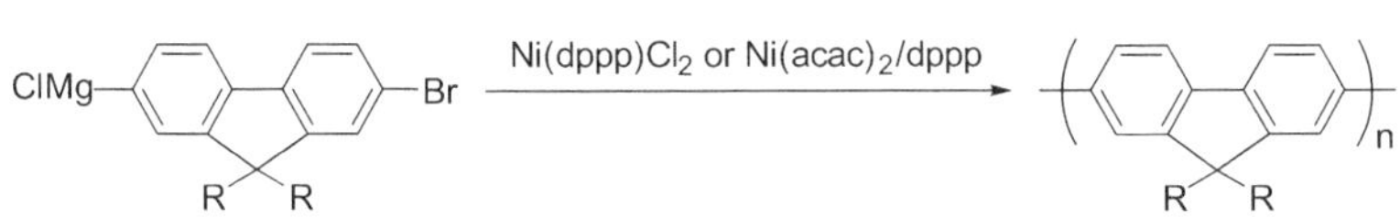

11

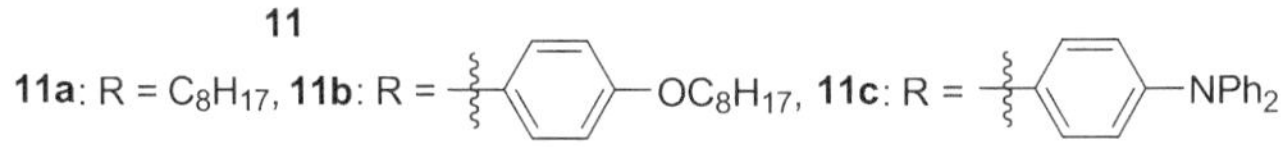

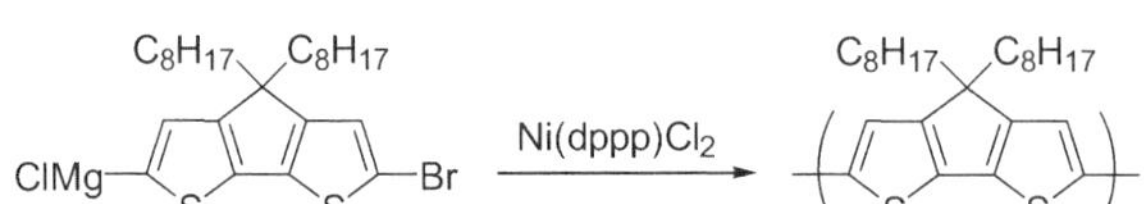

12

Figure 5.12 Synthesis of polycarbazole, polyfluorene, and poly(cyclopentadithiophene) by catalyst transfer condensation polymerization.

Geng and coworkers conducted polymerization of a bithienylmethane derivative **13**, in which two thiophene rings are connected via a saturated carbon, in the presence of Ni(dppe)Cl_2 (Fig. 5.13).[47] They found that catalyst transfer condensation polymerization was also applicable to the synthesis of well-defined polymers comprising nonconjugated aromatic repeating units. Bielawski and coworkers investigated polymerization of a biaryl monomer, 2-(4-bromo-2,5-bis(2-ethylhexyloxy)phenyl)-5-chloromagnesiothiophene **14**, in the presence of Ni(dppp)Cl_2 and LiCl.[48] They realized the synthesis of a conjugated alternating copolymer, poly(thiphene-*alt*-*p*-phenylene) with a narrow molecular weight distribution and controlled molecular weight by means of catalyst transfer condensation polymerization.

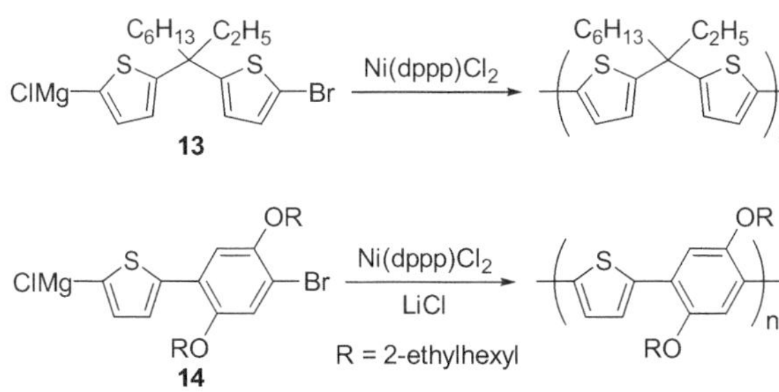

Figure 5.13 Synthesis of poly(bithienylmethylene) and poly(thiophene-*alt*-*p*-phenylene) by catalyst transfer condensation polymerization.

Polypyridine, which is an n-type polymer, was also synthesized by catalyst transfer Kumada–Tamao coupling polymerization (Fig. 5.14).[49] Polymerization of pyridine monomer **15**, generated by the reaction of 5-bromo-3-iodo-2-(2-(2-methoxyethoxy)ethoxy) pyridine with ^{i}PrMgCl in the presence of LiCl, with Ni(dppp)Cl_2 was carried out at ambient temperature to yield poly{2-[2-(2-methoxyethoxy)ethoxy]pyridine-3,5-diyl} with narrow molecular weight distribution and controlled molecular weight based on a feed ratio of monomer to Ni catalyst, indicating that polymerization proceeds in a chain-growth polymerization manner.

As another n-type π-conjugated polymer, polythiazole was also synthesized by catalyst transfer condensation polymerization (Fig. 5.15).[50] In this case, external initiation would allow to alleviate the steric hindrance of first cross-coupling step, and polymerization of **16** with the Ni initiator afforded polythiazole with narrow molecular weight distribution. The M_n value of the polythiazole increased in proportion to the monomer/catalyst ratio up to about 50. However, the GPC chromatograms typically showed bimodal peaks which corresponded to approximately twice the molecular weight of the main peak, implying that chain–chain coupling via disproportionation also occurred.

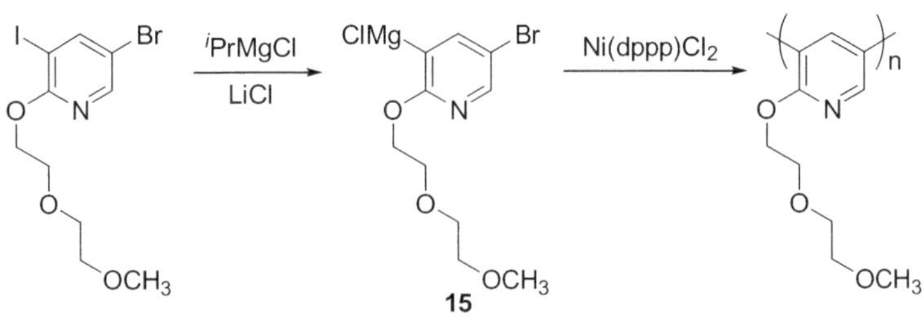

Figure 5.14 Synthesis of poly{2-[2-(2-methoxyethoxy)ethoxy]pyridine-3,5-diyl} by catalyst transfer condensation polymerization.

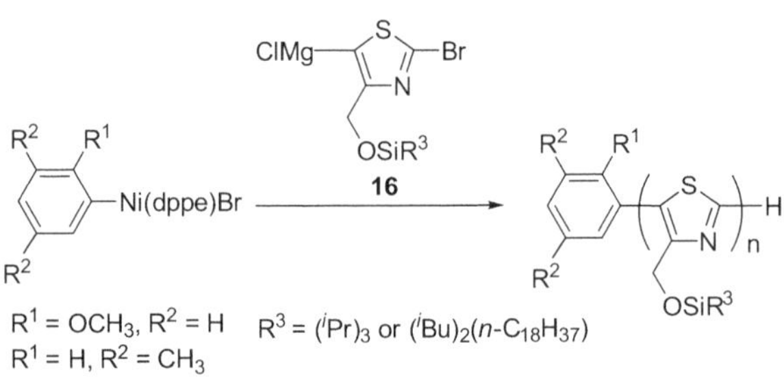

Figure 5.15 Synthesis of polythiazole by catalyst transfer condensation polymerization of **16** with the Ni initiator.

Seferos and coworkers carried out synthesis of polybenzotriazole, an electron-deficient π-conjugated polymer, by the polymerization of **17** with Ni(II)diimine catalyst (Fig. 5.16).[51] In the case of polymerization with **CF_3An** and ***t*-BuAn**, the GPC elution curve showed shoulders at the high-molecular-weight region, and molecular weight distribution was broad ($M_w/M_n > 2.0$). This result presumably stemmed from chain–chain coupling through disproportionation. On the other hand, in the case of polymerization with **OMeAn**, an electron-rich catalyst, the GPC elution curve showed a unimodal distribution ($M_w/M_n = 1.54$). Comparing the $\ln([M]_0/[M])$

versus time plots for each catalysts revealed that the electron-rich catalyst, **OMeAn**, displayed linearity up to 60% conversion, while the more electron-poor catalyst, **CF_3An** and ***t*-BuAn**, did not display a significant linear region. The M_n value of the polybenzotriazole increased in proportion to the monomer/catalyst ratio up to about 100. Furthermore, the chain extension polymerization with **OMeAn** was achieved. These results indicated polymerization of **17** with **OMeAn** proceeds in a chain-growth polymerization manner.

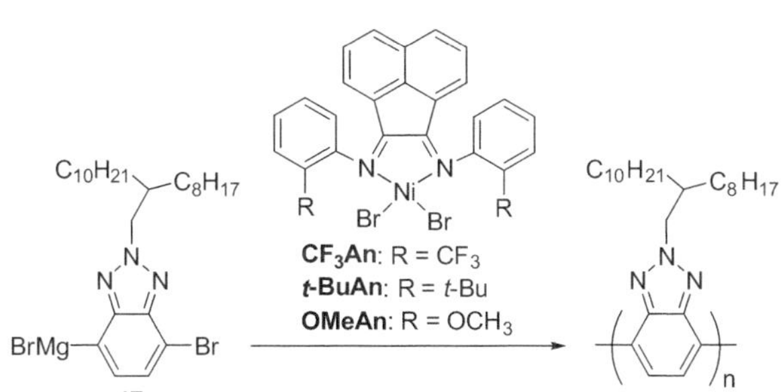

Figure 5.16 Synthesis of polybenzotriazole by catalyst transfer condensation polymerization of **17** with a Ni(II)diimine catalyst.

5.3 Suzuki–Miyaura Coupling Polymerization

It is important to reveal whether polymerization with a catalyst other than Ni proceeds in a catalyst transfer condensation polymerization mechanism. Yokozawa and coworkers conducted Suzuki–Miyaura coupling polymerization of a fluorene monomer **18** with arylpalladium(II) halide complex, $^tBu_3PPd(Ph)Br$, as an externally added initiator (Fig. 5.17). The obtained polymer had a Ph group at one end. Furthermore, conversion–M_n and feed ratio–M_n plots showed linear relationships and M_w/M_n values were low. These results indicated that this polymerization proceeds in a catalyst transfer condensation polymerization mechanism[52], as shown in the model reaction.[53-55] Poly(*p*-phenylene), P3HT, and poly(9,9'-dioctylfluorene-*alt*-benzothiadiazole) were also synthesized by means of chain-growth Suzuki–Miyaura coupling polymerization.[56-58]

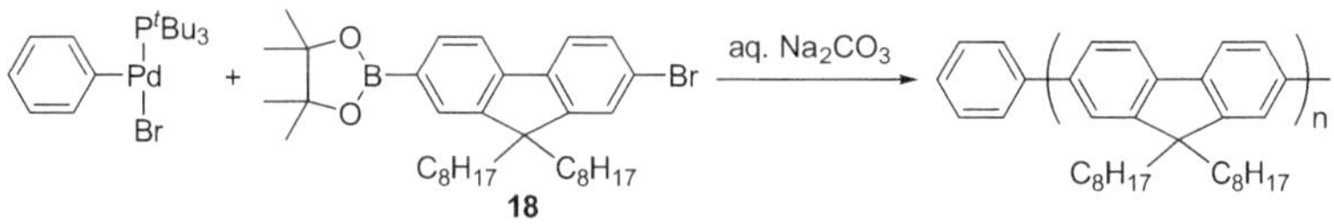

Figure 5.17 Catalyst transfer Suzuki–Miyaura coupling polymerization of **18** with ${}^{t}Bu_3PPd(Ph)Br$.

In comparison with Kumada–Tamao catalyst transfer condensation polymerization, Suzuki–Miyaura catalyst transfer condensation polymerization often afforded polymer with a broader molecular weight distribution and uncontrolled polymer ends in the high-molecular-weight region. For instance, when the molecular weight exceeded 8000, P3HT obtained by Kumada–Tamao catalyst transfer polymerization showed well-controlled polymer ends[8], while P3HT obtained by Suzuki–Miyaura catalyst transfer polymerization showed unexpected polymer ends.[57] Deboronation and dehalogenation were known as a side reaction of Suzuki–Miyaura coupling reaction. In particular, deboronation of five-membered 2-heterocyclic boronic acids was problematic.[59–62] The water and base in the reaction mixture contributed to the above side reaction, even though both components promote the coupling reaction. Yokozawa and coworkers conducted polymerization of triolborate halothiophene monomer **19** in the absence of water and base in dry tetrahydrofuran (THF) due to suppression of the above side reaction (Fig. 5.18).[63] However, the obtained polymer possessed a broad molecular weight distribution and many kinds of polymer ends. On the other hand, polymerization in water/THF gave a polymer with narrower molecular weight distribution and well-controlled polymer ends. These results imply that intramolecular transfer of the Pd catalyst in the Suzuki–Miyaura coupling reaction is hindered in the absence of water, whereas addition of a small amount of water suppresses intermolecular transfer of the catalyst.

Yokozawa and coworkers carried out polymerization of **20** with ${}^{t}Bu_3PPd(Ph)Br$ to extend the range of catalyst transfer condensation polymerization for synthesis of not only aromatic conjugated polymers but also aromatic and carbon–carbon double-bond (C=C) conjugated polymers (Fig. 5.19).[64] This polymerization gave higher-molecular-weight poly(phenylenevinylene) within a few minutes but molecular weight distribution was broad. Both polymers with and without initiator units were obtained, indicating that intermolecular chain transfer of the Pd catalyst occurred. Therefore, they investigated why ^{t}Bu$_3$PPd(0) did not undergo intramolecular transfer during polymerization of the monomer containing C=C by examining the model reaction of dibromostilbene derivatives with aryl boronic acid or boronic acid esters in the presence and absence of additives containing C=C.[65] They found that ^{t}Bu$_3$PPd(0) on the π-face of the substrate readily underwent intermolecular transfer to the C=C of another substrate or additive molecule. Furthermore, they also found that introduction of alkoxy groups at the *ortho*-positions of C=C was effective for suppression of the above intermolecular transfer of the catalyst. These results implied that intermolecular transfer of the ^{t}Bu$_3$PPd(0) to C=C of another monomer occurred in the polymerization of **20** with ^{t}Bu$_3$PPd(Ph)Br, leading to production of a polymer with a broad molecular weight distribution and uncontrolled end groups.

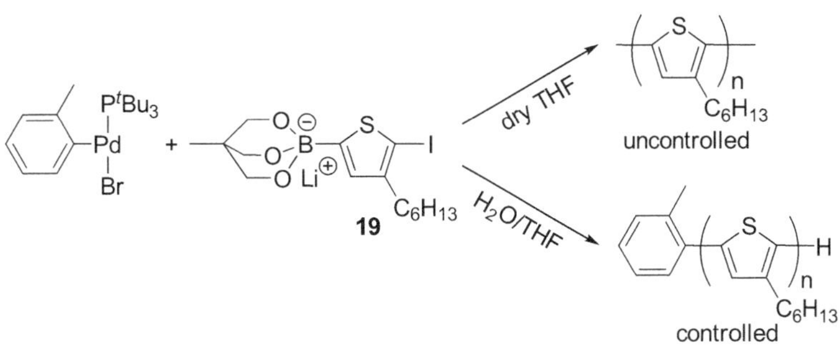

Figure 5.18 Catalyst transfer Suzuki–Miyaura coupling polymerization of **19** with ^{t}Bu$_3$PPd(*o*-tolyl)Br.

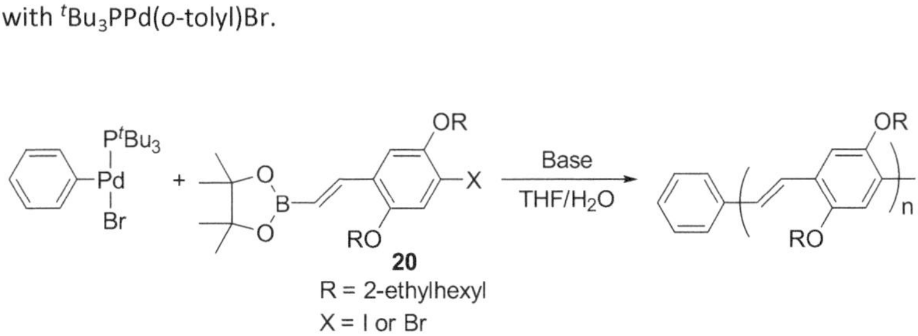

Figure 5.19 Suzuki–Miyaura coupling polymerization of **20** with ^{t}Bu$_3$PPd(Ph) Br.

5.4 Other Methods

Methods other than Kumada–Tamao and Suzuki–Miyaura catalyst transfer condensation polymerization for the synthesis of well-defined π-conjugated polymers are described here.

Bielawski and coworkers reported Stille coupling polymerization of **21** with t-$Bu_3PPd(Ph)Br$ for the synthesis of well-defined poly(*p*-phenyleneethynylene) (PPE) (Fig. 5.20).[66] To obtain a high-molecular-weight polymer in this polymerization, both CuI and additional Ph_3P were necessary. The M_n value of the polymer increased in proportion to the monomer conversion and feed ratio of monomer to catalyst, while the molecular weight distribution remained constant. A chain extension experiment was also conducted.

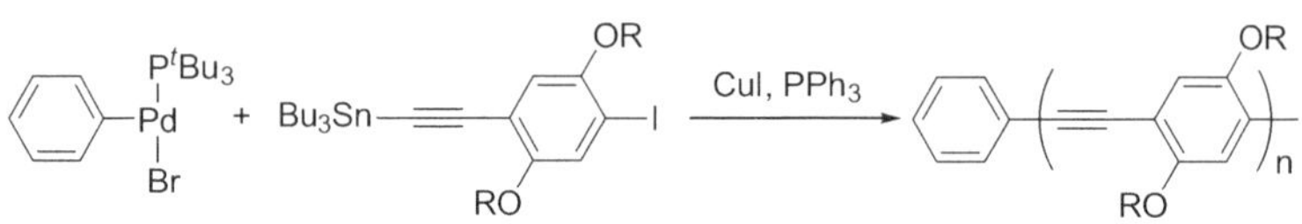

21 R = 2-ethylhexyl

Figure 5.20 Catalyst transfer Stille coupling polymerization of **21** for the synthesis of poly(*p*-phenyleneethnylene).

Senkovsky and coworkers reported polymerization of radical anion **22**, which was generated by the reaction of **23** with activated Zn, with PhNi(dppe)Br or $Ni(dppe)Br_2$ at ambient temperature (Fig. 5.21).[67,68] This polymerization involved reductive coupling and intramolecular catalyst transfer process. The single-electron redox process between PhNi(dppe)Br and **22** presumably afforded Ni(I) species, Ph-Ni(dppe), which oxidatively added to C–Br bond of the monomer residue to generate Ni(III) complex, Ph-Ni(dppe)Br-Ar-Br, followed by elimination of bromide anion, resulting in Ph-Ni(dppe)-Ar-Br. Reductive elimination of Ph-Ni(dppe)-Ar-Br occurred to generate a Ni(0) complex, Ni(dppe), followed by intramolecular transfer of the Ni(0) catalyst to the next C–Br bond to form Ph-Ar-Ni(dppe)-Br. Another **22** was reacted with Ph-Ar-Ni(dppe)-Br through single-electron redox process, and propagation continued in this manner. The M_n value of the obtained polymer increased with an increasing feed ratio of monomer to Ni catalyst, retaining relatively narrow molecular weight distribution (M_n = 25000–104000, M_w/M_n = 1.3–1.7).

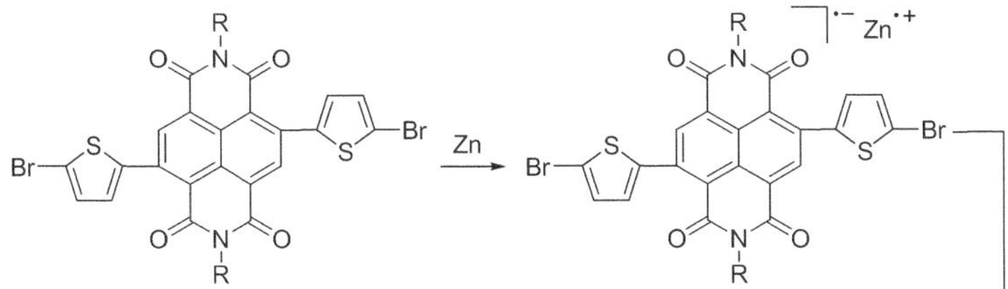

Figure 5.21 Chain-growth polymerization of **22**.

Koeckelberghs and coworkers conducted polymerization of a zinc monomer **24** with a Pd initiator **25** bearing a Ruphos phosphine ligand (Fig. 5.22).[69] Both $\ln([M]_0/[M])$ versus time and M_n versus conversion plots showed linear relationship. They proposed a polymerization mechanism as below. Transmetallation of **25** and **24** takes place, followed by reductive elimination, and the Pd(0)[Ruphos] produced is diffused in the reaction mixture. This Pd(0) catalyst does not react with **24** because the electron-donating organozinc of **24** deactivates the C–Br bond for oxidative addition from Pd(0)[Ruphos]. Thus, oxidative addition of C–Br bond at the propagating chain end to Pd(0)[Ruphos] occurs, and this Pd(II) complex reacts with **24**. Consequently, this polymerization proceeded not through the catalyst transfer mechanism but via change of reactivity between monomer and polymer end by virtue of the substituent effect.

Swager and coworkers reported the cationic chain-growth polymerization of 2-chloroalkylenedioxythiophenes **26** (Fig. 5.23).[70] The linear relationship between M_n and conversion of monomer was observed, retaining a relatively narrow molecular weight distribution. The proposed mechanism is as follows. The reaction of **26** with $SnCl_4$ afforded zwitterion **27**, which was subsequently reacted with another monomer to give dimer **28**. Elimination of $SnCl_5^-$ occurred from **28** to afford **29**. Growth continued from this sulfonium cation, resulting in polymer **30**. Finally, HCl elimination during work-up provided an objective polymer.

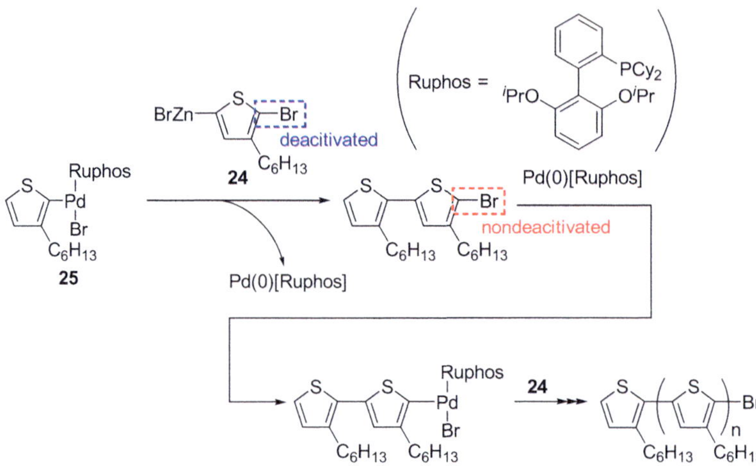

Figure 5.22 Chain-growth polymerization of **24** with **25** through substituent effect.

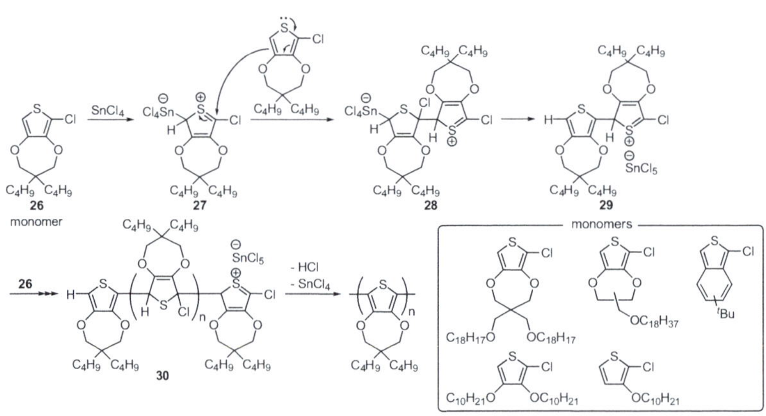

Figure 5.23 Lewis acid–promoted condensation polymerization of **26** with $SnCl_4$.

Sanji and Iyoda found that polymerization of conjugated monomers **31** and **32**, which contained trimethylsilyl and pentafluorophenyl groups, with a fluoride ion proceeded in a chain-growth polymerization manner (Fig. 5.24). This polymerization did not use a transition metal catalyst.[71,72] They proposed polymerization mechanism as below. A fluoride anion reacted with the trimethylsilyl group of the monomer to form a pentacoordinate silicate. The silicate was quite reactive and could regioselectively attack the 4-position of the pentafluorophenyl group of monomer to reproduce a fluoride anion. Then, the fluoride anion would transfer intramolecularly to the trimethylsilylthiophene or trimethylsilylethynyl group at the chain end, owing to anion–π interaction. The chain end reacted with another monomer and growth continued in such a way. The active polymer end could initiate polymerization of a different monomer to afford a block copolymer.

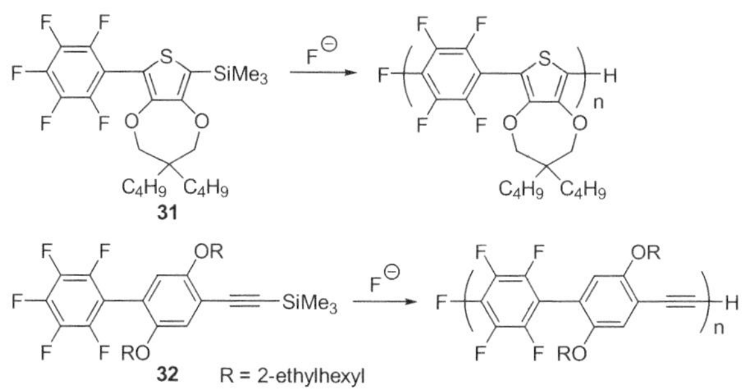

Figure 5.24 Transition-metal-free chain-growth polymerization of **31** and **32**.

5.5 Conclusions

In this chapter, we describe chain-growth polymerization from catalyst transfer Kumada–Tamao coupling polymerization to the recent progress involving a mechanism not via metal catalyst transfer. These methods enable us to synthesize many kinds of π-conjugated polymers: poly(3-alkylthiophene), poly(*N*-alkylpyrrole), poly(*p*-phenylene), polyfluorene, polybenzotriazole, poly(*p*-phenyleneethynylene), and so on. Although this chapter does not focus on architectures containing π-conjugated polymers such as block copolymers, star polymers, and graft copolymers, these architectures also can be prepared by the above methods. We hope that this chapter is helpful when design and synthesis of new high-performance materials, including π-conjugated polymers, are tried.

References

1. Grimsdale, A. C., Leok Chan, K., Martin, R. E., Jokisz, P. G., Holmes, A. B. *Chem. Rev.*, **109**, 897 (2009).
2. Arias, A. C., MacKenzie, J. D., McCulloch, I., Rivnay, J., Salleo, A. *Chem. Rev.*, **110**, 3 (2010).
3. Li, C., Liu, M., Pschirer, N. G., Baumgarten, M., Müllen, K. *Chem. Rev.*, **110**, 6817 (2010).
4. Beaujuge, P. M., Fréchet, J. M. J. *J. Am. Chem. Soc.*, **133**, 20009 (2011).
5. Yamamoto, T. *Bull. Chem. Soc. Jpn.*, **83**, 431 (2010).
6. Yokoyama, A., Miyakoshi, R., Yokozawa, T. *Macromolecules*, **37**, 1169 (2004).
7. Miyakoshi, R., Yokoyama, A., Yokozawa, T. *Macromol. Rapid Commun.*, **25**, 1663 (2004).
8. Miyakoshi, R., Yokoyama, A., Yokozawa, T. *J. Am. Chem. Soc.*, **127**, 17542 (2005).
9. Sheina, E. E., Liu, J., Iovu, M. C., Laird, D. W., McCullough, R. D. *Macromolecules*, **37**, 3526 (2004).
10. Iovu, M. C., Sheina, E. E., Gil, R. R., McCullough, R. D. *Macromolecules*, **38**, 8649 (2005).
11. Yokozawa, T., Yokoyama, A. *Chem. Rev.*, **109**, 5595 (2009).
12. Kiriy, A., Senkovskyy, V., Sommer, M. *Macromol. Rapid Commun.*, **32**, 1503 (2011).
13. Stefan, M. C., Bhatt, M. P., Sista, P., Magurudeniya, H. D. *Polym. Chem.*, **3**, 1693 (2012).
14. Yokozawa, T., Ohta, Y. *Chem. Commun.*, **49**, 8281 (2013).
15. Bryan, Z. J., McNeil, A. J. *Macromolecules*, **46**, 8395 (2013).
16. Wang, J., Higashihara, T. *Polym. Chem.*, **4**, 5518 (2013).
17. Handa, N. V., Serrano, A. V., Robb, M. J., Hawker, C. J. *J. Polym. Sci., Part A: Polym. Chem.*, **53**, 831 (2015).
18. Tkachov, R., Senkovskyy, V., Komber, H., Sommer, J.-U., Kiriy, A. *J. Am. Chem. Soc.*, **132**, 7803 (2010).
19. Verswyvel, M., Monnaie, F., Koeckelberghs, G. *Macromolecules*, **44**, 9489 (2011).
20. Senkovskyy, V., Khanduyeva, N., Komber, H., Oertel, U., Stamm, M., Kuckling, D., Kiriy, A. *J. Am. Chem. Soc.*, **129**, 6626 (2007).

21. Khanduyeva, N., Senkovskyy, V., Beryozkina, T., Bocharova, V., Simon, F., Nitschke, M., Stamm, M., Grötzschel, R., Kiriy, A. *Macromolecules*, **41**, 7383 (2008).

22. Bronstein, H. A., Luscombe, C. K. *J. Am. Chem. Soc.*, **131**, 12894 (2009).

23. Senkovskyy, V., Tkachov, R., Beryozkina, T., Komber, H., Oertel, U., Horecha, M., Bocharova, V., Stamm, M., Gevorgyan, S. A., Krebs, F. C., Kiriy, A. *J. Am. Chem. Soc.*, **131**, 16445 (2009).

24. Smeets, A., Van den Bergh, K., De Winter, J., Gerbaux, P., Verbiest, T., Koeckelberghs, G. *Macromolecules*, **42**, 7638 (2009).

25. Komber, H., Senkovskyy, V., Tkachov, R., Johnson, K., Kiriy, A., Huck, W. T. S., Sommer, M. *Macromolecules*, **44**, 9164 (2011).

26. Mao, Y., Wang, Y., Lucht, B. L. *J. Polym. Sci., Part A: Polym. Chem.*, **42**, 5538 (2004).

27. Miyakoshi, R., Yokoyama, A., Yokozawa, T. *J. Polym. Sci., Part A: Polym. Chem.*, **46**, 753 (2008).

28. Tamba, S., Fuji, K., Meguro, H., Okamoto, S., Tendo, T., Komobuchi, R., Sugie, A., Nishino, T., Mori, A. *Chem. Lett.*, **42**, 281 (2013).

29. Shi, X., Sui, A., Wang, Y., Li, Y., Geng, Y., Wang, F. *Chem. Commun.*, **51**, 2138 (2015).

30. Tamba, S., Shono, K., Sugie, A., Mori, A. *J. Am. Chem. Soc.*, **133**, 9700 (2011).

31. Hiorns, R. C., Khoukh, A., Gourdet, B., Dagron-Lartigau, C. *Polym. Int.*, **55**, 608 (2006).

32. Adachi, I., Miyakoshi, R., Yokoyama, A., Yokozawa, T. *Macromolecules*, **39**, 7793 (2006).

33. Vallat, P., Lamps, J. P., Schosseler, F., Rawiso, M., Catala, J. M. *Macromolecules*, **40**, 2600 (2007).

34. Boyd, S. D., Jen, A. K. Y., Luscombe, C. K. *Macromolecules*, **42**, 9387 (2009).

35. Wu, S., Huang, L., Tian, H., Geng, Y., Wang, F. *Macromolecules*, **44**, 7558 (2011).

36. Lamps, J. P., Catala, J. M. *Macromolecules*, **44**, 7962 (2011).

37. Yokoyama, A., Kato, A., Miyakoshi, R., Yokozawa, T. *Macromolecules*, **41**, 7271 (2008).

38. Hollinger, J., Jahnke, A. A., Coombs, N., Seferos, D. S. *J. Am. Chem. Soc.*, **132**, 8546 (2010).

39. Hollinger, J., Seferos, D. S. *Macromolecules*, **47**, 5002 (2014).

40. Miyakoshi, R., Shimono, K., Yokoyama, A., Yokozawa, T. *J. Am. Chem. Soc.*, **128**, 16012 (2006).
41. Ohshimizu, K., Takahashi, A., Higashihara, T., Ueda, M. *J. Polym. Sci., Part A: Polym. Chem.*, **49**, 2709 (2011).
42. Stefan, M. C., Javier, A. E., Osaka, I., McCullough, R. D. *Macromolecules*, **42**, 30 (2009).
43. Wen, H., Ge, Z., Liu, Y., Yokozawa, T., Lu, L., Ouyang, X., Tan, Z. *Eur. Polym. J.*, **49**, 3740 (2013).
44. Javier, A. E., Varshney, S. R., McCullough, R. D. *Macromolecules*, **43**, 3233 (2010).
45. Sui, A., Shi, X., Wu, S., Tian, H., Geng, Y., Wang, F. *Macromolecules*, **45**, 5436 (2012).
46. Willot, P., Govaerts, S., Koeckelberghs, G. *Macromolecules*, **46**, 8888 (2013).
47. Wu, S., Sun, Y., Huang, L., Wang, J., Zhou, Y., Geng, Y., Wang, F. *Macromolecules*, **43**, 4438 (2010).
48. Ono, R. J., Kang, S., Bielawski, C. W. *Macromolecules*, **45**, 2321 (2012).
49. Nanashima, Y., Yokoyama, A., Yokozawa, T. *Macromolecules*, **45**, 2609 (2012).
50. Pammer, F., Jäger, J., Rudolf, B., Sun, Y. *Macromolecules*, **47**, 5904 (2014).
51. Bridges, C. R., McCormick, T. M., Gibson, G. L., Hollinger, J., Seferos, D. S. *J. Am. Chem. Soc.*, **135**, 13212 (2013).
52. Yokoyama, A., Suzuki, H., Kubota, Y., Ohuchi, K., Higashimura, H., Yokozawa, T. *J. Am. Chem. Soc.*, **129**, 7236 (2007).
53. Sinclair, D. J., Sherburn, M. S. *J. Org. Chem.*, **70**, 3730 (2005).
54. Dong, C.-G., Hu, Q.-S. *J. Am. Chem. Soc.*, **127**, 10006 (2005).
55. Weber, S. K., Galbrecht, F., Scherf, U. *Org. Lett.*, **8**, 4039 (2006).
56. Yokozawa, T., Kohno, H., Ohta, Y., Yokoyama, A. *Macromolecules*, **43**, 7095 (2010).
57. Yokozawa, T., Suzuki, R., Nojima, M., Ohta, Y., Yokoyama, A. *Macromol. Rapid Commun.*, **32**, 801 (2011).
58. Elmalem, E., Kiriy, A., Huck, W. T. S. *Macromolecules*, **44**, 9057 (2011).
59. Tyrrell, E., Brookes, P. *Synthesis*, **2003**, 0469 (2003).
60. Billingsley, K. L., Anderson, K. W., Buchwald, S. L. *Angew. Chem. Int. Ed.*, **45**, 3484 (2006).
61. Billingsley, K., Buchwald, S. L. *J. Am. Chem. Soc.*, **129**, 3358 (2007).

62. Knapp, D. M., Gillis, E. P., Burke, M. D. *J. Am. Chem. Soc.*, **131**, 6961 (2009).

63. Kosaka, K., Ohta, Y., Yokozawa, T. *Macromol. Rapid Commun.*, **36**, 373 (2015).

64. Nojima, M., Ohta, Y., Yokozawa, T. *J. Polym. Sci., Part A: Polym. Chem.*, **52**, 2643 (2014).

65. Nojima, M., Ohta, Y., Yokozawa, T. *J. Am. Chem. Soc.*, **137**, 5682 (2015).

66. Kang, S., Ono, R. J., Bielawski, C. W. *J. Am. Chem. Soc.*, **135**, 4984 (2013).

67. Senkovskyy, V., Tkachov, R., Komber, H., Sommer, M., Heuken, M., Voit, B., Huck, W. T. S., Kataev, V., Petr, A., Kiriy, A. *J. Am. Chem. Soc.*, **133**, 19966 (2011).

68. Senkovskyy, V., Tkachov, R., Komber, H., John, A., Sommer, J.-U., Kiriy, A. *Macromolecules*, **45**, 7770 (2012).

69. Verswyvel, M., Verstappen, P., De Cremer, L., Verbiest, T., Koeckelberghs, G. *J. Polym. Sci., Part A: Polym. Chem.*, **49**, 5339 (2011).

70. Bonillo, B., Swager, T. M. *J. Am. Chem. Soc.*, **134**, 18916 (2012).

71. Sanji, T., Iyoda, T. *J. Am. Chem. Soc.*, **136**, 10238 (2014).

72. Sanji, T., Motoshige, A., Komiyama, H., Kakinuma, J., Ushikubo, R., Watanabe, S., Iyoda, T. *Chem. Sci.*, **6**, 492 (2015).

Biography

Yoshihiro Ohta was born in Kanagawa, Japan, in 1983. He received his BS (2006), MS (2008), and PhD (2011) in applied chemistry from Kanagawa University, Japan, under the direction of Prof. Tsutomu Yokozawa. After working as a research assistant professor in Prof. Atsushi Takahara's group at Kyusyu University (2011), he joined the faculty of Kanagawa University, where he is presently assistant professor. His research interests include the synthesis and characterization of hyperbranched polymers, functional block copolymers, and graft copolymers with well-defined structures.

Tsutomu Yokozawa was born in Chiba in 1957. He received his B.S. (1981), M.S. (1983), and Ph.D. (1987) in Organic Chemistry from Tokyo Institute of Technology under the direction of Professor Nobuo Ishikawa and Professor Takeshi Nakai. In 1985, he had already started an academic career in the Research Laboratory of Resources Utilization, Tokyo Institute of Technology as a Research Associate, and was promoted to Assistant Professor in 1988. He joined the Department of Applied Chemistry, Kanagawa University as a Lecturer in 1991, and was promoted to Associate Professor in 1993. During 1997-1998, he worked as a visiting scientist at the University of Illinois at Urbana-Champaign with Professor J. S. Moore. He was promoted to Full Professor in 1999. He was also a researcher for PRESTO, JST during 2001-2005. He received the Award of the Society of Polymer Science, Japan, in 2007. His research interest covers controlled synthesis of polymers, supramolecular chemistry of polymers, as well as synthetic organic chemistry.

Chapter 6

Synthesis of Polythiophene-Based Architectural Macromolecules: Recent Progress in Controlled Polymerization

Tomoya Higashihara
Department of Organic Materials Science, Graduate School of Materials Science, Yamagata University, 4-3-16 Jonan, Yonezawa, Yamagata 992-8510, Japan
thigashihara@yz.yamagata-u.ac.jp

In this chapter, the fundamental synthetic pathways for polythiophenes as well as recent progress in the synthesis of regioregular poly(3-alkylthiophene), well-defined chain-end-functional poly(3-alkylthiophene)s, and block and branched copolymers comprised of poly(3-alkylthiophene) segments are reviewed, focusing on the synthetic methodologies for controlling their regioregularities, molecular weights, dispersities, and block and branching structures.

Conjugated Objects: Developments, Synthesis, and Applications
Edited by Atsushi Nagai and Koji Takagi

ISBN 978-981-4774-03-1 (Hardcover), 978-1-315-22962-1 (eBook)
www.panstanford.com

6.1 Introduction

Since the discovery of polyacetylene films and the achievement of their excellent conductivity via a doping process reported by Shirakawa, MacDiarmid, and Heeger[1], much attention has recently focused on π-conjugated polymers, because they are unique materials due to their stiff rod-like structures in solution, melt, and/or solid states, specific crystalline/nanofibril structures, as well as many potentials for optoelectronic applications. Typical π-conjugated polymers include polyacetylenes, polyphenylenes, polyfluorenes, poly(phenylenevinylene)s, and polythiophenes (PTs) (Fig. 6.1). The investigation of various π-conjugated polymers currently provides a trend in which the principle, scope, and limitation in polymer synthetic chemistry are reconsidered to fulfill a wide range of requirements for the π-conjugated polymers.

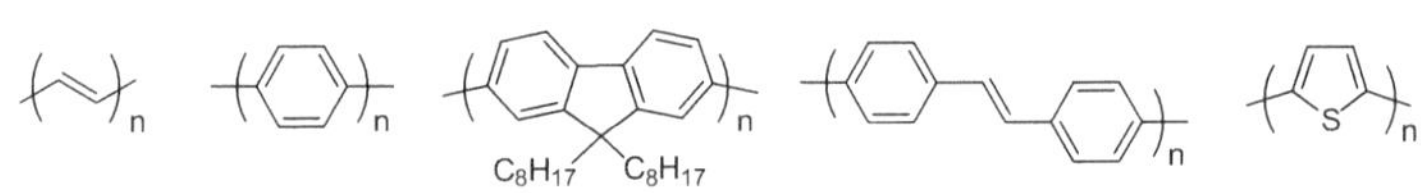

Polyacetylene Polyphenylene Polyfluorene Poly(phenylenevinylene) Polythiophene

Figure 6.1 Chemical structures of π-conjugated polymers.

Living/controlled polymerizations provide versatile methodologies for the synthesis of well-defined macromolecules with predictable molecular weights (MWs) and low dispersities (*Đ*). A wide range of other compositional and structural parameters could also be controlled, such as the chain-end functionality, copolymer composition, regioregularity, stereoregularity, and branching.[2–8] Recently, more complex architectural macromolecules, prepared by living/controlled polymerizations and efficient coupling/linking reactions of the building blocks, have rapidly emerged in polymer chemistry and have been well-reviewed elsewhere.[9–22] Among them, block copolymers (BCPs) have received special attraction due to their dissimilar phase structures that display unique behaviors in morphological studies, micelles, processing, material sciences, and functional device applications.[23–31] Under the current situation whereby the requirements for polymer properties and performances have become stricter and more versatile, the combinations of polymer segments in BCPs must be more versatile.

The breakthrough for a condensative chain-growth polymerization technique permits synthesizing well-defined π-conjugated polymers and BCPs containing low-dispersed π-conjugated polymer segments via living/controlled systems.[32–38] These efforts may contribute to the accelerated evolution of potential application areas such as polymer photonics materials, optoelectronic devices, etc.

In this chapter, the typical synthetic methods as well as their recent progress in the living/controlled synthesis in terms of regioregularity, MW, *Đ*, chain-end functionality, and block and branching structures are described.

6.2 Polythiophene

PT is a red powder possessing a bandgap of ca. 2 eV. Due to the strong intermolecular π–π stacking between the extended conjugation along the main chains, PT is also known as an insoluble and infusible material. To obtain a PT film, an electrolytic polymerization method has been applied. For instance, a thiophene monomer is dissolved in a solvent, such as acetonitrile, in the presence of an electrolyte, such as $LiBF_4$, followed by applying a voltage. A PT film can then be obtained on the anode electrode dipped into the solution.[39] A general polymerization mechanism for the electrolytic polymerization is shown in Scheme 6.1.

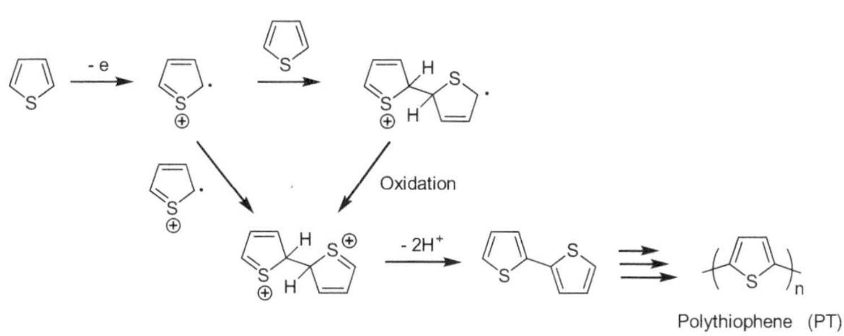

Scheme 6.1 Electrolytic polymerization of a thiophene monomer.

Jen and coworkers reported the synthesis of poly(3-alkylthiophene)s (P3ATs) (Fig. 6.2) to improve their solubility based on the electrolytic polymerization.[40] Sugimoto and coworkers successfully synthesized P3ATs by the $FeCl_3$-mediated oxidative coupling polymerization based on a solution polymerization method (Scheme 6.2).[41] Poly(3-alkylthiophene)s (P3ATs) are specifically high-performance materials as p-type semiconductors among all the π-conjugated polymers in terms of their solubility, chemical stability, charge mobility, and commercial availability. Therefore, very much attention has recently focused on these materials for potential applications in polymer electronic devices, such as organic field-effect transistors[42–47] and photovoltaic cells.[48–52]

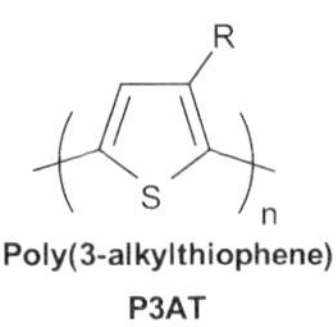

Figure 6.2 Chemical structure of poly(3-alkylthiophene).

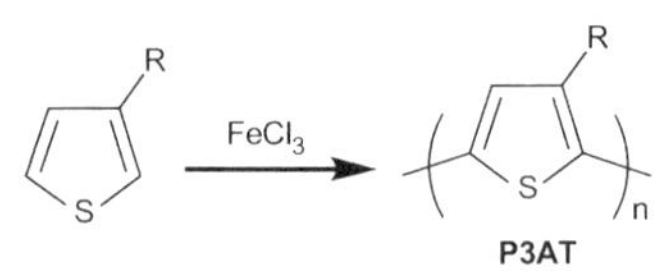

Scheme 6.2 Synthesis of poly(3-alkylthiophene) by $FeCl_3$-mediated oxidative coupling polymerization.

However, a significant problem has occurred from the substitution of an alkyl group at the 3-position of the thiophene rings due to the asymmetric monomer structure. Furthermore, a conventional condensative polymerization method limits the controllability of the MWs and *Đ* values; theoretically, the *Đ* value reaches 2 when the monomer conversion reaches 100% in the condensative polymerization.

6.3 Regioregularity

P3ATs have regiostructures along the main chains derived from the asymmetric-type monomers. As shown in Fig. 6.3a, there are three different diad regioisomers, that is, head-to-tail (HT), head-to-head (HH), and tail-to-tail (TT). Moreover, this produces four triad regioisomers in the polymer chain; that is, HT-HT, HT-HH, TT-HT, and TT-HH (Fig. 6.3b).

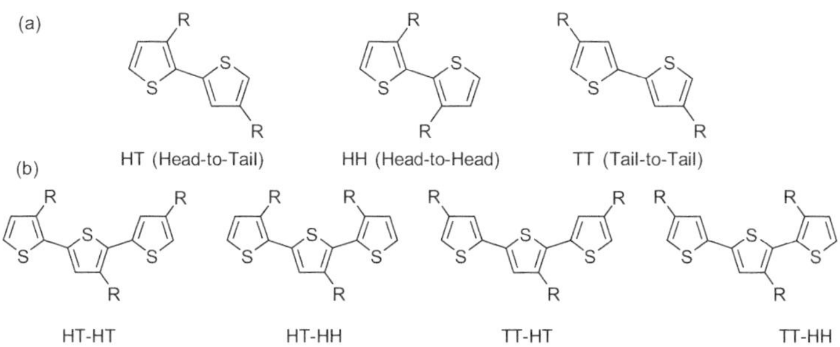

Figure 6.3 The (a) diad and (b) triad regioisomers of 3-hexylthiophene monomers.

McCullough and coworkers reported that the HT regioregularity of the P3ATs could be controlled based on the Kumada coupling reaction affording a regioregularity > 98% (Scheme 6.3a).[53,54] Rieke and coworkers reported the Negishi coupling polymerization to provide HT regioregular P3ATs (Scheme 6.3b) using a Rieke zinc.[55,56] Subsequently, the Stille coupling polymerization method (Scheme 6.3c)[57,58] and the Suzuki coupling polymerization method (Scheme 6.3d)[59,60] followed. Amou et al. optimized the conditions for the $FeCl_3$-mediated oxidative coupling polymerization (Scheme 6.2) to produce HT regioregular P3ATs with a 92% HT regioregularity.[61]

McCullough and coworkers developed an inexpensive methodology for the Grignard metathesis polymerization (GRIM) to afford HT regioregular P3ATs using the Grignard exchange reaction to produce the monomer intermediate. This system does not need to employ expensive reagents, such as lithium diisopropylamide (LDA) and magnesium bromide (Scheme 6.3e).[62,63]

The regioregularity could be calculated from the ^{1}H NMR spectra by comparing the signal intensity for the 4-H thiophene protons (HT-HT: 6.98 ppm, HT-HH: 7.00 ppm, TT-HT: 7.02 ppm, and TT-HH: 7.05 ppm) as well as the α-methylene protons (HT: 2.80 ppm, HH: 2.58 ppm) next to the thiophene ring.[55,56]

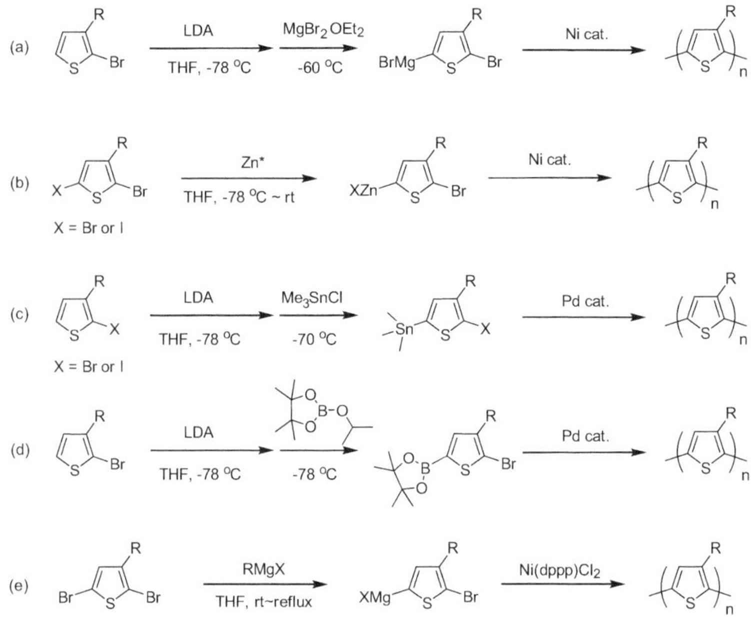

Scheme 6.3 Synthetic routes for regioregular poly(3-alkylthiophene)s.

6.4 Controlled Synthesis of Polythiophene

To control the MWs and $Đ$ values of the P3ATs, Yokozawa's group[62,63], shortly followed by McCullough's group[64,65], independently reported quasi-living Kumada catalyst transfer polycondensation (KCTP) systems. The mechanism has been proposed by Yokozawa[63] as follows (see Scheme 6.4): First, 2-bromo-5-chloromagnesio-3-hexylthiophene (**1**) is dimerized with Ni(dppp)Cl_2 (dppp=1,3-diphenylphosphinopropane) to afford the TT dimer. The Ni species then transfers to the next C–Br bonding, "walking" through the π-conjugation without diffusion of the Ni(0) species to afford the dimer (**2**). The dimer **2**, which is the virtual initiator, starts the polymerization of **1** via HT linkages by a sequence of transmetallation, reductive elimination of the Ni(0) species, and oxidative insertion of the Ni(0) species. The polymerization proceeds in a chain-growth manner due to prevention of the Ni(0) species diffusion, in which the reaction only takes place between the activated polymer chain end and the incoming monomer **1**. The resulting P3HTs possess controlled MWs with the feed ratio $[\mathbf{1}]_0/[Ni(dppp)Cl_2]_0$ as well as low $Đ$ values of less than 1.2. McCullough et al. proposed a different mechanism in which the associate pair between the generating polymer and Ni(0) species prevents diffusion of the Ni(0) species into the polymerization media to achieve the chain-growth polycondensation.[63]

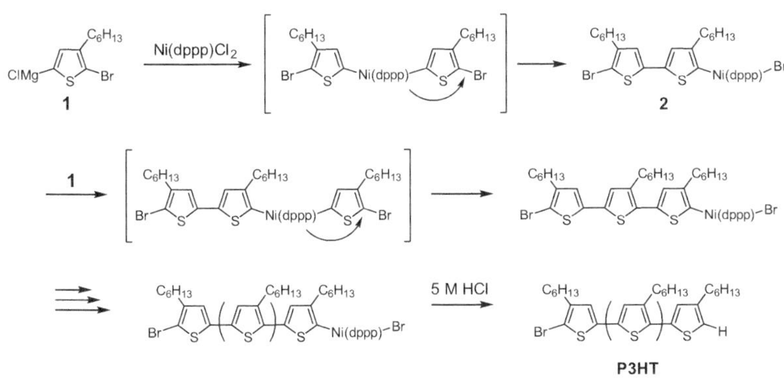

Scheme 6.4 Proposed mechanism of Kumada catalyst transfer polycondensation (KCTP) for synthesizing regioregular P3HT.

However, it is necessary to thoroughly purify the monomers and solvents for the KCTP system, because highly reactive and moisture-sensitive Grignard agents are employed. In addition, many functional groups, such as hydroxy, aldehyde, ketone, ester, amide, etc., cannot generally be tolerated in the presence of the Grignard agents. Higashihara and coworkers reported an alternative method based on the Negishi-type catalyst transfer polymerization (NCTP) using a zincate complex, $^tBu_4ZnLi_2$, instead of Grignard reagents. Actually, a living system for the NCTP using $^tBu_4ZnLi_2$ has been found to prepare regioregular P3HTs with controlled MWs and low $Đ$ values, as shown in Scheme 6.5a.[35,37,66]

The average MW value (M_n) proportionally increased by increasing the feed ratio of the monomer to Ni catalyst up to 30,700, while maintaining a relatively low $Đ$ of <1.2. Based on these successful results, the direct synthesis of poly(3-(6-hydroxyhexyl) thiophene) (P3HHT) could be performed without protection (Scheme 6.5b). The low nucleophilicity of the bulky zincate complex covered by the hydrophobic ^{t}butyl groups may possibly suppress the proton abstraction from the OH groups in P3HHT.

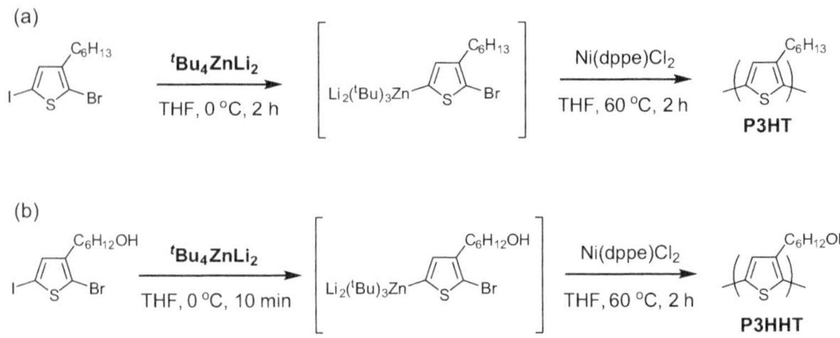

Scheme 6.5 Synthesis of (a) regioregular P3HT and (b) regioregular poly(3-(6-hydroxyhexyl)thiophene) (P3HHT) based on Negishi-type catalyst transfer polymerization (NCTP).

Very recently, P3HTs with extremely low *Đ* values (1.03–1.17) could also be prepared based on the NCTP by optimizing the phosphine ligand for the Ni catalyst to 1,2-dicyclohexylphosphinoethane (dcpe).[67] Another strategy involves a Suzuki–Miyaura CTP system. Indeed, Yokozawa et al. reported the $^{t}Bu_3PPd(Ph)Br$-catalyzed Suzuki–Miyaura coupling polymerization of 2-(4-hexyl-5-iodo-2-thienyl)-4,4,5,5-tetramethyl-1,3,2-dioxaborolane to prepare regioregular P3HTs (Scheme 6.6).[68]

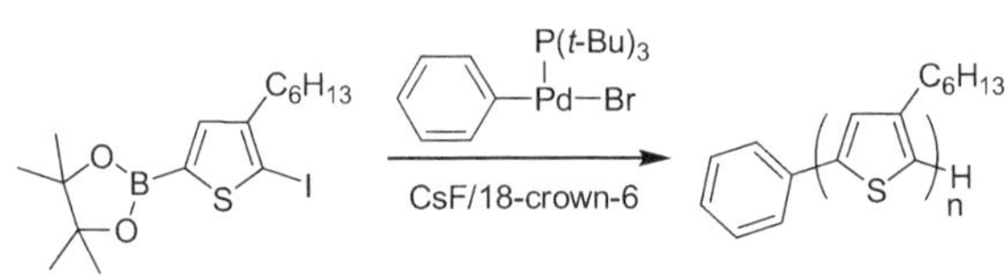

Scheme 6.6 Synthesis of regioregular P3HT by catalyst transfer Suzuki–Miyaura coupling polymerization.

6.5 Chain-End-Functional Polythiophene

The chain-end-functional PTs are very important materials for preparing BCPs, because the functional groups at the chain end(s) can be exploited for the introduction of the second block by the subsequent reactions or polymerizations. McCullough reported the preparation of a chain-end-functional P3HT by reacting P3HT-Ni(dppp)Br with a Grignard reagent (R-MgX) (see Scheme 6.7, termination method).[69–71] The functional groups include methyl, butyl, ^{t}butyl, vinyl, allyl, phenyl, tolyl, benzyl, formyl (acetal protected), alcohol/phenol (tetrahydropyran protected), and 3-aminopropyl/*p*-aminophenyl (trimethylsilyl protected) (see Scheme 6.7).

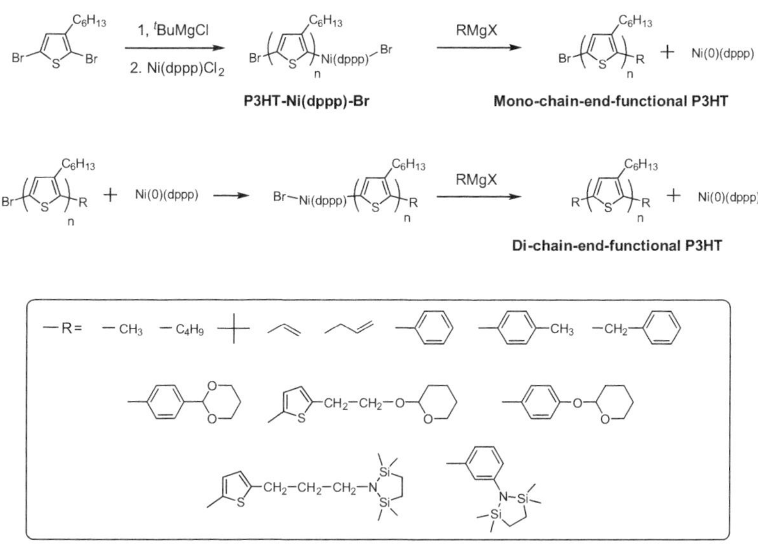

Scheme 6.7 Synthesis of chain-end-functional P3HTs by the termination method.

However, the type of R-MgX significantly affected the degree of functionalization at one chain end/both chain ends. As shown in Scheme 6.7, the di-chain-end-functional P3HT can be obtained by the following mechanism: (i) reductive elimination of Ni(0) catalyst after the monofunctionalization, (ii) the generated Ni(0) is oxidatively inserted into the α-terminal, and (iii) the metal exchange reaction with excess R-MgX. This mechanism suggested the difficulty in the selective synthesis of either the mono- or di-chain-end-functional P3HTs.

Luscombe and coworkers reported the efficient initiation method using the *o*-halotoluene precursor based on a ligand exchange approach (Scheme 6.8).[72] Koeckelberghs reported the synthesis of

α-chain-end-functional P3HTs with functional initiators having the OH (^{t}butyldimethylsilyl protected), SH (tri-isopropylsilyl protected), COOH (^{t}butylprotected), alkyne (trimethlsilyl protected), phosphonic ester, pyridine, and phenol (^{t}butyldimethylsilyl protected) groups based on the initiation method (see Fig. 6.4a).[73–75] Kiriy and coworkers developed a specially designed Ni initiator with an initiating fragment for the nitroxy-mediated radical polymerization (NMP) and one with an electronic functionality (see Fig. 6.4b).[76,77]

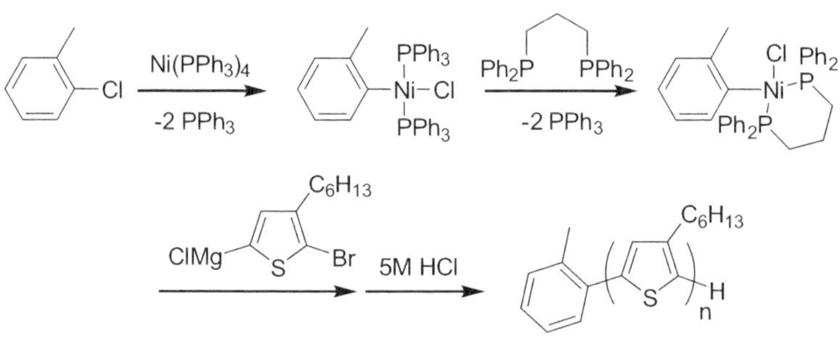

Scheme 6.8 Synthesis of α-chain-end-functional P3HTs by the initiation method.

On the other hand, Higashihara and coworkers succeeded in the direct synthesis of the ω-chain-end-functional P3HTs with monobromobutyl groups without protection by the combination of initiation and termination methods in a one-pot process (see Scheme 6.9).[78] In contrast, when Ni(dppp)Cl_2 was used as an initiator, α,ω-di-chain-end-functional P3HTs with 4-bromobutyl groups at both ends could be formed. This is the first report about the selective functionalization (ω-chain-end and α,ω-di-chain-ends) of P3HTs with the same functional groups. Furthermore, it should be mentioned that the bromoalkyl group can be transformed into a wide variety of other useful functionalities, such as azide, alkyne, amine, cyano, ether, alcohol, aldehyde, etc. This system is also quite important for the BCP synthesis using the chain-end-functional P3HT as a precursor, because the end functionality of P3HT is critical for the sequence control of the AB diblock or ABA triblock copolymers. The recent progress in chain-end-functional P3HTs has also been well reviewed by Hawker and coworkers.[79]

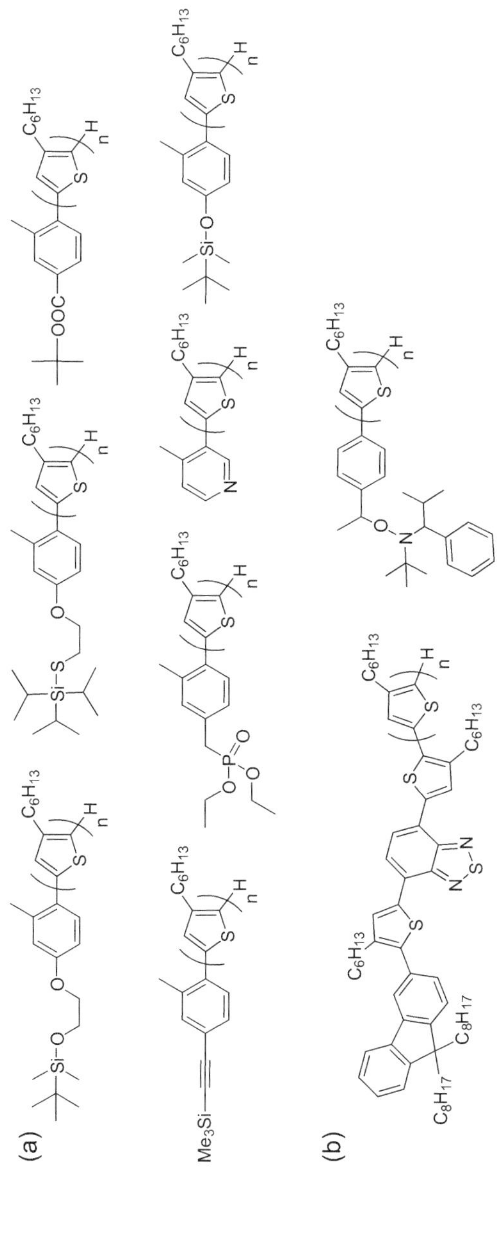

Figure 6.4 Chemical structures of α-chain-end-functional P3HTs prepared by the initiation method.

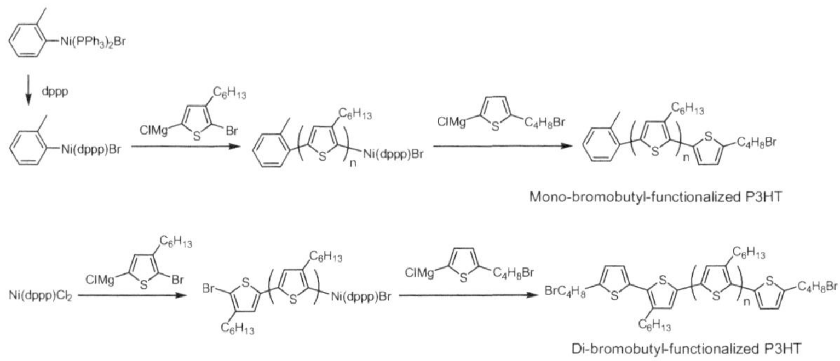

Scheme 6.9 Synthesis of mono- and di-bromobutyl chain-end-functional P3HTs.

6.6 All-Conjugated Block Copolymers

The all-conjugated block copolythiophene is generally defined as a BCP possessing a complete π-conjugation along the main chain. Indeed, a variety of block copolythiophenes could be synthesized based on KCTP utilizing a simple sequential monomer addition technique. McCullough and coworkers first reported the sequential polymerization of 2-bromo-5-chloromagnesio-3-dodecylthiophene (**3**) just after the polymerization of the first monomer **1** to afford P3HT-*b*-poly(3-dodecylthiophene) (**4**) (Scheme 6.10).[63] Soon after, Yokozawa and coworkers synthesized P3HT-*b*-poly(3-(2-(2-methoxyethoxy)ethoxy)thiophene) (**5**) based on a similar approach (see Fig. 6.5).[80] The Ueda and Tajima groups independently reported the synthesis of crystalline-amorphous P3HT-*b*-poly(3-phenoxymethylthiophene) (**6**)[81] and P3HT-*b*-poly(3-(2-ethylhexyl)thiophene) (**7**),[82] respectively (see Fig. 6.5). Many other all-conjugated block copolythiophenes have also been reported (see BCPs **8–17**, Fig. 6.5).[83–93] Higashihara and coworkers reported the synthesis of amphiphilic P3HT-*b*-poly(3-(2-(2-(2-methoxyethoxy)ethoxy)ethoxy)thiophene) (**18**)[94] and P3HT-*b*-poly(3-(4-(3,7-dimethyloctyloxy)-3-pyridyl)thiophene) (**19**), which could be used for organic solar cell devices.[95]

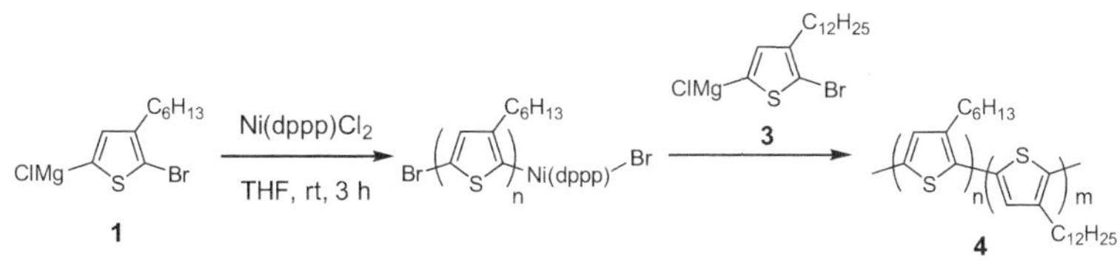

Scheme 6.10 Synthesis of all-conjugated block copolythiophene by the sequential monomer addition technique.

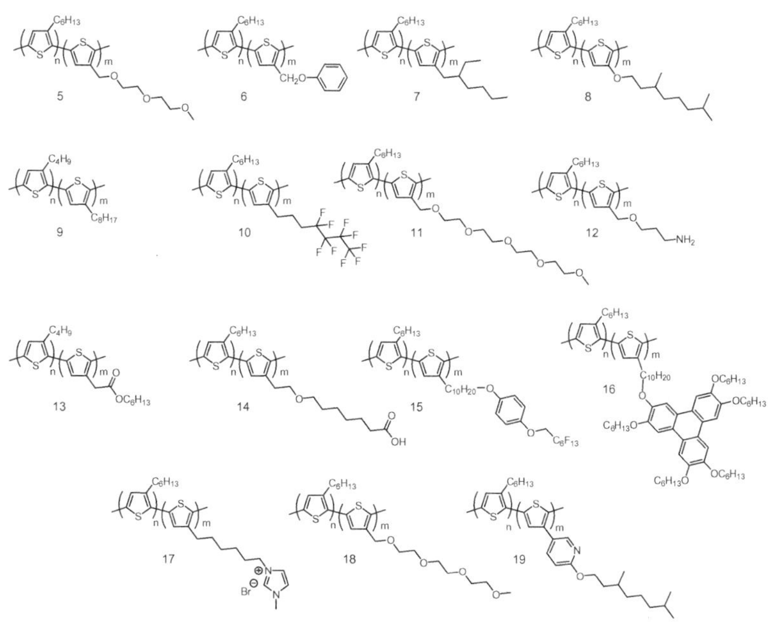

Figure 6.5 Chemical structures of all-conjugated block copolythiophenes.

All-conjugated BCPs containing PT and another main-chain polymer segment have also been reported. Scherf's group synthesized all-conjugated BCPs containing PT and polyfluorene segments using the H-/Br-terminated P3HT- and AB-type fluorine monomer.[96,97] On the basis of the recent progresses in the controlled synthesis of π-conjugated polymers, new types of all-conjugated BCPs have emerged. For instance, Yokozawa and coworkers reported the synthesis of poly(2,5-dihexyloxy-1,4-phenylene) (PPP)-*b*-P3HT based on the sequential monomer addition technique

(Scheme 6.11).[98] Other BCP examples (BCPs **20–22**) are shown in Fig. 6.6.[99–101] Recently, Higashihara and coworkers synthesized a novel soluble BCP having P3HT and graphene-like nanoribbon segments (**23**).[102] Furthermore, Koekelberghs and coworkers reported the first example of ABC-type BCPs containing P3HT, poly(3-octylselenophene), and poly(9,9-dioctylfluorene) segments (**24**) with all sequential combinations.[103]

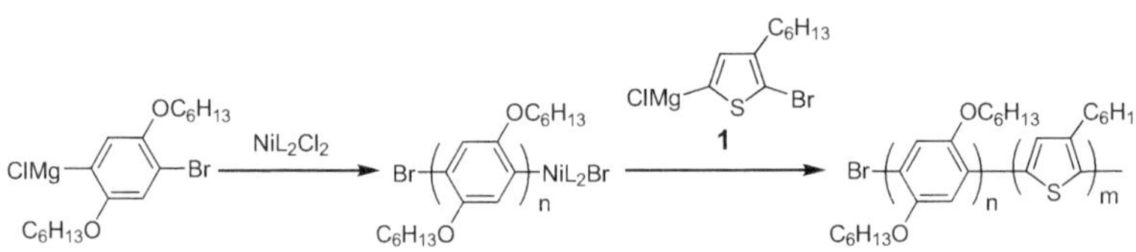

Scheme 6.11 Synthesis of poly(2,5-dihexyloxy-1,4-phenylene)-*b*-P3HT by the sequential monomer addition technique.

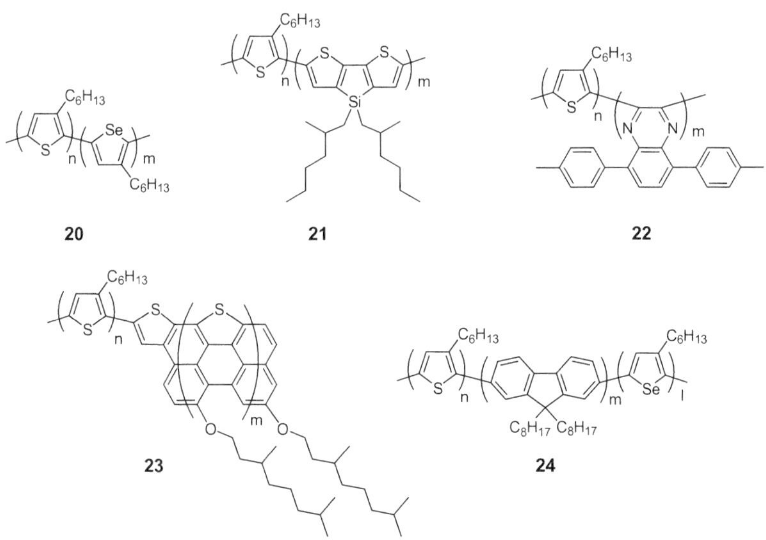

Figure 6.6 Chemical structures of all-conjugated BCPs containing polythiophene and another type of polymer segments.

Recently, much attention has been paid to the all-conjugated donor–acceptor BCPs due to their potential application to fullerene-free organic photovoltaics (OPVs). An all-conjugated donor-acceptor block copolythiophene was reported by Dagron-Lartigau's group[104] and Tajima's group modified the synthetic procedure involving the alkyne-azide click reaction.[105] They synthesized block copolythiophenes with P3HT and PT having the pendent C_{60} units at the side chains. Scherf, Greenham, and coworkers reported the synthesis of fullerene-free all-conjugated donor–acceptor BCPs by the polycondensation of the AB-type monomer initiated with H-/Br-terminated P3HT based on the Stille coupling reaction, as illustrated in Scheme 6.12a.[106] Hawker and coworkers reported the synthesis of another type of donor–acceptor BCP comprised of P3HT and a diketopyrrolopyrrole (DPP)-based narrow-bandgap block segment (Scheme 6.12b).[107]

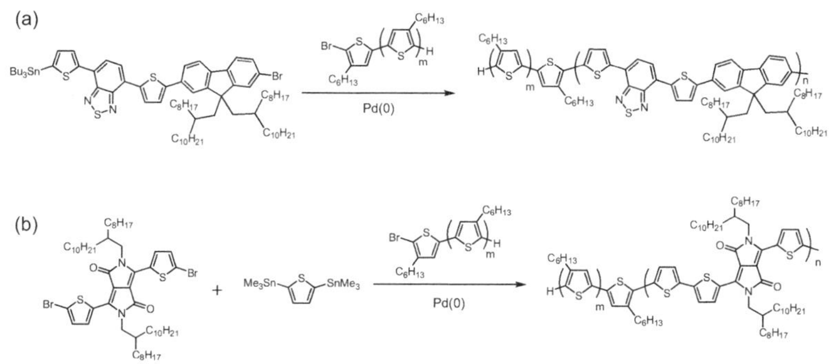

Scheme 6.12 Synthesis of all-conjugated donor–acceptor block copolymers.

Nakabayashi et al. reported a donor–acceptor BCP having P3HT and n-type naphthalene diimide (NDI)-based π-conjugated polymer segments (Scheme 6.13a) by the Yamamoto coupling between the dibrominated NDI monomer and Br-terminated P3HT.[108] Similarly, Wang, Higashihara, and coworkers synthesized a similar BCP (**25**) (Scheme 6.13b).[109] For the synthesis of BCP **25**, the Br-terminated NDI-based polymer was first prepared by the Stille coupling polymerization of bis(trimethylstannyl)thiophene and a slight excess of the dibrominated NDI monomer. It was then utilized as a macroinitiator for KCTP of the thiophene monomer **1** after insertion of the Ni catalyst with $Ni(COD)_2/PPh_3$, followed by the ligand exchange with dppp. The BCP **25** showed a clear phase separation of two distinct crystalline domains corresponding to the P3HT (edge-on rich) and NDI-based polymer segments (face-on rich) by grazing incidence wide-angle X-ray scattering (GIWAXS).[110] Quite recently, Koeckelberghs and coworkers reported the one-pot synthesis of the donor–acceptor P3HT-*b*-poly(didodecylthieno[3,4-b]pyrazine) by the sequential monomer addition technique.[111]

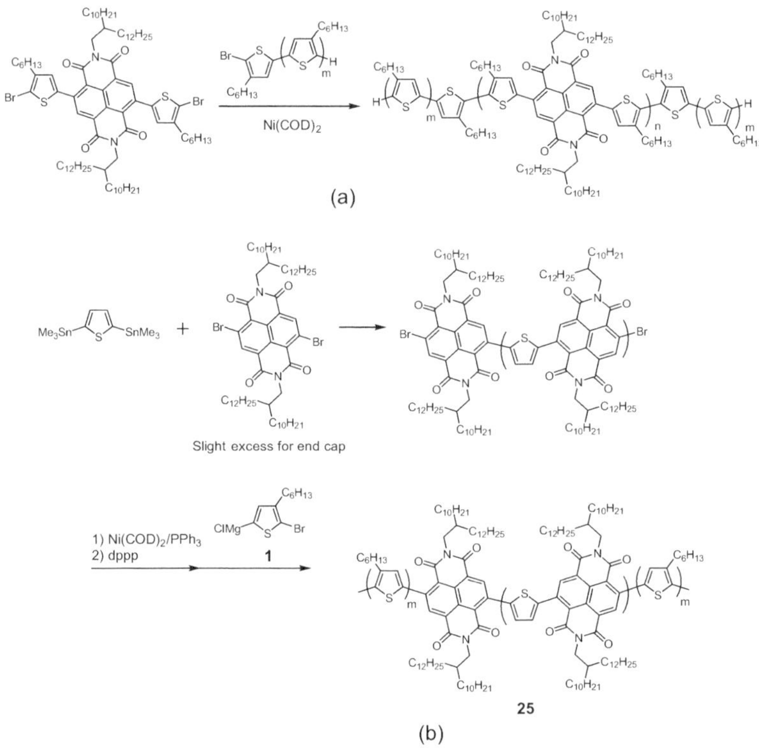

Scheme 6.13 Synthesis of all-conjugated donor–acceptor block copolymers containing naphthalene diimide-based n-type polymer segment.

6.7 Semiconjugated Block Copolymers

In the preceding section, the combination of block segments is limited to a π-conjugated system along the main chains. To extend the structural diversity, the second block segments can be changed to a non-π-conjugated system. To synthesize semiconjugated BCPs containing P3HT and poly(acrylate) (PMA) segments, McCullough combined KCTP with an atom transfer radical polymerization (ATRP) system (Scheme 6.14).[112] They also succeeded in synthesizing P3HT-*b*-polyisoprene(PI) (**26**) and P3HT-*b*-polystyrene(PS) (**27**) (see Fig. 6.7) using NMP and reversible addition-fragmentation chain transfer polymerization (RAFT), respectively.[113]

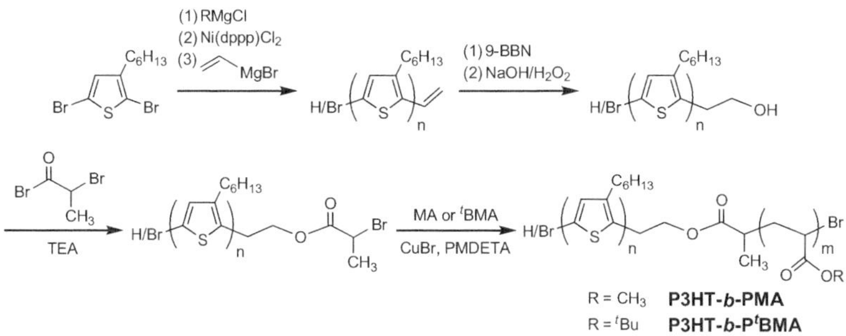

Scheme 6.14 Synthesis of P3HT-*b*-PMA and P3HT-*b*-P^{t}BMA by the combination of KCTP and ATRP.

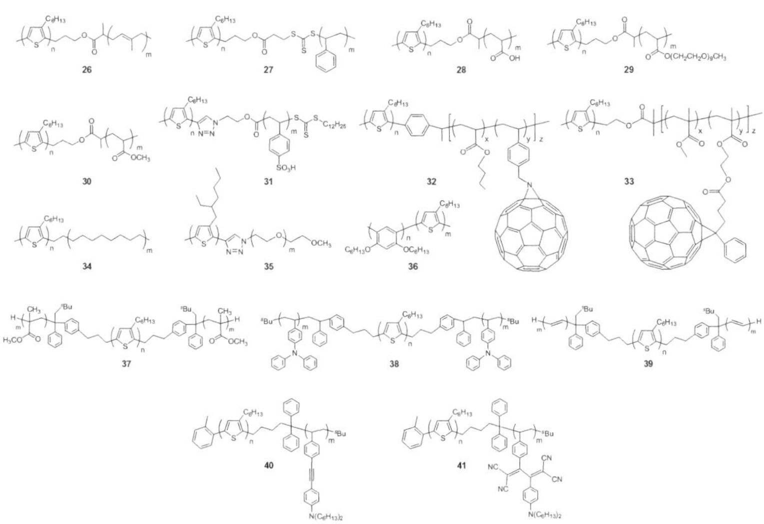

Figure 6.7 Chemical structures of semiconjugated BCPs containing polythiophene segments.

Furthermore, an amphiphilic P3HT-*b*-poly(acrylic acid) (PAA) (**28**) was synthesized by the hydrolysis of P3HT-*b*-P^{t}BMA.[114] Zou et al. synthesized a series of P3HT containing BCPs with PAA, poly(polyethylene glycol acrylate) (**29**), and PMMA (**30**) and used it as an efficient dispersant for carbon nanotubes.[115] Urien and coworkers reported an alternative synthetic approach based on click chemistry, combined with KCTP and ATRP, to afford AB and ABA BCPs containing P3HT and PS segments.[116] Another recent example based on click chemistry includes P3HT-*b*-poly(styrene sulfonate) (**31**).[117] Richard et al. synthesized P3HT-based donor–acceptor rod-coil BCPs with P3HT and fullerene-grafted PS segments (**32**) using the NMP system.[118] Soon after, Lee et al. synthesized P3HT-based rod-coil BCPs with fullerene-grafted PMMA segments (**33**).[119] However, these methodologies involve tedious site transformation reactions. In addition, the limited control nature of the radical polymerization usually causes high *Đ* values.

Besides radical polymerization cosystems, Jannssen and coworkers reported the synthesis of crystalline-crystalline BCPs comprised of P3HT and polyethylene segments (**34**) by a combination of KCTP and ring-opening polymerization.[120] Balsara and coworkers synthesized P3HT-*b*-poly(ethylene oxide) (**35**) based on click chemistry.[121] Higashihara and coworkers reported the synthesis of poly(4,6-dihexyloxy-1,3-phenylene)-*b*-P3HT with a low dispersity (*Đ* =1.15) by the sequential monomer addition technique (**36**).[122] Dai and coworkers first synthesized P3HT-*b*-poly(2-vinylpyridine) (P2VP) by a combination of KCTP and living anionic polymerization.[123]

A facile synthesis of PS-*b*-P3HT-*b*-PS was reported by Higashihara et al. based on a novel approach involving only two steps, that is, the in situ introduction of nonhomopolymerizable 1,1-diphenylethylene (DPE) moieties at the α,ω-ends of P3HT and a linking reaction with a living anionic polymer of styrene.[124] The synthetic routes are shown in Scheme 6.15. The key step is the in situ introduction of DPE moieties simultaneously at the α,ω-ends of P3HT, which readily and quantitatively link with the living PS without homopolymerization of the DPE units due to its steric hindrance and resonance effects. This spontaneous disubstitution permits skipping the extra steps of the transformation reactions at the chain end. In addition, the developed polymer–polymer linking method provides more reliable MWs and compositions of the BCPs than by conventional macroinitiator methods because both segments can be premade and separately characterized. The precursor α,ω-chain-end-DPE-difunctional P3HT was also used for synthesizing the well-defined PMMA-*b*-P3HT-*b*-PMMA with a low *Đ* value (1.10) (**37**).[125] On the other hand, J. K. Kim et al. synthesized the well-defined P3HT-*b*-PMMA based on the polymer–polymer linking approach which involves the use of the phenyl-acrylate(PA)-α-chain-end-functional P3HT.[126]

Semiconjugated BCPs normally show limited optoelectronic properties. Therefore, the introduction of optoelectronically active segments into the P3HT-based BCPs is essential. Higashihara and coworkers synthesized novel BCPs containing P3HT and the optoelectronically active poly(4-vinyltriphenylamine) (PVTPA) by a similar approach.[127] Indeed, the compatibilizing effect of **38** was discovered during the application to P3HT:[6,6]-phenyl-C_{61}-butyric acid methyl ester (PCBM) bulk heterojunction (BHJ) OPVs, showing a 13% improvement of the power conversion efficiency after the 1.5 wt% addition of **38**.[128] Other examples include polyacetylene-*b*-P3HT-*b*-polyacetylene (**39**),[129] P3HT-*b*-poly(4-(4′-*N*,*N*-dihexylaminophenylethynyl)styrene) (**40**), and the tetracyanoethylene (TCNE) adduct to **40** (**41**).[78]

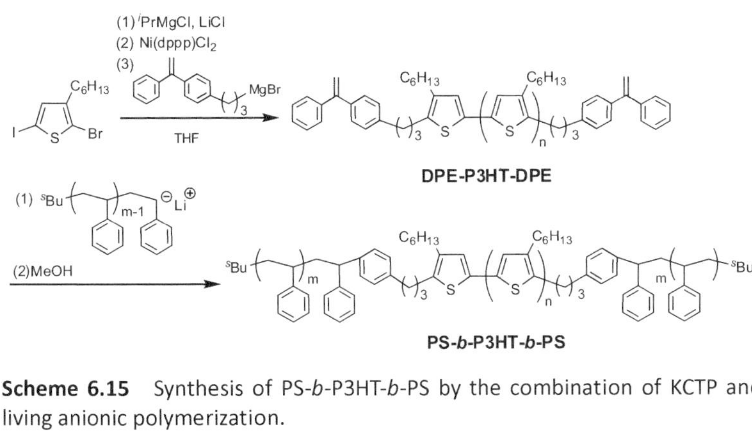

Scheme 6.15 Synthesis of PS-*b*-P3HT-*b*-PS by the combination of KCTP and living anionic polymerization.

6.8 Conjugated Branched Polymers

PT-based graft copolymers are divided into two types. The first one is comprised of PT side chains and another polymer backbone. Kiriy and coworkers reported the grafting-from method to prepare PS-*graft*-P3HT by the externally initiated KCTP.[130,131] On the other hand, the macromonomer method was developed by Kilbey II and coworkers using the ring-opening metathesis polymerization of a norbornenyl-functional P3HT macromonomer to synthesize poly(norbornene)-*graft*-P3HT.[132] The second one is comprised of a PT backbone and other polymer side chains. Hawker and coworkers prepared P3HT-*graft*-PS via a trityl ether linkage by the macromonomer method.[133] Higashihara and coworkers reported the first synthesis of the all-conjugated graft copolymers with a PT backbone and P3HT side chains as well as the ones with an n-type NDI-based polymer backbone and P3HT side chains by the grafting-from method (see Scheme 6.16).[134] Luscombe and coworkers synthesized the p-type donor–acceptor polymer backbone and P3HT side chains by the macromonomer method.[135]

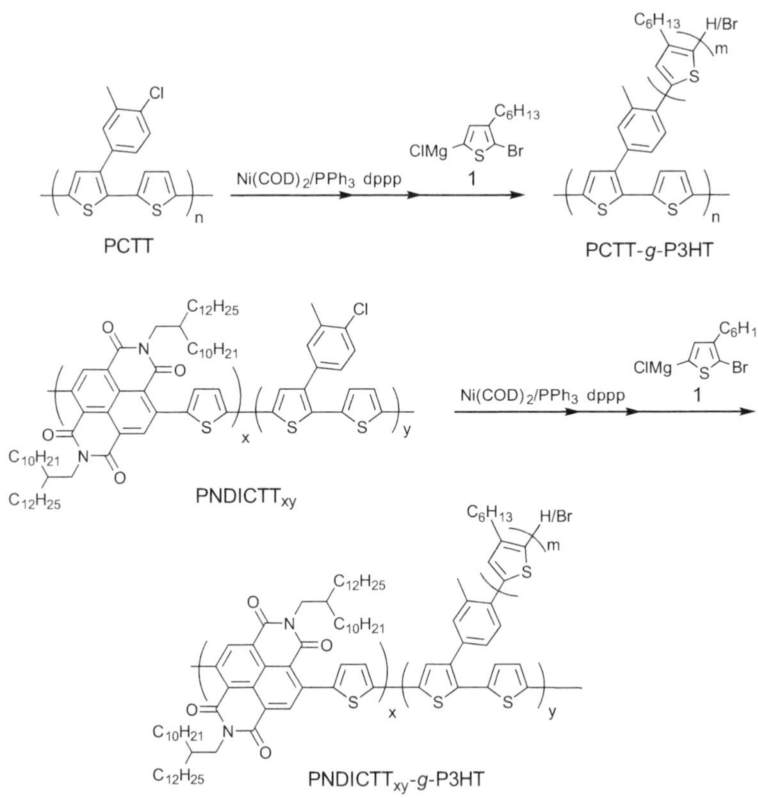

Scheme 6.16 Synthesis of all-conjugated graft copolymers with P3HT segments.

Compared to the PT-based graft copolymers, there are only a few reports about the synthesis of star P3HTs due to their synthetic difficulty.[136,137] Kim et al. reported the synthesis of the multiarm star P3HTs based on a divinyl compound technique.[138] However, the divinyl compound approach generally leads to ill-defined structures in terms of the arm number. Recently, Kiriy et al. utilized the core-first method based on the externally initiated KCTP to synthesize a 6-arm star P3HT.[139] Soon after, Luscombe and coworkers reported the first synthesis of a well-defined 3-arm star P3HT with a low dispersity ($Đ$ = 1.15) using a specially designed core compound.[140]

Asymmetric star polymers (also called *miktoarm star polymers*), in which different types of arm segments emanate from the core, have the potential for creating novel morphological nanostructures different from linear BCPs. However, it is normally quite difficult to synthesize such complicated branched polymers. To the best of our knowledge, Kim and coworkers first reported a well-defined AB_2-type asymmetric star polymer (A = PMMA, B = P3HT) by the copper (I)-catalyzed Huisgen 1,3-dipolar cycloaddition click reaction of the chain-end-functional PMMA with two azide moieties with a monoethynyl-chain-end-functional P3HT.[141] On the other hand, Higashihara and coworkers first reported the synthesis of well-defined ABC-type asymmetric star polymers (A = P3HT, B = PS, and C = P2VP) with controlled MWs, composition, and extremely low $Đ$ values ($Đ < 1.04$) based on the anionic linking reactions in conjunction with KCTP (see Scheme 6.17).[142] The synthetic methodology involves the following three-stage reaction: (i) an anionic linking reaction between the ω-chain-end-functional P3HT with a bromobutyl moiety (P3HT-C_4-Br) and living anionic PS end-capped with 1-(3-tbutyldim ethylsilyloxymethylphenyl)-1-phenylethylene, (ii) a transformation reaction of the 3-tbutyldimethylsilyloxymethylphenyl (TBDMS) moiety at the junction between the P3HT and PS segments into the PA moiety through the hydroxymethylphenyl moiety, and (iii) an anionic linking reaction between the PA-in-chain-functional P3HT-*b*-PS and living anionic P2VP. The grazing incidence X-ray scattering (GIXS) experiments of the ABC-type star polymer thin films confirmed their hierarchical nanofibril structures, in which the P3HT crystalline domains align with an edge-on orientation in the exclusive area of the microphase-separated domains from the PS+P2VP ones in the thin films. It was also found that the fibril period distance as well as the π–π stacking distance of the P3HT crystallites increased by increasing the length of the P2VP arm segments.

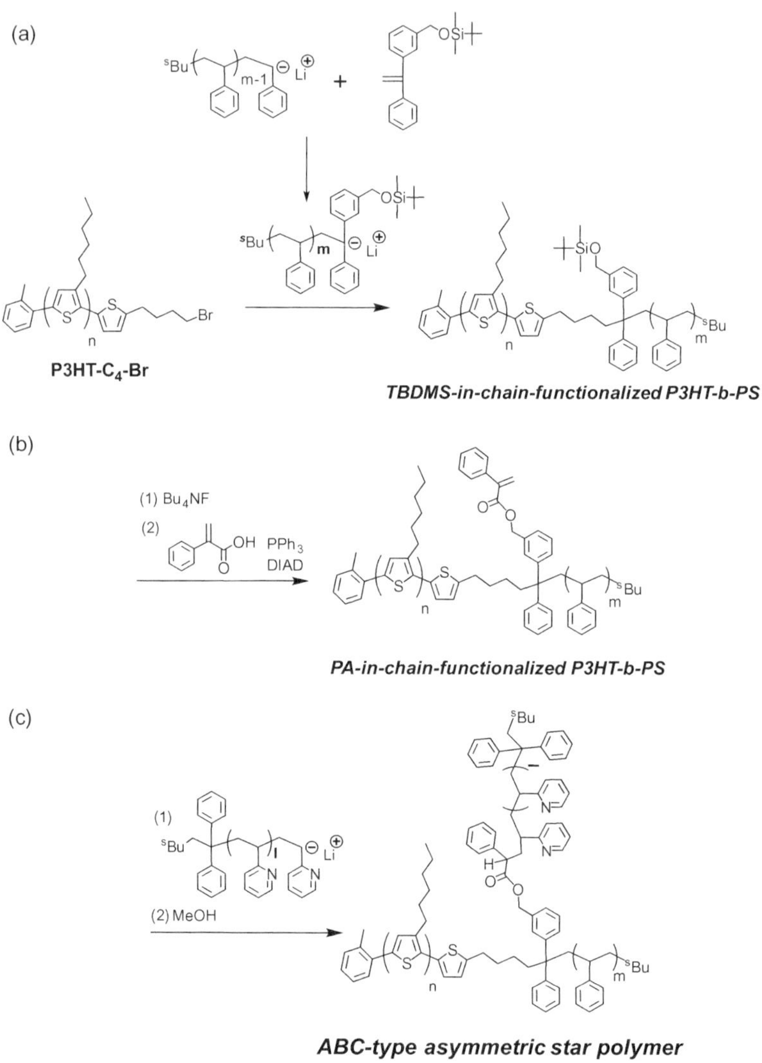

Scheme 6.17 Synthesis of all-conjugated graft copolymers with P3HT segments.

6.9 Conclusions

This chapter reviews in detail the fundamental studies of PTs and recent progress involving the controlled synthesis of chain-end-functional, block, and branched polymers containing PT segments. The versatile methodology and synthetic examples have been focused on, so this chapter should be useful, especially for beginning synthetic and material scientists who intend to exploit PT-based materials in optoelectronics. To design and synthesize PT-based polymers with predictable MWs and low $Đ$ values, the establishment of the controlled synthesis of the P3HT homopolymer as well as the combination with other living polymerization systems are essential. Therefore, understanding of not only the PT-based materials but also other polymers using other living polymerization systems has also become important. Although this review does not focus on morphological studies, such as self-assembly and micellation, these topics are also quite important for the PT-based materials. Unfortunately, the relationships among the primary PT-based polymer structures, morphologies, optoelectronic properties, and device performances are still not fully understood and they should be systematically clarified using a library of well-defined architectural polymer samples for providing reliable and reproducible conclusions in the near future.

References

1. Shirakawa, H., Louis, E. J., MacDiarmid, A. G., Chiang, C. K., Heeger, A. J. *J. Chem. Soc., Chem. Commun.*, 16, 578 (1977).
2. Morton, M. *Anionic Polymerization: Principles and Practice*, Academic Press: New York (1983).
3. Bywater, S. *Encyclopedia of Polymer Science and Engineering*, 2nd ed., Kroschwitz, J. I., ed., Wiley-Interscience: New York, Vol. 2, p. 1 (1985).
4. Rempp, P., Franta, E., Herz, J. E. *Adv. Polym. Sci.*, **86**, 145 (1988).
5. Grubbs, R. H., Tumas, W. *Science*, **243**, 907 (1989).
6. Webster, O. W. *Science*, **251**, 887 (1991).
7. Kenedy, J. P., Ivan, B. *Designed Polymers by Carbocationic Macromolecular Engineering: Theory and Practice*, Hanser Publishers, Munich (1992).

8. Hsieh, H. L., Quirk, R. P. *Anionic Polymerization: Principles and Practical Applications*, Hudgin, D. E., ed., Marcel Dekker: New York, Chapter 4, p. 71 (1996).
9. Pitsikalis, M., Pispas, S., Mays, J. W., Hadjichristidis, N. *Adv. Polym. Sci.*, **135**, 1 (1998).
10. Hadjichristidis, N. *J. Polym. Sci., Part A: Polym. Chem.*, **37**, 857 (1999).
11. Hadjichristidis, N., Pispas, S., Pitsikalis, M., Iatrou, H., Vlahos, C. *Adv. Polym. Sci.*, **142**, 71 (1999).
12. Hadjichristidis, N., Iatrou, H., Pispas, S., Pitsikalis, M. *J. Polym. Sci., Part A: Polym. Chem.*, **38**, 3211 (2000).
13. Hadjichristidis, N., Pitsikalis, M., Pispas, S., Iatrou, H. *Chem. Rev.*, **101**, 3747 (2001).
14. Hadjichristidis, N., Iatrou, H., Pitsikalis, M., Pispas, S., Avgeropoulos, A. *Prog. Polym. Sci.*, **30**, 725 (2005).
15. Hadjichristidis, N., Iatrou, H., Pitsikalis, M., Mays, J. W. *Prog. Polym. Sci.*, **31**, 1068 (2006).
16. Hirao, A., Hayashi, M., Tokuda, Y., Higashihara, T., Haraguchi, N., Ryu, S. W. *Polym. J.*, **34**, 633 (2002).
17. Hirao, A., Hayashi, M., Loykulnant, S., Sugiyama, K., Ryu, S. W., Haraguchi, N., Matsuo, A., Higashihara, T. *Prog. Polym. Sci.*, **30**, 111 (2005).
18. Hirao, A., Inoue, K., Higashihara, T., Hayashi, M. *Polym. J.*, **40**, 923 (2008).
19. Higashihara, T., Sugiyama, K., Yoo, H. S., Hayashi, M., Hirao, A. *Macromol. Rapid Commun.*, **31**, 1031 (2010).
20. Higashihara, T., Hayashi, M., Hirao, A. *Prog. Polym. Sci.*, **36**, 323 (2011).
21. Hirao, A., Hayashi, M., Higashihara, T., Hadjichristidis, N. *Complex Macromolecular Architecture*, eds. Hadjichristidis, N., Hirao, A., Tezuka, Y., Prez, F. D., John Wiley & Sons (Asia):, Singapore, Chapter 4, p. 97 (2011).
22. Higashihara, T., Segawa, Y., Sinananwanich, W., Ueda, M. *Polym. J.*, **44**, 14 (2012).
23. Krausch, G., Magerle, R. *Adv. Mater.*, **14**, 1579 (2002).
24. Ishizu, K., Tsubaki, K., Mori, A., Uchida, S. *Prog. Polym. Sci.*, **28**, 27 (2002).
25. Mori, H., Mueller, A. H. E. *Prog. Polym. Sci.*, **28**, 1403 (2003).
26. Park, C., Yoonand, J., Thomas, E. L. *Polymer*, **44**, 6725 (2003).
27. Hillmyer, M. A. *Adv. Polym. Sci.*, **190**, 137 (2005).

28. Olson, D. A., Chen, L., Hillmyer, M. A. *Chem. Mater.*, **20**, 869 (2008).

29. Hoeben, F. J. M., Jonkheijm, P., Meijer, E. W., Schenning, A. P. H. J. *Chem. Rev.*, **105**, 1491 (2005).

30. Tseng, W. H., Chen, C. K., Chiang, Y. W., Ho, R. M., Akasaka, S., Hasegawa, H. *J. Am. Chem. Soc.*, **131**, 1356 (2009).

31. Ho, R. M., Chen, C. K., Chiang, Y. W. *Macromol. Rapid Commun.*, **30**, 1439 (2009).

32. Miyakoshi, R., Yokoyama, A., Yokozawa, T. *J. Polym. Sci., Part A: Polym. Chem.*, **46**, 753 (2008).

33. Yokozawa, T., Ajioka, N., Yokoyama, A. *Adv. Polym. Sci.*, **217**, 1 (2008).

34. Yokozawa, T., Yokoyama, A. *Chem. Rev.*, **109**, 5595 (2009).

35. Higashihara, T., Ueda, M. *Macromol. Res.*, **21**, 257 (2013).

36. Wang, J., Higashihara, T. *Polym. Chem.*, **4**, 5518 (2013).

37. Higashihara, T., Goto, E. *Polym. J.*, **46**, 381 (2014).

38. Higashihara, T., Ueda, M. *Complex Macromolecular Architecture*, eds. Hadjichristidis, N., Hirao, A., Tezuka, Y., Prez, F. D., John Wiley & Sons (Asia): Singapore, Chapter 13, p. 395 (2011).

39. Kaneto, K., Yoshino, K., Inuishi, Y. *Jpn. J. Appl. Phys.*, **21**, L567 (1982).

40. Jen, K. Y., Miller, G. G., Elsembaumer, R. L. *J. Chem. Soc., Chem. Commun.*, **17**, 1346 (1986).

41. Sugimoto, R., Takeda, S., Gu, H. B., Yoshino, K. *Chem. Express*, **1**, 635 (1986).

42. Roncali, J. *Chem. Rev.*, **92**, 711 (1992).

43. Sirringhaus, H., Tessler, N., Friend, R. H. *Science*, **280**, 1741 (1998).

44. Sirringhaus, H., Brown, P. J., Friend, R. H., Nielsen, M. M., Bechgaard, K., Langeveld-Voss, B. M. W., Spiering, A. J. H., Janssen, R. A. J., Meijer, E. W., Herwig, P., de Leeuw, D. M. *Nature*, **401**, 685 (1999).

45. Zaumseil, J., Sirringhaus, H. *Chem. Rev.*, **107**, 1296 (2007).

46. Mas-Torrent, M., Rovira, C. *Chem. Rev.*, **111**, 4833 (2011).

47. Wang, C., Dong, H., Hu, W., Liu, Y., Zhu, D. *Chem. Rev.*, **112**, 2208 (2012).

48. Li, G., Shrotriya, V., Huang, J., Yao, Y., Moriarty, T., Emery, K., Yang, Y. *Nat. Mater.*, **4**, 864 (2005).

49. Kim, J. Y., Lee, K., Coates, N. E., Moses, D., Nguyen, T.-Q., Dante, M., Heeger, A. J. *Science*, **317**, 222 (2007).

50. Günes, S., Neugebauer, H., Sariciftci, N. S. *Chem. Rev.*, **107**, 1324 (2007).

51. Woo, C. H., Thompson, B. C., Kim, B. J., Toney, M. F., Fréchet, J. M. J. *J. Am. Chem. Soc.*, **130**, 16324 (2008).
52. Dang, M. T., Hirsch, L., Wantz, G. *Adv. Mater.*, **23**, 3597 (2011).
53. McCullough, R. D., Lowe, R. D. *J. Chem. Soc., Chem. Commun.*, 70 (1992).
54. McCullough, R. D., Lowe, R. D., Jayaraman, M., Anderson, D. L. *Org. Chem.*, **58**, 904 (1993).
55. Chen, T. A., Rieke, R. D. *J. Am. Chem. Soc.*, **114**, 10089 (1992).
56. Chen, T. A., Wu, X., Rieke, R. D. *J. Am. Chem. Soc.*, **117**, 233 (1995).
57. McCullough, R. D., Ewbank, P. C., Loewe, R. S. *J. Am. Chem. Soc.*, **119**, 633 (1997).
58. Iraqi, A., Barker, G. W. *J. Mater. Chem.*, **8**, 25 (1998).
59. Guillerez, S., Bidan, G. *Synth. Met.*, **93**, 123 (1998).
60. Liversedge, I. A., Higgins, S. J., Giles, M., Heeney, M., McCulloch, I. *Tetrahedron Lett.*, **47**, 5143 (2006).
61. Amou, S., Haba, O., Shirota, K., Hayakawa, T., Ueda, M., Takeuchi, K., Asai, M. *J. Polym. Sci., Part A: Polym. Chem.*, **37**, 1943 (1999).
62. Yokoyama, A., Miyakoshi, R., Yokozawa, T. *Macromolecules*, **37**, 1169 (2004).
63. Miyakoshi, R., Yokoyama, A., Yokozawa, T. *J. Am. Chem. Soc.*, **127**, 17542 (2005).
64. Sheina, E. E., Liu, J., Iovu, M. C., Laird, D. W., McCullough, R. D. *Macromolecules*, **37**, 3526 (2004).
65. Iovu, M. C., Sheina, E. E., Gil, R. R., McCullough, R. D. *Macromolecules*, **38**, 8649 (2005).
66. Higashihara, T., Goto, E., Ueda, M. *ACS Macro Lett.*, **1**, 167 (2012).
67. Goto, E., Nakamura, S., Kawauchi, S., Mori, H., Ueda, M., Higashihara, T. *J. Polym. Sci., Part A: Polym. Chem.*, **52**, 2287 (2014).
68. Yokozawa, T., Suzuki, R., Nojima, M., Ohta, Y., Yokoyama, A. *Macromol. Rapid Commun.*, **32**, 801 (2011).
69. Liu, J., McCullough, R. D. *Macromolecules*, **35**, 9882 (2002).
70. Jeffries-EL, M., Sauvé, G., McCullough, R. D. *Adv. Mater.*, **16**, 1017 (2004).
71. Jeffries-EL, M., Sauvé, G., McCullough, R. D. *Macromolecules*, **38**, 10346 (2005).
72. Bronstein, H. A., Luscombe, C. K. *J. Am. Chem. Soc.*, **131**, 12894 (2009).
73. Smeets, A., Van den Bergh, K., Winter, J. D., Gerbaux, P., Verbiest, T., Koeckelberghs, G. *Macromolecules*, **42**, 7638 (2009).

74. Smeets, A., Willot, P., Winter, J. D., Gerbaux, P., Verbiest, T., Koeckelberghs, G. *Macromolecules*, **44**, 6017 (2011).

75. Monnaie, F., Brullot, W., Verbiest, T., Winter, J. D., Gerbaux, P., Smeets, A., Koeckelberghs, G. *Macromolecules*, **46**, 8500 (2013).

76. Kaul, E., Senkovskyy, V., Tkachov, R., Bocharova, V., Komber, H., Stamm, M., Kiriy, A. *Macromolecules*, **43**, 77 (2010).

77. Senkovskyy, V., Sommer, M., Tkachov, R., Komber, H., Huck, W. T. S., Kiriy, A. *Macromolecules*, **43**, 10157 (2010).

78. Fujita, H., Michinobu, T., Tokita, M., Ueda, M., Higashihara, T. *Macromolecules*, **45**, 9643 (2012).

79. Handa, N. V., Serrano, A. V., Robb, M. J., Hawker, C. J. *J. Polym. Sci., Part A: Polym. Chem.*, **53**, 831 (2015).

80. Yokozawa, T., Adachi, I., Miyakoshi, R., Yokoyama, A. *High Perform. Polym.*, **19**, 684 (2007).

81. Ohshimizu, K., Ueda, M. *Macromolecules*, **41**, 5289 (2008).

82. Zhang, Y., Tajima, K., Hirota, K., Hashimoto, K. *J. Am. Chem. Soc.*, **130**, 7812 (2008).

83. Van den Bergh, K., Huybrechts, J., Verbiest, T., Koeckelberghs, G. *Chem. Eur. J.*, **14**, 9122 (2008).

84. Wu, P. T., Ren, G., Li, C., Mezzenga, R., Jenekhe, S. A. *Macromolecules*, **42**, 2317 (2009).

85. Yamada, I., Takagi, K., Hayashi, Y., Soga, T., Shibata, N., Toru, T. *Int. J. Mol. Sci.*, **11**, 5027 (2010).

86. Kim, J., Song, I. Y., Park, T. *Chem. Commun.*, **47**, 4697 (2011).

87. Hammer, B. A. G., Bokel, F. A., Hayward, R. C., Emrick, T. *Chem. Mater.*, **23**, 4250 (2011).

88. Ho, C. C., Liu, Y. C., Lin, S. H., Su, W. F. *Macromolecules*, **45**, 813 (2012).

89. Suspène, C., Miozzo, L., Choi, J., Gironda, R., Geffroy, B., Tondelier, D., Bonnassieux, Y., Horowitz, G., Yassar, A. *J. Mater. Chem.*, **22**, 4511 (2012).

90. Yao, K., Chen, L., Chen, X., Chen, Y. *Chem. Mater.*, **25**, 897 (2013).

91. Chen, X., Chen, L., Yao, K., Chen, Y. *Appl. Mater. Interfaces*, **5**, 8321 (2013).

92. Ghoos, T., Van den Brande, N., Defour, M., Brassinne, J., Fustin, C. A., Gohy, J. F., Hoeppener, S., Schubert, U. S., Vanormelingen, W., Lutsen, L., Vanderzande, D. J., Mele, B. V., Maes, W. *Eur. Polym. J.*, **53**, 206 (2014).

93. Thomas, A., Houston, J. E., Van den Brande, N., Winter, J. D., Chevrier, M., Heenan, R. K., Terry, A. E., Richeter, S., Mehdi, A., Mele, B. V. *Polym. Chem.*, **5**, 3352 (2014).

94. Higashihara, T., Ohshimizu, K., Ryo, Y., Sakurai, T., Takahashi, A., Nojima, S., Ree, M., Ueda, M. *Polymer*, **52**, 3687 (2009).

95. Lai, Y. C., Ohshimizu, K., Takahashi, A., Hsu, J. C., Higashihara, T., Ueda, M., Chen, W. C. *J. Polym. Sci., Part A: Polym. Chem.*, **49**, 2577 (2011).

96. Tu, G., Li, H., Forster, M., Heiderhoff, R., Balk, L. J., Sigel, R., Scherf, U. *Small*, **3**, 1001 (2007).

97. Scherf, U., Gutacker, A., Koenen, A. N. *Acc. Chem. Res.*, **41**, 1086 (2008).

98. Miyakoshi, R., Yokoyama, A., Yokozawa, T. *Chem. Lett.*, **137**, 1022 (2008).

99. Hollinger, J., Jahnke, A. A., Coombs, N., Seferos, D. S. *J. Am. Chem. Soc.*, **132**, 8546 (2010).

100. Erdmann, T., Back, J., Tkachov, R., Ruff, A., Voit, B., Ludwigs, S., Kiriy, A. *Polym. Chem.*, **5**, 5383 (2014).

101. Zhu, Y. Y., Yin, T. T., Yin, J., Liu, N., Yu, Z. P., Zhu, Y. W., Ding, Y. S., Yin, J., Wu, Z. Q. *RSC Adv.*, **4**, 40241 (2014).

102. Takahashi, A., Lin, C. J., Ohshimizu, K., Higashihara, T., Chen, W. C., Ueda, M. *Polym. Chem.*, **3**, 479 (2012).

103. Verswyvel, M., Steverlynck, J., Mohamed, S. H., Trabelsi, M., Champagne, B., Koeckelberghs, G. *Macromolecules*, **47**, 4668 (2014).

104. Ouhib, F., Khoukh, A., Ledeuil, J. B., Martinez, H., Desbrieres, J., Dagron-Lartigau, C. *Macromolecules*, **41**, 9736 (2008).

105. Miyanishi, S., Zhang, Y., Hashimoto, K., Tajima, K. *Macromolecules*, **45**, 6424 (2012).

106. Mulherin, R. C., Jung, S., Huetter, S., Johnson, K., Kohn, P., Sommer, M., Allard, S., Scherf, U., Greenham, N. C. *Nano Lett.*, **11**, 4846 (2011).

107. Ku, S. Y., Brady, M. A., Treat, N. D., Cochran, J. E., Robb, M. J., Kramer, E. J., Chabinyc, M. L., Hawker, C. J. *J. Am. Chem. Soc.*, **134**, 16040 (2012).

108. Nakabayashi, K., Mori, H. *Macromolecules*, **45**, 9618 (2012).

109. Wang, J., Ueda, M., Higashihara, T. *ACS Macro Lett.*, **2**, 506 (2013).

110. Wang, J., Ueda, M., Higashihara, T. *J. Polym. Sci., Part A: Polym. Chem.*, **52**, 1139 (2014).

111. Willot, P., Moerman, D., Leclere, P., Lazzaroni, R., Baeten, Y., Van der Auweraer, M., Koeckelberghs, G. *Macromolecules*, **47**, 6671 (2014).

112. Iovu, M. C., Jeffries-EL, M., Sheina, E. E., Cooper, J. R., McCullough, R. D. *Polymer*, **46**, 8582 (2005).

113. Iovu, M. C., Craley, C. R., Jeffries-EL, M., Krankowski, A. B., Zhang, R., Kowalewski, T., McCullough, R. D. *Macromolecules*, **40**, 4733 (2007).

114. Craley, C. R., Zhang, R., Kowalewski, T., McCullough, R. D., Stefan, M. *Macromol. Rapid Commun.*, **30**, 11 (2009).

115. Zou, J., Khondaker, S. I., Huo, Q., Zhai, L. *Adv. Funct. Mater.*, **19**, 479 (2009).

116. Urien, M., Erothu, H., Cloutet, E., Hiorns, R. C., Vignau, L., Cramail, H. *Macromolecules*, **41**, 7033 (2008).

117. Erothu, H., Kolomanska, J., Johnston, P., Schumann, S., Deribew, D., Toolan, D. T. W., Gregori, A., Dagron-Lartigau, C., Portale, G., Bras, W., Arnold, T., Distler, A., Hiorns, R. C., Mokarian-Tabari, P., Collins, T. W., Howse, J. R., Topham, P. D. *Macromolecules*, **48**, 2107 (2015).

118. Richard, F., Brochon, C., Leclerc, N., Eckhardt, D., Heiser, T., Hadziioannou, G. *Macromol. Rapid Commun.*, **29**, 885 (2008).

119. Lee, J. U., Cirpan, A., Emrick, T., Russell, T. P., Jo, W. H. *J. Mater. Chem.*, **19**, 1483 (2009).

120. Radano, C. P., Scherman, O. A., Stingelin-Stutzmann, N., Mueller, C., Breiby, D. W., Smith, P., Janssen, R. A. J., Meijer, E. W. *J. Am. Chem. Soc.*, **127**, 12502 (2005).

121. Patel, S. N., Javier, A. E., Beer, K. M., Pople, J. A., Ho, V., Segalman, R. A., Balsra, N. P. *Nano Lett.*, **12**, 4901 (2012).

122. Ohshimizu, K., Takahashi, A., Higashihara, T., Ueda, M. *J. Polym. Sci., Part A: Polym. Chem.*, **49**, 2709 (2011).

123. Dai, C. A., Yen, W. C., Lee, Y. H., Ho, C. C., Su, W. F. *J. Am. Chem. Soc.*, **129**, 11036 (2007).

124. Higashihara, T., Ohshimizu, K., Hirao, A., Ueda, M. *Macromolecules*, **41**, 9505 (2008).

125. Higashihara, T., Ueda, M. *React. Funct. Polym.*, **69**, 457 (2009).

126. Moon, H. C., Anthonysamy, A., Kim, J. K., Hirao, A. *Macromolecules*, **44**, 1894 (2011).

127. Higashihara, T., Ueda, M. *Macromolecules*, **42**, 8794 (2009).

128. Tsai, J. H., Lai, Y. C., Higashihara, T., Lin, C. J., Ueda, M., Chen, W. C. *Macromolecules*, **43**, 6085 (2010).

129. Higashihara, T., Liu, C. L., Chen, W. C., Ueda, M. *Polymers*, **3**, 236 (2011).

130. Khanduyeva, N., Senlovskyy, V., Beryozkina, T., Bocharova, V., Simon, F., Nitschke, M., Stamm, M., Grötzschel, R., Kiriy, A. *Macromolecules*, **41**, 7383 (2008).

131. Khanduyeva, N., Senlovskyy, V., Beryozkina, T., Horecha, M., Stamm, M., Uhrich, C., Riede, M., Leo, K., Kiriy, A. *J. Am. Chem. Soc.*, **131**, 153 (2009).

132. Ahn, S. K., Pickel, D. L., Kochemba, W. M., Chen, J., Uhrig, D., Hinestrosa, J. P., Carrillo, J. M., Shao, M., Do, C., Messman, J. M., Brown, W. M., Sumpter, B. G., Kilbey II, S. M. *ACS Macro Lett.*, **2**, 761 (2013).

133. Sivanandan, K., Chatterjee, T., Treat, N., Kramer, E. J., Hawker, C. J. *Macromolecules*, **43**, 233 (2010).

134. Wang, J., Lu, C., Mizobe, T., Ueda, M., Chen, W. C., Higashihara, T. *Macromolecules*, **46**, 1783 (2013).

135. Zeigler, D. F., Mazzio, K. A., Luscombe, C. K. *Macromolecules*, **47**, 5019 (2014).

136. Wang, F., Rauh, R. D., Rose, T. L. *J. Am. Chem. Soc.*, **119**, 11106 (1997).

137. Bras, J., Guillerez, S., Pépin-Donat, B. *Chem. Mater.*, **12**, 2372 (2000).

138. Kim, H., Lee, Y., Hwang, S., Choi, D., Yang, H., Baek, K. *J. Polym. Sci., Part A: Polym. Chem.*, **49**, 4221 (2011).

139. Senkovskyy, V., Beryozkina, T., Bocharova, V., Tkachov, R., Komber, H., Lederer, A., Stamm, M., Sevrin, N., Rabe, J. P., Kiriy, A. *Macromol. Symp.*, **291–292**, 17 (2010).

140. Yuan, M., Okamoto, K., Bronstein, H. A., Luscombe, C. K. *ACS Macro Lett.*, **1**, 392 (2012).

141. Park, J., Moon, H. C., Kim, J. K. *J. Polym. Sci., Part A: Polym. Chem.*, **51**, 2225 (2013).

142. Higashihara, T., Ito, S., Fukuta, S., Koganezawa, T., Ueda, M., Ishizone, T., Hirao, A. *Macromolecules*, **48**, 245 (2015).

Biography

Tomoya Higashihara was born in Kagawa, Japan, and received his BS, MS, and PhD degrees from the Tokyo Institute of Technology, Japan,

in 2000, 2002, and 2005, respectively, under the supervision of Prof. Akira Hirao. He won the Japan Society for the Promotion of Science (JSPS) fellowship DC2 and PD in 2004 and 2005, respectively. During 2005–2008, he joined the postdoctoral research program at the University of Massachusetts (UMASS) Lowell, USA, under the supervision of Prof. Rudolf Faust. He then returned to the Tokyo Institute of Technology promoted to assistant professor in Prof. Mitsuru Ueda's laboratory in 2008. During 2010–2013, he joined the Precursory Research for Embryonic Science and Technology (PRESTO) program "Photoenergy Conversion Systems and Materials for the Next-Generation Solar Cells" supported by the Japan Science and Technology Agency (JST). In 2013, he was promoted to associate professor of the Innovative Flex Course for Frontier Organic Material Systems (iFront) at Yamagata University. In 2015, he obtained the permanent position of associate professor at the Department of Organic Device Engineering at Yamagata University. He has published 198 papers/review articles, 12 books, and 30 patents. His research interests include living/controlled polymerization, polythiophene-based materials, π-conjugated polymers, and polymer solar cells.

Chapter 7

Through-Space-Conjugated Compounds

Yasuhiro Morisaki
Department of Applied Chemistry for Environment, School of Science and Technology, Kwansei Gakuin University, 2-1 Gakuen, Sanda, Hyogo 669-1337, Japan
ymo@kwansei.ac.jp

Charges and energy (excitons) are migrated and transferred through π-conjugated compounds by a hopping mechanism. A large number of π-conjugated compounds, σ-conjugated compounds, cross-conjugated compounds, and so on have been synthesized. However, few studies have focused on the synthesis of through-space-conjugated polymers and oligomers consisting of layered π-electron systems in their main chains. The synthesis of a through-space-conjugated polymer with a stacked structure of π-electron systems has rarely been carried out in the field of synthetic organic chemistry despite the fact that π-stacked structures are the common structure. Through-space-conjugated polymers and oligomers are π-stacked compounds, in which π-electron systems such as π-electron systems are layered in the face-to-face manner. This chapter focuses on the synthesis of through-space-conjugated compounds, in particular, by taking advantage of [2.2]paracyclophane compounds. First, synthetic routes of such [2.2]paracyclophane-based through-space-

Conjugated Objects: Developments, Synthesis, and Applications
Edited by Atsushi Nagai and Koji Takagi

ISBN 978-981-4774-03-1 (Hardcover), 978-1-315-22962-1 (eBook)
www.panstanford.com

conjugated polymers are shown. Then, through-space conjugation is discussed by the oligomeric approach. Application of the through-space-conjugated systems in single molecular wires, in which energy and charges are transferred, is introduced. Finally, the difference between pseudo-*para*- and pseudo-*ortho*-disubstituted [2.2]paracyclophane-based through-space-conjugated systems is discussed.

7.1 Introduction

Conjugated polymers and oligomers have attracted considerable attention in organic chemistry, polymer chemistry, and material chemistry, due to their potential for applications in optoelectronic devices, for example, electroluminescence (EL) devices[1], field-effect transistors (FETs)[2], and photovoltaic cells[3]. They exhibit the favorable characteristic of versatile processability, that is, thin-film-forming ability by spin-coating, spray-coating, and casting methods, in addition to the inherent features, such as light weight and flexibility, of organic polymers and oligomers. In these materials, charges and energy (excitons) are migrated and transferred in some layers from one side to the opposite side through a conjugated backbone by a hopping mechanism. A large number of π-conjugated compounds, σ-conjugated compounds, and cross-conjugated compounds[4] have been synthesized till date. However, few studies have focused on the synthesis of through-space-conjugated polymers and oligomers consisting of layered π-electron systems in the single chain. It is worth noting that layered or stacked π-electron systems are commonly found in nature. For example, in DNA, layered aromatic units can be found[5]; the π-stacked array of base pairs in double-stranded DNA plays an important role to stabilize their higher-ordered structures with hydrogen bonds. In photosynthetic systems, light-harvesting antenna complexes contain regular arrays of bacteriochlorophylls and chlorophylls[6]. Layered π-electron systems are found in various crystals as well as discotic liquid crystals[7,8]. Thus, the synthesis of a through-space-conjugated polymer with a stacked structure of π-electron systems has rarely been carried out in the field of synthetic organic chemistry despite the fact that π-stacked structures are the common structure.

Through-space-conjugated polymers and oligomers are π-stacked compounds, in which π-electron systems such as aromatic rings and/or π-electron systems are layered in the face-to-face manner. This chapter focuses on the synthesis of through-space-conjugated compounds, in particular, by taking advantage of [2.2] paracyclophane compounds[9–12].

7.2 Synthesis of [2.2]Metacyclophane-Based Through-Space-Conjugated Compounds

A through-space-conjugated polymer consisting of a cyclophane skeleton as a repeating unit in the main chain was reported in 1985[13,14]; 8,16-dihydroxy[2.2]metacyclophane **1** was incorporated as the repeating unit, as shown in Scheme 7.1. Metacyclophane **1** is electron rich; therefore, oxidative coupling with air (oxygen) under sunlight occurs to produce metacyclophane dimer **2**. The dimer was heated at 300°C–400°C to afford poly[2.2]metacyclophane **3**. The formed polymer **3** was insoluble in any organic solvents; thus, the structure was confirmed by Fourier transform infrared (FTIR) spectroscopy. The electrical conductivity of polymer **3** was estimated to be 10^{-9} S cm^{-1}, which increased dramatically to 0.25 S cm^{-1} by chemical doping with H_2SO_4.

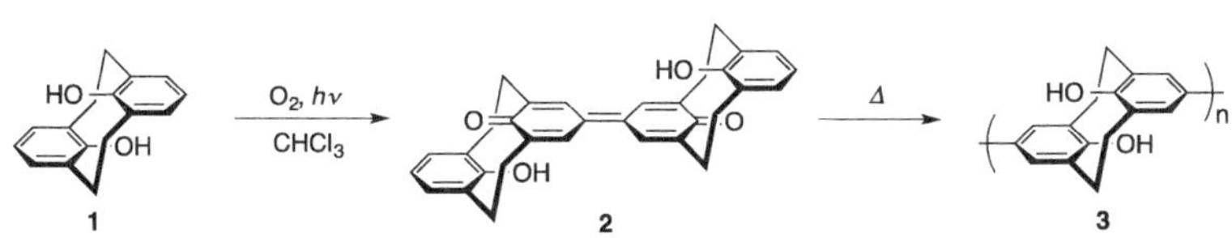

Scheme 7.1 Synthesis of poly[2.2]metacyclophane.

The synthesis of through-space-conjugated polymer **4**, in which diphenoquinone units were stacked repeatedly, was reported in 1985[14]. As shown in Scheme 7.2, the oxidative coupling polymerization of **1** using $FeCl_3$ produced diphenoquinone-stacked polymer **4**. This polymer was also characterized by FTIR spectroscopy as well as elemental analysis insoluble due to insolubility in common organic solvents. The electrical conductivity of **4** was measured, and the value for the pristine sample was 10^{-9} S cm^{-1}. Chemical doping with H_2SO_4 vapor led to increase of conductivity up to 10^{-4} S cm^{-1}.

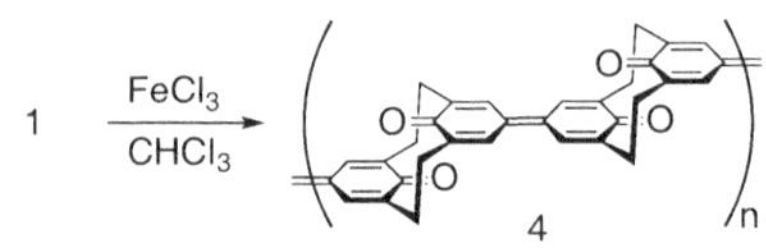

Scheme 7.2 Synthesis of diphenoquinone-stacked polymer.

7.3 Synthesis of [2.2]Paracyclophane-Based Through-Space-Conjugated Polymers

Pseudo-*p*-disubstituted [2.2]paracyclophanes can be employed as a building block for syntheses of through-space-conjugated polymers[15–19]. Palladium-catalyzed cross-coupling has been used for the synthesis of a wide variety of conjugated polymers; thus, Sonogashira–Hagihara cross-coupling, Heck cross-coupling, and Suzuki–Miyaura cross-coupling were used for polymer syntheses. Schemes 7.3–7.5 show the synthetic routes of [2.2]paracyclophane-containing poly(*p*-arylene-ethynylene) **7**[20], poly(*p*-arylenevinylene) **10**[21], and poly(*p*-arylene) **13**[22]. Various [2.2]paracyclophane-containing through-space-conjugated polymers[15–19] have been synthesized by using pseudo-*p*-disubstituted [2.2]paracyclophanes, such as pseudo-*p*-diethynyl [2.2]paracyclophane, pseudo-*p*-divinyl [2.2]paracyclophane, and pseudo-*p*-dibromo [2.2]paracyclophane. These through-space-conjugated polymers were well-soluble in common organic solvents due to the long alkyl chains, leading to thin-film-forming ability by casting and spin coating. Generally, the polymers exhibited good thermal stability[22].

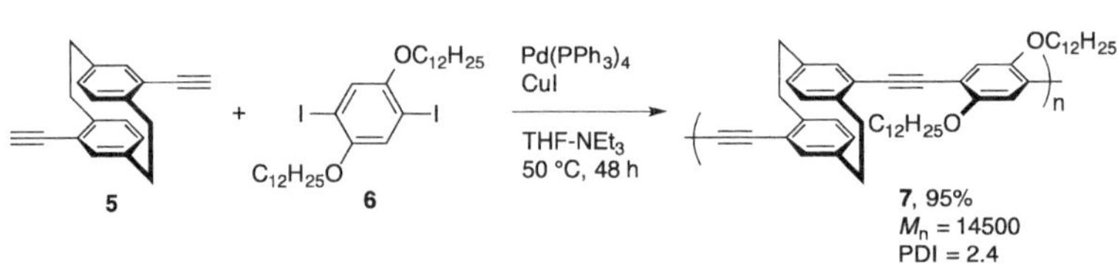

Scheme 7.3 Synthesis of [2.2]paracyclophane-containing poly(*p*-arylene-ethynylene).

Polymerizations of thiophene-substituted [2.2]paracyclophanes were independently reported by two groups[23–26]. π-Stacked polymers were synthesized by electrochemical polymerization;

for example, Audebert and coworkers reported the synthesis of polymer **15** by the electrochemical polymerization of monomer **14** (Scheme 7.6), which was precipitated on an electrode for cyclic voltammetry[23,24]. Collard and coworkers synthesized polymer **17** by the electrochemical oxidation as well as chemical oxidation using $FeCl_3$ (Scheme 7.7)[25,26]. The long alkyl chain at the thiophene rings provided the solubility of the polymer in organic solvents.

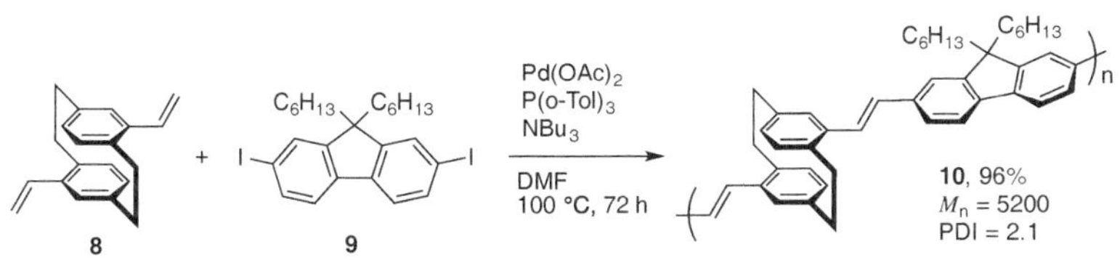

Scheme 7.4 Synthesis of [2.2]paracyclophane-containing poly(*p*-arylenevinylene).

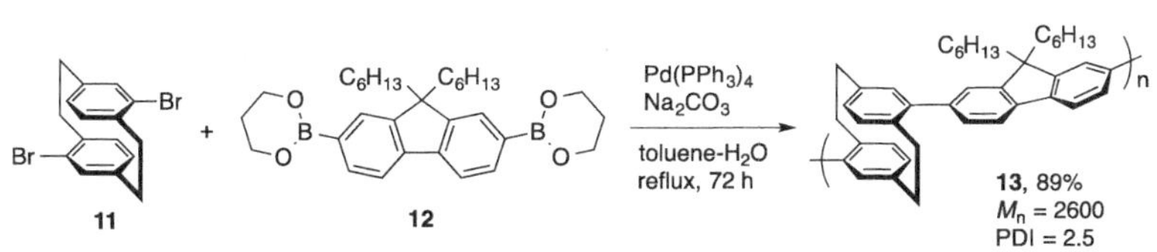

Scheme 7.5 Synthesis of [2.2]paracyclophane-containing poly(*p*-arylene).

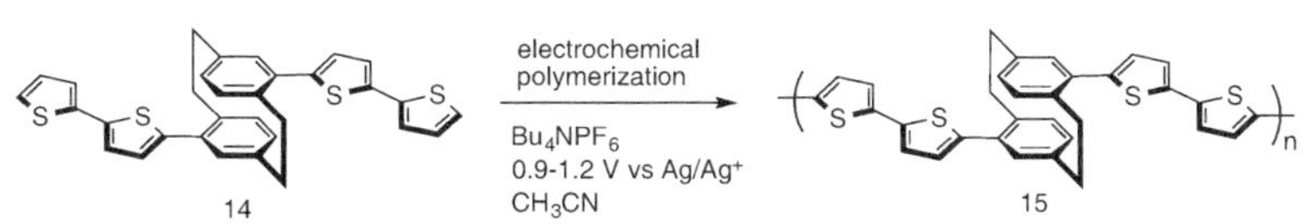

Scheme 7.6 Electrochemical polymerizations of thiophene-substituted [2.2] paracyclophane.

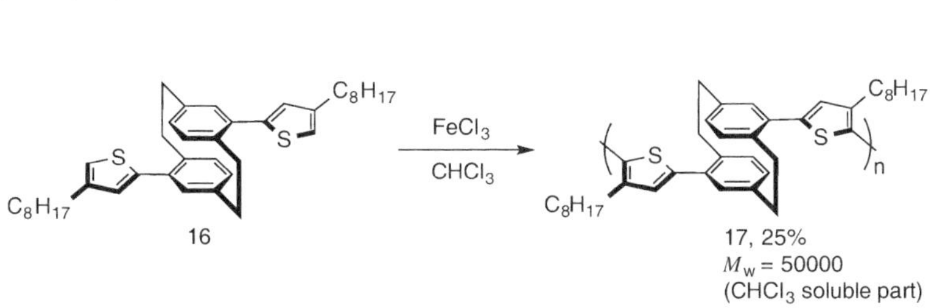

Scheme 7.7 Oxidative polymerizations of thiophene-substituted [2.2] paracyclophane.

7.4 [2.2]Paracyclophane-Based Conjugated Oligomers for Understanding Through-Space Conjugation

As shown in Fig, 7.1, [2.2]paracyclophane constructs new π-electron systems that are constrained to stack on each other in the oligomer and polymer chains. In general, almost all the conjugated polymers reported to date are through-bond conjugated polymers consisting of sp and/or sp^2 carbon frameworks. Through-space-conjugated polymers consisting of stacked π-electron systems have rarely been synthesized. Therefore, it is important to understand the characteristics of the through-space-conjugated system. From this viewpoint, [2.2]paracyclophane-containing through-space-conjugated oligomers as well as well-defined polymers with a narrow polydispersity index (PDI) based on poly(*p*-arylene-ethynylene) were prepared.

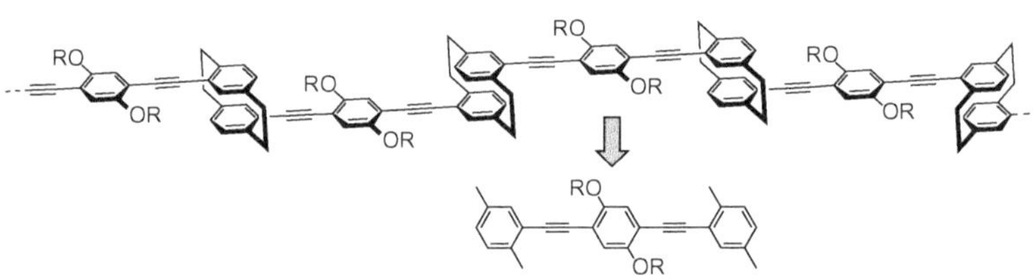

Figure 7.1 Structure of [2.2]paracyclophane-containing poly(*p*-arylene-ethynylene).

Through-space-conjugated oligomers **18–20** and a polymer were prepared by Sonogashira–Hagihara cross-coupling as the key reaction[20]. In addition, the polymer was separated into four fractions by liquid chromatography. Figure 7.2 shows the structures of oligomers **18–20**, separated polymers **21–24** with a narrow PDI, and polymer **7**. As shown in Fig. 7.2, compound **18** is partially stacked in oligomers **2**, **3**, and polymers (and also see Fig. 7.1).

Figure 7.3 shows the normalized UV-Vis absorption spectra of **18–20** in dilute $CHCl_3$ (1.0×10^{-5} M). All compounds exhibited typical π–π* transition bands of the *p*-arylene-ethynylene moiety. The peak top wavelengths of **18–20** were found to be 371, 378, and 381 nm, respectively, and those of the polymers were red-shifted as the number of the layered *p*-arylene-ethynylene units increased. Such a phenomenon is the same as the general through-bond conjugation system; therefore, we can conclude that the partially overlapped structure of the *p*-arylene-ethynylenes exhibited expansion of the conjugation, namely the through-space conjugation.

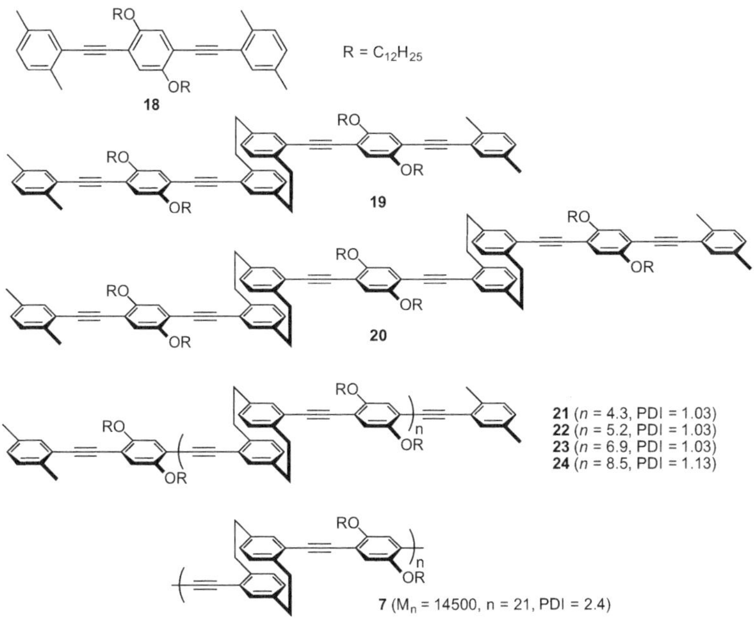

Figure 7.2 Structures of [2.2]paracyclophane-containing oligomers and polymers.

It is well-known that the absorption maximum λ_{max} (E_g cm^{-1}) of conjugated polymers with sufficiently long polymer chain is estimated from the relationship between the number of repeating units and λ_{max} (E_g cm^{-1}) for the well-defined π-conjugated oligomers[27]. The relationship was applied to the through-space-conjugated oligomers and polymer. As shown in Fig. 7.4, the peak top energies (E_g cm^{-1}) were plotted against the *n* inverse. The linear correlation was obtained with the coefficient of determination (R^2) of 0.99. When the polymer chain is sufficiently long (infinite layered units, $n = \infty$), the peak top energy of the through-space-conjugated poly(*p*-arylene-ethynylene) should reach unity E_g = 25,800 cm^{-1}. This result is in good agreement with the experimental peak top maximum (387 nm) for polymer **7** with M_n = 14,500 (approximately 21-layered *p*-arylene-ethynylenes). It is known that the electronic interactions of the stacked aromatic units in the π-stacked polymers extend to approximately 5 units for poly(dibenzofulvene)[28–31] and 7,7-diphenylnorbornane-based polymers[32–34]. In these polymers, π-stacked and unstacked conformations exist in a single polymer chain; therefore, the 4–5 aromatic units are uniformly stacked in solution. On the other hand, in the [2.2]paracyclophane-based through-space-conjugated compounds, *p*-arylene-ethynylene units are fixed in proximity, which leads to a conformationally stable π-stacked structure. The effective conjugation repeating unit number (n) was estimated to be 11.4 by the empirical power-law relationship ($E = 1/\lambda_{max} \approx n^{-x}$). The through-space-conjugated polymer improved through-space conjugation effects. The electronic interactions are effective through 11 stacked *p*-arylene-ethynylene moieties in the ground state.

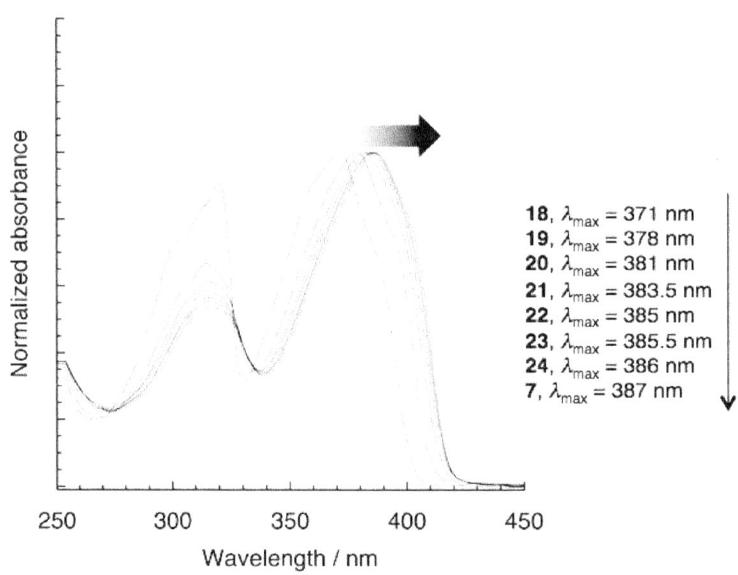

Figure 7.3 UV-Vis absorption spectra of [2.2]paracyclophane-containing oligomers and polymers in $CHCl_3$ solution.

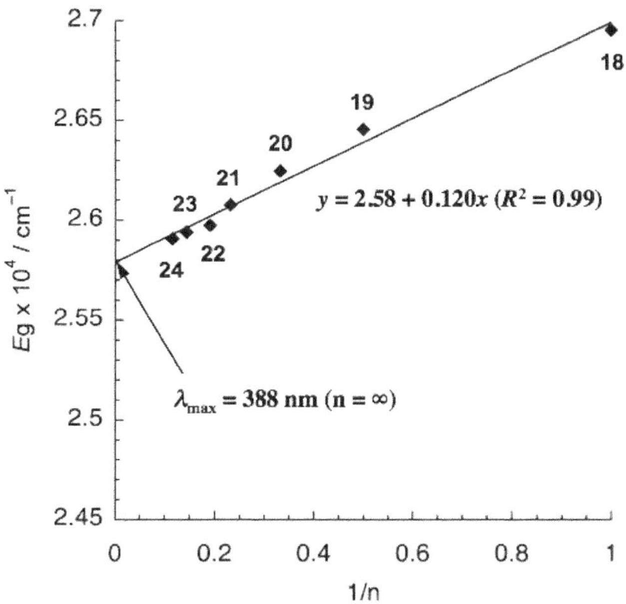

Figure 7.4 Correlation of the λ_{max} of oligomers and polymers with the inverse number of repeating units.

7.5 [2.2]Paracyclophane-Based Through-Space-Conjugated Polymers for Energy Transfer

Through-space-conjugated polymers **25** and **26** with xylene- and anthracene-end-capping groups, respectively, were prepared by Sonogashira–Hagihara coupling, as shown in Scheme 7.8[20]. The PL spectra of **25** and **26** in $CHCl_3$ solutions (1.0×10^{-5} M; excited at 385 nm) are shown in Fig. 7.5A. The emission peak of **25** showed a vibrational structure at around 410 nm, and the PL quantum efficiency (Φ_{PL}) was 0.82. This value is almost identical to the spectrum of **18** (Φ_{PL} = 0.86, Fig. 7.5B). The through-space-conjugated polymers emit from the monomer state (chromophore state) rather than the plane (excimer-like) state[35–37], and the emission of **25** occurs from each stacked π-electron system.

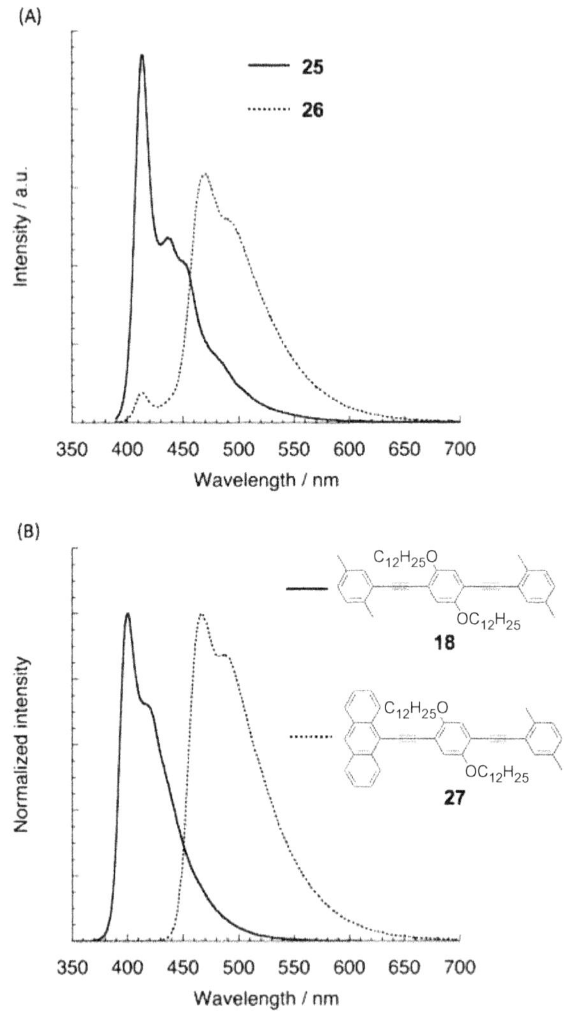

Figure 7.5 (A) PL spectra of polymers **25** and **26** in $CHCl_3$ excited at 385 nm. (B) PL spectra of compound **18** in $CHCl_3$ excited at 365 nm and compound **27** in $CHCl_3$ excited at 468 nm.

On the other hand, the spectrum of **26** with the anthracene-containing end-capping units was different from that of **25**. The emission at around 415 nm derived from the stacked π-electron systems clearly decreased, whereas an emission peak at around 470 nm with Φ_{PL} value of 0.70 was observed. This emission peak for **26** arose from the PL from the end-capping π-electron systems, which was identical to the PL spectrum of compound **27** (Fig. 7.5B). The time-resolved PL spectra of **26** exhibited that the emission from stacked π-electron systems decreased and that from the end-capping π-electron systems increased simultaneously (Fig. 7.6). Fluorescence resonance energy transfer (FRET)[38,39] occurred from the stacked π-electron systems to the end-capping π-electron systems.

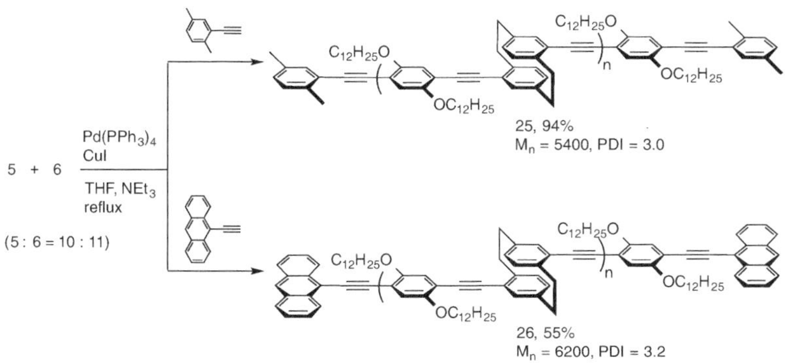

Scheme 7.8 Synthesis of [2.2]paracyclophane-containing poly(*p*-arylene-ethynylene) with end-capping groups.

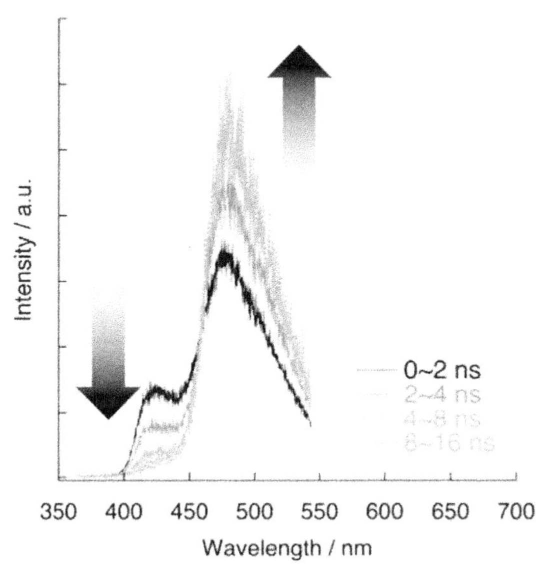

Figure 7.6 Time–resolved PL spectra of polymer **26** excited at 337 nm (N_2 laser) in $CHCl_3$.

7.6 [2.2]Paracyclophane-Based Through-Space-Conjugated Oligomers for Unidirectional Energy Transfer

In the case of polymer **26**, photoexcited energy was migrated and transferred to the both polymer chain ends. Through-space-conjugated compound **27** consisting of **M1**, **M2**, and **M3** moieties (Fig. 7.7) was synthesized to obtain a large overlap integral between an absorption band and a PL spectra of adjacent stacked π-electron systems[40–42]. It was designed for efficient FRET from the **M1** unit to the **M3** unit via the central **M2** moiety. Figure 7.8A shows the absorption and PL spectra of **27** in sufficiently diluted $CHCl_3$ solution. Compound **27** was excited at 290 nm considering that the **M1** moiety was mainly excited. The absorption spectrum of **27** consisted of the absorption spectra of three stacked π-electron systems (**18**, **28**, and **29**), whereas the PL spectrum was a single-component spectra with a broad peak top at 525 nm. This PL behavior was good agreement with that of **29**. Thus, the emissions from the **M1** and **M2** moieties were not observed, and only the **M3** moiety emitted.

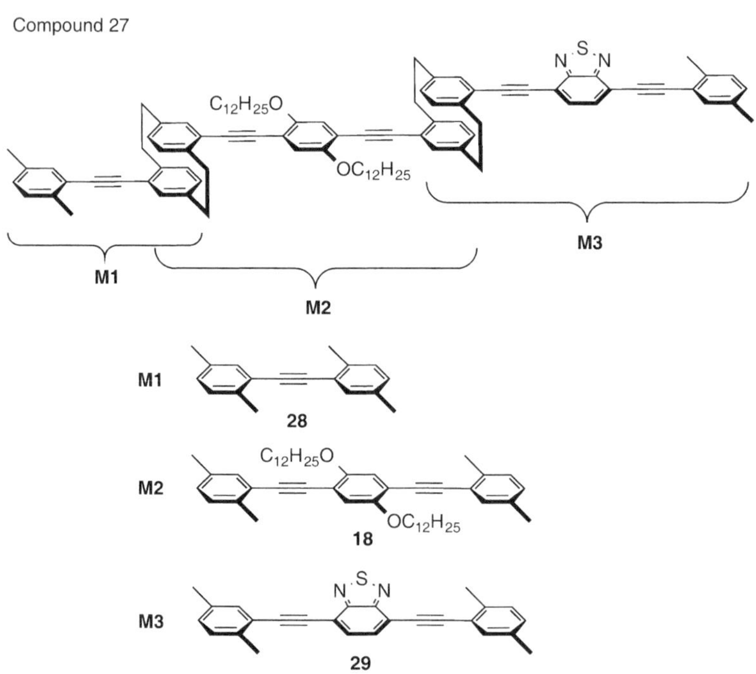

Figure 7.7 Structures compound **27** and stacked model compounds.

On the other hand, Fig. 7.8B shows the PL spectrum of a mixture of compounds 18, 28, and 29 in $CHCl_3$ solution (excitation wavelength = 290 nm), and each emission from the 18, 28, and 29 was clearly observed. By combining 18, 28, and 29 using the [2.2] paracyclophane skeleton, the emissions from the M1and M2 moieties were not observed from 27. These results suggest that photoexcited energy was transferred intramolecularly and unidirectionally from the **M1** and **M2** units to the **M3** moiety.

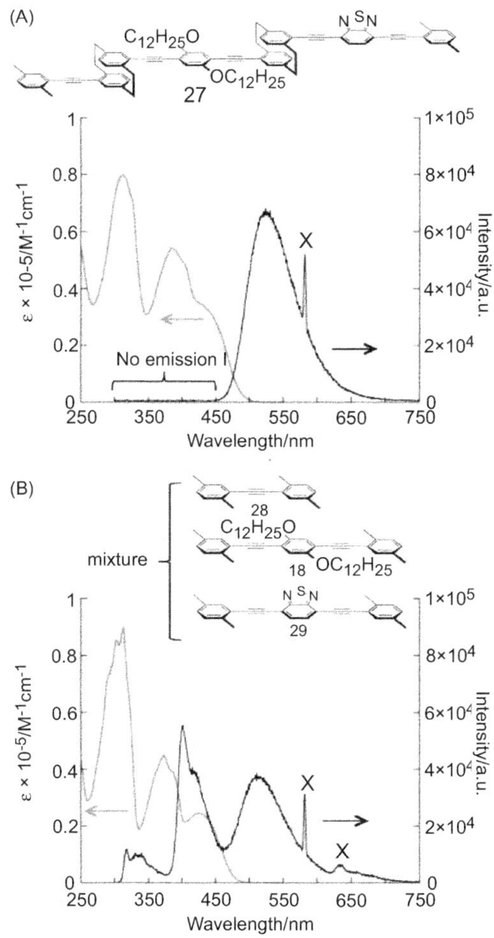

Figure 7.8 (A) UV-Vis absorption spectrum and PL spectrum of **27** in $CHCl_3$ (excited at 290 nm). (B) UV-Vis absorption and PL spectra of mixture of **28**, **18**, and **29** in $CHCl_3$ (excited at 290 nm). Letter X denotes scattered light and overtone peaks.

The time-dependent density functional theory (TD-DFT) revealed the transition dipole moments of each stacked π-electron unit in **T1**; all dipole moments were located on the long axis of the stacked π-electron system. The alignment of the dipole moments is theoretically favorable for FRET[38,39]. The orientation factors (κ^2) between the neighboring π-electron systems were calculated, and the values were 3.52 and 3.68 for the **M1/M2** and **M2/M3** pairs, respectively, which were sufficiently large values. This result indicates the dipole moments between the neighboring π-electron systems are ideally oriented in **T1**. The spectral overlap integrals for the **M1/M2** and **M2/M3** pairs were calculated from their spectra. In addition, the Förster radii (R_0) for the **M1/M2** and **M2/M3** pairs were estimated. From these results, the energy transfer efficiency (E_{ET}) and rate constants (k_{ET}) were calculated. Both the E_{ET} values (from **M1** to **M2** and from **M2** to **M3**) were over 0.999, and their k_{ET} values were calculated to be 1.10×10^{13} s^{-1} and 3.54×10^{12} s^{-1}, respectively[42].

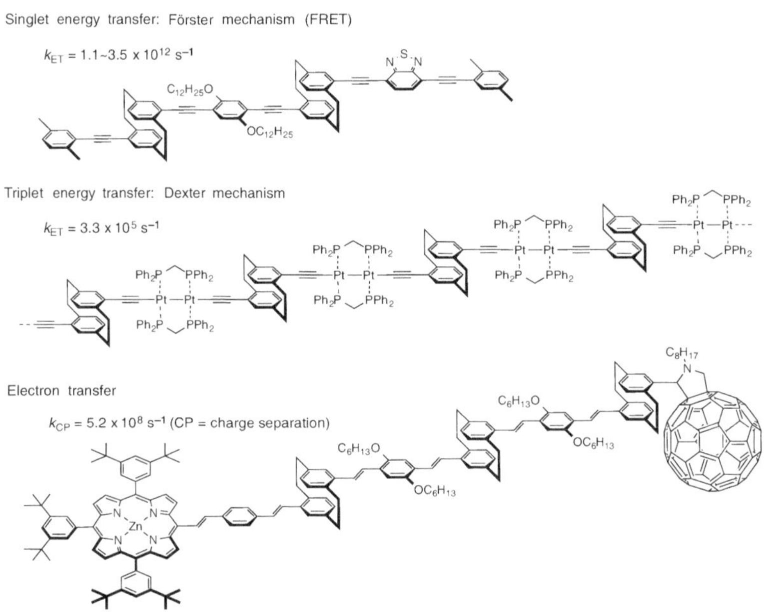

Figure 7.9 Structures of [2.2]paracyclophane-containing oligomers, including their energy transfer rate constants and charge separation rate constant.

Knorr and Harvey reported the Dexter-type energy transfer[43] using platinum-containing through-space-conjugated copolymers (Fig. 7.9)[44]. In this case, the k_{ET} value was calculated to be on the order of 10^4–10^5 s^{-1}. Guldi, Martín, and coworkers reported [2.2] paracyclophane-based through-space-conjugated oligomers (Fig. 7.9) containing zinc porphyrin and C60 at each oligomer chain end[45,46]. Photoexcited electron transfer (charge transfer) occurred with a rate constant of 10^8 s^{-1}.[6] According to these rate constant values, efficient FRET occurred mainly in **T1**. [2.2]Paracyclophane-based through-space-conjugated oligomers are promising candidates for single molecular wires that transfer energy as well as electron efficiently and unidirectionally.

7.7 Conjugation Length of Pseudo-*para*- and Pseudo-*ortho*-Disubstituted [2.2] Paracyclophanes

The absorption (UV) spectra of through-space-conjugated oligomers comprising pseudo-*ortho*-disubstituted [2.2]paracyclophane in dilute $CHCl_3$ (1.0×10^{-5} M) are shown in Fig. 7.10[47]. The absorption spectra were bathochromically shifted as the number of the stacked π-electron systems increased as a result of the through-space conjugation, which was the same manner as the oligomers containing pseudo-*para*-disubstituted [2.2]paracyclophane. The effect of the through-space conjugation of the pseudo-*ortho*-type oligomer was slightly smaller than that of the corresponding pseudo-*para*-type oligomer. For example, absorption maxima (λ_{max}) of **2** and **3** were 374.5 nm and 375.5 nm, respectively, whereas those of the pseudo-*para*-oligomers were 378 nm and 381 nm (Fig. 7.3), respectively. The DFT for the pseudo-*ortho*- and pseudo-*para*-dimer supported the results of UV spectral analysis (Fig. 7.11); namely, uniformity of phases as well as overlap of orbitals between face-to-face aromatic rings of the [2.2]paracyclophane unit reduced the HOMO/LUMO bandgap of the pseudo-*para*-dimer in comparison with the pseudo-*ortho*-dimer.

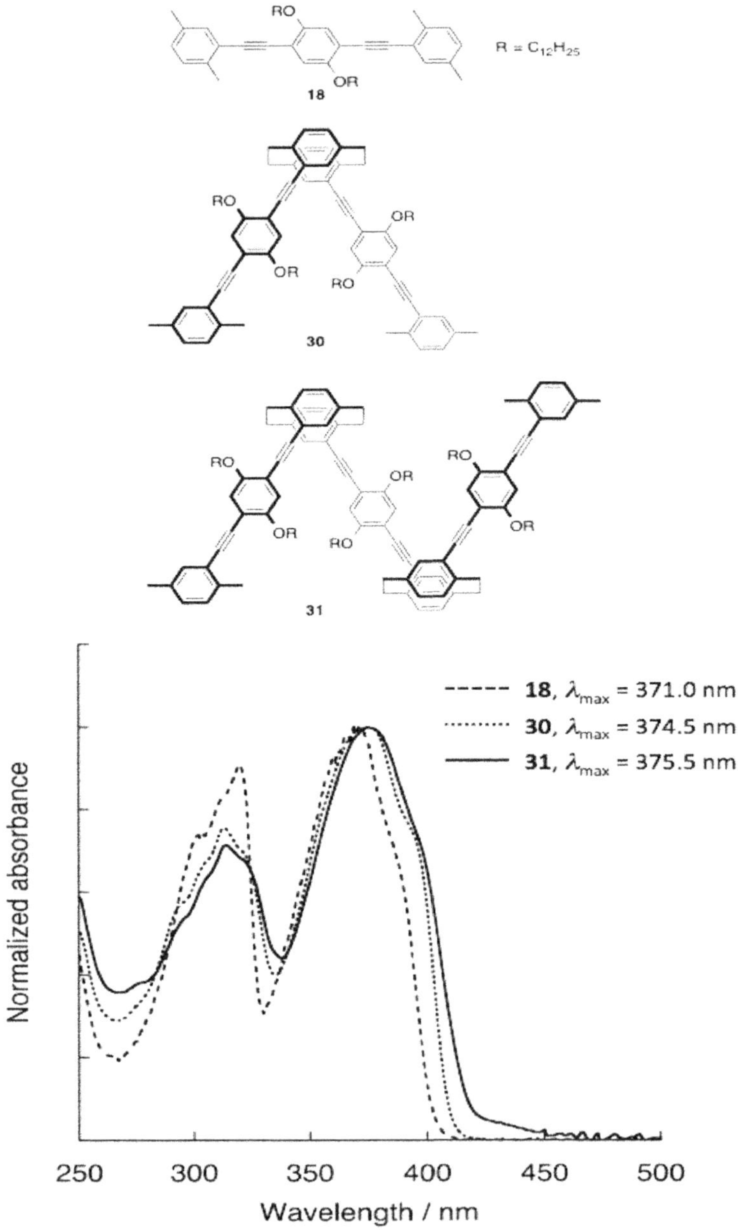

Figure 7.10 Structures of pseudo-ortho-di-substituted [2.2]paracyclophane-containing oligomers and their UV-Vis absorption spectra in $CHCl_3$.

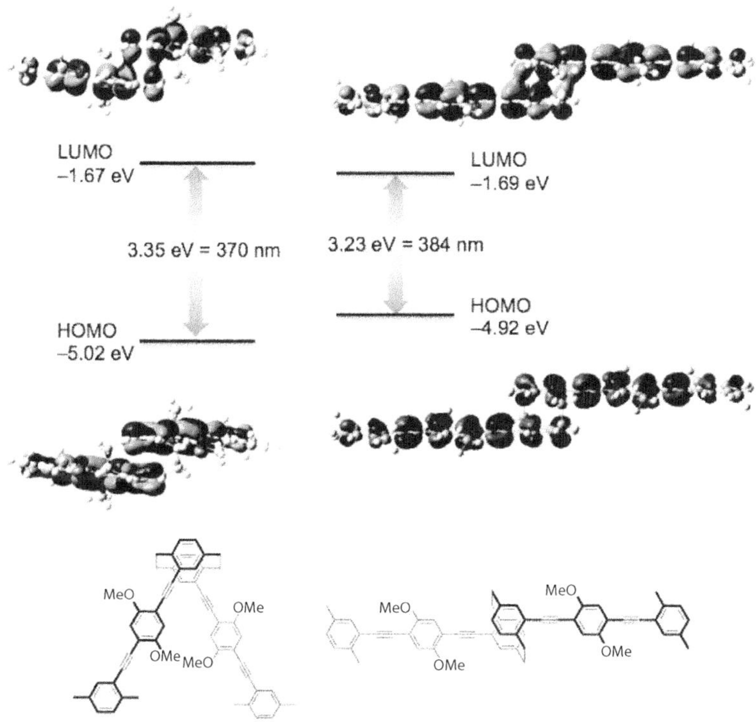

Figure 7.11 Density functional theory (DFT) calculations for pseudo-*ortho*- and pseudo-*para*-dimer at the B3LYP/6-31G(d) level.

7.8 Conclusions

This chapter mentioned the brief synthetic methods of through-space-conjugated polymers containing [2.2]paracyclophane in the polymer main chain and their fundamental properties, in addition to the oligomers. When the [2.2]paracyclophane unit is incorporated into the π-conjugated backbone, new π-electron systems are formed and stacked partially in the main chain in proximity. As the number of the stacked π-electron systems increased, absorption spectra are red-shifted, which is the same manner as the common through-bond π-conjugated compounds; this is called through-space conjugation. Through-bond π-conjugated compounds have a set of highest occupied molecular orbital (HOMO) (valence band) and lowest unoccupied molecular orbital (LUMO) (conduction band) energy bandgaps; on the other hand, in the through-space-conjugated compounds, each stacked π-electron system has a HOMO/LUMO energy bandgap. Various π-electron systems can be stacked using a [2.2]paracyclophane scaffold; therefore, photoexcited energy (Förster and Dexter mechanism) and charge transfer can be controlled by appropriate tuning of the bandgaps and energy levels of the stacked π-electron systems. Thus, [2.2]paracyclophane-containing through-space-conjugated molecules can act as a single molecular wire that flows energy and charges unidirectionally. This new class of through-space-conjugated polymers and oligomers make a fundamental contribution to the field of molecular electronics.

References

1. Müellen, K., Scherf, U., eds. *Organic Light Emitting Devices: Synthesis, Properties and Application*, Wiley-VCH: Weinheim, Germany (2006).
2. Groza, J. R., Locklin, J. J., eds. *Organic Field-Effect Transistors*, CRC Press Taylor & Francis: New York (2007).
3. Brabec, C., Dyakonov, V., Scherf, U., eds. *Organic Photovoltaics: Materials, Device Physics, and Manufacturing Technologies*, Wiley-VCH: Weinheim, Germany (2008).
4. Skotheim, T. A., Elsenbaumer, R. L., Reynolds, J. R., eds. *Handbook of Conducting Polymers*, 3rd ed., Marcel Dekker: New York (2006).
5. Wagenknecht, A., ed. *Charge Transfer in DNA: From Mechanism to Application*, Wiley: Weinheim, Germany (2005).
6. Roszak, A. W., Howard, T. D., Southall, J., Gardiner, A. T., Law, C. J., Isaacs, N. W., Cogdell, R. J. *Science*, **302**, 1969–1972 (2003).
7. Demus, D., Goodby, J. W., Gray, G. W., Spiess, H. W., Vill, V., eds. *Handbook of Liquid Crystals: Low Molecular Weight Liquid Crystals II*, Wiley-VCH: Weinheim, Germany (1998).
8. Herwig, P., Kayser, C. W., Müllen, K., Spiess, H. W. *Adv. Mater.*, **8**, 510–513 (1996).
9. Brown, C. J., Farthing, A. C. *Nature*, **164**, 915–916 (1949).
10. Cram, D. J., Steinberg, H. *J. Am. Chem. Soc.*, **73**, 5691–5704 (1951).
11. Vögtle, F. *Cyclophane Chemistry: Synthesis, Structures and Reactions*, John Wiley & Sons: Chichester (1993).

12. Gleiter, R., Hopf, H., eds. *Modern Cyclophane Chemistry*, Wiley-VCH: Weinheim (2004).

13. Mizogami, S., Yoshimura, S. *J. Chem. Soc. Chem. Commun.*, 427–428 (1985).

14. Mizogami, S., Yoshimura, S. *J. Chem. Soc. Chem. Commun.*, 1736–1738 (1985).

15. Morisaki, Y., Chujo, Y. *Angew. Chem. Int. Ed.*, **45**, 6430–6437 (2006).

16. Morisaki, Y., Chujo, Y. *Bull. Chem. Soc. Jpn.*, **82**, 1070–1082 (2009).

17. Morisaki, Y., Chujo, Y. *Prog. Polym. Sci.*, **33**, 346–364 (2008).

18. Morisaki, Y., Chujo, Y. *Polym. Chem.*, **2**, 1249–1257 (2011).

19. Morisaki, Y., Chujo, Y. *Chem. Lett.*, **41**, 840–846 (2012).

20. Morisaki, Y., Ueno, S., Saeki, A., Asano, A., Seki, S., Chujo, Y. *Chem. Eur. J.*, **18**, 4216–4224 (2012).

21. Morisaki, Y., Chujo, Y. *Macromolecules*, **37**, 4099–4103 (2004).

22. Morisaki, Y., Chujo, Y. *Bull. Chem. Soc. Jpn.*, **78**, 288–293 (2005).

23. Salhi, F., Lee, B., Metz, C., Bottomley, L. A., Collard, D. M. *Org. Lett.*, **4**, 3195–3193 (2002).

24. Salhi, F., Collard, D. M. *Adv. Mater.*, **15**, 81–85 (2003).

25. Guyard, L., Audebert, P. *Electrochem. Commun.*, **3**, 164–167 (2001).

26. Guyard, L., Audebert, P., Dolbier, W. R. Jr., Duan, J. X. *J. Electroanal. Chem.*, **537**, 189–193 (2002).

27. Müllen, K., Wegner, G., eds. *Electronic Materials: The Oligomer Approach*, Wiley-VCH: Weinheim (1998).

28. Nakano, T. *Polym. J.*, **42**, 103–123 (2010).

29. Nakano, T., Takewaki, K., Yade, T., Okamoto, Y. *J. Am. Chem. Soc.*, **123**, 9182–9183 (2001).

30. Nakano, T., Yade, T. *J. Am. Chem. Soc.*, **125**, 15474–15484 (2003).

31. Nakano, T., Yade, T., Yokoyama, M., Nagayama, N. *Chem. Lett.*, **33**, 296–297 (2004).

32. García Martínez, A., Osío Barcina, J., de Fresno Cerezo, A., Schlüter, A.-D., Frahn, J. *Adv. Mater.*, **11**, 27–31 (1999).

33. Osío Barcina, J., Colorado Heras, M. R., Mba, M., Gómez Aspe, R., Herrero-García, N. *J. Org. Chem.*, **74**, 7148–7156 (2009).

34. Caraballo-Martínez, N., del Rosario Colorado Heras, M., Mba Blázquez, M., Osío Barcina, J., García Martínez, A., del Rosario Torres Salvador, M. *Org. Lett.*, **9**, 2943–2946 (2007).

35. Oldham, W. J. Jr., Miao, Y. J., Lachicotte, R. J., Bazan, G. C. *J. Am. Chem. Soc.*, **120**, 419–420 (1998).
36. Bazan, G. C., Oldham, W. J. Jr., Lachicotte, R. J., Tretiak, S., Chernyak, V., Mukamel, S. *J. Am. Chem. Soc.*, **120**, 9188–9204 (1998).
37. Bazan, G. C. *J. Org. Chem.*, **72**, 8615–8635 (2007).
38. Förster, T. *Naturwissenschaften*, **33**, 166–175 (1946).
39. Lakowicz, J. R. *Principles of Fluorescence Spectroscopy*, 3rd ed., Springer: New York (2006).
40. Morisaki, Y., Kawakami, N., Nakano, T., Chujo, Y. *Chem. Eur. J.*, **19**, 17715–17718 (2013).
41. Morisaki, Y., Kawakami, N., Nakano, T., Chujo, Y. *Chem. Lett.*, **43**, 426–428 (2014).
42. Morisaki, Y., Kawakami, N., Shibata, S., Chujo, Y. *Chem. Asian J.*, **9**, 2891–2895 (2014).
43. Dexter, D. L. *J. Chem. Phys.*, **21**, 836–850 (1953).
44. Clément, S., Goudreault, T., Bellows, D., Fortin, D., Guyard, L., Knorr, M., Harvey, P. D. *Chem. Commun.*, **48**, 8640–8642 (2012).
45. Molina-Ontoria, A., Wielopolski, M., Gebhardt, J., Goulumis, A., Clark, T., Guldi, D. M., Martín, N. *J. Am. Chem. Soc.*, **133**, 2370–2373 (2011).
46. Wielopolski, M., Molina-Ontoria, A., Schubert, C., Margraf, J. T., Krokos, E., Kirschner, J., Gouloumis, A., Clark, T., Guldi, D. M., Martín, N. *J. Am. Chem. Soc.*, **135**, 10372–10381 (2013).
47. Morisaki, Y., Inoshita, K., Chujo, Y. *Chem. Eur. J.*, **20**, 8386–8390 (2014).

Biography

Yasuhiro Morisaki was born in 1972 in Osaka, Japan. He received his PhD from Kyoto University, Japan, in 2000. After his postdoctoral

research at Osaka National Research Institute for six months, he became assistant professor (2000–2007) and then associate professor (2008–2014) at Kyoto University. During his research at Kyoto University, he carried out postdoctoral work at the University of Alberta (2004–2005). He became a full professor at Kwansei Gakuin University in 2015. His research interests focus on synthetic organic chemistry and polymer chemistry.

Chapter 8

Synthesis, Structures, and Functions of Helical π-Conjugated Polymers

Hiromitsu Sogawa, Kazuko Nakazono, and Toshikazu Takata
Department of Organic and Polymeric Materials, Tokyo Institute of Technology, 2-12-1, Ookayama, Meguro, Tokyo 152-8552, Japan
ttakata@polymer.titech.ac.jp

The development of artificial helical polymers has attracted the interest of researchers from the viewpoint of exploration and mimicry of biomolecular helices. It has also prompted the development of novel properties and functional materials such as liquid crystals, nonlinear optical materials, chiral sensing and separation materials, and asymmetric catalysts. This chapter focuses on the synthesis, secondary structure, and functions of polyacetylene and foldamers known as dynamic helical π-conjugated polymers.

8.1 Introduction

Proteins, amylose, and DNA typically form well-defined higher-order structures. These biomacromolecules exhibit sophisticated

Conjugated Objects: Developments, Synthesis, and Applications
Edited by Atsushi Nagai and Koji Takagi

ISBN 978-981-4774-03-1 (Hardcover), 978-1-315-22962-1 (eBook)
www.panstanford.com

functions such as in vivo exquisite reactions, molecular recognition, and replication, depending on their structures. Helices, which are common molecularly asymmetric higher-order structures, adopt various configurations in biomacromolecules, such as α-, 3_{10}-, and β-helical proteins; amylose helix; DNA double helix; and collagen triple helix.[1] Artificial polymers form helical structures when each monomer unit is regulated uniformly in one direction. These polymers are important from the viewpoint of fundamental research as well as potential applications. In their pioneering work in 1955, Natta et al. found that isotactic polypropylene generated a helical structure in the crystalline state but instantly turned into a random coil in solution because of the lack of steric repulsion between methyl side groups.[2] Since then, various artificial helical polymers such as polymethacylates, polyisocyanates, polyisocyanides, polysilanes, polyacetylenes, polyquinoxaline, polyarylene ethylenes, and polythiophenes have been synthesized and their secondary structures have been investigated along with their unique optical properties.[3] In particular, polyacetylenes[3d,f,h,j,k] and conjugated foldamers[3b,c,h] such as poly(m-phenylene ethynylene)s are classified as dynamic helical π-conjugated polymers because their small helix inversion barriers promote reversible helix–helix and helix–coil transitions in response to external stimuli such as heating and polar solvent addition. Therefore, these polymers exhibit useful stimuli-responsive properties such as molecular recognition, chiral catalysis, and chemical sensing in addition to electronic and optical properties. This chapter focuses on the synthesis, secondary structure, stimuli-responsiveness, and optical properties of polyacetylenes and conjugated foldamers.

8.2 Polymerization Catalysts and Configurations of Substituted Acetylenes

The polymerization of various mono- and disubstituted acetylene monomers has been extensively investigated using numerous transition metal catalysts. It follows two reaction mechanisms, depending on the polymerization catalysts (Fig. 8.1a): the insertion mechanism, which involves alkyl metals bearing a metal–carbon single bond as active species, and metathesis, which relies on metal carbenes containing a metal–carbon double bond. Polymerization proceeds via insertion in the presence of Ti, Fe, Rh, and Pd catalysts. Group 5 and 6 transition metal chlorides such as WCl_6, $MoCl_5$, $TaCl_5$, $NbCl_5$, and Ru catalysts promote metathesis.[3g,j,k] Polyacetylenes comprise carbon–carbon alternating double bonds (which adopt *cis*- or *trans*-geometry) in their main chain, resulting in four conformational isomers (*cis-cisoidal*, *cis-transoidal*, *trans-cisoidal*, and *trans-transoidal*; Fig. 8.1b). Polymerization mechanisms strongly affect this stereoregularity. Rhodium catalysts exhibit high activity and relatively high tolerance toward functional groups, making them suitable for the polymerization of various monosubstituted polyacetylenes (Fig. 8.1c). Furthermore, the obtained polymers display highly *cis*-stereoregular structures with high head-to-tail content. *Cis*-contents are usually calculated by ^{1}H NMR spectroscopy.[4] The discovery of performance of Wilkinson's catalyst $[RhCl(PPh_3)_3]$ in the polymerization of the monosubstituted phenylacetylene[5] has led to the synthesis of various Rh catalysts such as neutral dinuclear complexes {$[(nbd)RhCl]_2$, $[(cod)RhCl]_2$, and $[(tfb)RhCl]_2$ (nbd = 2,5-norbornadiene, cod = 1,5-cyclooctadiene, tfb = tetrafluorobenzobarrelene)} and zwitterionic complex {$(nbd)Rh^+[\eta^6\text{-}C_6H_5B^-(C_6H_5)_3]$} (Fig. 8.2).[6] Relatively low-polarity solvents such as toluene, benzene, and tetrahydrofuran (THF) were particularly useful for producing high-molecular-weight polymers. Rh-catalyzed polymerization has proceeded even in water.[7] Organometallic Li, Zn, B, and Al compounds such as *n*-BuLi, MeLi, *t*-BuLi, PhLi, Et_2Zn, Et_3B, Et_3Al, and alkali metal amides act as effective co-catalysts in these systems. The polymerization of phenylacetylene using $[(nbd)RhCl]_2$ in the presence of alkali metals such as $KN(SiMe_3)_2$ and $LiN(i\text{-}Pr)_2$was completed almost instantaneously in toluene at moderate temperature.[8] In addition, Et_3N effectively assisted $[(nbd)RhCl]_2$-catalyzed polymerizations.[6b] In 1994, the living polymerization of phenylacetylene was achieved using $(Ph_3P)_2(nbd)RhC{\equiv}CPh$ in the presence of 4-(*N*,*N*-dimethylaminopyridine) (DMAP).[9a] The polymers with narrow polydispersity ($M_w/M_n \approx 1.1$) were obtained in quantitative yields despite a relatively low initiation efficiency (<35%). The initiation efficiency increased to 70% when the $[(nbd)RhOMe]_2$–Ph_3P–DMAP catalytic system was used. End functionalization and block polymerization were achieved using living Rh-catalyzed systems.[9b] In addition to these experiments, detailed theoretical studies of the reaction mechanism, regioselectivity, and stereoregularity of Rh-catalyzed polymerizations have been conducted by various methods such as density function theory (DFT), our own *n*-layered integrated molecular orbital and molecular mechanics (ONIOM), and ONIOM molecular dynamics.[10]

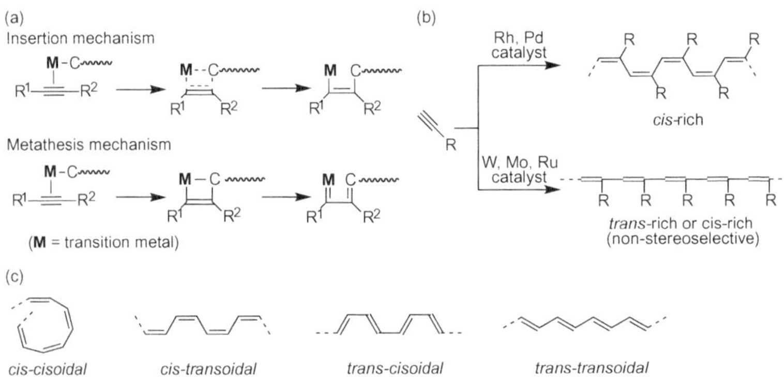

Figure 8.1 (a, b) Transition metal–catalyzed polymerization of acetylenes and (c) polymer configurations.

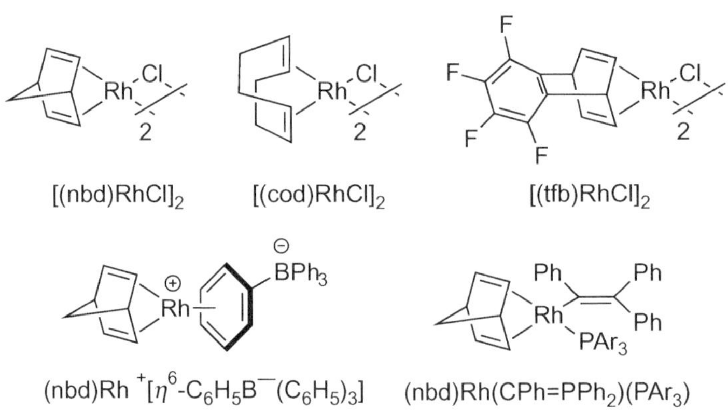

Figure 8.2 Conventional Rh catalysts.

8.3 Dynamic Secondary Structures of π-Conjugated Polyacetylenes

Numerous π-conjugated helical polyacetylenes exhibiting predominant one-handedness have been synthesized. Their preparation methods include (i) polymerization of acetylene monomers in an asymmetric reaction field using chiral nematic liquid crystals, (ii) Rh-catalyzed polymerization of achiral monomers in the presence of chiral ligands or co-catalysts, (iii) helicity induction by combination of achiral polyacetylenes with low-molecular-weight compounds, and (iv) polymerization of optically active monomers (Fig. 8.3). The helix-sense-selective polymerization of acetylene using the Ziegler–Natta catalyst in the chiral nematic liquid crystals produced insoluble helical polyacetylene films, which formed clockwise or counterclockwise nanofibers.[11] Other helix-sense-polymerization systems gave poly(*p*-alkoxy-*m*,*m*-dihydoryxy-phenylacetylene) (**1**) using an optically active amine as a co-catalyst (Fig. 8.4).[12] Asymmetric polymerization of the same monomer was achieved in the presence of a Rh catalyst bearing a C_2-chiral tfb ligand.[13] The opposite chiral catalyst isomer generated a polymer featuring the opposite helical handedness. Achiral poly(phenylacetylene)s containing carboxylic acid (**2**), sulfonic acid (**3**), phosphoric acid (**4**), tertiary amine (**5**), crown ether (**6**), and cyclodextrin moieties (**7**) induced predominantly one-handed helical structures upon the addition of optically active compounds such as carboxylic acids, esters, amines, and amino acid ammonium salts (Fig. 8.5).[14] The preferred handedness was detected immediately, creating characteristic circular dichroism (CD) signals in the π-conjugated main-chain absorption regions. A good relationship was observed between the signs of CD signals corresponding to the polymer helicity and the absolute configurations of chiral additives. In this method, the helicity was tuned by the chirality of the small molecules after polymerization. Finally, several optically active monosubstituted acetylene derivatives produced one-handed helical π-conjugated polymers using Rh catalysts (Fig. 8.6). This method has been widely applied because of its simplicity. Many π-conjugated helical polyacetylenes obtained by this method are described more specifically below.

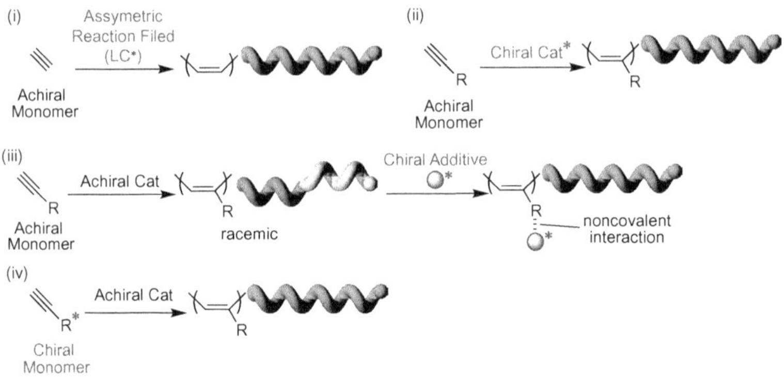

Figure 8.3 Typical preparation of helical polyacetylenes with a predominant screw sense.

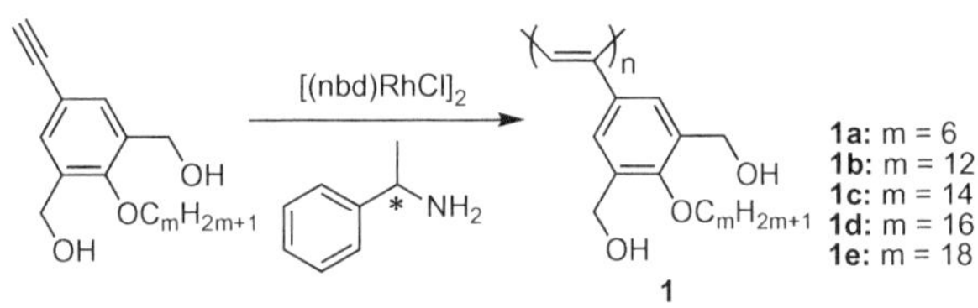

Figure 8.4 Helix-sense-selective polymerization systems using chiral amine co-catalysts.

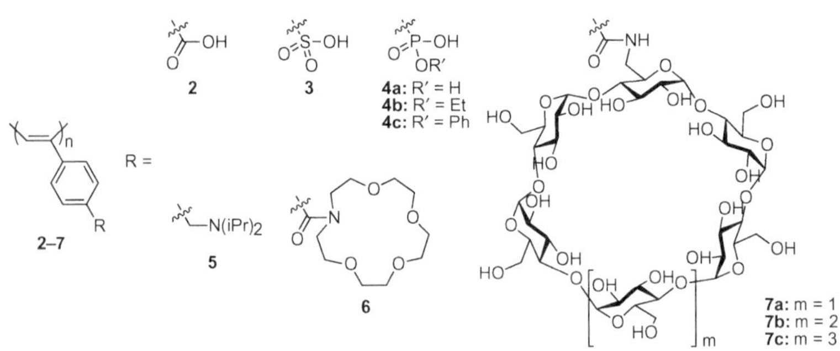

Figure 8.5 Substituted polyacetylenes forming helical structures with a predominant one-handedness biased by noncovalent interactions.

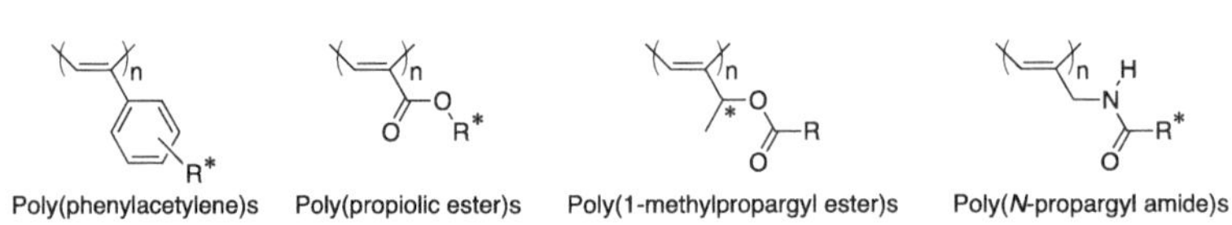

Figure 8.6 Helical polyacetylenes obtained using chiral monomers.

Intramolecular hydrogen bonding and steric repulsion between neighboring units play important roles in stabilizing the predominantly one-handed helical structures. *Cis*-stereoregular poly(*N*-propargyl amide)s exhibit intense CD signals and large optical rotations in $CHCl_3$, indicating a predominantly one-handed helical conformation resulting from intramolecular N–H•••O=C hydrogen bonding and steric repulsion between side chains. This helical conformation collapsed in polar solvents such as MeOH and dimethyl formamide (DMF) or upon heating, which introduced disorder into the hydrogen-bonded networks (Fig. 8.7).[15] In contrast, poly(1-methylpropargyl alcohol) and ester derivatives exhibited intense CD signals even in polar solvents such as MeOH, dimethyl sulfoxide, and DMF. The methyl branch adjacent to the main chain plays an important role in helicity induction. In other words, the presence of a chiral group in close proximity to the main chain significantly impacts helical structure formation.[16] Stereoregular poly(*N*-alkynylamide)s containing several alkylene spacers between amide groups and chiral centers or the main chain have also been synthesized. Their secondary structure strongly depended on the chiral center and amide group positions. Specifically, the helical pitch decreased when the chiral center was located away from the main chain.[17]

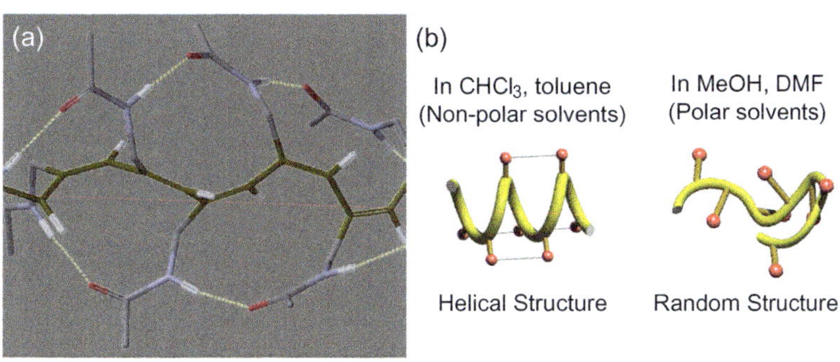

Figure 8.7 (a) Possible conformations of helical poly(*N*-propargyl amide)s and (b) solvent-dependent conformational change.

Amino acids and peptide residues, which represent a versatile source of chirality in organic synthesis, have been incorporated in several helical π-conjugated polyacetylenes as pendant groups (Fig. 8.8).[18–20] These polymers showed unique chiroptical properties including helical sense inversion in response to external stimuli such as temperature, solvent, and pH. Aspartic (**8**) and glutamic acid-based poly(*N*-propargylamide)s (**9**) adopted right-handed helical structures under acidic and neutral conditions but left-handedness in basic environments. Their helical pitch also changed upon inversion.[19d] Furthermore, serine-derived poly(*N*-propargylamide) s (**10**) exhibited a distinctive chiroptical behavior. In particular, **10b** adopted a helical conformation in polar solvents such as MeOH but not in $CHCl_3$-like nonpolar solvents. The hydroxy groups likely prevented polar solvents from disturbing the intramolecular hydrogen bonding between amide groups. Sanda and coworkers reported that *α*-propargyl amino acid–derived polyacetylenes (**11** and **12**) exhibited intense CD signals in the absorption region of the conjugated polyacetylene backbone, consistent with the formation of a predominantly one-handed helical structure. The helical structure of **11b** extended upon the addition of alkali hydroxides and tetra-*n*-butylammonium fluoride. The degree of extension increased with the cation size. The existence of cations close to the carboxylate moieties may create high steric repulsion between the side chains, loosening the helix during main-chain elongation.[19i] Anion recognition systems relying on noncovalent electrostatic interactions were generated using leucine- and aspartic acid–derived poly(phenylacetylene)s bearing urea (**13**) and sulfonamide (**14**) groups, respectively.[20c–e] *N*-methylvaline-derived poly(phenylacetylene) (**15**) catalyzed the asymmetric reduction of aromatic ketimines.[19f] Asymmetric aldol reactions, in which chiral interaction between the helix main chain and the intermediate may affect the product enantioselectivity, were achieved using proline-derived poly(phenylacetylene) (**16**).[19h] Various poly(phenylacetylene)s bearing oligo and polypeptide pendant groups (**17** and **18**) were also produced. Pendant conformations resulted in unique catalytic activity and conformational changes in the polymer backbone.[20a,b]

As mentioned above, polyacetylenes present small helix inversion barriers, promoting chiral amplification processes (Fig. 8.9).[14c,e,16b,c,21] The excess of biased helical conformation is enhanced by the copolymerization of a chiral monomers with a small amount of optically active comonomers according to the "sergeants and soldiers effect." Alternatively, this enhancement may stem from the "majority rule," which involves copolymerization of nonracemic monomers exhibiting a small enantiomeric excess. In addition, various chiral amplification systems have been developed using "noncovalent" chiral interactions. For example, a poly(phenylacetylene)-bearing bulky aza-18-crown (**6b**), a typical host molecule in host–guest chemistry, adopted a predominantly one-handed helical conformation in the presence of 0.1 equiv. of alanine in acetonitrile. The addition of only 5% enantiomeric excess (ee) of alanine generated a CD signal. Moreover, this polymer detected less than 0.005% ee of alanine. It also exhibited induced CD signals upon complexation with all common amino acids, suggesting its applicability to highly sensitive and practical synthetic receptors for amino acid chirality detection.[14e]

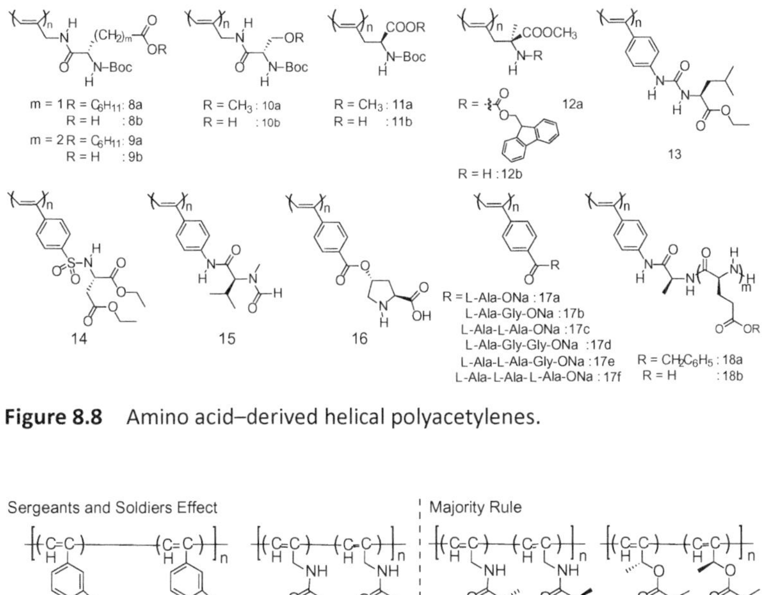

Figure 8.8 Amino acid–derived helical polyacetylenes.

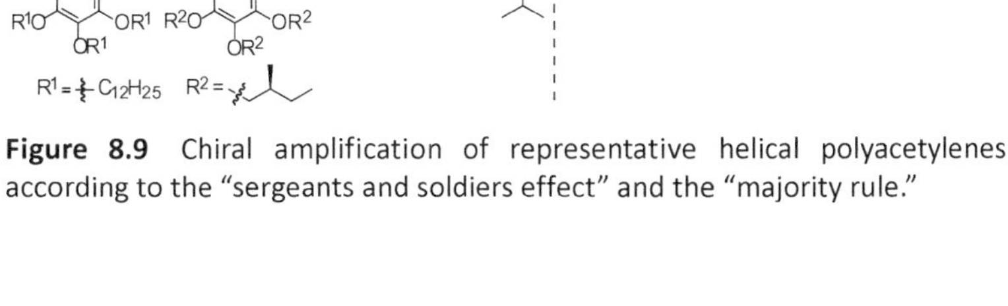

Figure 8.9 Chiral amplification of representative helical polyacetylenes according to the "sergeants and soldiers effect" and the "majority rule."

In addition to synthetic improvements, the elucidation of helical polymer handedness provides valuable insight into the helix formation mechanism and functions (Fig. 8.10). CD and optical rotation measurements offer useful information on preferred helix handedness but cannot definitely solve a helical structure including its helical pitch, diameter, and sense. The exciton-coupled CD method and CD simulations are often used to predict preferred helical senses but display a degree of uncertainty when applied to helical polymers (**19–22**).[22] Artificial helical polymer structures are usually difficult to determine accurately at the molecular level by X-ray diffraction (XRD) except when the corresponding uniform oligomers are isolated and crystals suitable for single-crystal X-ray analysis are formed. Detailed helical structures of poly(pheylacetylene) derivatives (**23**) have been identified and visualized by atomic force microscopy (AFM) and scanning tunneling microscopy (STM).[23b–f] Secondary structures of chiral and achiral phenylacetylene copolymers (**25**) have also been elucidated.[23g] These studies provide a direct evidence of macromolecular helicity in these dynamic polymer sunder various conditions.

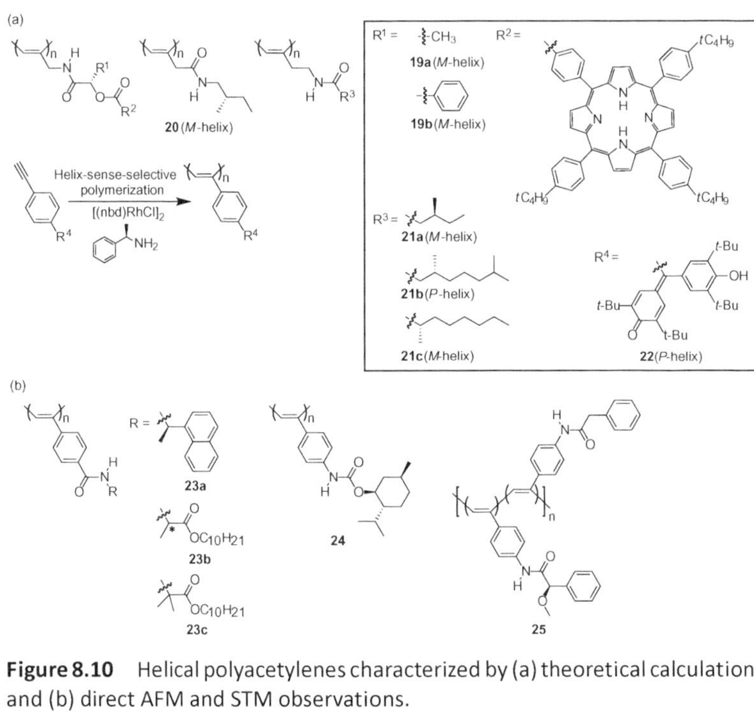

Figure 8.10 Helical polyacetylenes characterized by (a) theoretical calculations and (b) direct AFM and STM observations.

Helical polyacetylenes have also been applied to chiral stationary phases (CSPs) for high-performance liquid chromatography (HPLC) similarly to other synthetic helical polymers such a poly(triphenylmethyl methacrylate)s and polyisocyanides.[24] Copolymerization of the chiral monomer in the presence of monomer-bearing silica gel gave a coated-type CSP, which resolved several racemates such as alcohols, amines, and epoxides. A unique switchable enantioseparation system has also been developed using 2,2′-bisphenol-derived polyacetylene.[24f] Polymer main-chain helicity and axial chirality were amplified and switchable in the solid state, allowing elution order and enantioselectivity to be switched at will.

8.4 Foldamer-Type π-Conjugated Helical Polymers

This section describes recent advances in foldamer-type π-conjugated polymers (Fig. 8.11). According to Gellman[25], foldamers are "any polymer with a strong tendency to adopt a specific compact conformation." Oligo(*m*-phenylene ethynylene) foldamers were first developed by Moore and coworkers in 1997. Optically active oligo(*m*-phenylene ethynylene)s bearing ethylene glycol units (**26**) folded into predominantly one-handed helices in polar solvents because of their amphiphilicity, which originated from their hydrophilic side chains and hydrophobic main chains (Fig. 8.11).[26] The *para*-linked poly(phenylene ethynylene)s bearing chiral side chains exhibited strong bisignate CD signals as a result of aggregation, demonstrating that the *meta* connectivity of the monomer units played an important role in the oligomer folding into a helix.[27] Optically active polymers with analogous main chains also adopted a helical conformation with a biased screw sense (Fig. 8.12).[28] Polymer **27** exhibited strong bisignate CD signals in methanol, water, and their mixtures, consistent with the formation of folded helical structures presenting a preferred handedness.[28c] A poly(*m*-phenylene ethynylene)-based polyradical bearing galvinoxyl and chiral pinanyl pendant groups (**28**) showed a weak but characteristic CD in methanol/chloroform mixtures.[28d] Poly(*m*-phenylenedisilanylene) (**29**) comprising a σ-π-conjugated main chain exhibited positive Cotton effects in the absorption region of the main chain in methanol but no intense CD signals in CH_2Cl_2.

Interestingly, this polymer maintained its helical handedness in CH_2Cl_2 with appropriate metal addition.[28e] Poly(*m*-ethynylpyridine) containing chiral pendant groups (**30**) also formed a predominantly one-handed helix.[28j] The introduction of an optically active binaphtol moiety into the backbone of oligo(*m*-phenylene ethynylene)s (**31**) gave a bias to the folded helical structure, causing typical solvent-dependent conformational changes.[28a] Photo switchable helical foldamers (**32**) were developed using azobenzene repeat units in the main chain. The *trans*-to-*cis* azobenzene isomerization upon UV irradiation unfolded the helical conformation. In contrast, visible light irradiation readily, quantitatively, and reversibly reorganized the initial helical conformation.[28i] Optically active triazole-based polymers (**33**) obtained by Cu-catalyzed 1,3-dipolar cycloaddition also exhibit folded helix-forming ability. The kinked *anti–anti* conformation is key for the formation of helices with a biased screw sense.[28g] Silver-containing metallosupramolecular polymers (**34**) produced tunable secondary structures such as folded helices, depending on the counteranions. However, well-defined linear structures were difficult to generate. A combination of Ag ions and analogous dipyridyl ligands gave folded helical coordination polymers with a switchable pitch. The helical pitch displayed unique temperature-triggered reversible extension–contraction motions.[28f] Furthermore, the helical sense, pitch, and other properties of typical helical poly(*m*-phenylene ethynylene)s (**35**) were elucidated by direct AFM observation. A comparison of CD, UV-Vis spectra, and other data provided information on the secondary structure of these polymers.[28k]

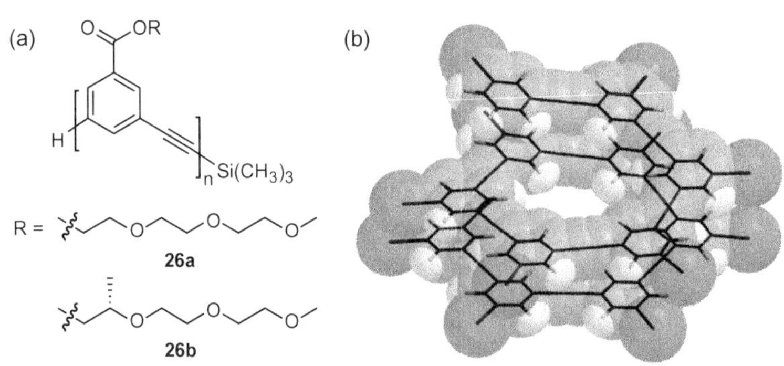

Figure 8.11 Oligo(*m*-phenylene ethynylene)s folded into a predominantly one-handed helical conformation in polar solvents.

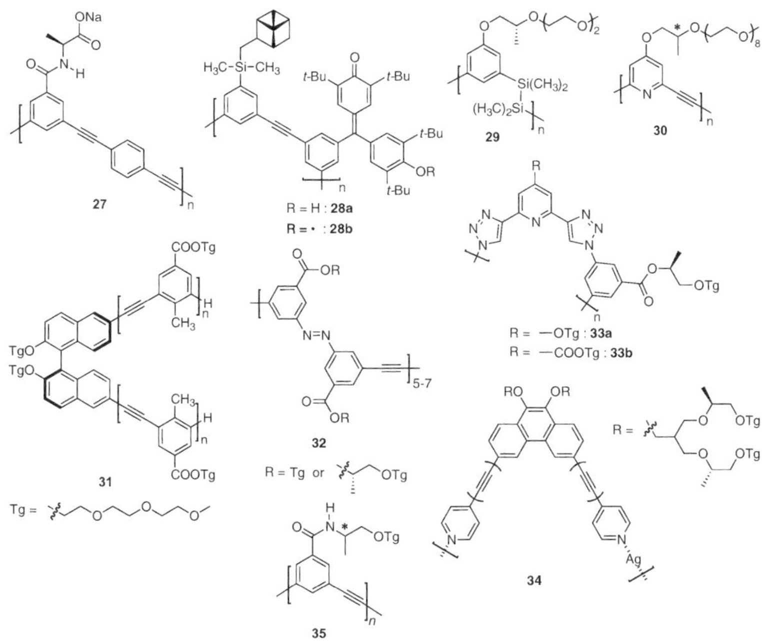

Figure 8.12 π-Conjugated helical foldamers.

Unlike most of the other polyphenylene arylenes forming folded helical structures as mentioned above, Sanda and coworkers have reported novel amino acid–derived optically active poly(*m*-phenylene arylene)s bearing hydroxy groups (Fig. 8.13).[29] D-hydroxyphenylglycine-derived poly(*m*-phenylene ethynylene-*p*-phenylene ethynylene)s (**36**) formed a helix with an excess of one-handed screw sense by the opposite amphiphilic balance based on a hydrophobic exterior and a hydrophilic interior in nonpolar solvents such as $CHCl_3$.[29a] The formation of regulated intramolecular hydrogen bonding between the amide groups at the *n*th and $(n + 6)$th units at the side chains, confirmed by solution-state IR spectroscopic and molecular modeling studies, supported to stabilize its helical conformations. The addition of MeOH and H_2O induced the collapse of the folded helical structure with biased screw sense. CD, UV-Vis, and fluorescence spectra, together with the molecular mechanic and detailed DFT calculations, elucidated the formation of predominantly one-handed folded helical structures in nonpolar solvents. The tuning of π-conjugated arylene units of main chains including biphenyl, naphthalene, pyrene, and azobenzene enabled to control the conjugation lengths of obtained helical foldamers.[29f,g] Similar to Hecht's work, azobenzene-derived polymers (**40** and **41**) exhibited reversible conformational changes in nonpolar solvents between folded-helical structure and random coil upon light irradiation, which was induced by the photoisomerization of the azobenzene units from the planar *trans*-form to the bent *cis*-form. The introduction of azobenzene units in the π-conjugated main chains instead of the side chains allowed to induce large conformational changes.[29c,e] Meanwhile, the stabilization of preferred one-handed helical conformation in polar solvents was achieved via the intramolecular metathesis reaction. **38** exhibited bisignated CD signals in $CHCl_3$ and THF, but no CD in DMF. The intramolecular acyclic diene metathesis reaction at high dilute conditions in $CHCl_3$ resulted in the fixation of folded helical structures without the formation of networked polymers. The obtained polymers exhibited intense CD signals in not only nonpolar solvents but also DMF and even THF/H_2O mixture, indicating the formation of covalently fixed helical conformations.[29d] Moreover, the successful energy transfer from the conjugated main chain of helical foldamer to perylene moiety at the side chain was observed for **39**. Although the reason is not clear yet, the energy transfer efficiency of helical foldamer is high in some content compared with that in analogous polymer, which could not form clear helical structures. [29h]

Meanwhile, Takata and coworkers developed precise control of the higher-order structure of polymers by connecting C_2-chiral units including biaryl derivatives and spirobifluorene with a rigid spacer unit (Fig. 8.14).[30] The obtained polymers formed one-handed helical structure, which is a priori organized by the structure and chirality of the C_2-chiral unit. For example, the C_2-chiral spirobifluorene-derived polymer containing covalently two-point linkage (**44**) was developed by the stepwise reductive annulation between spirobifluorene and anthraquinone spacer units.[30d] In addition, the unique planarity-switchable spacer units consisting of an oligothiophene moiety were applied to achieve redox reversible transformation of exact higher-order structures of C_2-chiral polymers (**45**). Oxidizations of polymers using $FeCl_3$ resulted in a large red shift of the absorption band and CD signals due to the generation of a radical cation. The thermal stability of spacer structures was enhanced compared with that of neutral states, resulting in the formation of thermally stable folded helical structures. The addition of sodium bisulfite as a reducing agent implied the efficient reversibility of a redox-driven folding system.[30e]

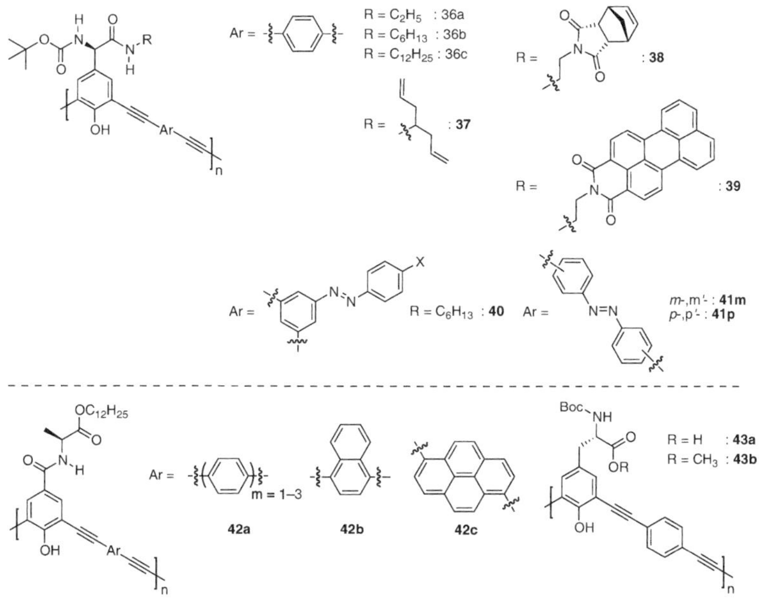

Figure 8.13 Poly(*m*-phenylene arylene)s bearing hydroxy groups folded into a helical structure in nonpolar solvents.

The conjugated foldamers have a unique nature to encapsulate specific guest molecules into its cavity under certain conditions (Fig. 8.15).[31] As mentioned above, **26a** bearing achiral groups folded into helices with the same equivalent of right- and left-handed ones in acetonitrile. Meanwhile, the induced CD signals were observed at the regions of the main-chain chromophore when nonracemic hydrophobic and rod-like guests were added.[31a,b] The encapsulation within the hydrophobic inner cavity of foldamers took place based on the hydrophobic interaction in acetonitrile. This phenomenon is regarded as chiral recognition by the helical π-conjugated polymers. Tew and coworkers have successfully induced chirality in achiral polymers bearing mono-ammonium groups by the addition of chiral mandelic acid.[31c] Poly(*m*-ethynylpyridine)s, which were developed by Inouye's groups, exhibited high selectivity toward chirality of saccharides.[31d] On the other hands, Moore et al. demonstrated the practical usage of *m*-phenylene ethynylene oligomers modified with a dimethylaminopyridine unit as reactive sieves. Foldamer bearing reactive sites in its cavity exhibited variable reactivity toward substrates, even though the substrates differ minimally in size and shape. The methylation of pyridine moiety proceeded much faster with alkyl sulfonates bearing "ideal" linear chains compared with short ones or branched ones. The foldamer-based covalent catalyst can be realized by developing this system.[31e]

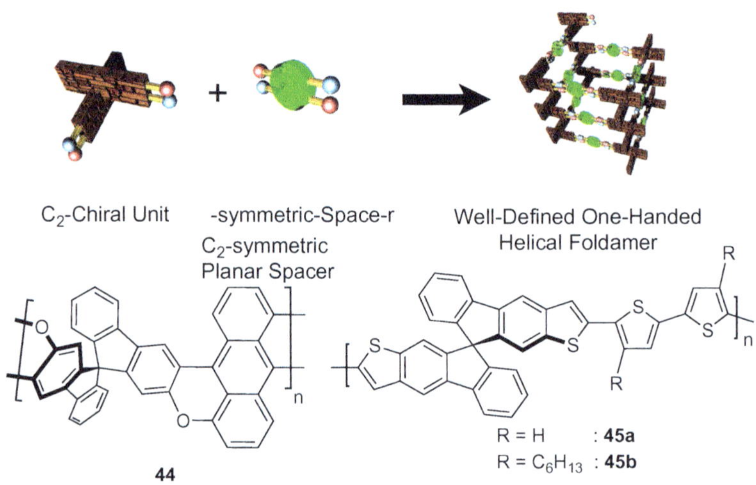

Figure 8.14 Synthesis of a π-conjugated C_2-chiral polymer-forming exact helical structure.

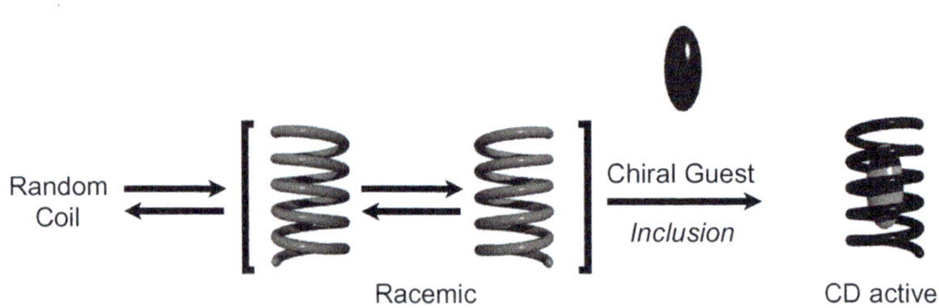

Figure 8.15 Schematic of a predominantly one-handed helix induction in foldamers via the guest inclusion process.

8.5 Application of π-Conjugated Helical Polymers

As described above, how to control the structure of π-conjugated helical polymers has been well studied. On the other hand, the dynamic helical property of π-conjugated helical polymers is attractive for supramolecular chemistry as the functional device and scaffold. Polymers having interlocked structures as key skeletons, that is, polyrotaxanes, promise the characteristic property owing to the unique mobility of the mechanical bond. Takata et al. have recently developed a class of side-chain-type polyrotaxane having π-conjugated helical polymers as the backbone. Rotaxane is one of the typical supramolecule studies as a molecular switch attribute to the shuttling property of the interlocked structure. The components, wheel and axle, essentially move relatively due to Brownian motion. Such a mobile property can be controlled by introducing interaction sites on the axle as the station for the wheel. Takata et al. took advantage of rotaxane's characteristic properties to control polymer structures. A combination of a rotaxane switch and π-conjugated helical polymers achieved not only a rational control of the helical structure of the polymer but also the visualization of rotaxane's switching behavior. Thus, they applied the conjugated helical polymers as the detector or amplifier of microscopic structural changes in molecules. The rotaxane side chains can switch the steric influence around the polymer main chain such as bulkiness and chirality by tuning the position of the wheel of the rotaxane moiety (Fig. 8.16).

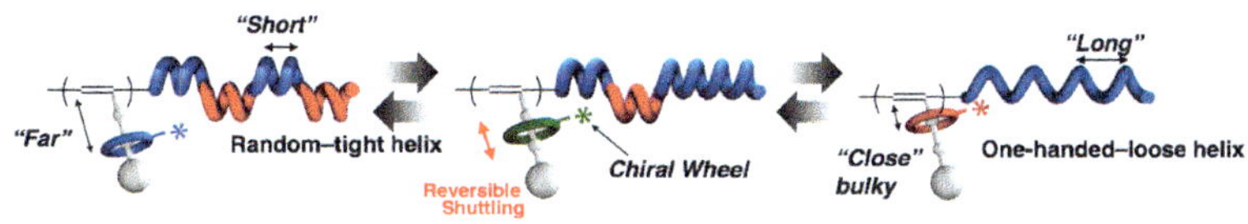

Figure 8.16 Helical pitch control by a pendant rotaxane switch.

First, how the distance between the wheel and the main chain influences the effective conjugation length and helical pitch of polyacetylene was clarified. The color of polyacetylene possessing pendant rotaxanes with the wheel (crown ether) locating far from the main chain (**46a**) was yellow; however, that of the polymer with the wheel locating near the main chain (**46b**) was red (Fig. 8.17a).[32] The significant color change occurred during the *N*-acetylation of ammonium moiety of polymer **46a**, that is, the removal of the extraordinary strong hydrogen bonds between the crown ether and ammonium moiety permitted the crown ethers to approach the main chain. According to the Percec's loose and tight helix theory[21c], the red color indicates a loose and long pitch of the polyacetylene structure, while the yellow color indicates a tight and short pitch. Thus, the plausible reason for the difference in the helical pitch is owing to the degree of steric hindrance between the bulky wheels. Namely, the location of wheels near the main chain inhibits shrinking of the helix; however, the wheels localize far from the main chain and the steric hindrance around the main chain is eliminated.

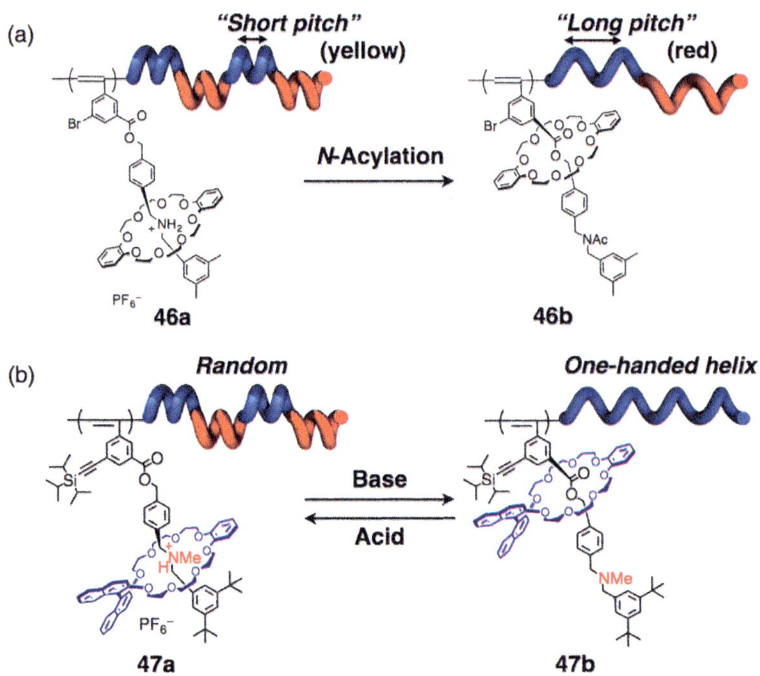

Figure 8.17 Positional change in the chiral rotaxane side chain and the through-space chirality transfer on the helix sense of the polyacetylene main chain.

Furthermore, Takata et al. achieved the chirality control of the helix sense of polyacetylene by pendant chiral rotaxane switches. Polyacetylene, which possess an optically active side chain, can reportedly form a one-handed helix when the optically active group is close to the main chain. Masuda et al. showed the importance of such an effect on the distance between the chiral point and the polyacetylene main chain.[17] According to their report, if the chiral center is placed in the seven atoms far from the main chain, the CD spectrum will show no Cotton effect to deny the one-handed helix induction of the main chain. On the basis of such a phenomenon, Takata et al. synthesized polyacetylene possessing pendant chiral rotaxane switches (Fig. 8.17b).[33] The pendant rotaxane consists of an optically active binaphthol-containing crown ether and *tert*-ammonium or *tert*-amine-containing axle component (**47a**,**b**). *tert*-Amine is the key functional group for the reversible positional switching of crown ether-based rotaxane. *tert*-Amine (**47b**) keeps the crown ether away on it in rotaxane because of the electrostatic repulsion, while *tert*-ammonium (**47a**) can localize the crown ether on it by hydrogen bonding. Note that the *tert*-nitrogen center easily and certainly modulates the interaction between the crown ether and axle to enable good positional switching performance driven by acid/base stimuli. In fact, polyacetylenes with *tert*-amine subjected to localize near its main chain exhibited the CD signal attribute to the one-handed helical main-chain structure. On the other hand, with treatment of the amine-type polyacetylene with a strong acid like trifluoroacetic acid, the Cotton effect disappeared to support the generation of a racemic helix. The treatment of the protonated polymer with a base such as 1,8-diazabicyclo[5.4.0]undec-7-ene (DBU) regenerated the Cotton effect similarly to the initial amine-type polymer. These results lead to two important influences: first is the rotaxane switch promising the helicity control of the polymer structure via through-space chirality transfer, and the other is that the positional switching of the pendant rotaxane has sufficient ability to control the higher-order structure of the polymer. Note also that the obtained polymer is a molecular architecture consisting of high-density and highly integrated molecular devices. The rotaxane switch and the structure of the polymer backbone are cooperatively operated, and the microscopic structural transformation of the rotaxane side chain was amplified through the polymer backbone to the more macroscopic change such as polymer color or helix sense.

A similar tequnique has demonstrated even in the control of poly(*m*-phenylene diethynylene) as the example of foldamer by Takata and coworkers (Fig. 8.18).[34,35] Although the structural control of foldamers using poly(*m*-phenylene diethynylene) has not been reported so far, their studies proved that poly(*m*-phenylene diethynylene) (**48a** and **48b**) is an attractive foldamer with a large helical conformation and high polymerization degree. The synthesis of poly(*m*-phenylene diethynylene) possessing rotaxane switch can be achieved under the typical oxidative coupling polymerization condition with CuCl-O_2 catalyst system in pyridine.

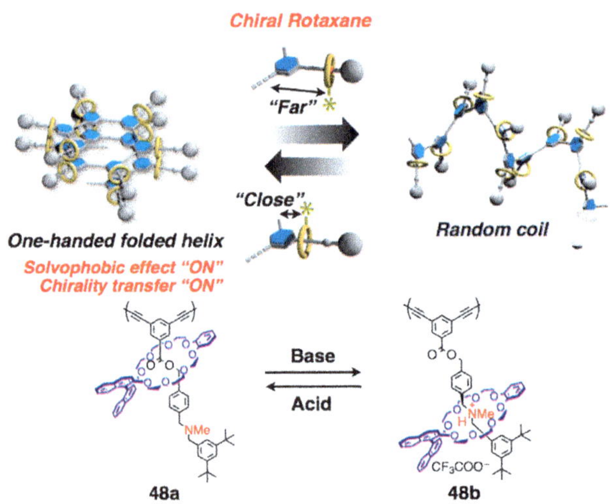

Figure 8.18 Second structure control of poly(*m*-phenyelene diethynylene) by chiral rotaxane switching in the side chain.

The solvophovic effect is the key driving force to induce the folded helical conformation in the foldamer system, the pendant rotaxane switches of the poly(*m*-phenylene diethynylene) can induce the solubility and polarity conversion of polymer. In fact, the CD spectra in a mixed solvent, $CH_2Cl_2/CH_3CN = 1/1$ (v/v), induced an intense CD that was completely different from that of the monomer to indicate that the polymer forms a one-handed helix via through-space chirality transfer. In addition, the UV-Vis spectrum showed a remarkable hypochromism, and the red shift of the main-chain absorption strongly supported the formation of the folded conformation of the polymer as a result of the solvophobic effect. Hypochromic effect is a powerful indicator of the oriented chromophores commonly observed in DNA, RNA, and other foldamers, and is sensitive to the distance between chromophores and their relative orientations.[36,37] Therefore, the 1,4-diphenyl-1,3-butadiyne units were arranged by π–π stacking interactions, which strongly suggest the formation of a folded conformation of the polymer in a mixed solvent. Meanwhile, the polymer treated with trifluoroacetic acid showed CD spectra same as those of a monomer in a similar mixed solvent. The UV-Vis spectrum also denied the folded helical conformation of the acidified polymer. However, the acidified ionic polymer can take the folded structure in a CH_3CN-H_2O mixed solvent as a result of increasing solvophobic effect. The UV-Vis spectrum in this mixed solvent showed a remarkable hypochromism and red shift of the main-chain absorption, but its CD signal coincided well with that of CH_3CN and should form a racemic helical conformation. This is consistent with the position of the optically active wheel component being far from the main chain, as indicated from the chemical shift of NMR in such a polar solvent.

These results show that poly(*m*-phenylene diethynylene) is a foldamer conformation can be controlled by solvents. The rotaxane switch plays a key role in determining the helix sense, while drastically changing the affinity of the polymer with solvent. These results consolidated the possibility of the rotaxane switch as a molecular device and through-space chirality transfer.

8.6 Conclusions

In this chapter, we focused and reviewed the synthesis, secondary structures, and functions of polyacetylene and π-conjugated foldamers. The research area reviewed here has remarkably developed over the past decade, and a large number of helical polyacetylenes and foldamers have been synthesized. For instance, some of them have been successfully applied to the commercializable materials. There are ever so many more interesting and fascinating reports concerning about the artificial helical polymers. We just focused on specific subjects. Many chemists have been challenged and attracted to develop artificial helical π-conjugated polymers in terms of exploring the origin of life and creating novel functional materials. It will be our great pleasure if the concepts and understanding introduced in this review will be useful for learning the related fields.

References

1. Branden, C., Tooze, J. *Introduction to Protein Structure*, 2nd ed., Garland: New York (1999).
2. Natta, G., Pino, P., Corradini, P., Danusso, F., Mantitca, E., Nazzanti, G., Moraglio, G. *J. Am. Chem. Soc.*, **77**, 1708–1710 (1955).
3. For reviews, see: (a) Green, M. M., Peterson, N. C., Sato, T., Teramoto, A., Cook, R., Lifson, S. *Science*, **268**, 1860–1866 (1995); (b) McQuade, D. T., Pullen, A. E., Swager, T. M. *Chem. Rev.*, **100**, 2537–2574 (2000); (c) Hill, D. J., Mio, M. J., Prince, R. B., Hughes, T. S., Moore, J. S. *Chem. Rev.*, **101**, 3893–4011 (2001); (d) Nakano, T., Okamoto, Y. *Chem. Rev.*, **101**, 4013–4038 (2001); (e) Sato, T., Terao, K., Teramoto, A., Fujiki, M. *Polymer*, **44**, 5477–5495 (2003); (f) Aoki, T., Kaneko, T., Teraguchi, M. *Polymer*, **47**, 4867–4892 (2006); (g) Masuda, T. *J. Polym. Sci., Part A: Polym. Chem.*, **45**, 165–180 (2007); (h) Yashima, E., Maeda, K., Iida, H., Furusho, Y., Nagai, K. *Chem. Rev.*, **109**, 6102–6211 (2009); (i) Akagi, K. *Chem. Rev.*, **109**, 5354–5401 (2009); (j) Liu, J., Lam, J. W. Y., Tang, B. Z. *Chem. Rev.*, **109**, 5799–5867 (2009); (k) Shiotsuki, M., Sanda, F., Masuda, T. *Polym. Chem.*, **2**, 1044–1058 (2011); (l) Suginome, M., Yamamoto, T., Nagata, Y., Yamada, T., Akai, Y. *Pure Appl. Chem.*, **84**, 1759–1769 (2012).
4. (a) Simionescu, C., Dumitrescu, S., Percec, V. *J. Polym. Sci., Part C: Polym. Symp.*, **2**, 209–227 (1978); (b) Lam, J. W. Y., Tang, B. Z. *Acc. Chem. Res.*, **38**, 745–754 (2005).
5. Kern, R J. *J. Polym. Sci., Part A: Polym. Chem.*, **7**, 621–631 (1969).
6. (a) Furlani, A., Licoccia, S., Russo, M. V., Camus, A., Marsich, N. *J. Polym. Sci., Part A: Polym. Chem.*, **24**, 991–1005 (1986); (b) Yang, W., Tabata, M., Kobayashi, S., Yokota, K., Shimizu, A. *Polym. J.*, **23**, 1135–1138 (1991); (c) Saeed, I., Shiotsuki, M., Masuda, T. *Macromolecules*, **39**, 8567–8573 (2006); (d) Saeed, I., Shiotsuki, M., Masuda, T. *Macromolecules*, **39**, 8977–8981 (2006); (e) Onishi, N., Shiotsuki, M., Sanda, F., Masuda, T. *Macromolecules*, **42**, 4071–4076 (2009); (f) Onishi, N., Shiotsuki, M., Masuda, T., Sano, N., Sanda, F. *Organometallics*, **32**, 846–853 (2013).

7. Tang, B. Z., Poon, W. H., Leung, S. M., Leung, W. H., Peng, H. *Macromolecules*, **30**, 2209–2212 (1997).

8. (a) Furlani, A., Napoletano, C., Russo, M. V., Camus, A., Marsich, N. *J. Polym. Sci., Part A: Polym. Chem.*, **27**, 75–86 (1989); (b) Kanki, K., Misumi, Y., Masuda, T. *Macromolecules*, **32**, 2384–2386 (1999); (c) Saeed, I., Shiotsuki, M., Masuda, T. *Macromolecules*, **39**, 5347–5351 (2006).

9. (a) Kishimoto, Y., Eckerle, P., Miyatake, T., Ikariya, T., Noyori, R. *J. Am. Chem. Soc.*, **116**, 12131–12131 (1994); (b) Kishimoto, Y., Eckerle, P., Miyatake, T., Ikariya, T., Noyori, R. *Macromoleucles*, **29**, 5054–5055 (1996).

10. (a) Ke, Z., Abe, S., Ueno, T., Morokuma, K. *J. Am. Chem. Soc.*, **133**, 7926–7941 (2011); (b) Sanda, F., Shiotsuki, M., Masuda, T. *Macromol. Symp.*, **350**, 67–75 (2015).

11. (a) Akagi, K., Piano, G., Kaneko, S., Sakamaki, K., Shirakawa, H., Kyotani, M. *Science*, **282**, 1683–1686 (1998); (b) Goh, M., Kyotani, M., Akagi, K. *J. Am. Chem. Soc.*, **129**, 8519–8527 (2007).

12. (a) Aoki, T., Kaneko, T., Maruyama, N., Sumi, A., Takahashi, M., Sato, T., Teraguchi, M. *J. Am. Chem. Soc.*, **125**, 6346–6347 (2003); (b) Kaneko, T., Umeda, Y., Yamamoto, M., Teraguchi, M., Aoki, T. *Macromolecules*, **38**, 9420–9426 (2005); (c) Kaneko, T., Umeda, Y., Jia, S., Hadano, M., Teraguchi, M., Aoki, T. *Macromolecules*, **40**, 7098–7102 (2007).

13. Nishimura, T., Ichikawa, Y., Hayashi, T., Onishi, N., Shiotsuki, M., Masuda, T. *Organometallics*, **28**, 4890–4893 (2009).

14. (a) Yashima, E., Matsushima, T., Okamoto, Y. *J. Am. Chem. Soc.*, **117**, 11596–11597 (1995); (b) Yashima, E., Maeda, Y., Okamoto, Y. *Chem. Lett.*, 955–966 (1996); (c) Yashima, E., Matsushima, T., Okamoto, Y. *J. Am. Chem. Soc.*, **119**, 6345–6359 (1997); (d) Yashima, E., Matsushima, T., Okamoto, Y. *J. Am. Chem. Soc.*, **123**, 8159–8160 (2001); (e) Nonokawa, R., Yashima, E. *J. Am. Chem. Soc.*, **125**, 1278–1283 (2003); (f) Onouchi, H., Hasegawa, T., Kashiwagi, D., Ishiguro, H., Maeda, K., Yashima, E. *Macromolecules*, **38**, 8625–8633 (2005); (g) Maeda, K., Mochizuki, H., Watanabe, M., Yashima, E. *J. Am. Chem. Soc.*, **128**, 7639–7650 (2006); (h) Yashima, E., Maeda, K. *Macromolecules*, **41**, 3–12 (2008); (i) Maeda, K., Miyagawa, T., Furuko, A., Onouchi, H., Yashima, E. *Macromolecules*, **48**, 4281–4293 (2015).

15. (a) Nomura, R., Tabei, J., Masuda, T. *J. Am. Chem. Soc.*, **123**, 8340–8341 (2001); (b) Deng, J., Tabei, J., Shiotsuki, M., Sanda, F., Masuda, T. *Macromol. Chem. Phys.*, **205**, 1103–1107 (2004); (c) Tabei, J., Nomura, R., Sanda, F., Masuda, T. *Macromolecules*, **37**, 1175–1179 (2004).

16. (a) Suzuki, Y., Shiotsuki, M., Sanda, F., Masuda, T. *Macromolecules*, **40**, 1864–1867 (2007); (b) Suzuki, Y., Shiotsuki, M., Sanda, F., Masuda, T. *Chem. Asian J.*, **3**, 2075–2081 (2008); (c) Shirakawa, Y., Suzuki, Y., Terada, K., Shiotsuki, M., Masuda, T., Sanda, F. *Macromolecules*, **43**, 5575–5581 (2010).

17. Tabei, J., Shiotsuki, M., Sanda, F., Masuda, T. *Macromolecules*, **38**, 5860–5867 (2005).

18. (a) Li, B. S., Cheuck, K. K. L., Ling, L., Chen, J., Xiao, X., Bai, C., Tang, B. Z. *Macromolecules*, **36**, 77–85 (2003); (b) Cheuck, K. K. L., Lam, J. W. Y., Chen, J., Lai, L. M., Tang, B. Z. *Macromolecules*, **36**, 5947–5959 (2003); (c) Lai, L. M., Lam, J. W. Y., Qin, A., Dong, Y., Tang, B. Z. *J. Phys. Chem. B*, **110**, 11128–11138 (2006); (d) Cheuck, K. K. L., Li, B. S., Lam, J. W., Xie, Y., Tang, B. Z. *Macromolecules*, **41**, 5997–6005 (2008).

19. (a) Sanda, F., Araki, H., Masuda, T. *Macromolecules*, **37**, 8510–8516 (2004); (b) Sanda, F., Terada, K., Masuda, T. *Macromolecules*, **38**, 8149–8154 (2005); (c) Sanda, F., Araki, H., Masuda, T. *Macromolecules*, **38**, 10605–10608 (2005); (d) Zhao, H., Sanda, F., Masuda, T. *J. Polym, Sci. Part A: Polym. Chem.*, **45**, 1691–1698 (2007); (e) Liu, R., Sanda, F., Masuda, T. *Polymer*, **48**, 6510–6518 (2007); (f) Terada, K., Masuda, T., Sanda, F. *J. Polym. Sci. Part A: Polym. Chem.*, **47**, 4971–4981 (2009); (g) Terada, K., Masuda, T., Sanda, F. *Macromolecules*, **42**, 913–920 (2009); (h) Ikeda, A., Terada, K., Shiotsuki, M., Sanda, F. *J. Polym. Sci. Part A: Polym. Chem.*, **49**, 3783–3796 (2011); (i) Sogawa, H., Shiotsuki, M., Sanda, F. *J. Polym. Sci. Part A: Polym. Chem.*, **50**, 2008–2018 (2012).

20. (a) Maeda, K., Kamiya, N., Yashima, E. *Chem. Eur. J.*, **10**, 4000–4010 (2004); (b) Maeda, K., Tsukui, H., Matsuhita, Y., Yashima, E. *Macromolecules*, **40**, 7721–7726 (2007); (c) Kakuchi, R., Nagata, S., Sakai, R., Otsuka, I., Nakade, H., Satoh, T., Kakuchi, T. *Chem. Eur. J.*, **14**, 10239–10266 (2008); (d) Kakuchi, R., Tago, Y., Sakai, R., Satoh, T., Kakuchi, T. *Macromolecules*, **42**, 4430–4435 (2009); (e) Sakai, R., Sakai, N., Satoh, T., Li, W., Zhang, A., Kakuchi, T. *Macromolecules*, **44**, 4249–4257; (f) Ohsawa, S., Sakurai, S., Nagai, K., Bannno, M., Maeda, K., Kumaki, J., Yashima, E. *J. Am. Chem. Soc.*, **133**, 108–114 (2011); (g) Song, J., Zhang, H., Deng, J. *React. Funct. Polym.*, **93**, 10–17 (2015).

21. (a) Nomura, R., Tabei, J., Masuda, T. *Macromolecules*, **35**, 2955–2961 (2002); (b) Tabei, J., Nomura, R., Masuda, T. *Macromolecules*, **35**, 5405–5409 (2002); (c) Percec, V., Obata, M., Rudick, J. G., De, B. B., Glodde, M., Bera, T. K., Magonov, S. N., Balagurusamy, V. S. K., Heiney, P. A. *J. Polym. Sci., Part A: Polym. Chem.*, **40**, 3509–3533 (2002); (d) Gao, G. Z., Sanda, F., Masuda, T. *Macromolecules*, **36**, 3938–3948 (2003); (e) Morino, K., Watase, N., Maeda, K., Yashima, E. *Chem. Eur. J.*, **10**, 4703–4707; (f)

Deng, J., Luo, X., Zhao, W., Yang, W. *J. Polym. Sci., Part A: Polym. Chem.*, **46**, 4112–4121 (2008).

22. (a) Tabei, J., Nomura, R., Sanda, F., Masuda, T. *Macromolecules*, **38**, 9448–9454 (2005); (b) Kaneko, T., Umeda, Y., Yamamoto, T., Teraguchi, M., Aoki, T. *Macromolecules*, **38**, 9420–9426 (2005); (c) Suzuki, Y., Tabei, J., Shiotsuki, M., Inai, Y., Sanda, F., Masuda, T. *Macromolecules*, **41**, 1086–1093 (2008); (d) Suzuki, Y., Miyagi, Y., Shiotsuki, M., Inai, Y., Masuda, T., Sanda, F. *Chem. Eur. J.*, **20**, 15131–15143 (2014).
23. (a) Shinohara, K., Yasuda, S., Kato, G., Fujita, M., Shigekawa, H. *J. Am. Chem. Soc.*, **123**, 3619–3620 (2001); (b) Sakurai, S., Kuroyanagi, K., Morino, K., Kunitake, M., Yashima, E. *Macromolecules*, **36**, 9670–9674 (2003); (c) Sakurai, S., Okoshi, K., Kumaki, J., Yashima, E. *J. Am. Chem. Soc.*, **128**, 5650–5651 (2006); (d) Yashima, E. *Polym. J.*, **42**, 3–16 (2010); (e) Ohsawa, S., Sakurai, S., Nagai, K., Bannno, M., Maeda, K., Kumaki, J., Yashima, E., Sakurai, S. *J. Am. Chem. Soc.*, **133**, 108–114 (2011); (f) Bergueiro, J., Freire, F., Wendler, E. P., Seco, J. M., Quiñoá, E., Riguera, R. *Chem. Sci.*, **5**, 2170–2176 (2014).
24. (a) Yashima, E., Huang, S., Okamoto, Y. *J. Chem. Soc., Chem. Commun.*, 1811–1812 (1994); (b) Okamoto, Y., Yashima, E. *Angew. Chem. Int. Ed.*, **37**, 1020–1043 (1998); (c) Haupt, K., Mosbach, K. *Chem. Rev.*, **100**, 2495–2504 (2000); (d) Nakano, T. *J. Chromatogr. A*, **906**, 201–205 (2001); (e) Okamoto, Y., Ikai, T. *Chem. Soc. Rev.*, **37**, 2593–2608 (2008); (f) Shimomura, K., Ikai, T., Kanoh, S., Yashima, E., Maeda, K. *Nat. Chem.*, **6**, 429–434 (2014).
25. Gellman, S H. *Acc. Chem. Res.*, **31**, 173–180 (1998).
26. (a) Nelson, J. C., Saven, J. G., Moore, J. S., Wolynes, P. G. *Science*, **277**, 1793–1796 (1997); (b) Prince, R. B., Saven, J. G., Wolynes, P. G., Moore, J. S. *J. Am. Chem. Soc.*, **121**, 3114–3121 (1999); (c) Prince, R. B., Brunsveld, L., Meijer, E. W., Moore, J. S. *Angew. Chem. Int. Ed.*, **39**, 228–230 (2000).
27. (a) Fiesel, R., Scherf, U. *Macromol. Rapid Commun.*, **19**, 427–431 (1998); (b) Wilson, J. N., Steffen, W., McKenzie, T. G., Lieser, G., Oda, M., Neher, D., Bunz, U. H. F. *J. Am. Chem. Soc.*, **124**, 6830–6831 (2002); (c) Resta, C., Pescitelli, G., Bari, L. D. *Macromolecules*, **47**, 7052–7059 (2014).
28. (a) Gin, M. S., Yokozawa, T., Prince, R. B., Moore, J. S. *J. Am. Chem. Soc.*, **121**, 2643–2644 (1999); (b) Kim, H. J., Zin, W. C., Lee, M. *J. Am. Chem. Soc.*, **126**, 7009–7014 (2004); (c) Zhao, X. Y., Schanze, K. S. *Langmuir*, **22**, 4856–4862 (2006); (d) Kaneko, T., Yoshimoto, S., Hadano, S., Teraguchi, M., Aoki, T. *Polyhedron*, **26**, 1825–1829 (2007); (e) Sanji, T., Sato, Y., Kato, N., Tanaka, M. *Macromolecules*, **40**, 4747–4749 (2007); (f)

Kim, H. J., Lee, E., Park, H., Lee, M. *J. Am. Chem. Soc.*, **129**, 10994–10995 (2007); (g) Kobayashi, S., Itomi, K., Morino, K., Iida, H., Yashima, E. *Chem. Commun.*, 3019–3021 (2008); (h) Liu, R., Shiotsuki, M., Masuda, T., Sanda, F. *Macromolecules*, **42**, 6115–6122 (2009); (i) Yu, Z., Hecht, S. *Angew. Chem. Int. Ed.*, **50**, 1640–1643 (2011); (j) Abe, H., Okada, K., Makida, H., Inouye, M. *Org. Bioorg. Chem.*, **10**, 6930–6936 (2012); (k) Banno, M., Yamaguchi, T., Nagai, K., Kaiser, C., Hecht, S., Yashima, E. *J. Am. Chem. Soc.*, **134**, 8718–8728 (2012); (l) Ousaka, N., Yamaguchi, T., Yashima, E. *Chem. Lett.*, **43**, 512–514 (2014); (m) Burkhalter, M., Kandre, R., Lukin, O. *Tetrahedron*, **71**, 4132–4136 (2015).

29. (a) Liu, R., Shiotsuki, M., Masuda, T., Sanda, F. *Macromolecules*, **42**, 6115–6122 (2009); (b) Liu, R., Sogawa, H., Shiotsuki, M., Masuda, T., Sanda, F. *Polymer*, **51**, 2255–2263 (2010); (c) Sogawa, H., Shiotsuki, M., Matsuoka, H., Sanda, F. *Macromolecules*, **44**, 3338–3345 (2011); (d) Hashimoto, A., Sogawa, H., Shiotsuki, M., Sanda, F. *Polymer*, **53**, 2559–2563 (2012); (e) Sogawa, H., Shiotsuki, M., Sanda, M. *Macromolecules*, **46**, 4378–4387 (2013); (f) Sogawa, H., Shiotsuki, M., Hirao, T., Haino, T., Sanda, M. *Macromolecules*, **46**, 8161–8170 (2013); (g) Sogawa, H., Shiotsuki, M., Sanda, M. *Macromolecules*, **46**, 8896–8904 (2013); (h) Hashimoto, A., Miyagi, Y., Sogawa, H., Yamamoto, S., Sanda, M. *Chem. Lett.*, **43**, 1622–1644 (2014).

30. (a) Takata, T., Furusho, Y., Murakawa, K., Endo, T., Matsuoka, H., Hirasa, T., Matsuo, J., Sisido, M. *J. Am. Chem. Soc.*, **120**, 4530–4531 (1998); (b) Takata, T., Murakawa, K., Furusho, Y. *Polym. J.*, **31**, 1051–1056 (1999); (c) Seto, R., Maeda, T., Konishi, G., Takata, T. *Polym. J.*, **39**, 1351–1359 (2007); (d) Yi, Z., Okuda, H., Koyama, Y., Seto, R., Uchida, S., Sogawa, H., Kuwata, S., Takata, T. *Chem. Commun.*, **51**, 10423–10426 (2015); (e) Okuda, H., Koyama, S., Uchida, S., Michinobu, T., Sogawa, H., Takata, T. *ACS Macro Lett.*, **4**, 462–466 (2015).

31. (a) Prince, R. B., Barnes, S. A., Moore, J. S. *J. Am. Chem. Soc.*, **122**, 2758–2762 (2000); (b) Tanatani, A., Mio, M. J., Moore, J. S. *J. Am. Chem. Soc.*, **123**, 1792–1793 (2001); (c) Arnt, L., Tew, G. N. *Macromolecules*, **37**, 1283–1288 (2004); (d) Inouye, M., Waki, M., Abe, H. *J. Am. Chem. Soc.*, **126**, 2022–2027 (2004); (e) Smaldone, R. A., Moore, J. S. *J. Am. Chem. Soc.*, **129**, 5444–5450 (2007); (f) Suk, J-m., Naidu, V. R., Liu, X., Lah, M. S., Jeong, K-S. *J. Am. Chem. Soc.*, **133**, 13938–13941 (2011); (g) Akine, S., Sairenji, S., Taniguchi, T., Nabeshima, T. *J. Am. Chem. Soc.*, **135**, 12948–12951 (2013).

32. Nakazono, K., Fukasawa, K., Sato, T., Koyama, Y., Takata, T. *Polym. J.*, **42**, 208–215 (2010).

33. Ishiwari, F., Nakazono, K., Koyama, Y., Takata, T. *Chem. Commun.*, **47**, 11739–11741 (2011).
34. Suzuki, S., Ishiwari, F., Nakazono, K., Takata, T. *Chem. Commun.*, **48**, 6478–6480 (2012).
35. Suzuki, S., Matsuura, K., Nakazono, K., Takata, T. *Polym. J.*, **46**, 355–365 (2014).
36. Bloomfield, V. A., Crothers, D. M., Tinoco, I. *Physical Chemistry of Nucleic Acids*, Harper & Row: New York, NY, USA, pp. 83–102 (1974).
37. Cantor, C. R., Schimmel, P. R. in *Biophysical Chemistry Part II, Techniques for the Study of Biological Structure and Function*, W. H. Freeman: San Francisco, pp. 399–404 (1980).

Biography

Hiromitsu Sogawa was born in Tokushima, Japan, in 1984. He received his BS (2008), MS (2010), and PhD (2013) degrees from Kyoto University, Japan, under the supervision of Prof. Fumio Sanda. He joined the group of Prof. Toshikazu Takata at the Tokyo Institute of Technology, Japan, as assistant professor in 2013. His research interests are the synthesis, design, and application of functional conjugated polymers and the development of supramolecular materials based on interlocked molecules such as (poly)rotaxanes.

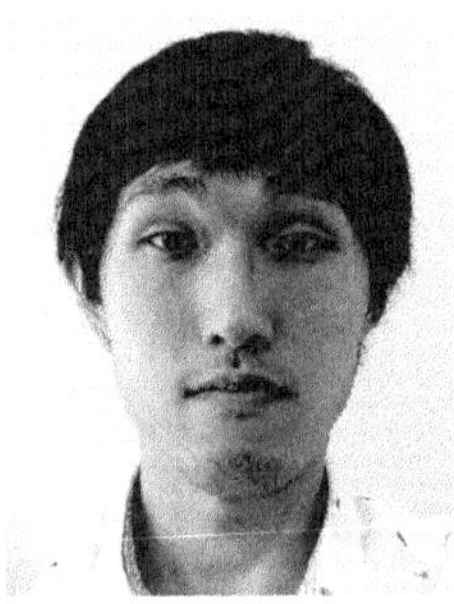

Kazuko Nakazono is a CREST scientist at the Tokyo Institute of Technology in Prof. Takata's lab. She earned her BS in chemistry from the Tokyo University of Science. She performed graduate studies on organic chemistry under the supervision of Prof. Shinichi Saito at the Tokyo University of Science. In 2005, she went to the Tokyo Institute of Technology and received her PhD in chemistry under the supervision of Prof. Toshikazu Takata in 2008. She became assistant professor in 2008 and subsequently a CREST scientist with the research group of Prof. Takata. Her research is directed toward polymer functionalization utilizing rotaxane's dynamic property.

Toshikazu Takata, professor of chemistry at the Tokyo Institute of Technology, was born in Toyama Prefecture of Japan in 1953. He received his BS degree from Okayama University (1976) and PhD from Tsukuba University (1981) under the supervision of Prof. Shigeru Oae. He was promoted to professor of the Japan Advanced Institute of Science and Technology in 1994 and then moved to Osaka Prefecture University in 1995 and the present Tokyo Tech in 2003. He has published more than 400 original papers,

160 reviews and books, and 160 patents in organic, polymer, and supramolecular chemistry areas. He has been awarded the Molecular Chirality Award; the CSJ Award for Technical Development; the Commendation of the Ministry of Education, Culture, Sports, Science and Technology; the Award of the Society of Rubber Science and Technology, Japan; the Award of the Society of Polymer Science, Japan; and the CSJ Award 2016 following Polymer Science, Japan.

Chapter 9

Control of Self-Assembling Behavior of Organic Polymers via Charge Transfer (CT) Interaction of π-Conjugated Planes

Kenta Kokado
Department of Chemistry, Faculty of Science, Hokkaido University, Kita10 Nishi8, Kita-ku, Sapporo, Hokkaido 060-0810, Japan
kokado@sci.hokudai.ac.jp

A molecule with rich π-electrons can donate its electrons to a molecule with poor π-electron, and they form a charge transfer (CT) interactions between their π-conjugated planes. Until now, the use of CT interaction in supramolecular chemistry has been fewer than that of hydrogen bonding or metal–ligand coordination. The major drawback of CT interaction for supramolecular chemistry should be the low association constant. However in recent, the surpass directionality and complementarity of CT interaction, in addition to durability toward solvents or simplicity of probing, have prompted chemists to explore various supramolecular architectures. In this chapter, the author reviews the utilization of CT interaction from the viewpoint of polymer chemistry, such as linear polymers connected

Conjugated Objects: Developments, Synthesis, and Applications
Edited by Atsushi Nagai and Koji Takagi

ISBN 978-981-4774-03-1 (Hardcover), 978-1-315-22962-1 (eBook)
www.panstanford.com

via CT interaction, and secondary structure variation of synthetic polymer chains via CT interaction, in detail.

9.1 Introduction

Charge transfer (CT) interactions of π-conjugated planes can be defined as intermolecular interactions between π-electron-rich (donor molecules) and π-electron-deficient species (acceptor molecules). The combination of electron donor and acceptor provides supramolecular complex formation with high specificity and complementarity. The resulting CT complex exhibits a distinctive absorption band in the visible region, offering important information on the association. As the electron-donating species, hexamethylbenzene (**1**), dimethylaniline (**2**), dialkoxynaphthalene (DAN, **3**), anthracene (**4**), pyrene (**5**), carbazole (**6**), tetrathiafulvalene (TTF, **7**) triphenylene (**8**), or truxene (**9**) are often employed, and as the electron-accepting species, trinitrobenzene (**10**), dichlorocicyanobenzoquinone (DDQ, **11**), tetracyanoquinonedimethane (TCNQ, **12**), viologen (**13**), trinitrofluorenone (TNF, **14**), pyromellitic diimide (PDI, **15**), naphthalene diimide (NDI, **16**), mellitic triimide (**17**), and truxenone (**18**) are often employed (Fig. 9.1). Traditionally, CT interactions have drawn much attention in the field of organic solid state chemistry for a long time, from the viewpoint of CT cocrystals or CT salts and consequent remarkable properties of their superconductivity or conductivity.[1–9]

On the other hand, recent extensive advancement of supramolecular chemistry has extensively explored the immense diversity in self-organizing matters.[10–25] In these reports, hydrogen bonding or metal–ligand coordination is mainly employed as the driving force. Compared to the utilization of hydrogen bonding or metal–ligand coordination in supramolecular chemistry, fewer examples have been reported for self-assembling process driven by CT interaction, probably because of a modest association constant and double-faced interacting moiety. However, a rational molecular design to improve the inherently modest association constant of CT interaction has been recently explored to accomplish a sophisticated supramolecular architecture, for example[24,26–28], often with involvement of another interaction such as hydrogen bonding, metal–ligand coordination, or solvophobic interactionas the aid. Additionally, the double-faced feature of CT interaction can offer an advantageous point from the perspective to construct multiple π-stacking structures with long-range order. In this chapter, the author sheds light on CT interaction from the viewpoint of polymer chemistry, thus linear polymers linked by CT interaction, and control of secondary structure of polymeric chains via CT interaction. The formation of gels or organic electronics by CT interaction is reviewed elsewhere.[20,24]

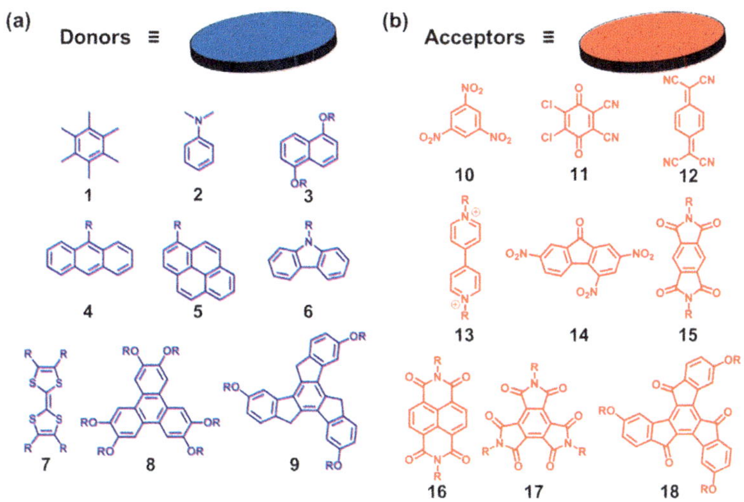

Figure 9.1 Molecular structures of representative donors and acceptors performing CT interaction of the π-conjugated plane.

9.2 Linear Polymers Linked by CT Interaction

Substantial efforts have been recently devoted to develop supramolecular polymers, because they can offer healable, stimuli-responsive, or switchable materials due to the reversible connectivity.[14,29] Therein, chemists have usually resorted to hydrogen-bonding motifs or metal–ligand coordination rather than CT interaction. This is because CT interaction can typically provide a modest association constant, and the linear growth of a polymer chain by CT interaction requires a laborious molecular design. Additionally, although the traditional measurement techniques such as UV/Vis or nuclear magnetic resonance (NMR) spectroscopy can visualize the CT interaction at each interacting point, capturing the entire structure of supramolecular polymer chain have remained difficult. Recent drastic advancement in measurement such as scattering, mass analysis, or microscopy has overcome above situation, thus now we can know how often and how large a supramolecular polymer forms. In this context, chemists have started to explore supramolecular polymers linked by CT interaction, exhibiting a characteristic features for conventional polymers such as viscosity or spinnability. In this section, the author reviews examples for preparation of supramolecular polymers by utilizing CT interaction as the main connectivity.

9.2.1 Porphyrin-Based Linear Polymers

Porphyrins are π-conjugated planar macrocyclic compounds having 26 π-electrons, and with regard to CT interaction, they can act as electron donors.[30] Haino et al. prepared a supramolecular polymer by using CT interaction of porphyrin as the main connectivity.[31] In previous reports, they found that a bisporphyrin "cleft" linked by 2,6-pyridinedicarboxamide efficiently bound effectively binds an electron-deficient flat aromatic guest such as 2,4,7-trinitrofluorenone (TNF), or 2,6-pyridinedicarboxamide linker itself.[32,33] The dimerization constant of 1 was found to be 5500 M^{-1} in toluene-d_8 and 310M^{-1} in $CDCl_3$, and the binding constant for TNF was determined to be 410,000 M^{-1}.[33] o build a linear polymer from the cleft molecule, they adopted a tetrakis porphyrin derivative (**19**), connecting the two clefts by a diyne group via Eglinton coupling (Fig. 9.2a). The obtained molecule **19** showed temperature and concentration-dependent adsorption variation of the single Soret band, and the association constant was found to be 1,500,000 M^{-1} in toluene from a nonlinear curve-fitting analysis by applying isodesmic model produced. The ^{1}H NMR spectroscopy of **19** in toluene exhibited broadened features even at a temperature as high as 80°C, indicating the existence of a polymeric structure. A tapping mode atomic force microscopy (AFM) observation of the drop-cast sample on a mica substrate showed the formation of fibrous structure with more than 1 μm length. The addition of an electron-deficient guest molecule TNF resulted in disruption of the polymeric assembly revealed from ^{1}H NMR.

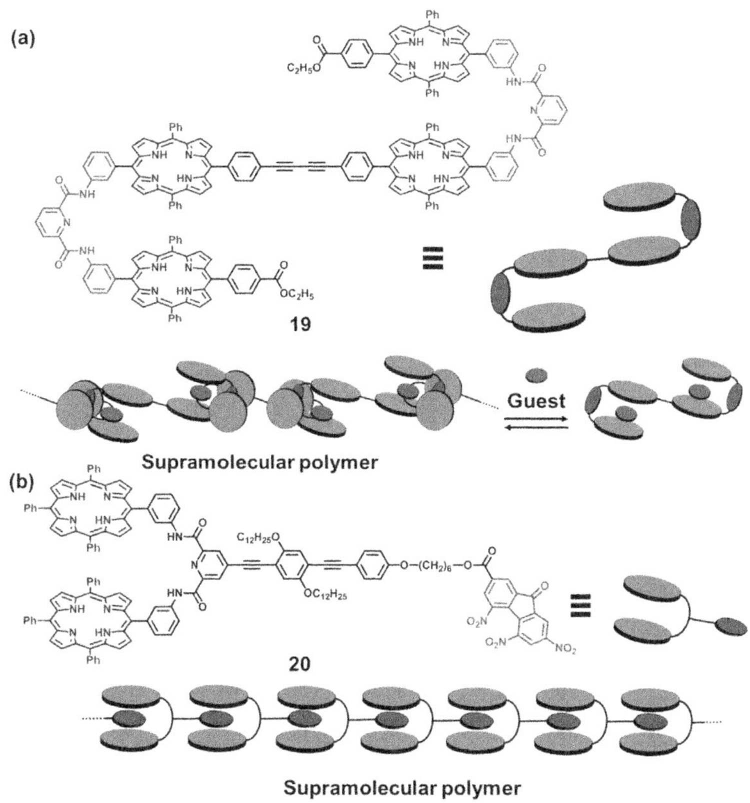

Figure 9.2 (a) Tetrakisporphyrin (**19**) forming linear polymers and accepting guest-induced depolymerization of **19**. (b) Heteroditopic monomer **20**.

In 2012, Haino and coworkers reported supramolecular polymerization of a heteroditopic monomer having both bisporphyrin cleft and TNF moiety (**20**, Fig. 9.2b).[34] The obtained monomer **20** showed concentration-dependent UV/Vis absorption behavior in toluene and chloroform, and the curve-fitting analysis based on isodesmic model determined the association constant as 42,000 M^{-1}. The intermolecular close contact of the cleft and TNF moiety was also confirmed by nuclear Overhauser effects

(NOE), indicating the formation of head-to-tail supramolecular polymer. Diffusion-ordered spectroscopy (DOSY) NMR technique revealed the reduction of the diffusion constant upon the increase of concentration of **20** in $CDCl_3$, due to the increase of the hydrodynamic radii. A viscometry also showed chain propagation upon the increase of concentration. The fibrous network structure of the supramolecular polymer was proved by AFM observation. Note that an analogue molecule having acetyl group instead of TNF never show such features characteristic for polymers. Very recently, Haino and coworkers explored well-defined network formation of the above supramolecular polymer via metalation of porphyrin and following complexation with bispyridine derivatives.[35] The obtained elastic freestanding film with a high Young's modulus as 1 GPa was fabricated by the networking technique.

9.2.2 TTF-Based Linear Polymers

Tetrathiafulvalene (TTF) is π-conjugated organosulfur cyclic compound possessing 14 π-electrons, and it can act as an electron donor to present CT interaction.[36] Martin and coworkers reported synthesis of supramolecular polymers by using a heteroditopic monomer consisting of a fullerene (C_{60}) and exTTF (2-[9-(1,3-dithiol-2-ylidene]anthracen-10(9*H*)-ylidene)-1,3-dithiole),[37] which was found to be a building block of receptor for fullerene.[38] An isophthalic acid derivative having two exTTF moieties showed a high association constant with the fullerene by CT interaction in $CHCl_3$ (3560 M^{-1}) in the previous report. To expand the connectivity, they adopted the heteroditopic monomer (**21**) synthesized via condensation between PCBA and an isophthalic acid derivative. The increase of concentration in $CDCl_3$ solution directly caused the broadening of the ^1H NMR signal, indicating the formation of high-molecular-weight aggregates. The variable-temperature NMR measurement showed a preferential association of the fullerene on the aromatic side of exTTF at lower temperature. The signal from α methylene proton of the PCBA ester unit exhibited an up-field shift, which was not observed in the concentration experiment. In MALDI-TOF mass spectrometry (MS) measurement, the peaks derived from pentamer or less were detected. Contrastingly in DLS measurement, a broad distribution was observed with the maximum at 378 nm corresponding to 160-mer, indicating the presence of nonspecific oligomer aggregates. The cyclic voltammetry (CV) measurement showed the variation of electronic properties depending on the concentration; thus the increase of concentration caused the more energetic and irreversible feature of both the reduction process for the C_{60} moiety and the oxidation process of the exTTF moiety. From this result, the electronic communication took place in the ground state, reflected the data from UV/Vis measurement. AFM observation of the drop-cast sample illustrated the formation of a long (15–300 nm) and winding necklace-like tube with 2 nm height, which was greatly different from PCBM alone.

Consequently, Martin and coworkers extended the concept to synthesis of dendrimer by tethering two exTTF moieties on one PCBA moiety.[39] In this case, the ^{1}H NMR patterns also exhibited broadening and slight shielding upon the increase of concentration (1.2–20 mM), likewise to VT-NMR measurement upon cooling. Significant change of the signal was observed on the proton of isophthalate at the periphery, indicating the intermolecular binding of C_{60} moiety within the receptor cavity having two exTTF. The MALDI-TOF MS spectra showed the formation of a hexamer ($m/z \approx 18{,}000$), although DLS measurement disclosed the existence two kinds of distribution, thus a small assembly with around 10 nm diameter and large aggregates with around 1 μm diameter. AFM observation displayed the size and morphology of the assembly with uniform height (~1 nm). The appearance of a CT band was confirmed in UV/Vis spectroscopy at 487 nm, implying that the CT interaction acted as the driving force. As can be seen in the linear polymer described above, the obtained dendrimer also exhibited a more energetic and irreversible feature in electronic properties upon an increase of concentration, revealed by CV measurement.

Martin and coworkers recently reported great improvement of binding constant and consequent molecular weight of the above C_{60}-exTTF linear supramolecular polymer system by connecting the two exTTF via a covalent linkage.[40] They previously found the improvement of the associating constant by adopting macrocyclic scaffolds in low-molecular-weight compounds.[41–43] The UV/Vis titration experiment of PC61BM and a model exTTF derivative indicated high K_a (126,000 M^{-1}), proved by an emerging CT absorption band at 480 nm with a definite isosbectic point at 451 nm. As the supramolecular polymerization result, the MALDI-TOF MS measurement showed a very high molecular weight up to 90 kDa corresponding to 46-mer. The measurement of 1H NMR did not display any significant changes upon the variation of concentration (0.05–0.5 mM) as well as UV/Vis absorption spectroscopy (0.007–0.4 mM), meaning the formation of supramolecular polymer at the concentration in this range. The square-wave voltammetry (SWV) showed a higher energetic shift of the electronic properties. The DLS measurement disclosed the detailed profile of the growing polymer; thus at higher concentration (1 mM), only a very narrow distribution was observed at around 1000 nm, while at lower concentration (0.01 mM), several peaks were detected at a full observable size range, including the monomeric peak at around 1 nm. These data suggested that this polymerization does not follow an isodesmic growth model, and at 1mM, the molecular weight should be >150 kDa calculated from the hydrodynamic radius. AFM observation showed the production of remarkably long and straight fibers with 2 μm length and 1.5 nm height.

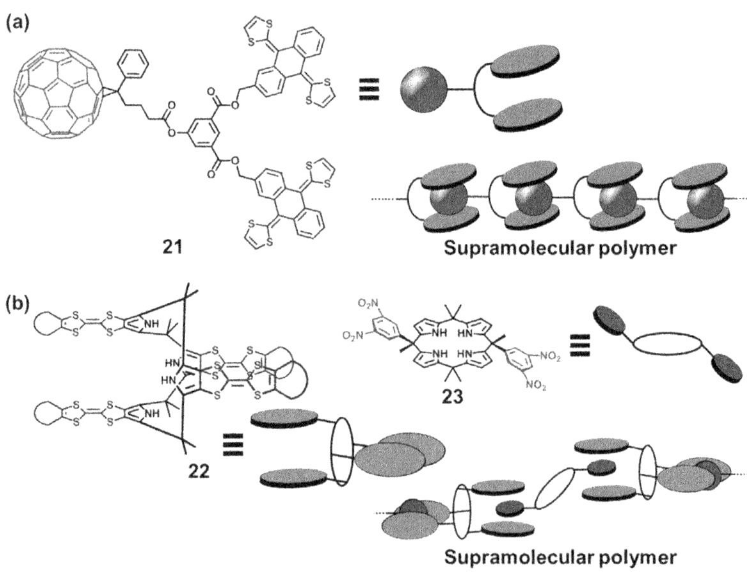

Figure 9.3 (a) Heteroditopic monomer having fullerene and exTTF. (b) Calix[4] pyrroles with TTF (**22**) and dinitrobenzene (**23**).

Sessler and coworkers reported the formation of supramolecular polymers from tetra-TTF-functionalized calix[4]pyrrole as the electron donor (**22**, Fig. 9.3b) and bis(dinitrophenyl)-meso-functionalized calix[4]pyrrole as the acceptor (**23**).[44] Therein, two TTF sandwiched one dinitrophenyl group regardless of diastereomeric isomers of **23**, revealed by single-crystal X-ray analysis. The 1:1 complexation was also confirmed by Job's plot, and the association constant between **22** and **23** was found to be 79,000 M^{-1} in $CHCl_3$, thus at 20 mM, 90-mer of the supramolecular polymer could be obtained. Furthermore, the supramolecular polymer could undergo a deaggregation triggered by addition of 2 equiv. of chloride anion or trinitrobenzene (TNB). In the case of chloride anion, the conformational change of the core calix[4]pyrrole from flat to cone shape was responsible for the deaggregation, while in the case of TNB, the competitive complexation with tetra-TTF-functionalized calix[4]pyrrole was responsible for the deaggregation.

9.2.3 Host-Stabilized CT-Based Linear Polymers

As the author described in the introductive part, CT interaction usually presents modest strength of connection, thus reinforcing the interaction that utilizing host stabilization is one of the rational designs to achieve supramolecular polymerization. From this point of view, Kim and coworkers reported the formation of supramolecular polymers by the aid of host-stabilized CT interaction in a macrocyclic compound, cucurbit[8]uril (CB[8]).[45] They previously found the enhanced CT interaction in CB[8],[46] and presented formation of a "molecular necklace", [6]MN, in which a small rings are threaded onto a large ring, from 10 components, by using a molecule having both electron-donating and electron-accepting groups.[47] As the donor group, naphthalene or benzene with electron-donating groups were employed, and as the acceptor group, viologen or pyridinium derivatives were used. The formation of a CT complex was confirmed by UV/Vis absorption spectra, for example, the combination of dihydroxy naphthalene and methylviologen showed emergence of absorption band at 580 nm.[46] To form a supramolecular polymer, they adopted growth of polymer chain from the surface of gold. At first, a guest molecule was synthesized in which both 2-hydroxynaphthalene and 4,4'-bipyridinium moiety connected by rigid p-xylene linker to prevent intramolecular complexation (**24**, Fig. 9.4a). In H_2O, **24** showed strong absorption at 482 nm upon the addition of CB[8]. The formation of 2:2 complex and polymeric complex was confirmed by NMR spectroscopy technique, and the degree of polymerization was estimated to be ~4 at the concentration of 2 mM. By using thiol-terminated molecule having dipyridiniumethylene unit (**25**), a stronger electron acceptor than 4,4'-pyridinium, a "seed" for polymerization was immobilized on a gold surface as a self-assembled monolayer (SAM). Soaking the modified substrate in an aqueous solution of CB[8] and **24** resulted in an increase of IR absorption peaks corresponding to C=O and C–N stretching band of CB[8] up to 4times compared to the initial SAM. The depolymelization was also possible just by washing the substrate by pure water. A surface plasmon resonance (SPR) technique also revealed the stepwise growth of supramolecular polymer on the surface. After evaporation, the growth was also confirmed by AFM observation, and the height was increased by 3.9 nm average compared to that of initial SAM.

The successful synthesis of supramolecular polymers by using HSCT spurs chemists to explore other polymers. For example, Zhang and coworkers reported elongation of the obtainable supramolecular polymers by utilizing multiple HSCT of DADV (1',1"-(butane-1,4-diyl) bis(1-(anthracen-2-ylmethyl)-4,4'bipyridine-1,1'di-ium)bromide, **26**) in CB[8] (Fig. 9.4b).[48] The molecule contains two anthracene units as an electron donor, and two viologen units as an electron acceptor, linked by short alkyl chain. Thus the resulting double HSCT interaction can construct head-to-tail connection. The short alkyl chain prevents the formation of 1:1 or 2:2 cyclic complex, thus a long polymer chain can be produced. Utilization of AFM-based single-molecule force microscopy (SMFS) showed slightly larger Kuhn length as 2.2 nm compared to conventional covalent polymers, and much smaller segment elasticity as 1.25 $pN\cdot nm^{-1}$. Thus the chain is not flexible but acts as a spring thanks to the noncovalent bonding. DLS measurement showed a hydrodynamic radius at around 30 nm, indicating the formation of polymeric compounds. The data from ^1H NMR and UV/Vis spectroscopy exhibited the progress supramolecular polymerization. In 1H NMR spectra, the peaks at aromatic regions were broadened after addition of CB[8]. In UV/Vis spectra, firstly the peak at 540 nm was emerged, where as it was hypsochromically shifted after addition of enough CB[8], indicating two step complexation, that is, single HSCT and subsequent double HSCT. With increased concentration (4 mM) of **26**, a gelation was observed, probably due to interaction between carbonyl groups on CB[8] or the formation of branch. Actually, the use of monofunctionalized model compound did not caused the gel formation, and the addition of MV2+ or K+ resulted in fast destruction of formed gel.

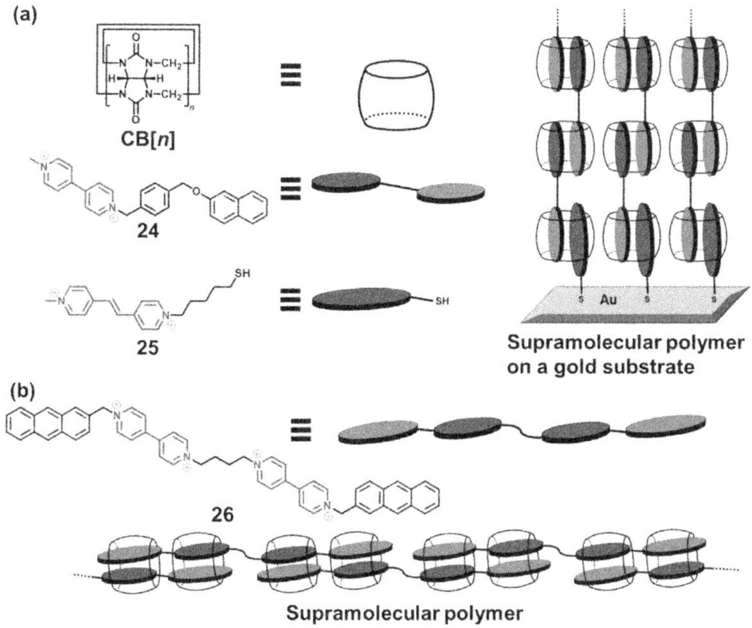

Figure 9.4 (a) Supramolecular polymerization of **24** on a gold surface driven by host-stabilized CT interaction (HSCT). (b) Supramolecular polymerization of **25** driven by HSCT.

As another example, Pang et al. presented formation of linear supramolecular polymers from thiazole orange, an organic dye, and CB[8].[49] Upon addition of CB[8] to aqueous solution of thiazole orange, ESI-MS measurement illustrated the formation of 2:2 complex. The absorption spectra showed a blue shift from 500 nm to 484 nm with a small distribution of a red shift to 520 nm, indicating the formation of an H-dimer and a J-aggregate. In addition, a drastic enhancement of fluorescence intensity was also observed. A two-dimensional ESI traveling wave ion mobility mass spectroscopy (TWIM-MS) plot showed the existence of $[1:1]^{+}$,$[2:2]^{2+}$,$[3:3]^{3+}$, and$[4:4]^{4+}$ complexes from the peak as m/z = 1634, meaning the formation of linear supramolecular polymers with different sizes. The formation of fibrous objects was observed by AFM, having the length of 40 to 200 nm and the height of 1.3 nm. DLS measurement showed the presence of aggregates with 130 nm and 2,800 nm diameter. As the concentration of the complex increased, the viscosity was gradually increased with enhancement of fluorescence intensity. The use of CB[7], a smaller host compound, resulted in only moderate fluorescence enhancement, and ESI-MS measurement pointed out the formation of only 1:2 complex of thiazole orange and CB[7].

As reviewed in this section, the preparation of linear polymers connected with CT interaction has been accomplished by rational molecular designs. These researches suggest the possibility of a unique strategy for controlling the size and morphology of supramolecular polymeric assemblies with sophisticated structures at the nanometer scale by the aid of CT interaction. The sandwich-type CT complex formation or the use of host molecule are essential to grow the linear polymer chains, both for accomplishing extremely high association constant and preventing infinite growth of alternating stacks to the vertical direction of the π-conjugated planes.

9.3 Control of Secondary Structure of Polymeric Chains via CT Interaction

Polymers generally present their characteristics and functions not only from the primary structure but also the secondary or assembling structure, as can be seen in DNA or proteins in biological systems. For example, well-designed functional spaces in enzymes can offer their smart and precise catalytic activity, derived from their secondary or more highly ordered structure constructed by accumulation of weak inter or intramolecular interactions, including hydrogen-bonding or hydrophobic interactions. Recently, the generation of synthetic polymers with precisely controlled conformations have attracted many chemists, because of the possibility for emulation of nature systems and design of smart artificial materials. On the basis of this point of view, the term "foldamer" has gained prominence in the field of polymer and supramolecular chemistry, which points out synthetic polymers "adopting a specific compact conformation" in solution.[50–52] The most reported examples for foldamers have predominantly exploited the intramolecular interactions such as hydrogen bonding, solvophobic interaction, or steric hindrance to govern the polymer conformation. In this circumstance, the use of CT interaction for a foldamer has received a certain level of attention because it can provide a specific, soft, and sometimes double-faced interaction. In this section, the author reviews examples for controlling the secondary or more highly ordered structure of a polymer chain via CT interaction.

9.3.1 Folding Copolymer System with Alternating Donor and Acceptor Units

In 1995, Iverson and coworkers reported the folding phenomenon of synthetic polymers driven by CT interaction between 1,5-dialkoxynaphthalene (DAN)and 1,4,5,8-naphthalene diimide (NDI).[53] They firstly prepared a cocrystal of DAN as the electron donor and NDI as the electron acceptor (Fig. 9.5a), and confirmed the proper connectivity from the single-crystal X-ray analysis. Then they synthesized oligomeric compounds alternatively having the donor and the acceptor units linked by L-aspartic acid via a solid-phase method. The tetramer (**28**) or hexamer (**29**) showed 18 nm of bathochromic shift for the CT absorption band in 20 mM sodium phosphate at pH 7.0 compared to that of dimer (**27**) at 522 nm, indicating more than two aromatic groups are stacked simultaneously.[54] The observed hypochromism at the UV region revealed that the aromatic rings are electronically coupled via the stacked arrangement with a parallel configuration between the rings—thus the folding nature. The measurement of ^{1}H NMR in 20 mM sodium phosphate/D_2O clearly showed the stacking formation by the up-field shift of diimide protons in **28** and **29**, while the shift of **27** was found to be half, thus being partially stacked. In addition, COSY NMR presented diastereotopic signals for the methylene groups in **28** and **29**, meaning a restriction of rotational motion of the backbone. NOE measurement showed an enhancement between protons adjacent to the aromatic rings.

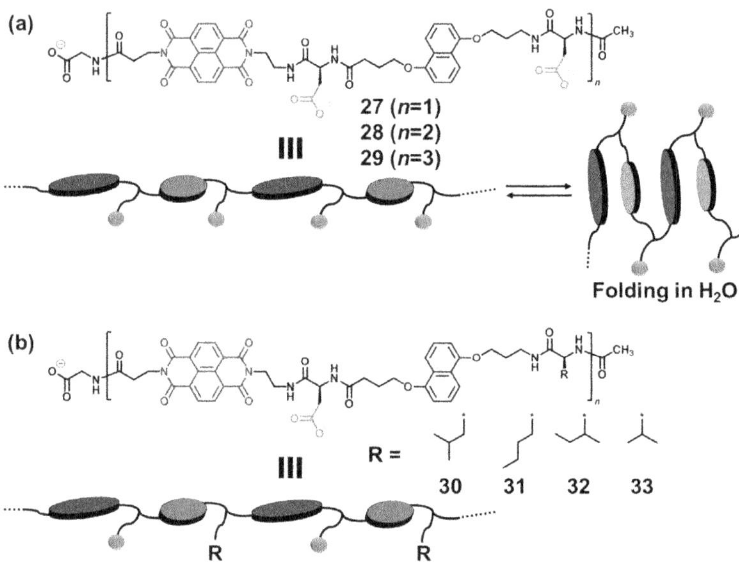

Figure 9.5 (a) Folding oligomers **27–29** with alternating donor and acceptor units. (b) Folding oligomers **30–33** with various side chains.

By altering the Asp unit to Leu in the oligomer, the oligomer showed more aggregating fashion in H_2O, found in ^1H NMR spectroscopy.[55] The addition of 2% CTAB resulted in unfolding of the oligomer. More importantly, a solution of the oligomer in H_2O presented irreversible transition to a viscous gel upon heating at 80°C, and the gel was not reverted to the original; solution even after 3 weeks. On the other hand, purification by reverse phase column chromatography can recover the original oligomer. By a kinetic study of the CT band on UV/Vis absorption, the existence of two growing phases was revealed—thus a slow phase and a fast phase. The fast phase was also able to induce by adding 10% of an already gelated sample to the oligomer solution before heating. The presumable mechanism of this is as follows: (1) First, the unfolding of the oligomer was induced by heating, exposing the hydrophobic surface of aromatic moiety. (2) Then, tangled aggregation grew and subsequently captured a number of unstacked molecules. A concentrate solution of peptide often behaves similarly, wherein the interior hydrophobic residues in initially folded protein are exposed to solvent by heat denaturation. The systematic study for solvent effect on the complexation of the oligomer by using a model compound unveiled the linear relationship between free energy and solvent polarity, and the association constant was higher in solvents with a higher solvent polarity parameter—meaning primarily driven by hydrophobic interaction.[56] In addition, the control of linkage sequence could finely tune the conformational modularity, that is, the sequence of DAN, NDI, and other residues.[57]

A thorough study using various dimer units by using spectroscopy and molecular modeling revealed that the oligomers predominantly showed a folded structure and thus face-to-face stacking of the electron donor and electron acceptor.[58] The offset and dihedral angles were significantly dependent on the length of linkage, disclosed by the molecular modeling. The molecule with longest linkage consisting of four amino acid residues showed the widest variety of the offset and angles. Nevertheless, the prepared dimers with flexible linkage could form the intramolecular folding structure, evidenced from ^{1}H NMR and UV/Vis spectroscopy.

The topology control of stacking is also achievable by changing the sequence of donor and acceptor in an oligomer chain. [59] Iverson and coworkers examined the formation of different type of folding from the pleated folding, thus showing intercalative or turn-type trimer folding by employing a trimer with DAN-DAN-NDI sequence. In this research, they employed β-peptide as the polymer backbone, and a decrease of only one methylene unit resulted in a dimer stack.

The phenomenon of heat-induced gelation was also deeply investigated by incorporating the hydrophobic amino acid residue such as Leu (**30**, Fig. 9.5b), Nle (**31**), Ile (**32**), and Val (**33**).[60] As the result, the Val derivative (**33**) was not gelated upon heating at 80°C, even at the maximum concentration (5 mM), while all samples showed the increase of viscosity at 80°C. The other three samples formed hydrogels at 80°C, and the hydrogel was irreversibly retained even after cooling at room temperature. The measurement of circular dichroism (CD) spectra exhibited the increase of ellipticity after heating at 80°C for all four samples, whereas a sample from Asp exhibited no change in the CD signal. The observation by scanning electron microscopy (SEM) demonstrated the formation of a fibrous structure for samples **30**, **31**, and **32**, while **33** showed a large mass of materials. These results indicated that the amyloid-like behavior is not unique to protein, but relatively a general property of amphiphilic foldable molecules, in which a stable folding state as well as irreversible formation of highly ordered aggregates are existing.

Ramakrishnan and coworkers presented a folding of synthetic polymers assisted by metal ion complexation.[61,62] They used a synthetic polymer having DAN as an electron donor and pyromellitic diimide (PDI) as an electron donor (**36**, Fig. 9.6), therein they were linked by hexa(oxyethylene) as a flexible chain with coordinating ability, synthesized via simple imidation polycondensation (M_n = 17,000).[61] The obtained polymer showed a folding structure derived from the intramolecular CT complexation, revealed by appearance of a CT band at 450 nm. Addition of a poor solvent MeOH resulted in an increase of the CT band. The formation of a CT complex was also confirmed by an up-field shift of aromatic protons in ^{1}H NMR, and more importantly, the addition of alkali metal ions such as K^+, Na^+, and Li^+. For example, the addition of KSCN caused a significant up-field shift of the aromatic protons and downfield shift of the aliphatic protons, indicating the enhancement of CT complexation in the presence of alkali metal ions. A systematic study by changing the spacer length from tetra(oxyethylene) to hexa(oxyethylene) (**34**–**36**, Fig. 9.6) unveiled the relationship between the spacer lengths.[62] In pristine polymers in solution, **34** showed the highest up-field shift, corresponding to the largest CT absorption in UV/Vis spectra. The polymer **34** also exhibited the highest thermosensitive up-field shift at lower temperature, meaning a greater tendency of sandwiched structure formation, thus the fully folded state, than those with a longer spacer. On the other hand upon the addition of alkali metal ion such as K^+, Na^+, and Li^+, **34** showed no significant change in ^{1}H NMR or UV/Vis spectra, while those of **35** and **36** exhibited a considerable shift. Interestingly, **35** showed the largest response to the addition of the smallest Li^+, whereas **36** showed the largest response with K^+, reflecting the compatibility between the ionic diameter and the spacer length. The observation by SMFS enabled direct stretching of the foldamer with hexa(oxyethylene).[63] Compared to the freely coiled structure of the polymer in DMSO, the K^+-coordinated foldamer required extra energy to stretch, although the folding was found to be far from perfect (~8%) in this case.

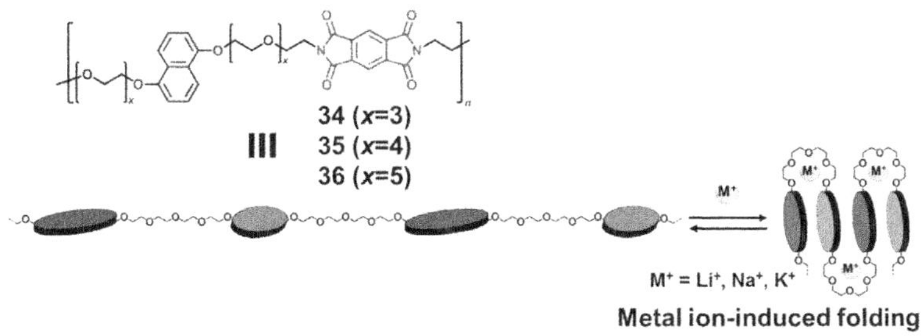

Figure 9.6 Folding of copolymers **34–36** induced by metal ions.

Instead of the poly(oxyethylene) chain, an tetraalkylammonium derivative was also employed as the space between donor and acceptor units by Ramakrishnan et al.[64] A di-*tert* amine monomer having DAN donor and a dibromo monomer having PDI acceptor were mixed to obtain the desired ionene (M_w = 45,000). The obtained ionene showed solvent-dependent CT absorption, thus the band became larger in a highly polar solvent such as H_2O due to acceleration of intramolecular CT complex formation, accompanying with a decrease of fluorescence intensity. The temperature decrease or ionic strength increase could also enhance the formation of a CT complex. The AFM observation directly illustrated the collapsed ionene aggregates having ~3 nm thickness, well corresponded to the distance between two ammonium groups, with the diameter of 50 to 160 nm, indicative of postulated accordion-type folding.

9.3.2 Folding Homopolymer System Affected by External Molecules

The addition of a small molecule to a polymer, both having electron-donating or electron-accepting ability, can directly tune the polymer conformation. Ramakrishnan and coworkers presented a small-molecule-induced folding process by employment of electron-accepting PDI linked by hexa(oxyethylene) chain as the polymer (**37**, Fig. 9.7) and electron-donating DAN derivative as the small molecule (**38**).[65] A small molecule having both DAN and ammonium units underwent the desired stacking, while the separate addition of DAN and equimolar ammonium salt resulted in only modest stacking, revealed by ^1H NMR and UV/Vis spectroscopy. The folding could be released by the addition of [18] crown-6 via extraction of the small molecule from the polymer. By using a DAN and PDI copolymer instead of the PDI polymer, two-step folding was presented.[66] In this case, the polymer was synthesized from PD anhydride and DAN diamine, and a second diamine, diaminohexa(oxyethylene), as a stack breaker. At high content of the stack breaker, the pristine polymer exhibited larger extent of isolated PDI unit in ^{1}H NMR spectra. The addition of NH_4SCN reinforced the stacking of originally sandwiched PDI, but the isolated PDI did not show any change, even with an excess amount of the ammonium salt. On the other hand, the further addition of a stackable DAN derivative having ammonium caused an up-field shift, even on isolated PDI. A reverse design was also possible by using a polymer having a DAN and PDI derivative as the small molecule.[67] An alkylamino moiety was incorporated in the main chain of the polymer; thus PDI tethering a carboxylic acid was used to form an ionic interaction with the amino group. The addition of a small molecule (PDI-carboxylic acid) to the polymer solution caused an appearance of CT absorption and inherent viscosity. Note that the addition of propionic acid gave no change; thus the two kinds of interactions, ionic and CT interactions, provided successful folding behavior.

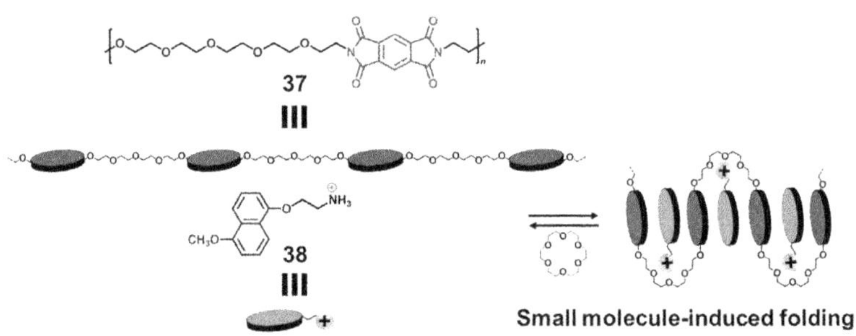

Figure 9.7 Folding of polymer (**37**) induced by a small molecule (**38**).

Rotello and coworkers presented restriction of the polymer solution structure by CT interaction between the polystyrene-tethering anthracene and the coexistent small molecule, picric acid. The polymer was firstly self-complexed via π-staking of the side chain; however, it underwent strong complexation with picric acid, resulting in improvement of thermostablity in the globular polymer structure.[68]

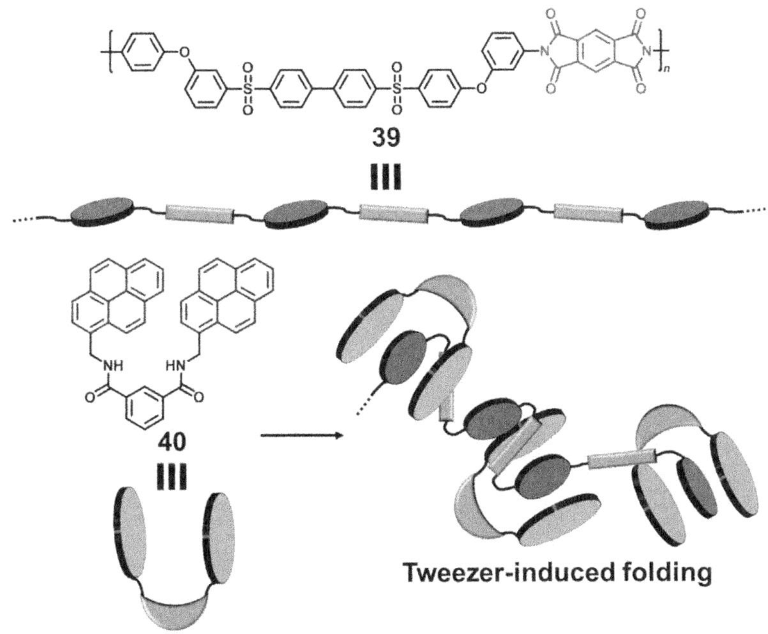

Figure 9.8 Folding of polymer (**39**) induced by a molecular tweezer (**40**).

Colquhoun and coworkers presented a folding control of the polymer by using a molecular tweezer.[69–71] The pyrene-based tweezer molecule **40** could bind the PDI moiety in the polymer **39** prepared from sulfone-ether-derived diamine and PD anhydride, revealed by ^{1}H NMR study of the foldamer and single-crystal X-ray analysis of the model compounds. Interestingly, the tweezer could recognize the difference of triplet in the polymer; thus it could effectively bind the unhindered PDI moiety than the hindered PDI moiety having adjacent only one methyl group, similarly to enzymes. The tough mechanical property and self-healing behavior was also explored by this group, with mixing a polyamide having an electron-donating pyrene unit at the chain end and an electron-accepting polymer having PDI connected by diethylene glycol.[72,73]

The use of a rotaxane structure is also an effective way to control polymer conformation through a CT interaction. Stoddart and coworkers reported the synthesis of a series of oligorotaxanes consisting of a oligomer of DAN and tetraethyleneglycol (TEG) units as the electron-donating thread (**41**) and cyclo-bis(paraquat-*p*-

phenylene) ($CBPQT^{4+}$, **42**) as the electron-accepting ring, forming a rotaxane-type CT complex (Fig. 9.9).[74,75] The thread chain end was capped by bulky diisopropylphenyl unit via a click reaction. Lengthening the thread oligomer resulted in the increase of stability in the secondary structure through the two kinds of CT interactions, revealed from the ^{1}H NMR analysis.[75] Further elongation of the donor thread was achieved by Stoddart et al.[74], in which the longest one had 15 DAN units with 8 $CBPQT^{4+}$ rings. From the thorough investigation, they found a well-defined folding structure of the oligorotaxanes, and an empirical rule forbidding $CBPQT^{4+}$ rings to occupy adjacent DAN sites. By combining the rotaxane with a hydrogen-bonding foldamer having two types of electron-donating units, DAN and TTF were synthesized to achieve a molecular switch.[76] The hydrogen-bonding aril amide linker acted as both a linker and a steric barrier for the ring movement. Thanks to this molecular design, the metastable state co-conformation (MSCC), and thus complexation of DAN and PDI, was highly stabilized and the lifetime of postoxidation MSCC was increased.

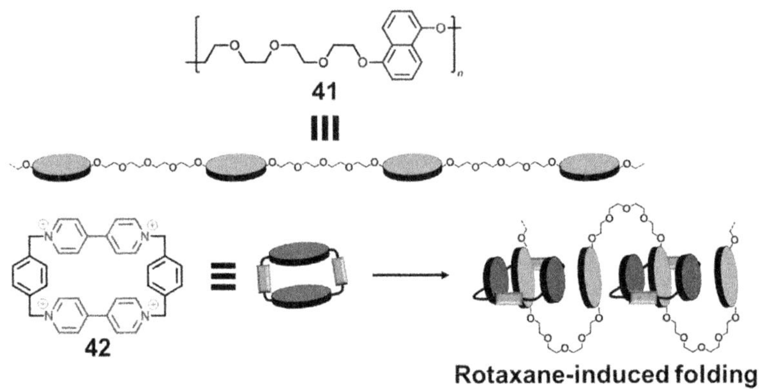

Figure 9.9 Rotaxane-induced folding of polymer (**41**) threaded through a macrocyclic compound (**42**).

Recently, the author and coworkers have shown tunable lower critical solution temperature (LCST) properties of a polymer containing pendant donor units (pyrene) in organic solvents through CT interactions (Fig. 9.10a).[77] In 1,2-dichloroethane, the polymer **43** was not soluble because of the presence of the π-stacking of aromatic units. However, the addition of acceptors having branched alkyl chains (**44–46**, named effector) resulted in the formation of a CT complex, changing the insoluble polymer to soluble. Interestingly, the elevation of temperature caused disassembly of the CT complex, resulting in macroscopic precipitation of the polymer, and thus LCST-type phase transition. The LCST-type phase transition is usually observed in H_2O, because it is governed by the hydrogen bonding between the polymer and H_2O. The present example illustrated a rational design for LCST phase transition of a polymer solution through a noncovalent complexation such as CT interaction with an effect or tethering a solubilizing group. In this sense, they synthesized a gel (**47**) by chemically crosslinking the above polymer (Fig. 9.10b).[78] The obtained gel was not swollen in organic solvents such as toluene, but mixing of electron-accepting small molecules drastically improved the swelling degree. Furthermore, the swollen gel by the aid of an effector showed volume phase transition temperature (VPTT); thus the gel was collapsed at higher temperature through disassembly of the CT complex.

The use of two kinds of oligomer sequences, donor oligomer and acceptor oligomer, enabled the formation of a heteroduplex in an aqueous solution like a DNA duplex. Li and coworkers reported a zipper-styled heteroduplex from two oligomers based on ethyleneimine, wherein one tethered DAN (**48**, Fig. 9.11a) and the other tethered PDI (**49**).[79–81] The combination of the longest oligomers having four aromatic rings exhibited the highest association constant of 11,000 M^{-1} in $CDCl_3$ from ^1H NMR.[79] Later, they presented the combination of two kinds of oligomers: the acceptor is the above PDI-tethering oligoethyleneimine and the donor is tetraethyleneglycol and DAN oligomer.[80] Regardless of the difference in the oligomer main chain, these two oligomers could form a heteroduplex, and the longest oligomer also showed the highest association constant (10,000 M^{-1}). Another accepting oligomer having PDI linked by diethyleneglycol analogously showed a high association constant (2650 M^{-1}). The other main chain such as δ-peptide based on L-ornithine was also applicable to express the above zipper feature, which was a kind of a block oligomer of the L-ornithine-tethering donor DAN and acceptor PDI.[81] The block oligomer showed a hairpin-like folded structure in one oligomer chain driven by CT interaction. The longest one having two DAN and three PDI units exhibited the highest CT absorption, and it did not vanish even at 150°C.

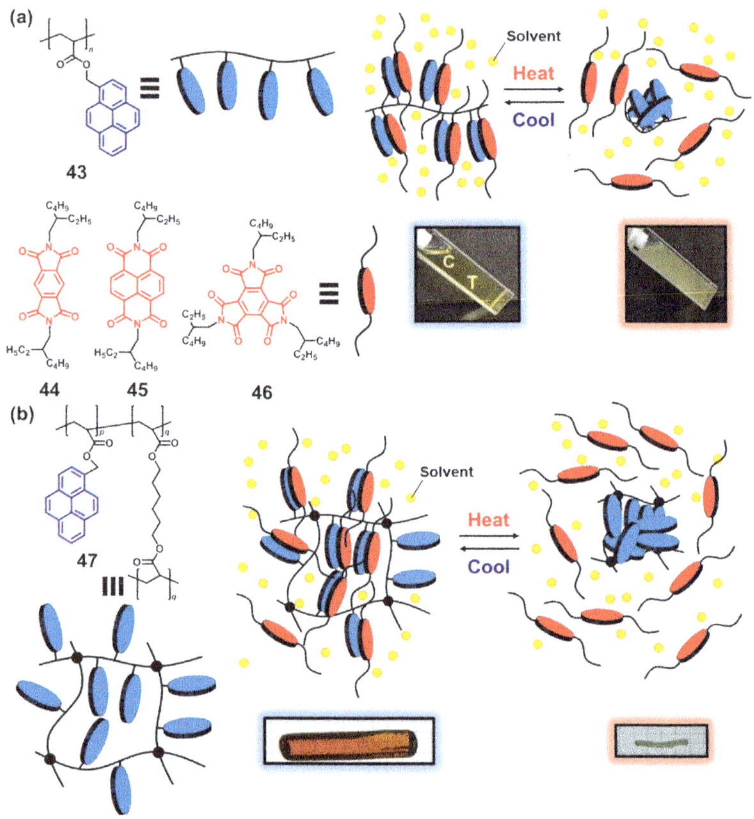

Figure 9.10 (a) LCST-type thermoresponsive behavior of polymer **43**, driven by association with effectors **44–46**. (b) Thermoresponsive volume phase transition behavior of gel **47**, driven by association with effectors **44–46**.

Iverson and coworkers reported two kind of oligomers, one with only DAN (**50**, Fig. 9.11b) and the other with only NDI (**51**).[82] They prepared eight oligomers with four different chain lengths, monomer to tetramer. The donor oligomer and acceptor oligomers with the same length formed a 1:1 complex when mixed together in H_2O, and NMR and isothermal titration calorimetry (ITC) revealed the increase of the association constant with an increase in the chain length from 130 M^{-1} (monomer) to 350,000 M^{-1} (tetramer, **50**+**51**) despite the larger electrostatic repulsion between the longer chains. The 1:1 complex formation was also confirmed by polyacrylamide gel electrophoresis (PAGE) besides Job's plot, wherein only 1:1 concentration provided one spot, otherwise an unbound oligomer was detected. Thanks to the intramolecular charge repulsion, the chain was spread and the duplex was consequently formed.

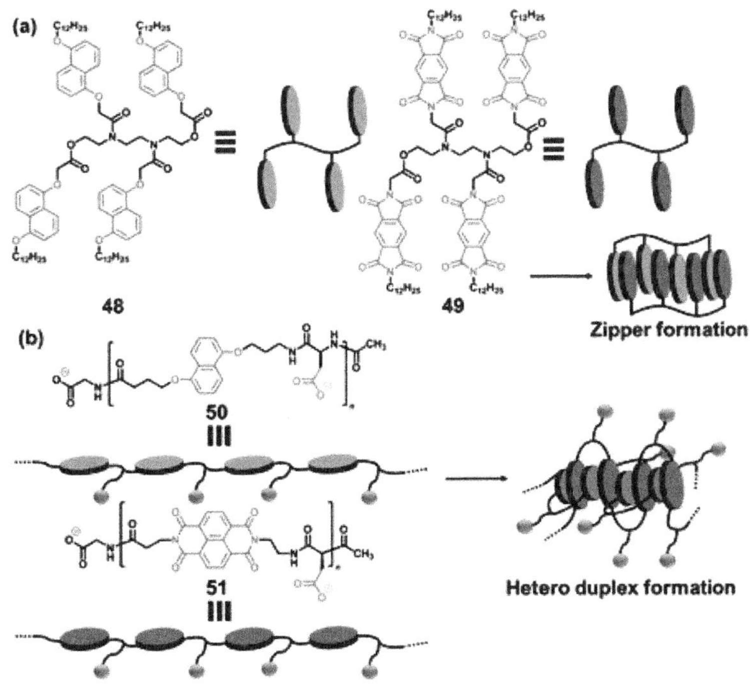

Figure 9.11 (a) Zipper formation through CT interaction between oligomers **48** and **49**. (b) Heteroduplex formation through CT interaction between oligomers **50** and **51**.

The use of CT interaction can also affect the morphology of solid state materials. The earliest example of mixing electron donor and acceptor polymers was reported by Percec et al., to fabricate polymer blends.[83] The glass transition temperature (T_g) of the polymer blends shifted to a higher temperature compared to a simple arithmetic average of those for original polymers tethering carbazole as the donor and dinitrobenzene as the acceptor. Recently, Iverson and coworkers tried controlling the solid state morphology via CT interaction between DAN and NDI.[84,85] In the first example, they employed polyethylene-*alt*-maleic anhydride as the polymer backbone, and two independent polymers were prepared by tethering DAN or NDI units.[84] The polymer was essentially insoluble in water, but mixing it with equimolar the other polymer resulted in a very viscous purple liquid. After drying the viscous solution on a glass substrate, AFM observation showed structures approximately 700 nm high, 1.6 μm across, and 14 μm long. The high viscosity could provide a long fiber when injected the solution to 1 M HCl. In the second example, they utilized poly(arylene-ethynylene) as the polymer backbone.[85] The π-conjugated polymers were synthesized via Migita–Kosugi–Stille coupling copolymerization between a dibromo NDI derivative and distannylethynylarylenes catalyzed by $Pd(PPh_3)_4$. As the aryl groups, benzene, naphthalene, DAN, and anthracene were employed. These polymers showed a large absorption coefficient in the wavelength region from 500 to 700 nm, derived from the intramolecular CT interaction, and also exhibited concentration-dependent fluorescence quenching, especially in the polymers from DAN and anthracene. The XRD measurement of the precipitated bulk polymers from MeOH represented strong diffraction from a lamellar structure with interdigitated side chains (23 Å) and also moderate diffraction from ~12 Å, 4.5 Å, and 3.5 Å *d*-spacing, attributed to an adjacent donor moiety in one polymer chain and interchain π–π stacking. Though the optoelectronic properties were dependent on the intramolecular CT interaction in this example, the intermolecular CT complexation would present the well-organized architecture via a proper molecular design.

Therefore to summarize this section, the employment of CT interaction for folding successfully achieved the similarity with the function of natural protein or DNA, and stimuli responsivity is also applicable through a rational molecular design, enabling smart and complex polymeric materials.

9.4 Conclusions

In this chapter, the author reviewed the utilization of CT interaction from the viewpoint of polymer chemistry, such as linear polymers connected via CT interaction, and secondary structure variation of synthetic polymer chains via CT interaction. Although the utilization of CT interaction in supramolecular chemistry has been limited compared to hydrogen bonding or metal–ligand coordination, the excel directionality and complementarity is highly suitable for construction of supramolecular architectures as introduced. The superior usability of CT interaction, including durability toward solvents or simplicity of probing, will allow us to fabricate more complexed structures and functionality simulating biological systems.

References

1. Mulliken, R. S., Person, W. B. *Annu. Rev. Phys. Chem.*, **13**, 107 (1962).
2. Bent, H. A. *Chem. Rev.*, **68**, 587 (1968).
3. Soos, Z. G. *Annu. Rev. Phys. Chem.*, **25**, 121 (1974).
4. Bryce, M. R. *Chem. Soc. Rev.*, **20**, 355 (1991).
5. Yan, Y.-K., Mingos, D. M. P. *Chem. Soc. Rev.*, **24**, 203 (1995).
6. Kurmoo, M., Graham, A. W., Day, P., Coles, S. J., Hursthouse, M. B., Caulfield, J. L., Singleton, J., Pratt, F. L., Hayes, W. *J. Am. Chem. Soc.*, **117**, 12209 (1995).
7. Powell, B. J., McKenzie, R. H. *J. Phys.: Condens. Matter*, **18**, R827 (2006).
8. Saito, G., Yoshida, Y. *Bull. Chem. Soc. Jpn.*, **80**, 1 (2007).
9. Bredas, J. L., Beljonne, D., Coropceanu, V., Cornil, J. *Chem. Rev.*, **104**, 4971 (2004).
10. Desiraju, G. R. *Angew. Chem. Int. Ed.*, **34**, 2311 (1995).
11. Ikeda, A., Shinkai, S. *Chem. Rev.*, **97**, 1713 (1997).
12. Zeng, F., Zimmerman, S. C. *Chem. Rev.*, **97**, 1681 (1997).
13. Balzani, V., Credi, A., Raymo, F. M., Stoddart, J. F. *Angew. Chem. Int. Ed.*, **39**, 3348 (2000).
14. Brunsveld, L., Folmer, B. J. B., Meijer, E. W., Sijbesma, R. P. *Chem. Rev.*, **101**, 4071 (2001).
15. Holliday, B. J., Mirkin, C. A. *Angew. Chem. Int. Ed.*, **40**, 2022 (2001).
16. Lehn, J.-M. *Polym. Int.*, **51**, 825 (2002).
17. Lee, J. W., Samal, S., Selvapalam, N., Kim, H. J., Kim, K. *Acc. Chem. Res.*, **36**, 621 (2003).
18. Hofmeier, H., Schubert, U. S. *Chem. Soc. Rev.*, **33**, 373 (2004).

19. Sangeetha, N. M., Maitra, U. *Chem. Soc. Rev.*, **34**, 821 (2005).
20. Chen, Z., Lohr, A., Saha-Moller, C. R., Wurthner, F. *Chem. Soc. Rev.*, **38**, 564 (2009).
21. Steed, J. W. *Chem. Commun.*, **47**, 1379 (2011).
22. Zhang, X., Wang, C. *Chem. Soc. Rev.*, **40**, 94 (2011).
23. Appel, E. A., del Barrio, J., Loh, X. J., Scherman, O. A. *Chem. Soc. Rev.*, **41**, 6195 (2012).
24. Babu, S. S., Prasanthkumar, S., Ajayaghosh, A. *Angew. Chem. Int. Ed.*, **51**, 1766 (2012).
25. Erbas-Cakmak, S., Leigh, D. A., McTernan, C. T., Nussbaumer, A. L. *Chem. Rev.*, **115**, 10081 (2015).
26. Das, A., Ghosh, S. *Angew. Chem. Int. Ed.*, **53**, 2038 (2014).
27. Klosterman, J. K., Yamauchi, Y., Fujita, M. *Chem. Soc. Rev.*, **38**, 1714 (2009).
28. Meyer, E. A., Castellano, R. K., Diederich, F. *Angew. Chem. Int. Ed.*, **42**, 1210 (2003).
29. Folmer, B. J. B., Sijbesma, R. P., Versteegen, R. M., van der Rijt, J. A. J., Meijer, E. W. *Adv. Mater.*, **12**, 874 (2000).
30. Lash, T. D. *J. Porphyrins Phthalocyanines*, **15**, 1093 (2011).
31. Haino, T., Fujii, T., Watanabe, A., Takayanagi, U. *Proc. Natl. Acad. Sci. U.S.A.*, **106**, 10477 (2009).
32. Haino, T., Fujii, T., Fukazawa, Y. *Tetrahedron Lett.*, **46**, 257 (2005).
33. Haino, T., Fujii, T., Fukazawa, Y. *J. Org. Chem.*, **71**, 2572 (2006).
34. Haino, T., Watanabe, A., Hirao, T., Ikeda, T. *Angew. Chem. Int. Ed.*, **51**, 1473 (2012).
35. Kinjo, K., Hirao, T., Kihara, S. I., Katsumoto, Y., Haino, T. *Angew. Chem. Int. Ed.*, **54**, 14830 (2015).
36. Bendikov, M., Wudl, F., Perepichka, D. F. *Chem. Rev.*, **104**, 4891 (2004).
37. Fernandez, G., Perez, E. M., Sanchez, L., Martin, N. *Angew. Chem. Int. Ed.*, **47**, 1094 (2008).
38. Perez, E. M., Sanchez, L., Fernandez, G., Martin, N. *J. Am. Chem. Soc.*, **128**, 7172 (2006).
39. Fernandez, G., Perez, E. M., Sanchez, L., Martin, N. *J. Am. Chem. Soc.*, **130**, 2410 (2008).
40. Isla, H., Perez, E. M., Martin, N. *Angew. Chem. Int. Ed.*, **53**, 5629 (2014).
41. Isla, H., Gallego, M., Perez, E. M., Viruela, R., Orti, E., Martin, N. *J. Am. Chem. Soc.*, **132**, 1772 (2010).

42. Canevet, D., Gallego, M., Isla, H., de Juan, A., Perez, E. M., Martin, N. *J. Am. Chem. Soc.*, **133**, 3184 (2011).
43. Canevet, D., Perez, E. M., Martin, N. *Angew. Chem. Int. Ed.*, **50**, 9248 (2011).
44. Park, J. S., Yoon, K. Y., Kim, D. S., Lynch, V. M., Bielawski, C. W., Johnston, K. P., Sessler, J. L., *Proc. Natl. Acad. Sci. U.S.A.*, **108**, 20913 (2011).
45. Kim, K., Kim, D., Lee, J. W., Ko, Y. H., Kim, K., *Chem. Commun.*, 848 (2004).
46. Kim, H.-J., Heo, J., Jeon, W. S., Lee, E., Kim, J., Sakamoto, S., Yamaguchi, K., Kim, K. *Angew. Chem. Int. Ed.*, **40**, 1526 (2001).
47. Ko, Y. H., Kim, K., Kang, J.-K., Chun, H., Lee, J. W., Sakamoto, S., Yamaguchi, K., Fettinger, J. C., Kim, K. *J. Am. Chem. Soc.*, **126**, 1932 (2004).
48. Liu, Y., Yu, Y., Gao, J., Wang, Z., Zhang, X. *Angew. Chem. Int. Ed.*, **49**, 6576 (2010).
49. Xu, Y., Guo, M., Li, X., Malkovskiy, A., Wesdemiotis, C., Pang, Y. *Chem. Commun.*, **47**, 8883 (2011).
50. Gellman, S. H. *Acc. Chem. Res.*, **31**, 173 (1998).
51. Hill, D. J., Mio, M. J., Prince, R. B., Hughes, T. S., Moore, J. S. *Chem. Rev.*, **101**, 3893 (2001).
52. Schmuck, C. *Angew. Chem. Int. Ed.*, **42**, 2448 (2003).
53. Lokey, R. S., Iverson, B. L. *Nature*, **375**, 303 (1995).
54. Benniston, A. C., Harriman, A., Lynch, V. M. *Tetrahedron Lett.*, **35**, 1473 (1994).
55. Nguyen, J. Q., Iverson, B. L. *J. Am. Chem. Soc.*, **121**, 2639 (1999).
56. Cubberley, M. S., Iverson, B. L. *J. Am. Chem. Soc.*, **123**, 7560 (2001).
57. Zych, A. J., Iverson, B. L. *Helv. Chim. Acta*, **85**, 3294 (2002).
58. Zych, A. J., Iverson, B. L. *J. Am. Chem. Soc.*, **122**, 8898 (2000).
59. Gabriel, G. J., Sorey, S., Iverson, B. L. *J. Am. Chem. Soc.*, **127**, 2637 (2005).
60. Bradford, V. J., Iverson, B. L. *J. Am. Chem. Soc.*, **130**, 1517 (2008).
61. Ghosh, S., Ramakrishnan, S. *Angew. Chem. Int. Ed.*, **43**, 3264 (2004).
62. Ghosh, S., Ramakrishnan, S. *Macromolecules*, **38**, 676 (2005).
63. Liu, K., Zheng, X., Samuel, A. Z., Ramkumar, S. G., Ghosh, S., Tan, X., Wang, D., Shuai, Z., Ramakrishnan, S., Liu, D., Zhang, X. *Langmuir*, **29**, 14438 (2013).
64. De, S., Ramakrishnan, S. *Macromolecules*, **42**, 8599 (2009).
65. Ghosh, S., Ramakrishnan, S. *Angew. Chem. Int. Ed.*, **44**, 5441 (2005).
66. Ramkumar, S. G., Ramakrishnan, S. *Macromolecules*, **43**, 2307 (2010).

67. De, S., Koley, D., Ramakrishnan, S. *Macromolecules*, **43**, 3183 (2010).
68. Ilhan, F., Gray, M., Blanchette, K., Rotello, V. M. *Macromolecules*, **32**, 6159 (1999).
69. Colquhoun, H. M., Zhu, Z. *Angew. Chem. Int. Ed.*, **43**, 5040 (2004).
70. Colquhoun, H. M., Zhu, Z., Cardin, C. J., Gan, Y. *Chem. Commun.*, 2650 (2004).
71. Colquhoun, H. M., Zhu, Z., Cardin, C. J., Gan, Y., Drew, M. G. B. *J. Am. Chem. Soc.*, **129**, 16163 (2007).
72. Burattini, S., Colquhoun, H. M., Fox, J. D., Friedmann, D., Greenland, B. W., Harris, P. J., Hayes, W., Mackay, M. E., Rowan, S. J. *Chem. Commun.*, 6717 (2009).
73. Burattini, S., Greenland, B. W., Hayes, W., Mackay, M. E., Rowan, S. J., Colquhoun, H. M. *Chem. Mater.*, **23**, 6 (2011).
74. Zhu, Z., Bruns, C. J., Li, H., Lei, J., Ke, C., Liu, Z., Shafaie, S., Colquhoun, H. M., Stoddart, J. F. *Chem. Sci.*, **4**, 1470 (2013).
75. Basu, S., Coskun, A., Friedman, D. C., Olson, M. A., Benitez, D., Tkatchouk, E., Barin, G., Yang, J., Fahrenbach, A. C., Goddard, W. A., III, Stoddart, J. F. *Chem. Eur. J.*, **17**, 2107 (2011).
76. Zhang, K. D., Zhao, X., Wang, G. T., Liu, Y., Zhang, Y., Lu, H. J., Jiang, X. K., Li, Z. T. *Angew. Chem. Int. Ed.*, **50**, 9866 (2011).
77. Amemori, S., Kokado, K., Sada, K. *Angew. Chem. Int. Ed.*, **52**, 4174 (2013).
78. Gharib, D. H., Amemori, S., Naya, M., Kokado, K., Sada, K. *RSC Adv.*, **5**, 89319 (2015).
79. Zhou, Q. Z., Jiang, X. K., Shao, X. B., Chen, G. J., Jia, M. X., Li, Z. T., *Org. Lett.*, **5**, 1955 (2003).
80. Zhou, Q.-Z., Jia, M.-X., Shao, X.-B., Wu, L.-Z., Jiang, X.-K., Li, Z.-T., Chen, G.-J. *Tetrahedron*, **61**, 7117 (2005).
81. Zhao, X., Jia, M. X., Jiang, X. K., Wu, L. Z., Li, Z. T., Chen, G. J. *J. Org. Chem.*, **69**, 270 (2004).
82. Gabriel, G. J., Iverson, B. L. *J. Am. Chem. Soc.*, **124**, 15174 (2002).
83. Rodriguez-Parada, J. M., Percec, V. *Macromolecules*, **19**, 55 (1986).
84. Reczek, J. J., Iverson, B. L. *Macromolecules*, **39**, 5601 (2006).
85. Alvey, P. M., Ono, R. J., Bielawski, C. W., Iverson, B. L. *Macromolecules*, **46**, 718 (2013).

Biography

Kenta Kokado was born in Osaka, Japan, in 1982. He obtained his

MSc (2007) and PhD (2010) in polymer chemistry from Kyoto University (supervisor, Prof. Y. Chujo). During 2009–2010, he was a JSPS research fellow in Kyoto University. Then, he joined the group of Prof. K. Sada as an assistant professor in 2010. His current research interests focus on smart materials having network structures.

Chapter 10

Liquid Crystalline Materials in Mesoscopic Scale

Tomomichi Itoh
Department of Materials Science and Biotechnology, Ehime University, 3 Bunkyo-cho, Matsuyama, Ehime 790-8577, Japan
itoh@ehime-u.ac.jp

A liquid crystal is a fluid exhibiting anisotropic properties such as birefringence and elasticity. This chapter outlines fundamentals of liquid crystalline properties and focuses on micromechanical properties of mesoscopic liquid crystalline materials.

10.1 Introduction

There has been a growing interest in liquid crystalline (LC) materials because of their great potential for use in attractive applications. Since LC molecules form an oriented structure with keeping a flow property, the ordered structure is arranged by external stimuli such as temperature, an electric field, a magnetic field, a shear

Conjugated Objects: Developments, Synthesis, and Applications
Edited by Atsushi Nagai and Koji Takagi

ISBN 978-981-4774-03-1 (Hardcover), 978-1-315-22962-1 (eBook)
www.panstanford.com

flow, and light. This switching property is widely applied in liquid crystalline displays (LCDs) which are essential for our modern lives. In addition, the molecular rearrangement can make a change of a material morphology so that LC materials have been applied for micromechanical devices.

This chapter outlines the fundamentals of LC materials, including molecular structures, phases, anisotropic properties, measurements, and elastic energy, and then describes the strategies for micromechanical manipulation induced by LC properties.

10.2 What Is a Liquid Crystal?

It is well known that materials take three phases of solid (crystal), liquid (isotropic), and gas. Crystals are characterized by three-dimensional order, while molecules in liquids are free from a position with a short-ranged interaction. Some materials can take another stable phase between the crystal and the liquid, namely *liquid crystal*. LC molecules have intermediate order and molecular mobility. We must carefully define *order*, which includes an orientational and a positional order. The most important property of a liquid crystal is a long-range orientational order, which is caused by the anisotropy of molecular shape. Therefore, the molecule taking LC phases has a rod-like or disc-like shape.

Anisotropic properties of LC materials provide many interesting features to lead to useful applications. Although the most famous application would be LCDs, LC actuators have received considerable attention in these days. In this chapter, we survey essential properties of a liquid crystal and briefly review mesoscopic morphologies of liquid crystalline polymers (LCPs) in an attempt to provide micromechanical actuators. Details of LC properties are provided in other books.[1,2]

10.2.1 Director Vector and Orientational Order of LC Molecules

Here, we consider the order of rod-like molecules. Figure 10.1 represents the molecular arrangement for an isotropic and an LC phase. The molecules in a crystalline phase have three-dimensional positional order and highly aligned orientational order. In the isotropic phase, the molecules are completely random and free from any order. In the LC phase, molecules are aligned along a preferred axis on average, although the long axis of the molecules is somewhat fluctuated. The preferred direction is called *director*, which is denoted by a unit vector **n**.

The director of the rod-like molecules is caused by an excluded-volume effect. When you pack matchsticks in a match box as many as possible, you have to align them in one direction. This is the excluded-volume effect of rod-like materials. When the free volume of the molecules is large in the isotropic phase, the long axis of each molecule can take random direction. As the system is condensed, on the other hand, molecular alignment is required in order to pack the molecules in a restricted space.

The degree of orientational order, S, is determined as follows:

$$S = \left\langle \frac{1}{2}(3\cos^2\theta - 1) \right\rangle$$

Here <> denotes the thermal average, and θ is the angle of the long axis for each molecule with respect to the director vector **n**. Perfect alignment of LC molecules gives $S = 1$, while the average of $\cos^2\theta$ is 1/3 to lead to $S = 0$ in the isotropic phase. The orientational order S is typically determined by nuclear magnetic resonance or a polarized optical measurement.

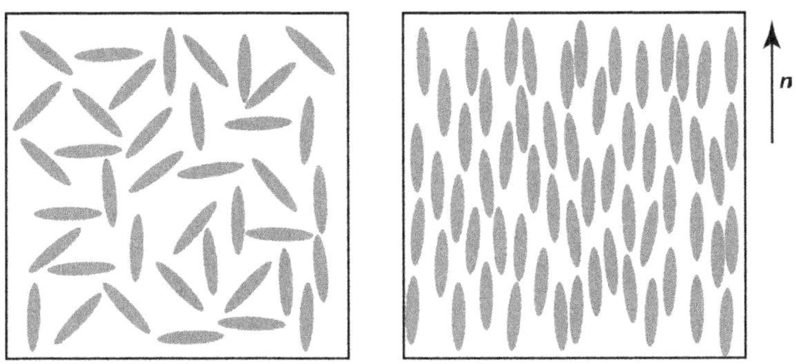

Isotropic Phase **Liquid Crystalline Phase (Nematic)**

Figure 10.1 Basic structure of LC phases.

10.3 LC Molecules

10.3.1 Molecular Structure of a Liquid Crystal

Either a rod-like (or *calamic*) or a disc-like (or *discotic*) part, which is called *mesogen*, is a core of an LC molecule to provide an orientational and a positional order in an LC phase as illustrated in Fig. 10.2. The mesogen has a rigid and highly symmetric structure (i.e. biphenyl, azobenzene, triphenylene) so that neat mesogenic molecules are liable to form a crystalline structure. To possess molecular mobility, an increase in the free volume of the mesogens is provided by adding a solvent or attaching a flexible aliphatic spacer to the mesogens.

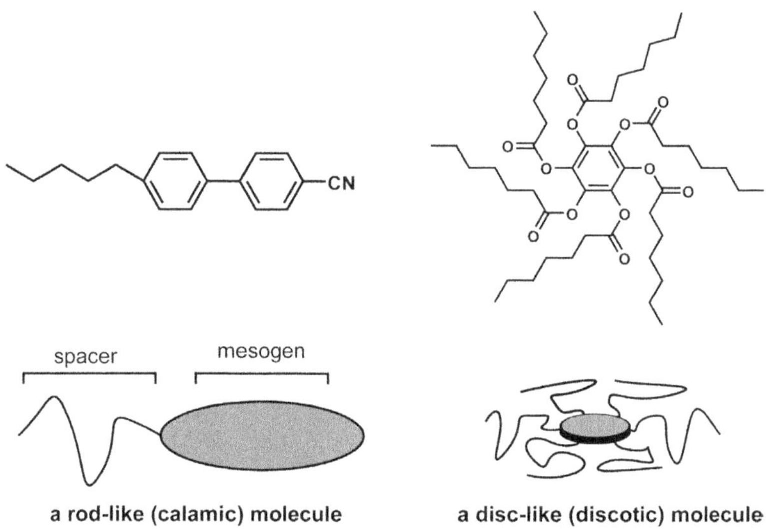

Figure 10.2 Molecular structure of liquid crystals.

10.3.2 Lyotropic and Thermotropic Liquid Crystals

According to the way of appearance, LC phases are classified into a lyotropic phase or a thermotropic one. The former is an LC phase formed in a solution and the latter exhibits that formed without any solvent.

Lyotropic LC phases are further divided into two classes: one is formed by mesogens, and the other is by amphiphilic molecules. The former is typically observed in a condensed solution of rigid-rod polymers such as hydroxypropyl cellulose, and the latter is caused by phase segregation between a hydrophilic and a hydrophobic part of amphiphiles in an aqueous solution. Although amphiphiles do not have mesogens, they aggregate to form anisotropic-fluid structures of spherical micellar lattice, hexagonal-packed cylindrical micelles, bilayer lamellae, and bicontinuous structures. This chapter does not cover the lyotropic LC phase any more. Details are provided in other books.[1,2] Hereafter, we discuss a thermotropic LC phase.

In a thermotropic LC phase, thermal fluctuation of mesogens is caused by the melt of spacers which have a solvent effect above a certain temperature.

10.3.3 LC Polymers

Polymers can take an LC phase as well as low-molecular-weight LC molecules. LCPs are classified into three types of main-chain, side-chain, and rigid-rod LCPs as shown in Fig. 10.3. A main-chain LCP is composed of alternatively connected mesogens and spacers, while each mesogen is attached to a backbone via a spacer in a side-chain LCP. A rigid-rod LCP is a polymer forming a macromolecular mesogen such as α-helical poly(γ-benzyl-L-glutamate), hydroxypropyl cellulose, tobacco mosaic virus, and DNA. Some rigid-rod LCPs can form a highly ordered crystalline structure on cooling, providing a high-modulus fiber material such as Vectra and Kevlar.

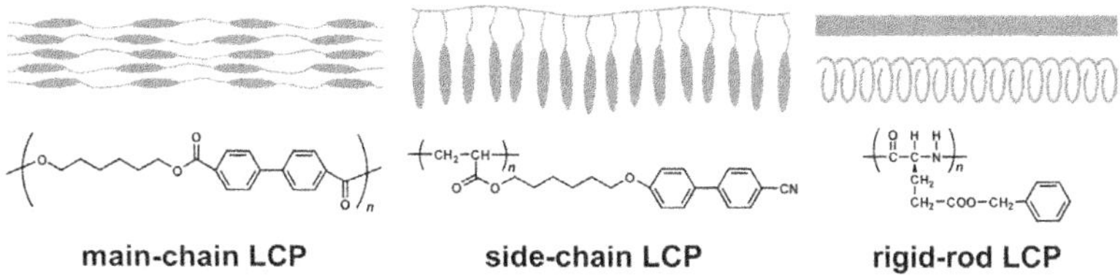

Figure 10.3 LC polymers (LCPs).

10.4 LC Phases

An LC phase is classified into various types according to a translational and a rotational symmetry of an LC structure. The most common type of liquid crystal is a calamic liquid crystal, which is composed of rod-like calamic molecules. In contrast, disc-like discotic molecules take a discotic LC phase. The calamic LC phase illustrated in Fig. 10.1 is a *nematic* phase, in which molecules are free to move but subjected to a long-range orientational order. Another typical LC phase, a *smectic* phase, has a one-dimensional positional order of mesogens to form a periodically layered structure. When an LC molecule has a chiral center, in addition, a nematic and a smectic phase can form a twisted structure, the so-called chiral nematic (or *cholesteric*) and chiral smectic phase, respectively.

10.4.1 Nematic Phases

Nematic (N) is the simplest LC phase where molecules are completely free from a positional order but have a long-range orientational order. Typical molecules are shown in Fig. 10.4. Mesogens in the chiral nematic phase (N*) form a twisted structure as illustrated in Fig. 10.5. The chiral nematic phase is the firstly observed LC phase in the nineteenths century for cholesteryl benzoate and also known as a cholesteric phase. Directors twist in the direction perpendicular to the long axis of the mesogens. The helical structure often takes a characteristic pitch in the range of the 100 nm order. This size is near to wavelength of visible light so that the chiral nematic phase can show structural colors by Bragg reflection. The pitch is responsive to thermal stimuli to change the scattered colors. For example, it is known that a highly concentrated aqueous solution of hydroxypropyl cellulose exhibits a chiral nematic phase and that its structural color is tuned by changing its temperature. While a chiral nematic phase commonly has a single twisted director, an LC phase with a double-twisted structure has also been observed in a small temperature range just below the transition to the isotropic phase, called *blue phase*.

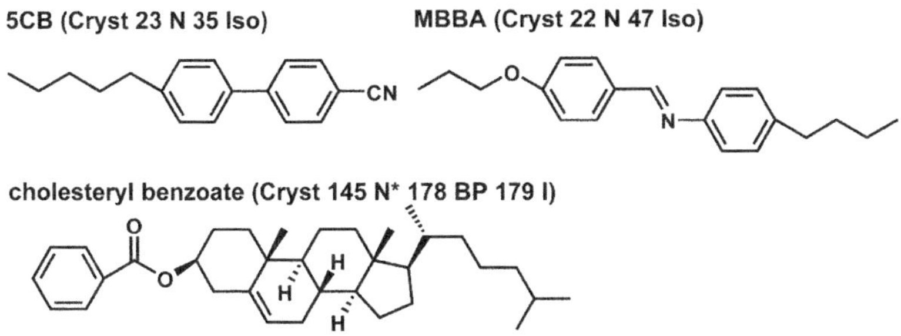

Figure 10.4 Typical molecules showing nematic phases.

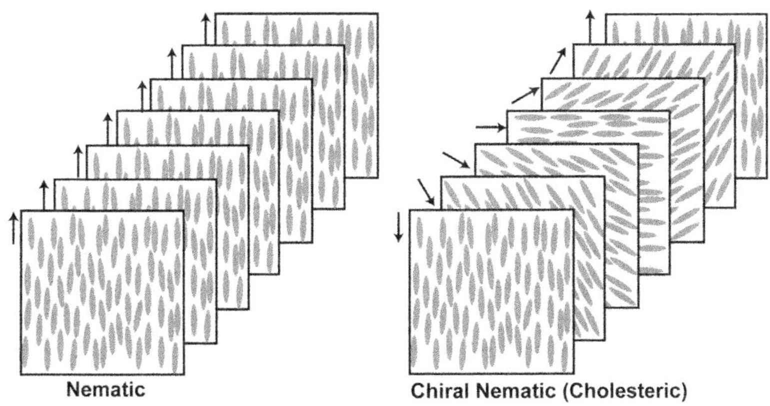

Figure 10.5 Director alignment of molecules in nematic phases.

10.4.2 Smectic Phases

A smectic phase is characterized by a one-dimensional positional order to result in the formation of a periodic layered structure. Typical molecules are illustrated in Fig. 10.6. A smectic phase is classified into more than 10 different types, depending on the tilting director and further positional orders. There are two typical smectics of smectics A (SmA) and C (SmC) in which the in-plane mesogenic position is free. The long axis of mesogens is parallel to the layer normal in smectic A, while the director is tilted to the layer normal in smectic C.

8CB (Cryst 22 SmA 34 N 41 Iso)

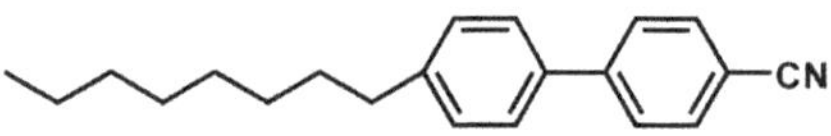

racemic MHPOBC (Cryst 60 SmC_A 113 SmC 121 SmA 145 Iso)

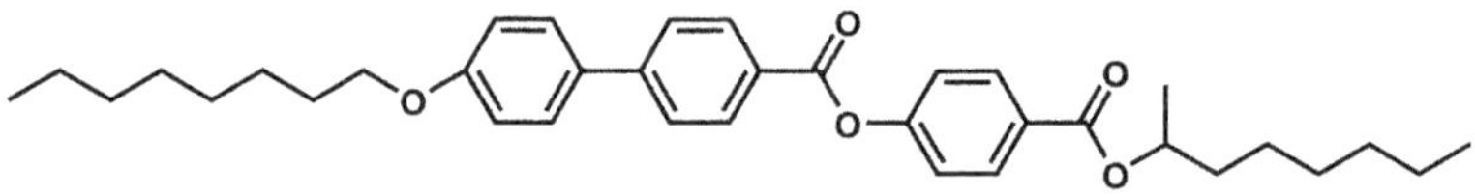

DOBAMBC (Cryst 76 SmC* 93 SmA 117 Iso)

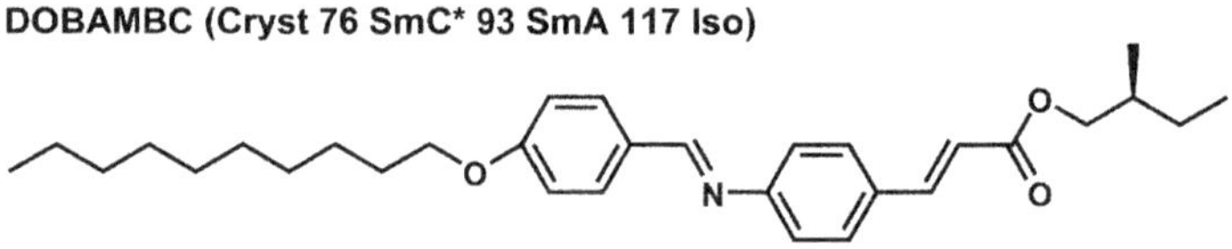

Figure 10.6 Typical molecules showing smectic phases.

The periodic distance of smectic layers (d) is strongly related to molecular length (l). A smectic A phase has its layered periodicity between l and $2l$, as shown in Fig. 10.7. At $d = l$, molecules are aligned within a layer regardless of their head–tail structure, which is called a smectic A_1 (SmA_1) phase. When a smectic layer is composed of interdigitated mesogens, the periodic distance is larger than the molecular length, $d > l$, the so-called smectic A_d (SmA_d) phase. The periodic distance is twice as large as the molecular length, $d = 2l$, in a smectic A_2 (SmA_2) phase, where molecules are arranged antiparallel in a smectic layer. The smectic A_d and the smectic A_2 phases are often observed in an LC molecule with a polar mesogen such as cyanobiphenyl.

In a typical smectic C phase, the director vectors in the smectic layers are tilted in the same direction. When director vectors are alternatively tilted in the opposite direction, the phase is called a smectic C_A (SmC_A) phase. A chiral version of smectic C is called a chiral smectic C (SmC*) phase, where director vectors rotate around a cone, as shown in Fig. 10.8. Similarly to the chiral nematic phase, the SmC* phase forms a helical structure. This helix is the key factor to cause ferroelectricity.

The in-plane mesogenic position is free in the smectic A and the smectic C. On the other hand, other types of smectic phases have a positional order in a smectic layer. Director in a smectic B phase is aligned parallel to the layer normal as with a smectic A phase, but molecules form a hexagonal-packed structure in a smectic layer. When director is tilted to a smectic layer normal, a smectic F or a smectic I phase appears. A higher-ordered smectic phase, which was called a smectic B_{cryst}, G, J, H, or K phase, is classified into a soft-crystal phases. A cubic (Cub) phase, which was formerly known as a smectic D phase, does not have any layered structure but forms a bicontinuous structure.

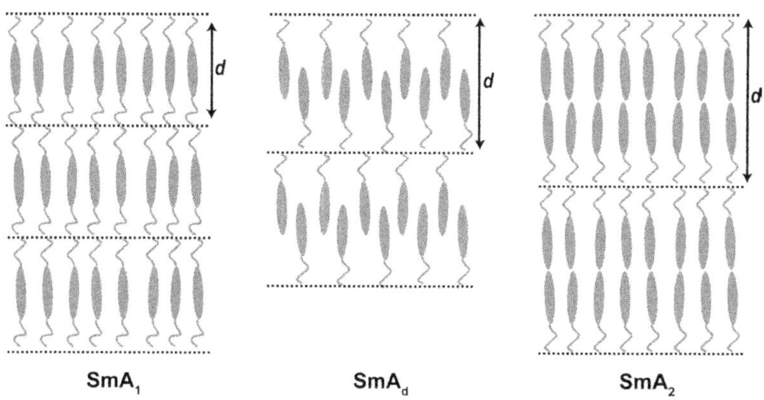

Figure 10.7 Smectic A phases.

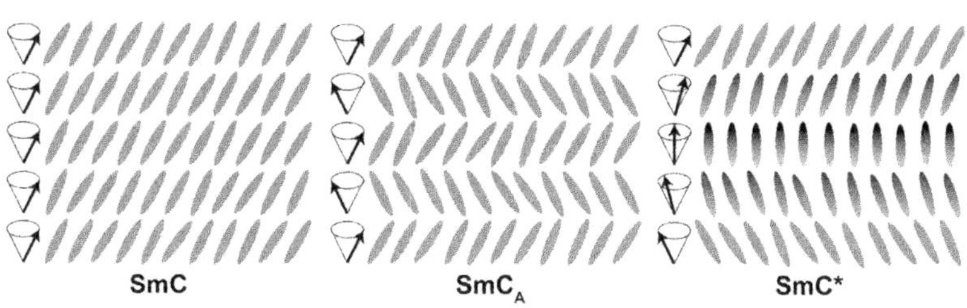

Figure 10.8 Smectic C phases.

10.4.3 Columnar Phases

Similarly to rod-like molecules, disc-like molecules form a nematic (N_D) phase. When the disc-like molecules are stacked, column structures are formed to show various columnar (Col) phases. The columns behave as rod-like objects. In a columnar nematic (N_C) phase, the columns have a long-range orientational order but are free from the column position. Other columnar phases are classified according to types of packing structures of the disc-like molecules and stacking structures of the columns. In a column, normal vectors of disc-like molecules are parallel or tilted to the long axis of the column. The columns are stacked in a hexagonal, a rectangular, or an oblique structure.

10.5 Anisotropic Properties of Liquid Crystals

The alignment of LC molecules causes many interesting anisotropic properties. The director vectors are responsive to external fields so that liquid crystals show anisotropic interactions with external fields.

10.5.1 Effect of Substrate Surface and Shear Flow

An LC director is strongly affected by a contact surface of a substrate to be aligned in a parallel or a perpendicular direction to the surface. The parallel alignment is called *homogeneous* or *planar*, while the perpendicular one is *homeotropic*. Even when the LC alignment is changed by some external fields, the molecules at the surface keep their director.

By rubbing a substrate, the director is aligned along the rubbing direction. This alignment is achieved by rubbing a glass plate with a paper filter, a polymer film with nylon cloth, and so on. This technique is very easy to obtain a highly ordered alignment of LC molecules.

When LC molecules are sandwiched between two flat plates, the director vector can be aligned by a shear process. This is also a simple technique since the shear field is provided by the parallel shift of one plate of the sandwiched cell. The shear alignment is caused by the anisotropic viscosity of LC materials. Because the viscosity is lowest when the mesogenic director is parallel or perpendicular to the shear direction, the mesogens rotate to minimize the flow viscosity to result in the director alignment.

10.5.2 Switch of LC Alignment by Magnetic and Electric Fields

A rod-like molecule, including a benzene ring, possesses a characteristic diamagnetism. In general, the diamagnetism in the perpendicular direction to the long axis of the molecule is larger than that in the parallel direction. Therefore, the director vector is aligned along the magnetic field.

When an electric field is applied to LC molecules, a dipole moment, which is proportional to the electric field, is induced. The anisotropic response to the electric field is expressed by anisotropic permittivity ($\Delta\varepsilon$), which is a difference between the permittivity parallel to the director vector ($\varepsilon_\perp$) and that perpendicular to it ($\varepsilon_{||}$).

$$\Delta\varepsilon = \varepsilon_{||} - \varepsilon_\perp$$

The anisotropic permittivity takes either positive or negative, depending on the chemical structure of an LC molecule. When $\Delta\varepsilon$ is negative, the director is aligned in the perpendicular direction to the electric fields, while parallel alignment is caused for an LC molecule with $\Delta\varepsilon < 0$.

10.5.3 Optical Anisotropy

The reflective index is proportional to the square of permittivity. Since an LC molecule has anisotropic permittivity, as described above, the reflective index in the parallel direction to the director vector ($n_{||}$) has a different value from that in the perpendicular ($n_\perp$). This optical anisotropy is called *birefringence* (Δn):

$$\Delta n = n_{||} - n_\perp$$

An LC director is controlled by an electric field, leading to a switch of the birefringence. The switching property is the principle mechanism of LCDs. Nematic LC molecules are sandwiched between two transparent electrodes such as indium tin oxide (ITO), giving an LC cell. The surface of the electrodes is rubbed to provide a highly ordered planar alignment of the LC molecules. The LC cell is further sandwiched with two polarizers in the perpendicular directions each other. When linearly polarized light enters from the cell bottom, the polarization changed to elliptic one through the LC molecules due to their birefringence. Hence, the light can penetrate the upper crossed polarizer to lead to a bright field. As an electric field is applied to the cell, the nematic alignment is switched from planar to homeotropic. Since an LC molecule does not have optical anisotropy along the long molecular axis, polarized light retains its linearity through the homeotropic liquid crystal. Therefore, the light cannot penetrate the crossed polarizer, resulting in a dark field. Today, the mechanism of the optical shutter has developed into twisted nematic (TN) and supertwisted nematic (STN).

10.6 Observation and Measurements of Liquid Crystals

10.6.1 Polarized Optical Microscope

When an LC structure has a defect, a characteristic texture is observed under a polarized optical microscopy (POM) observation, which is widely used to identify LC phases. Details with POM photographs are provided in the book by Dierking.[3]Although molecules have a long-range orientational order in a nematic phase, the defect breaks a director continuity, which is called a *disclination*. Figure 10.9 shows examples of director configuration with a disclination. The strength of the disclination (m) is defined by rotation of the director vector around the disclination.

Under a POM observation, the director configuration around a disclination shows a *schlieren* texture, which is characterized by dark brushes in a bright field. When a nematic is placed between crossed polarizers, linearly polarized light changed into elliptically polarized one due to birefringence, resulting in a bright field. On the other hand, Δn is zero in the case of parallel and perpendicular alignments of the director vectors to the polarizer, leading to no birefringence. Therefore, the incident light cannot penetrate the crossed polarizers, resulting in a dark field. Since director vectors around a disclination are partially parallel or perpendicular to the analyzer, a schlieren texture is observed as a dark brushes. In general, $4|m|$ brushes are observed from a disclination with the strength m. Whether positive or negative of the strength is determined by the rotation of the sample stage of the POM. Positive strength leads to rotation of the brushes to the same direction to the stage rotation, while an inverse rotation is observed with a negative *m* value.

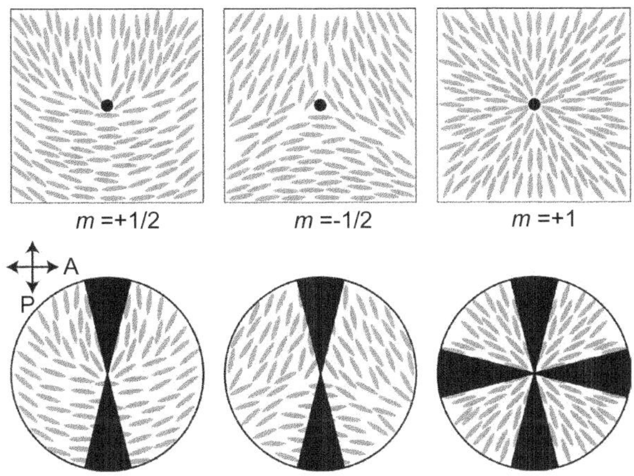

Figure 10.9 Director configuration around a disclination with various strengths (upper) and corresponding schlieren textures (bottom) in nematic liquid crystals.

In a smectic phase, smectic layers bend to form onion-like structures, which is called *focal conic*. Although the focal conics show various textures, the most common textures are fan shape and broken-fan shape for smectic A and C phases, respectively. For a chiral nematic and a chiral smectic C phase, a fingerprint texture results from a periodic structure of helical structures.

10.6.2 X-Ray Diffraction

X-ray diffraction is a powerful tool to study LC structures. For powder samples, the diffraction pattern is divided into inner and outer rings. Outer reflections are placed at large angles around 20° by using a common wavelength such as CuK_α (λ = 0.154 nm). These come from the average spacing of mesogens in the lateral direction to their long axis about 0.5 nm. Inner reflections at small angles about 3° are accounted for periodic structures along director vectors.

Diffraction patterns for alignment samples are shown in Fig. 10.10. When alignment direction is placed vertically to this page, nematic, smectic A, and smectic C phases show a broad halo at the large angles in the horizontal direction because mesogens have a liquid-like short-ranged positional order. The pattern of inner reflections in nematic phases is differently from that in smectic phases. Nematics are characterized by a broad halo, indicating a liquid-like interaction between the molecules along their director. In contrast, sharp reflections are observed for smectic phases due to a long-range periodicity of smectic layers. The periodic distance is measured from the spot angle of the inner reflections via Bragg's law.

For the smectic phases, detailed information is obtained by using the alignment samples. Typical patterns of Bragg reflections are illustrated in Fig. 10.10. An outer halo locates in the horizontal direction for smectic A and C as with nematic. In the case of a smectic C_A phase, the outer reflections are split at the angle of alternately tilted directors. In the small-angle region, sharp reflections appear owing to the periodicity of the smectic layers. A conventional smectic C shows tilted reflections according to the azimuthal angle of the smectic layers, while sharp reflections appear in the vertical direction in smectic A and C_A.

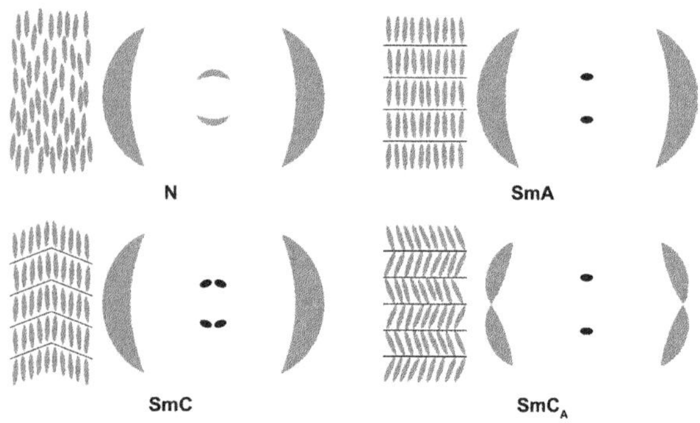

Figure 10.10 Typical X-ray diffraction patterns for oriented samples in various LC phases.

10.6.3 Differential Scanning Calorimetry

Differential scanning calorimetry (DSC) is a common method to investigate phase transitions of LC materials. In general, an LC phase transition is detected as a first-order transition showing a peak. A peak-top temperature and the peak area provide a transition temperature and corresponding enthalpy change, respectively.

Compared with a crystal–LC transition, the enthalpy change of an LC–isotropic phase transition is small but detectable. However, that of an LC–LC phase transition is too small to be recognized as a peak. In that case, other techniques such as POM and X-ray measurements equipped with a hot stage are useful to determine the phase transition temperatures.

10.7 Elastic Energy of Liquid Crystals

Deformations of molecular alignment in nematic phases are classified into splay, twist, and bend deformations, as shown in Fig. 10.11. On the basis of liquid crystals where all molecules are oriented in one direction, elastic energy is caused by the deformation and expressed as follows:

$$F = \frac{1}{2}K_1[\nabla \cdot \mathbf{n}]^2 + \frac{1}{2}K_2[\mathbf{n} \cdot (\nabla \times \mathbf{n})]^2 + \frac{1}{2}K_3[\mathbf{n} \times \nabla \times \mathbf{n}]^2$$

Here K_1, K_2, and K_3 are elastic constants for splay, twist, and bend deformations, respectively. Since the complete alignment is the most stable, the elastic energy repairs the deformations. When LC molecules are enclosed in a tiny domain, however, the LC structure is altered by surface curvature to cause an elastic energy. This results in two interesting properties. One is an alteration of LC phase behavior, and the other is a change of the domain morphology. In the forthcoming sections, the interplay between LC phase behavior and domain morphologies will be discussed.

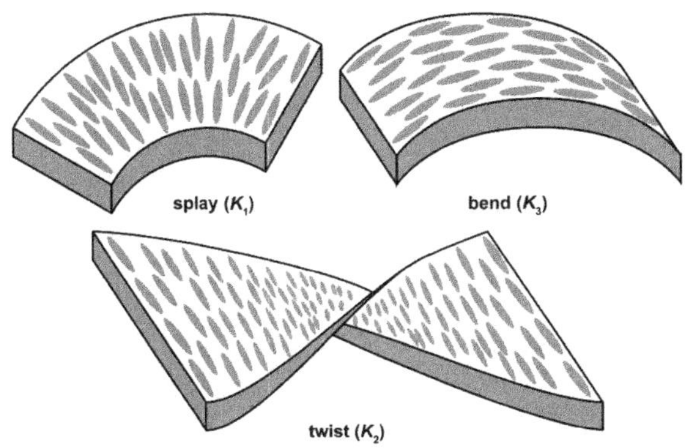

Figure 10.11 Deformations of nematic liquid crystals.

10.8 Mesoscopic LCP Morphologies: LC Block Copolymers

A phase transition of an LCP changes main-chain conformation, as illustrated in Fig. 10.12, leading to a change of mesoscopic morphologies of the LCP materials. This is applied for manipulation of microdomain structures and actuation of mesoscopic materials.

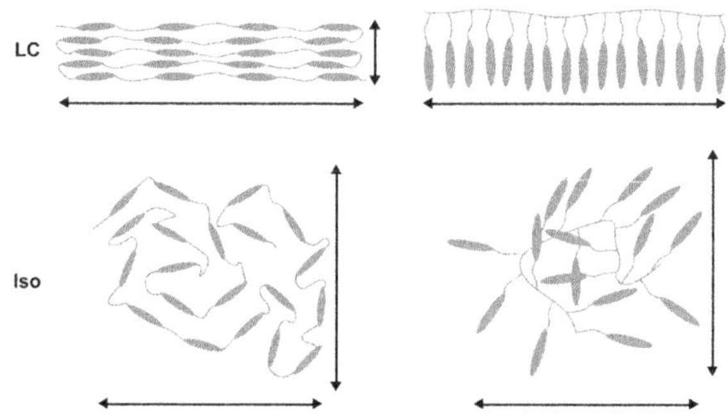

Figure 10.12 Conformation of LCPs in LC and isotropic phases.

10.8.1 Synthesis and Microdomain Structures of LC Block Copolymers

A block copolymer is composed of different polymers which are covalently attached at their chain termini each other.[4] Different polymers are generally immiscible so that block copolymers cause microphase segregation with a periodic-ordered structure at several tens of nanometers. Size and morphology of the microdomain structure depends on segmental composition and segregation strength. Lamellar, hexagonal-packed cylinder, and body-centered cubic phases are typical morphologies.

A side-chain LCP has been extensively used as one segment in an LC block copolymer. There are two approaches for the preparation of a side-chain LC block copolymer. One is the attachment of a low-molecular LC molecule with a functional group to a side-chain functional group in one segmental chain by polymer-analogous reaction.[5–7] The other is direct block copolymerization of a conventional and a side-chain mesogenic monomer. Compared with the polymer-analogous reaction, the direct approach is preferred to produce a well-defined LC block copolymer. Although a side-chain LC block copolymer was formerly prepared by anionic polymerization[8–11], living radical polymerization is widely used these days. Living radical polymerization retains the advantages of conventional radical polymerization of simplicity, robustness, versatility, and a wide variety of available monomers, resulting in the production of various side-chain LC block copolymers.[12–16] Directly synthesized side-chain LCPs are often composed of a (meth)acrylic derivative because a conjugated vinyl monomer is easy to undergo a radical polymerization to give side-chain LCPs with various properties in the wide range of molecular weight. Recently, a main-chain LC block copolymer has been developed by a combination of polycondensation and living radical polymerization.[17–19]

An LC block copolymer forms a hierarchical-ordered nanostructure where an LC structure is formed in a microphase-segregated structure, as illustrated in Fig. 10.13. Mesogenic alignment is affected by the interface of the microdomain and vice versa.[15]

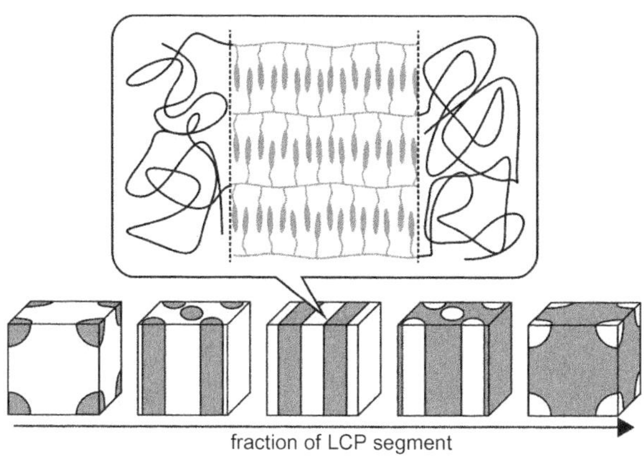

Figure 10.13 Microdomain morphologies of LC block copolymers.

10.8.2 LC Phase Behavior in Microdomain Structures

Effects of LC phase behavior on a microdomain structure have been intensively investigated. Watanabe et al. prepared narrowly distributed LC block copolymers (PS-*b*-PM6B1s) composed of polystyrene and poly(6-[4-(4′-methoxyphenyl)phenoxy]hexyl methacrylate) (PM6B1) segments by anionic polymerization.[8] The LC block copolymer with 50% of the LCP segment formed a lamellar-type microdomain structure. The LCP segment took crystalline–SmA_d and SmA_d–isotropic phase transitions at 100°C and 130°C, respectively. Their transition enthalpies were lower than those of corresponding PM6B1 homopolymer in addition to a decrease in the crystal–SmA_d phase transition temperature, indicating that the microdomain structure caused difficulty in the formation of the crystalline and LC structures. When poly(α-methyl styrene) whose glass transition temperature is 160°C was used as an amorphous segment instead of the polystyrene, the microdomain was verified even above the SmA_d–isotropic transition temperature of the PM6M1 segment to lead to a further decrease in crystallinity, liquid crystallinity, and the crystal–SmA_d phase transition temperature.[20]

Microdomain morphology has an effect on phase behavior of an LCP segment. Fischer et al. synthesized LC block copolymers composed of polystyrene and poly[2-(3-cholesteryloxycarbonyloxy) ethyl methacrylate] segments with various compositions of the LCP segment by a polymer-analogous reaction.[21] The LCP segment took SmA phase when the LC block copolymer formed an alternative lamellar morphology, polystyrene cylinders in the LCP matrix, or polystyrene spheres, while only a nematic phase was observed in LCP spheres embedded in the polystyrene matrix. In this case, LCP cylinders were not observed. They explained that thermodynamically stable smectic layers could not be formed within LC-cylindrical and LC spherical microdomains. On the other hand, counterresults have been reported by Watanabe et al. synthesized a block copolymer (PS-*b*-PM6CB) composed of a polystyrene and a poly(6-[4-(4′-cyanophenyl)phenoxy]hexyloxy methacrylate) (PM6CB) segment and observed the SmA_2 phase in LCP cylinders, even when the weight fraction of the LCP segment was 50%.[11] Ober et al. synthesized a series of LC block copolymers by a polymer-analogous reaction of polystyrene-*block*-poly(hydrogenated isoprene) and cyanoazobenzene-contained side-chain groups and observed an increase in the SmA–isotropic phase transition temperature in cylindrical microdomains by 20°C compared to that in lamellae.[22]

10.8.3 Manipulation of Microdomain Structures

Microdomain structures have an effect on orientational behavior of an LCP segment. Watanabe et al. reported the orientational behavior in a fiber specimen.[9] When a fiber of a PM6B1 homopolymer was prepared by pulling up from an isotropic melt, methacrylic backbones were aligned along the fiber axis. As the fiber was cooled to ambient temperature via SmA_d, the director of the side-chain mesogenic groups which were attached to the main chains were aligned in the perpendicular direction to the fiber axis. For a PS-*b*-PM6B1 fiber specimen which was also pulled up from an isotropic melt, on the other hand, the smectic layer normal was parallel to the fiber axis. This alignment is caused as follows, in turn; the lamellar microdomains were aligned along the fiber axis, the main-chain backbones of the PM6B1 segment were distributed perpendicular to the lamellar interfaces, and the smectic layer normal was aligned parallel to the fiber axis. This behavior indicates that the mesogenic alignment was strongly coupled to the main-chain conformation via six methylene spacers in the microdomain structure.

As methylene unit of side-chain spacers increased, mesogenic groups were decoupled with the main-chain conformation. Hamley et al. observed that smectic layer normal was aligned in the perpendicular direction to the long axis of LCP cylindrical microdomains for an LC block copolymer composed of a polystyrene and a poly[(*S*)-(−)-methylbutyl-4'-(12-methacryloyloxydodecyloxy) biphenyl-4-carboxylate] with 12 methylene spacers.[13] When LCP segments filled a matrix of a microphase-segregated structure, in addition, both parallel and perpendicular alignments of smectic layer normal to the long axis of the polystyrene cylinders have been allowed.

It has been reported that LC phase behavior sometimes has an effect on microdomain size in an LC block copolymer. Watanabe et al. found that periodic spacing of a lamellar microdomain morphology in a PS-*b*-PM6B1 gradually decreased by 20–25% with increasing temperature in the SmA_d temperature region, accompanied by a decrease in the smectic order of the PM6B1 segment.[9,23] The periodic spacing and the smectic order were reversible at each temperature, indicating thermal equilibrium of both the size and the smectic order. The thermodynamically reversible change is caused by the interplay between smectic-layered enthalpy in the side chain and conformational entropy in the main chain. The formation the smectic layer is more stable for the side-chain mesogenic groups in the SmA_d phase than layer deformation. The interdigitated structure of the side-chain mesogens in SmA_d requires a chain extension of the PM6M1 main-chain backbone, while a random coil is the most stable conformation for the methacrylic main chain. Therefore, the chain extension results in an entropic loss of the main-chain conformation. Since the main chain and the side chain are coupled via the rather short spacer with six methylene units, temperature elevation leads to an entropic gain of the main-chain conformation accompanied by an enthalpic loss of the layered structure of the side-chain mesogenic groups. The size change of the LCP microdomain is characteristic to SmA_d-forming LC block copolymers. In contrast, a PM6CB segment which formed SmA_2 phase in a PS-*b*-PM6CB did not show such a temperature dependence of microdomain size.[11] The SmA_2 phase does not require chain extension of the main-chain backbone, although the main chain is located between smectic layers.

A microdomain morphology in an LC block copolymer is affected by LC phase behavior. Microphase segregation is related to the degree of polymerization and the Flory–Huggins χ parameter.[4] The latter is determined by the chemical structure of each segment and inversely proportional to temperature. Therefore, a thermoreversible order–disorder transformation (ODT) of microphase segregation has been observed in a low-molecular-weight amorphous block copolymer at elevated temperature.[4] For an LC block copolymer, Hammond et al. reported that an ODT took place accompanied by the LC–isotropic phase transition of the LCP segment, indicating that formation of an LC structure enhanced microphase segregation.[24] Even when the polymer segment in the LC block copolymer is miscible above the LC–isotropic temperature, the mesogenic groups are segregated to form an LC structure on cooling.

It has also been reported that an order–order transformation (OOT) of a microdomain structure can be caused by an LC phase transition. Gronski et al. found an OOT where polystyrene spheres in an LCP matrix changed into polystyrene cylinders accompanied by the isotropic–nematic phase transition of the LCP segment on cooling.[25] Hammond et al. reported an OOT from a perforated layered LCP microdomain to alternative lamellae.[26] Finkelmann et al.

observed that a gyroid structure with a bicontinuous LCP microdomain was transformed into lamellae along with the isotropic–smectic phase transition.[27] These can be explained by LC elastic energy causing loss of microdomain curvature and prevention from disclination. Conformational anisotropy of an LCP segment is also responsible for an OOT of LC block copolymers. Watanabe et al. prepared a series of a symmetrical PS-*b*-PM6B1s and found that the lamellar morphology in the isotropic phase of the PM6B1 segment was transformed into an LCP cylindrical one in SmA_d phase at 30–32 wt% of the PM6B1 composition.[28] While conformation of each segment was relaxed in the isotropic phase, the formation of the SmA_d structure requires chain extension of the PM6B1 segment. This imposed a conformational anisotropy on the polystyrene segment because the both segments shared the same interface of the microdomain. When the polystyrene segment cannot keep the imposed anisotropy, the entropic loss of the conformational free energy might be recovered by making the interfacial curvature of the microdomain structures to transform their morphology from lamellae to LCP cylinders.

A magnetic field has been applied for alignment of an LC block copolymer. Hamley et al. found that LCP cylinders and the director of side-chain mesogens were aligned parallel to the magnetic field when a sample was slowly cooled down to a smectic phase in the presence of a magnetic field.[13] Thomas et al. observed polystyrene cylinders in an LCP matrix aligned perpendicular to a magnetic field at a smectic phase temperature.[29] In contrast, Watanabe et al. reported a parallel alignment of polystyrene cylinders to a magnetic field in a nematic phase.[16] In this case, it is noteworthy that the orientational order of the polystyrene cylinders was $S = 0.99$, indicating perfect alignment of the microdomain. In these reports, the magnetic alignment of LCP microdomain structures has not been observed for a lamellar morphology. The magnetic alignment of microdomain structures could depend on an LC phase, magnetic susceptibility, and elastic energy of mesogenic alignment as well as a microdomain morphology.

Here, an interrelation between microphase segregation and LC phase behavior in a hierarchical-ordered structure of an LC block copolymer has been described. The interrelation causes manipulation of size, morphology, and alignment of a microphase-segregated structure as well as alteration of LC phase behavior.

Recently, an LC block copolymers composed of a photochromic LCP has been investigated to realize a light-controllable microdomain structure, which is a candidate material for application of ultrafine resolution lithography.[30–32]

10.9 Mesoscopic LCP Morphologies: LCP Micro- and Nanoparticles

The interrelation between morphology and LC phase behavior has also been observed in LCP particles, which are larger than the microdomain structures in an LC block copolymer by one or more order.

10.9.1 Preparation of LCP Particles

Yoshida et al. carried out suspension, emulsion, and dispersion polymerizations of a mixture of an LC monomer and styrene, respectively.[33] The suspension polymerization was a polymerization in water-dispersed hydrophilic droplets which were consisted of the LC monomer and AIBN and stabilized by poly(vinyl alcohol), giving 45 μm LCP microspheres. The emulsion system involved water medium, water-soluble potassium persulfate, and LC-monomer droplets stabilized by sodium dodecyl sulfate, affording 80 nm LCP microspheres. The dispersion polymerization started from a solution of the LC monomer, AIBN, poly(vinyl pyrrolidone) as a steric stabilizer, and methanol medium, providing 1.5 μm LCP microspheres by polymerization-induced phase segregation.

Zentel et al. obtained narrowly distributed LCP microspheres in the size range from submicron to several micrometers by dispersion polymerization.[34] They observed optical anisotropy in the resultant LCP microspheres. The LCP microsphere with 1.8 μm showed uniform birefringence as a spot, and its brightness fluctuated accompanied by rotation of the LCP microsphere under a POM observation. For LCP microspheres in the size range of 2–4 μm, a Maltese cross and a baseball-patterned texture were observed, indicating a bipolar director configuration with $m = +1$ disclinations at the north and the south pole. More complexed patterns were observed in larger LCP microspheres. Interestingly, these bipolar LCP microspheres were able to be rotated by optical tweezers irradiating circularly polarized light and, moreover, the rotation was changed between clockwise and counter clockwise according to the rotational sense of the circularly polarized light.[35] Electric fields were also useful to rotate bipolar LCP microspheres.[36]

Huck et al. prepared main-chain LCP particles by miniemulsion techniques.[37]A chloroform solution of a main-chain LCP was dispersed in water by ultrasonication in the presence of a surfactant. Then, a stable dispersion of the main-chain LCP particles was obtained after evaporation of the chloroform to precipitate the LCP particles. Interestingly, the particles had an ellipsoidal morphology with their aspect ratio of 1.0–1.5, depending on the particle size in the range of 30–1000 nm. The ellipsoidal morphology was changed into spherical one by elevation of temperature to the isotropic phase and recovered by slowly cooled down. This must be caused by a conformational transformation of the main-chain LCP along with the LC phase behavior.

10.9.2 LC Elastomer Particles

A reversible change of particle morphology has been observed in an LC elastomer system. A polymeric elastomer is a crosslinked polymer in a rubber state showing rubber elasticity in a dry condition. An LC elastomer possesses a stimuli-responsive reversible change of their shape, induced by an LC phase transition. Therefore, an LC elastomer has been expected as a candidate material for an actuator. Compared with a piezo element, an LC elastomer can undergo a significant strain change and can be made in a small size to several tens of nanometers.

Cairns et al. reported a template approach for the preparation of LC elastomer particles.[38] A ceramic filter membrane was filled with a mixture of an LC diacrylate and a photoinitiator, exposed to UV in order to polymerize the LC diacrylate, and removed by an alkaline etching. Tubes, toroids, and yarn balls of a crosslinked LCP in the size range of several micrometers were obtained, depending on monomer concentration and surface treatment of the template. Zentel et al. also carried out a similar approach using a template to produce crosslinked LCP nanowires.[39] Mesogens were highly aligned along the nanowires, leading to a strong change of their length during the LC phase transition.

A miniemulsion technique has also been applied for preparation of LC elastomer particles. Zentel et al. used a side-chain LC polysiloxane with acrylate end groups for crosslinking.[40] They dispersed a chloroform solution of the polymer and AIBN to water, and then elevated temperature of the dispersion to undergo crosslinking of the LCP and evaporation of the chloroform from the LCP emulsions, giving spherical LC elastomer particles with 100 nm in size. Interestingly, they observed the smectic layer structure by cryo–transmission electron microscopy (cryo-TEM) without any stain and found two types in the particles. One is a fringe pattern with unbent smectic layers which were found for a side-chain LCP with three-ring mesogens possessing a high bending modulus. The other is onion-like smectic layers along the curved surface of the spheres as well as the unresolved center part, which was observed for LCP with short mesogens.

Another method for preparation of LC elastomer particles is a continuous flow synthesis, which was developed by Zentel et al.[41,42] Droplets of a mixture of an LC monomer, a crosslinker, and a photoinitiator were put into a microfluidics and polymerized by UV exposure in the LC phase temperature range, giving a large number of narrowly distributed microspheres in the size range of several hundred micrometers. The microfluidic made a shear effect on the mesogens in the droplets so that mesogenic director was aligned in one direction in a particle. The particles showed a reversible change of their morphology between spheres and ellipsoids accompanied by the LC–isotropic phase transition, reflecting a large change in the conformation of the crosslinked LCP chains. This was applied for preparation of micropumps composed of a glycerol core and a crosslinked nematic-LCP shell.[43] When the core-shell LC elastomers were punctured with a capillary, the glycerol was pumped out from the core on heating and moved back into it on cooling by the actuation of the LC elastomer shell.

Highly anisotropic LC elastomer particles in the size range of tens micrometers were prepared by Smoukov et al. with a mechanical stretch.[44] Droplets containing an LC precursor, a crosslinker, and a photoinitiator were embedded into a poly(vinyl alcohol) film specimen. Then, the firm was stretched, exposed to UV, and immersed in water to give ellipsoidal LC elastomer particles. Because their mesogenic directors were aligned in one direction by the stretch, the LC–isotropic phase transition caused a reversible morphological transformation of the particles.

Abbot et al. reported preparation of ellipsoidal LC elastomer particles by miniemulsion polymerization.[45] A mixture of an LC diacrylate, 4-cyano-4'-pentylbiphenyl (5CB), and a photoinitiator was dispersed in water/glycerol medium by using a homogenizer and polymerized by UV exposure. After the extraction of the 5CB, 10 μm ellipsoidal LC elastomer particles with their aspect ratio of 1.4 were obtained.

10.10 Conclusions

In this chapter, we have reviewed micromechanical manipulation of LC materials as well as fundamentals of LC properties. Manipulation of a mesoscopic morphology using LC phase behavior is a promising candidate for application in ultrafine resolution lithography, microelectromechanical systems (MEMS), and so on.

As described above, LCP particles showing micromechanical actuation have been prepared by several techniques. Recently, surface modification of polymer particles receives large attention for advanced materials such as a sensing material showing a stimuli-responsive dispersion/aggregation behavior. Dispersion polymerization is an attractive technique to prepare surface-functional narrowly distributed polymer particles. In addition, precise control of surface structure is becoming possible.[46–49] By a combination of these techniques, we expect surface-functional LCP particles possessing multistimuli responsibility, in addition to the micromechanical manipulation in near future.

References

1. Chandrasekhar, S. *Liquid Crystals*, Cambridge University Press (1992).
2. de Gennes, P. G., Prost, J. *The Physics of Liquid Crystals*, Oxford University Press (1995).

3. Dierking, I. *Textures of Liquid Crystals*, John Wiley & Sons (2006).
4. Hamley, I. W. *The Physics of Block Copolymers*, Oxford University Press (1998).
5. Adams, J., Gronski, W. *Makromol. Chem. Rapid Commun.*, **10**, 553–557 (1989).
6. Zaschke, B., Frank, W., Fischer, H., Schmutzler, K., Arnold, M. *Polym. Bull.*, **27**, 1–8 (1991).
7. Iyengar, D. R., Perutz, S. M., Dai, C.-A., Ober, C. K., Kramer, E. J. *Macromolecules*, **29**, 1229–1234 (1996).
8. Bohnert, R., Finkelmann, H. *Macromol. Chem. Phys.*, **195**, 689–700 (1994).
9. Yamada, M., Hirao, A., Nakahama, S., Iguchi, T., Watanabe, J. *Macromolecules*, **28**, 50–58 (1995).
10. Zheng, W. Y., Hammond, P. T. *Macromol. Rapid Commun.*, **17**, 813–824 (1996).
11. Yamada, M., Itoh, T., Nakagawa, R., Hirao, A., Nakahama, S., Watanabe, J. *Macromolecules*, **32**, 282–289 (1999).
12. Kasko, A. M., Heintz, A. M., Pugh, C. *Macromolecules*, **31**, 256–271 (1998).
13. Hamley, I. W., Castelletto, V., Lu, Z. B., Imrie, C. T., Itoh, T., Al-Hussein, M. *Macromolecules*, **37**, 4798–4807 (2004).
14. Hamley, I. W., Castelletto, V., Parras, P., Lu, Z. B., Imrie, C. T., Itoh, T. *Soft Matter*, **1**, 355–363 (2005).
15. Ansari, I. A., Castelletto, V., Mykhaylyk, T., Hamley, I. W., Lu, Z. B., Itoh, T., Imrie, C. T. *Macromolecules*, **36**, 8898–8901 (2003).
16. Tomikawa, N., Lu, Z., Itoh, T., Imrie, C. T., Adachi, M., Tokita, M., Watanabe, J. *Jpn. J. Appl. Phys.*, **44**, L711–L714 (2005).
17. Ishige, R., Ishii, T., Tokita, M., Koga, M., Kang, S., Watanabe, J. *Macromolecules*, **44**, 4586–4588 (2011).
18. Koga, M., Ishige, R., Sato, K., Ishii, T., Kang, S., Sakajiri, K., Watanabe, J., Tokita, M. *Macromolecules*, **45**, 9383–9390 (2012).
19. Koga, M., Abe, K., Sato, K., Koki, J., Kang, S., Sakajiri, K., Watanabe, J., Tokita, M. *Macromolecules*, **47**, 4438–4444 (2014).
20. Yamada, M., Itoh, T., Hirao, A., Nakahama, S., Watanabe, J. *High Perform. Polym.*, **10**, 131–138 (1998).
21. Fischer, H., Poser, S., Arnold, M. *Liq. Cryst.*, **18**, 503–509 (1995).

22. Mao, G., Wang, J., Clingman, S. R., Ober, C. K., Chen, J. T., Thomas, E. L. *Macromolecules*, **30**, 2556–2567 (1997).

23. Yamada, M., Iguchi, T., Hirao, A., Nakahama, S., Watanabe, J. *Polym. J.*, **30**, 23–30 (1998).

24. Zheng, W. Y., Hammond, P. T. *Macromolecules*, **31**, 711–721 (1998).

25. Sänger, J., Gronski, W., Maas, S., Stühn, B., Heck, B. *Macromolecules*, **30**, 6783–6787 (1997).

26. Anthamatten, M., Hammond, P. T. *Macromolecules*, **32**, 8066–8076 (1999).

27. Schneider, A., Zanna, J.-J., Yamada, M., Finkelmann, H., Thomann, R. *Macromolecules*, **33**, 649–651 (2000).

28. Itoh, T., Yamada, M., Hirao, A., Nakahama, S., Watanabe, J. *Mol. Cryst. Liq. Cryst.*, **347**, 211–220 (2000).

29. Osuji, C., Ferreira, P. J., Mao, G., Ober, C. K., Vander Sande, J. B., Thomas, E. L. *Macromolecules*, **37**, 9903–9908 (2004).

30. Priimagi, A., Shevchenko, A. *J. Polym. Sci., Part B: Polym. Phys.*, **52**, 163–182 (2014).

31. Seki, T. *Polym. J.*, **46**, 751–768 (2014).

32. Sano, M., Hara, M., Nagano, S., Shinohara, Y., Amemiya, Y., Seki, T. *Macromolecules*, **48**, 2217–2223 (2015).

33. Yoshida, M., Fan, X., Uemura, Y., Hatate, Y. *J. Appl. Polym. Sci.*, **81**, 2490–2499 (2001).

34. Vennes, M., Zentel, R. *Macromol. Chem. Phys.*, **205**, 2303–2311 (2004).

35. Vennes, M., Martin, S., Gisler, T., Zentel, R. *Macromolecules*, **39**, 8326–8333 (2006).

36. Haseloh, S., Zentel, R. *Macromol. Chem. Phys.*, **210**, 1394–1401 (2009).

37. Yang, Z., Huck, W. T. S., Clarke, S. M., Tajbakhsh, A. R., Terentjev, E. M. *Nat. Mater.*, **4**, 486–490 (2005).

38. Cairns, D. R., Eichenlaub, N. S., Crawford, G. P. *Mol. Cryst. Liq. Cryst.*, **352**, 275–282 (2000).

39. Ohm, C., Haberkorn, N., Theato, P., Zentel, R. *Small*, **7**, 194–198 (2011).

40. Vennes, M., Zentel, R., Rössle, M., Stepputat, M., Kolb, U. *Adv. Mater.*, **17**, 2123–2127 (2005).

41. Ohm, C., Serra, C., Zentel, R. *Adv. Mater.*, **21**, 4859–4862 (2009).

42. Ohm, C., Fleischmann, E.-K., Kraus, I., Serra, C., Zentel, R. *Adv. Funct. Mater.*, **20**, 4314–4322 (2010).

43. Fleischmann, E.-K., Liang, H.-L., Kapernaum, N., Giesselmann, F., Lagerwall, J., Zentel, R. *Nat. Commun.*, **3**, 1178 (2012).

44. Marshall, J. E., Gallagher, S., Terentjev, E. M., Smoukov, S. K. *J. Am. Chem. Soc.*, **136**, 474–479 (2014).

45. Mondiot, F., Wang, X., de Pablo, J. J., Abbott, N. L. *J. Am. Chem. Soc.*, **135**, 9972–9975 (2013).

46. Itoh, T., Fukutani, K., Hino, M., Ihara, E., Inoue, K. *J. Colloid Interface Sci.*, **330**, 292–297 (2009).

47. Itoh, T., Komada, S., Ihara, E., Inoue, K. *J. Colloid Interface Sci.*, **388**, 112–117 (2012).

48. Itoh, T., Abe, I., Tamamitsu, T., Shimomoto, H., Inoue, K., Ihara, E. *Polymer*, **55**, 3961–3969 (2014).

49. Itoh, T., Tamamitsu, T., Shimomoto, H., Ihara, E. *Polymer*, **70**, 183–193 (2015).

Biography

Tomomichi Itoh studied polymer chemistry at the Tokyo Institute of Technology. He received his PhD in liquid crystalline behavior the

and microdomain morphology of liquid-crystalline block copolymers, under the guidance of Prof. Junji Watanabe, in 2002. He worked as a JSPS research fellow at the Tokyo Institute of Technology and as an honorary researcher at the University of Aberdeen. The liquid crystalline behavior in the microdomain structure and the manipulation of microdomain morphology with liquid crystalline phase behavior, which is described in this chapter, were mainly studied in this period. Then, he worked at Materials Science and Biotechnology, Ehime University, from 2002, and his current position there is senior assistant professor. Throughout his research career, he has been interested in control over mesoscopic structures in polymer materials, including block copolymers, liquid crystalline polymers, polypeptides, and polymer particles. Recently, he is studying two subjects, although neither is reviewed in this chapter. One is fine control over the size and surface structure of functional polymer particles, which are prepared by dispersion polymerization in the presence of functional block copolymers or functional macromonomers. He has obtained detailed information on the formation mechanism of particle surfaces through dispersion polymerization, which is a basic aspect of surface control of functional polymer particles. The second is the significance of the chain terminus in a polypeptide during a conformational transformation, which is investigated by using block copolymers composed of an amorphous polymer and a polypeptide segment. In addition to laboratory teaching, he gives classroom lectures on polymer science and laboratory work in polymer chemistry and in general engineering, as well as on English grammar in chemical papers and writing for academic reports in Japanese. It is noteworthy that he designes and manages a teaching program on how to make an academic report for chemistry students, which was selected for an award by the Faculty of Engineering, Ehime University, in 2014. He is contributing, and will in the future, to the promotion of scientific and technological development through both his research and his education in the area of polymer science.

Chapter 11

Bulk Supramolecular Assemblies Constructed from Macrocyclic Compounds

Tomoki Ogoshi[a,b,*], **Takahiro Kakuta**[a], **and Tada-aki Yamagishi**[a]

[a]*Graduate School of Natural Science and Technology, Kanazawa University, Kakuma-machi, Kanazawa, Ishikawa 920-1192, Japan*

[b]*Japan Science and Technology Agency, PRESTO, 4-1-8 Honcho, Kawaguchi, Saitama 332-0012, Japan*

ogoshi@se.kanazawa-u.ac.jp

Macrocyclic compounds have been key players in the field of supramolecular chemistry because of their interesting features such as their highly symmetrical structures, nanosized uniform building blocks, cavity-size-dependent host–guest interactions, and versatile functionality. In this chapter, we discuss the assembly of bulk supramolecular materials using macrocyclic compounds, such as cyclodextrins, crown ethers, calix[*n*]arenes, cucurbit[*n*]urils, pillar[*n*]arenes, and other macrocyclic compounds. Several examples are provided of bulk supramolecular assemblies such as liquid crystals and guest uptake and storage porous materials using the original features of the macrocyclic hosts.

Conjugated Objects: Developments, Synthesis, and Applications
Edited by Atsushi Nagai and Koji Takagi

ISBN 978-981-4774-03-1 (Hardcover), 978-1-315-22962-1 (eBook)
www.panstanford.com

11.1 Introduction

Macrocyclic compounds have been key players in the field of supramolecular chemistry because of their interesting features, including (i) highly symmetrical structures owing to the lack of terminal groups, (ii) a uniform size at the nanometer scale, (iii) cavity-size-dependent host–guest interactions, and (iv) facile functionalization.[1] Various supramolecular assemblies with highly uniform structures can be constructed using macrocyclic compounds as building blocks because of their highly symmetrical structures and uniform sizes. The host–guest ability of the macrocyclic hosts allows separation and storage of guest molecules, and supramolecular structural switching upon addition of guests. Functionalization of the supramolecular assemblies can be useful for adding new functions to the supramolecular assemblies.

In this chapter, we describe bulk supramolecular assemblies using macrocyclic compounds such as cyclodextrins (CDs), crown ethers, calix[*n*]arenes, cucurbit[*n*]urils and pillar[*n*]arenes. We discuss the original features of the respective macrocyclic hosts such as their structures, functionalities, and host–guest properties. We also describe bulk supramolecular assemblies of other macrocyclic compounds. Several examples are provided of bulk supramolecular assemblies such as liquid crystals and guest uptake and storage porous materials using the original features of the macrocyclic hosts.

11.2 Cyclodextrins

CDs, which are a family of macrocyclic oligosaccharides, were discovered by A. Villiers in 1891.[2] Therefore, CDs have the longest history and are the most widely used macrocyclic hosts. CDs consist of glucopyranose units attached by α-1,4-linkages (Fig. 11.1a). The most common and commercially available CDs are cyclic pentamers, hexamers, and octamers, which are named α-, β-, and γ-CD, respectively. The cavity size of the CDs increases with an increasing number of repeating units. The cavity diameters of α-, β-, and γ-CDs are 4.9 Å, 6.2 Å, and 8.0 Å, respectively. Their structure resembles a nonsymmetric bucket with a wide and a narrow rim on either side. CDs have many hydroxyl groups on both faces owing to the constituent glucopyranose units. Owing to the high reactivity of the OH groups, various functionalized CDs can be obtained.

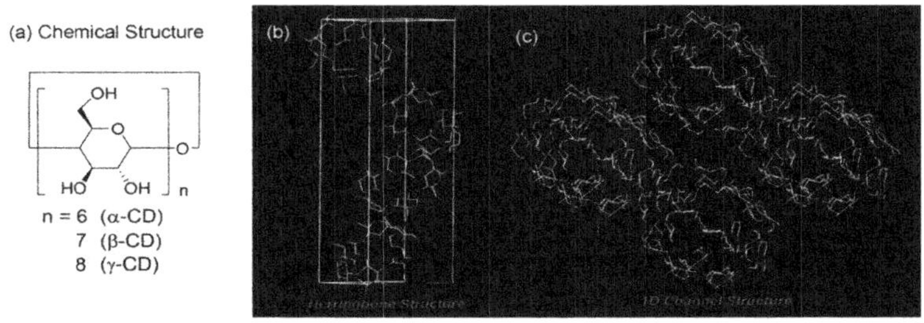

Figure 11.1 (a) Chemical structure of CDs. Crystal structures of CDs with guests. (b) Herringbone structure of α-CD⊃methanol complexes and (c) 1D channel structure of β-CD⊃poly(trimethylene oxide) complexes.

CDs have nonsymmetric structures and hydroxyl groups at both rims; thus, channel-type assemblies are obtained by connecting CD molecules by intermolecular hydrogen bonds between the hydroxyl groups on the rims of the CDs. However, when small molecules are incorporated into the CD cavity, CDs crystalize in a herringbone fashion, where by both ends of the CD cavity are blocked by adjacent molecules.[3] The guest molecule is enclosed within an isolated cavity; this type of packing is called cage type. When 1-propanol, methanol, iodine, and krypton are used as guests for α-CD, the size of these guest molecules is small enough to be fully included in the cavity (Fig. 11.1b). Channel-type packing is induced in many cases when large-sized guests, which are too large to be included into the cavity, are used because CDs need to rearrange to cover the un-included parts of the guest. Harada et al. reported the formation of 1D channels from β-CD molecules by inclusion of a polymeric chain. Inclusion of poly(trimethylene oxide) into β-CD cavities triggers the formation of a 1D channel structure (Fig. 11.1c). It is believed that large guests induce 1D channel structures.[4]

Tonelli et al. reported the formation of channel-type CD assemblies in the absence of guest molecules.[5] They discovered that α- and γ-CDs in columnar structures with one water molecule included were successfully produced by recrystallization from their aqueous solutions. α-CDs in the columnar structure showed host–guest ability toward small molecules and polymers. Akashi et al. found that channel-type assemblies of γ-CDs formed unique cubic microstructures.[6] They investigated the size control of the microcubes. The size of γ-CD cubes was successfully controlled by changing the concentration of γ-CD. By decreasing the γ-CD concentration, the size of the γ-CD cubes became larger (Fig. 11.2A–D). The size of γ-CD cubes could also be controlled by using other poor solvents instead of acetone. When 2-propanol was used as a poor solvent, the size of the γ-CD cubes became larger than those prepared in acetone, whereas the size was smaller when 2-butanol was used (Figs. 11.2E and 11.2F). They also found that evaporating 1,1,1,3,3,3-hexafluoro-2-propanol, which has been used to dissolve insoluble peptides and polymers in common organic solvents, from CD solutions gave crystalline solids composed of channel-type CD assemblies.[7]

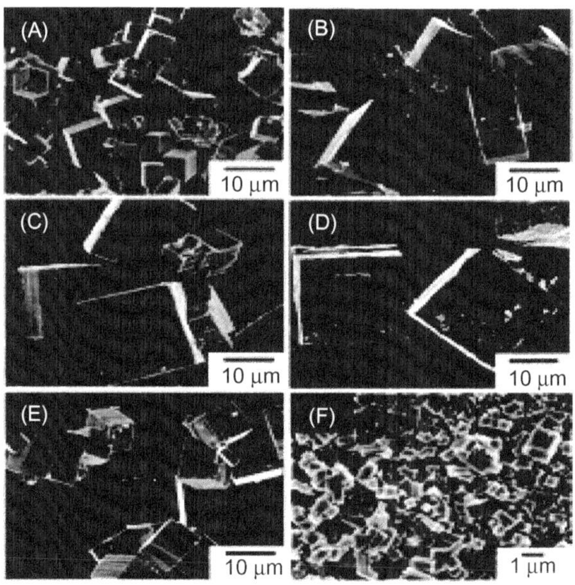

Figure 11.2 SEM images of γ-CD cubes formed by dropping aqueous γ-CD solutions—(A) 0.17, (B) 1.7×10^{-2}, (C) 3.5×10^{-3}, and (D) 1.7×10^{-3} M—into acetone or aqueous γ-CD solution (170 mM) into (E) 2-propanol or (F) 2-butanol. Reprinted with permission from Ref. [6]. Copyright (2010) American Chemical Society.

They also reported the application of the channel-type CD assemblies as organogelators.[8] Channel assemblies consisting of

γ-CD cubes, which were produced by following the preparation method by Tonelli et al.[5], formed organogels in a variety of oils and organic solvents (Fig. 11.3d). In contrast, native γ-CD with a cage-type packing structure (Fig. 11.3c) and α-CDs with a channel (Fig. 11.3b) and cage-type (Fig. 11.3a) structures showed no gelation ability. The construction of 3D networks by noncovalent forces such as London dispersion forces and van der Waals interactions should be the main driving force for the formation of the organogels. The organogelation ability of the CDs depended on their sizes and shapes; the smallest cubic nanoparticles showed the greatest gelationability.

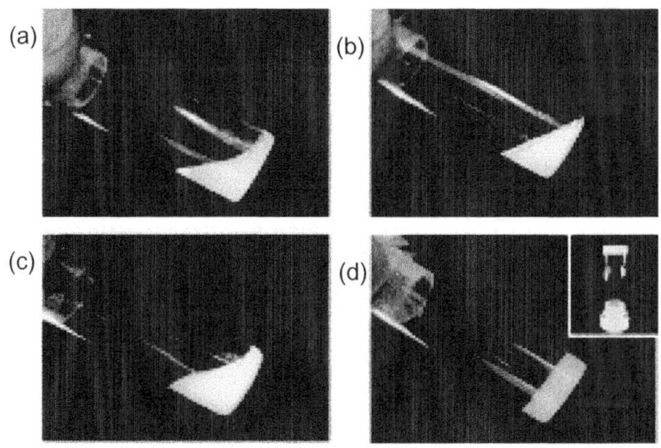

Figure 11.3 Photographs of mixtures of insulating oil (150 mL) and various assemblies (100 mg): (a) native α-CD (cage-type α-CD assembly), (b) channel-type α-CD assembly, (c) native γ-CD (cage-type γ-CD assembly), and (d) channel-type γ-CD assembly. Reproduced from Ref. [8] with permission from the Royal Society of Chemistry.

Liquid crystal materials based on CDs have also been reported.[9] A straightforward approach for producing liquid crystal materials using CDs is installing mesogenic pendants on the rims of CDs. For example, Zhang et al. synthesized β-CD with seven biphenyl pendant groups on the small face of the β-CD by copper(I)-catalyzed alkyne–azide cycloaddition (CuAAC) reaction between β-CD with seven azides and a biphenyl moiety with one alkyne group (Fig. 11.4a). The β-CD functionalized with seven biphenyl moieties showed an ordered smectic E phase at low temperature, but changed to a less ordered smectic A phase in the angstrom scale. In the nanoscale, it formed an ordered lamellar structure, and the structure did not change even after phase transition from smectic E to a less ordered smectic A phase (Fig. 11.4b).

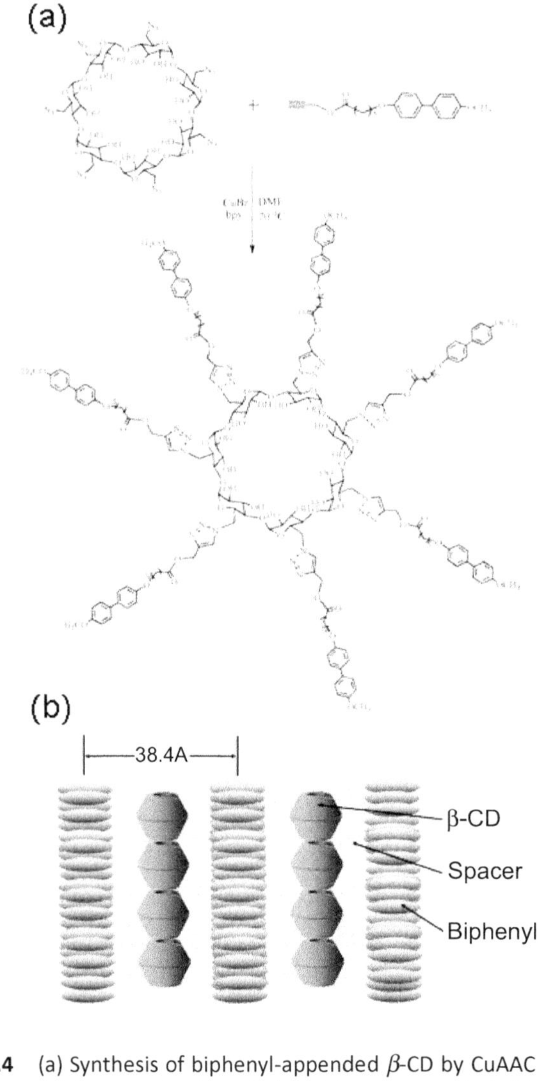

Figure 11.4 (a) Synthesis of biphenyl-appended β-CD by CuAAC reaction. (b) Proposed model for organization of the biphenyl-appended β-CD within the lamellar structure. Reproduced with permission from Ref. [9]. Copyright 2010 Wiley Periodicals, Inc.

Polyrotaxanes, in which many cyclic molecules are mechanically incorporated onto a polymer, have attracted much attention as new polymeric materials.[10] Harada et al. first reported CD-based polyrotaxanes.[11] Ito et al. synthesized CD-based polyrotaxanes showing liquid crystalline properties by covalently introducing mesogenic pendants to the polyrotaxanes (Fig. 11.5a).[12] The polyrotaxanes with mesogenic pendants showed glass transition at 70°C and mesophase until 129°C (Fig. 11.5b).

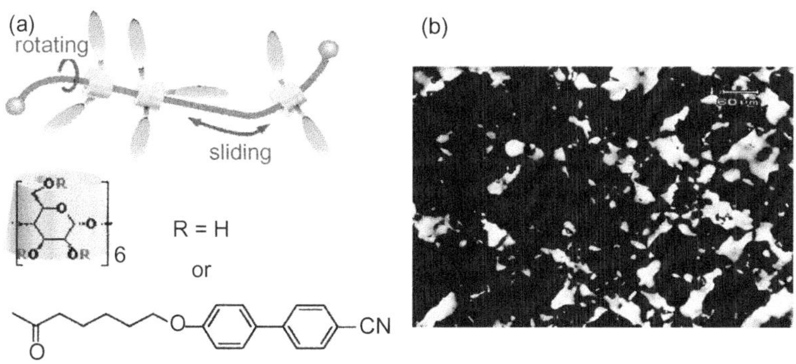

Figure 11.5 (a) Structure of mesogen-substituted polyrotaxane and (b) polarizing optical micrograph of the mesogen-substituted polyrotaxane at 100°C. Reprinted with permission from Ref. [12]. Copyright (2007) American Chemical Society.

Terao et al. reported unique liquid crystalline properties of rotaxanes resulting from the rigidity of the main chain of the polyrotaxane structure.[13] The main-chain-type polyrotaxanes (Fig. 11.6a), in which rigid poly(phenylene ethynylene) polymeric chain are efficiently covered by CDs, formed a cholesteric phase in a concentrated solution (Fig. 11.6c), whereas an uncovered polymer (Fig. 11.6b) only formed an isotropic liquid even at high concentrations. Covering of the rigid π-conjugated polymers by chiral CD wheels could contribute to the observation of the cholesteric phase.

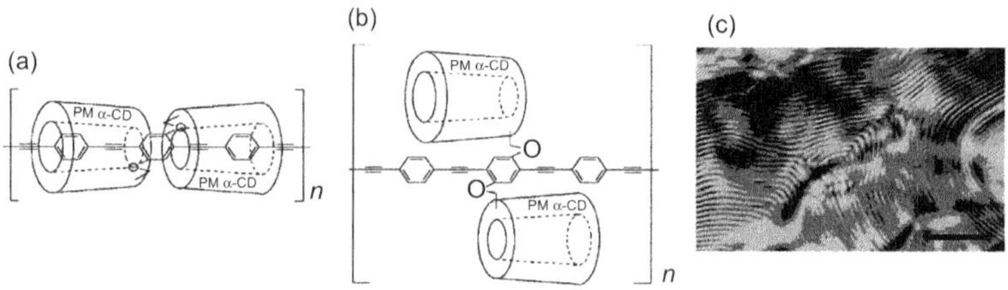

Figure 11.6 Chemical structures of (a) fully covered and (b) uncovered polymers by CDs. (c) Polarized optical micrograph of the fully covered polymer in $CHCl_3$ solution (~15 wt%) sealed in a glass capillary tube; the micrograph was obtained at 25°C under crossed polarizers. The LC phase was separated from the isotropic phase in equilibrium, and its cholesteric pitch was maintained for more than 3 months. Black scale bar: 50 μm. Reprinted with permission from Ref. [13]. Copyright (2009) American Chemical Society.

11.3 Crown Ethers

Crown ethers were the first synthetic macrocyclic compounds reported by Pedersen in 1967.[14] The repeating units of crown ethers are flexible and amphiphilic ethylene oxide moieties. Crown ethers show ion coordination ability because the unpaired electron pairs of the oxygen atoms coordinate with metal cations.

Laschat et al. reported the switching of liquid crystalline phases from smectic to discotic phases by complexation of crown ethers with metal cations.[15] They synthesized unsymmetrical benzo[15]crown-5 ethers with one lateral *ortho*-terphenyl unit bearing alkoxyside chains. Without a metal cation, they formed smectic mesophases resulting from an antiparallel orientation of the wedge-shaped mesogens (Fig. 11.7a). Conversely, with alkali metal cations, they exhibit textures typical of columnar mesophases (Fig. 11.7b). Alkali metal ions form sandwich-type 1:2 complexes with crown ethers. This supramolecular structural change triggered the rearrangement of the liquid crystalline structures from smectic to discotic mesophases.

Loeb and Eichhorn et al. synthesized liquid crystalline [2]rotaxanes.[16] They synthesized [2]rotaxanes using a dibenzo[24]crown-8 ether macrocycle that wraps around a dumbbell-shaped axle containing cationic pyridinium moieties and bulky stoppers with long alkyl chains. The [2]rotaxanes displayed a smectic A phase, whereas the dumbbells did not show liquid crystal phases (Fig. 11.8). A reduction in ionic interactions between the cores (sheltering of the ions) could be a major reason for the observation of smectic A liquid crystal phases for the [2]rotaxanes.

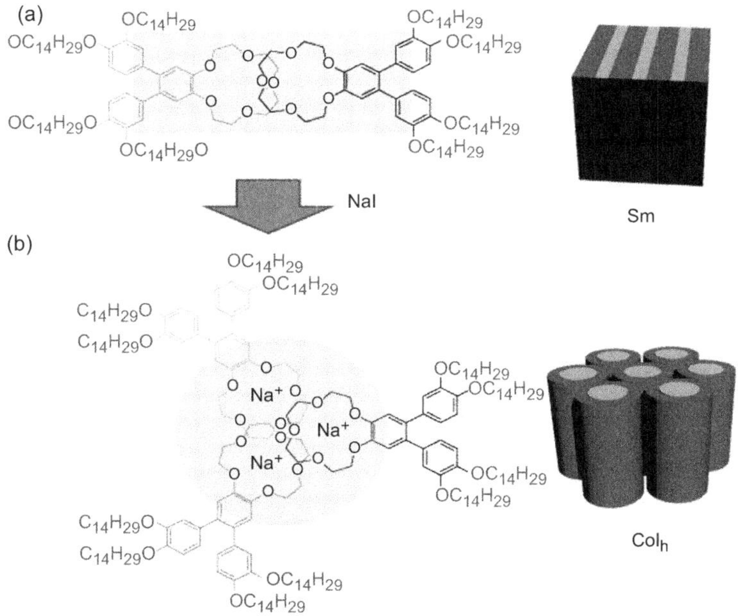

Figure 11.7 Sodium-ion-responsive chemical switching between (b) columnar and (a) smectic phases based on crown ethers containing a lateral *ortho*-terphenyl unit bearing alkoxy side chains.

Schurko and Loeb et al. constructed a molecular organic framework containing mechanically interlocked crown ether rings.[17] The macrocycles in the molecular organic framework underwent different degrees and rates of rotation depending on the size and shape of the macrocycle, as determined by ^2H solid-state nuclear magnetic resonance. In the case of a large macrocyclic ring, the ring had three distinct motions: two-site jump of CD_2 groups (Fig. 11.9b), partial rotation of the macrocyclic ring (Fig. 11.9c), and full rotation of the ring (Fig. 11.9d), at 251, 324, and 423 K, respectively. When a small ring was used, the two-site jump and partial rotation occurred at higher temperatures compared with the large rings, and no full rotation occurred. The two-site jump was only observed for the large ring containing bulky benzo substituents.

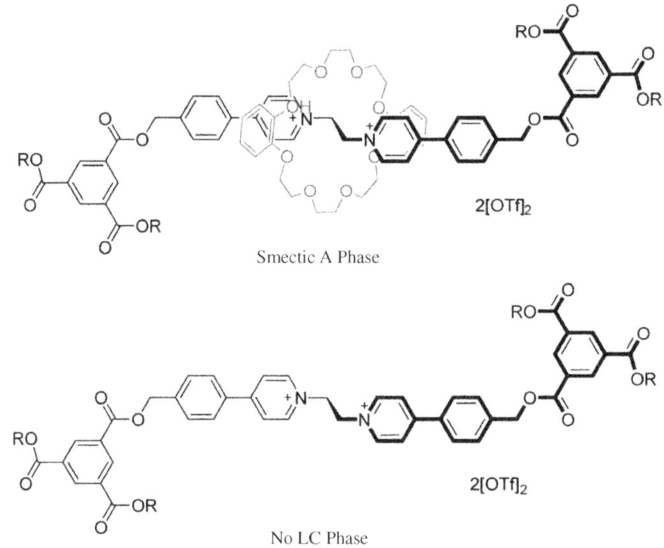

Figure 11.8 A [2]rotaxane with liquid crystalline properties (top) and the dumbbell structure of the axle (bottom).

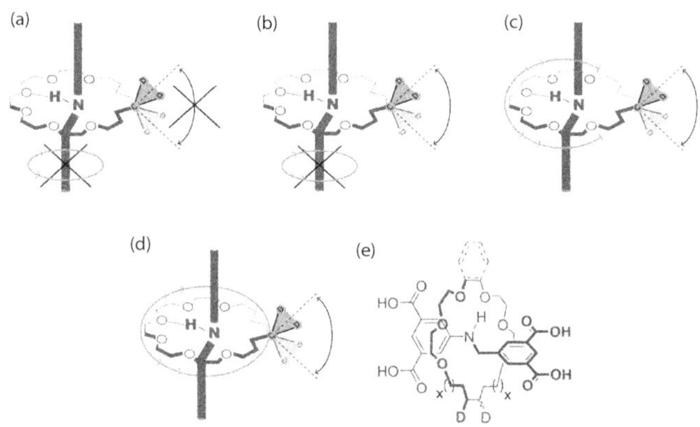

Figure 11.9 Schematic representation of the possible modes of motion of the macrocyclic rings within [2]rotaxanes in molecular organic frameworks: (a) no motion of the macrocyclic ring; (b) two-site jump of the CD_2 groups; (c) partial rotation of the macrocycle around the six ether oxygen atoms within the ring, in addition to the two-site jump of the CD_2 groups; and (d) full rotation of the macrocyclic ring, in addition to the two-site jump of the CD_2 groups. (e) Chemical structure of [2]rotaxanes incorporated in molecular organic frameworks. Reprinted with permission from Ref. [17]. Copyright (2015) American Chemical Society.

Fyles and Barboiu et al. synthesized columnar self-assembled ureido crown ethers (Fig. 11.10e).[18] The driving force for the self-assembly was head-to-tail hydrogen bonding between the urea groups (Fig. 11.10a). Antiparallel association of the ureas produced columnar channels composed of face-to-face crown ethers (Fig. 11.10b-d).

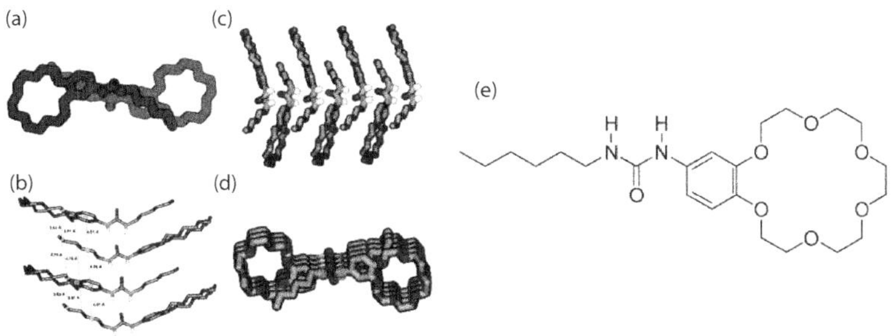

Figure 11.10 Crystal structure of the receptor stick representation of (a) a single dimer, (b) dimer packing distances, and (c and d) two side views of the crystal packing of the heteroditopic receptor in the hydrogen-bonding channel-type superstructures. (e) Chemical structure of a heteroditopic receptor. Reprinted with permission from Ref. [18]. Copyright (2006) American Chemical Society.

11.4 Calixarenes

Calix[*n*]arenes, which are phenolic macrocycles, were popularized by Gutsche in 1978.[19] The phenolic units are linked by methylene bridges at their 2- and 6-positions (meta positions). Owing to the meta-bridge linkage, they form unique calix-shaped structures, which is why they were named calix[*n*]arenes. Phenolic moieties are placed at the lower rims. Therefore, functional groups can be installed in the calix[*n*]arene scaffolds. A straightforward approach to obtain liquid crystal materials using calix[*n*]arenes is to introduce mesogenic moieties using the high reactivity of the phenolic groups. Yang et al. studied the synthesis of a mesophase of a series of calix[4]arene-based liquid crystals. Calix[4]arenes with cholesterol and gallic acid moieties on the lower rim displayed columnar liquid crystals by stacking of calix[4]arenes with a cone conformation (Fig. 11.11a).[20, 21]

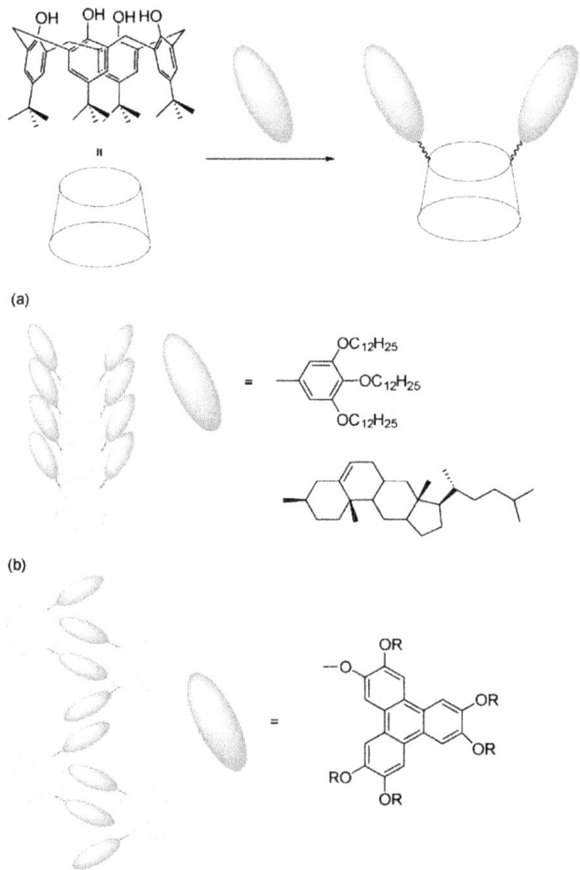

Figure 11.11 Two kinds of schematic representation of the columnar layered molecular arrangement for calix[4]arene with cholesterol, gallic, and triphenylenemoieties.

Conversely, cyclic hexamer calix[6]arene with only two gallic acid moieties did not show liquid crystalline properties. Calix[6]arenes do not have a stable conformation compared with calix[4]arenes, and therefore did not show liquid crystalline properties. These results indicate that the cone conformation of the calix[4]arenes is very important for the presence of liquid crystalline properties.

When discotic triphenylene moieties were incorporated in the calix[4]arene scaffold, the compounds showed the triphenylene column with calix[4]arene units on ancillary lateral sides in the liquid crystalline state (Fig. 11.11b).[22] The columnar assembly of planar triphenylene moieties is stronger than that of bowlic calix[4] arenes. Dendronized cyclotriveratrylene derivatives have cone conformations, thus self-assembled structures of these compounds form columns by intermolecular stacking between cone-shaped cyclotriveratrylenes.

Crego-Calama and Reinhoudt et al. produced self-assembled molecular boxes using calix[4]arene as building blocks (Fig. 11.12).[23] The molecular box consists of two rosette motifs connected through three calix[4]arene dimelamine molecules. The molecular box was constructed by 36 hydrogen bonds; thus it was thermodynamically stable even at low concentrations (5 μM). In the solid state, the molecular boxes were stacked in an ordered columnar fashion.

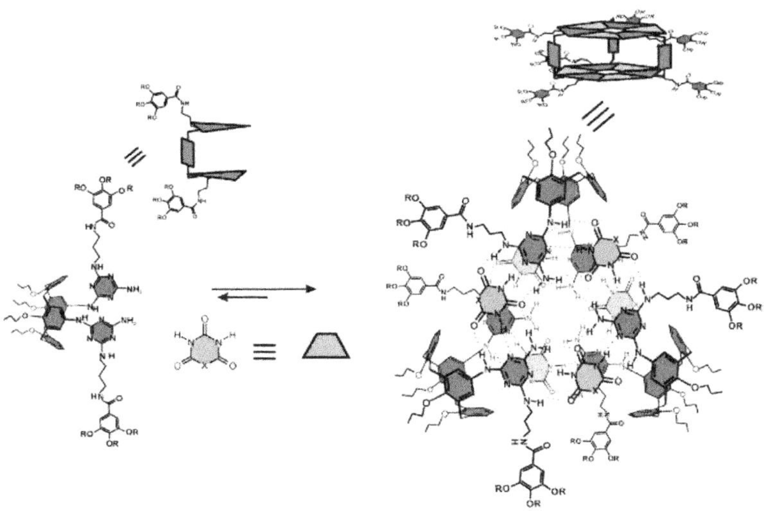

Figure 11.12 Molecular box consisting of two rosette motifs connected through three calix[4]arene dimelamine molecules. Reproduced with permission from Ref. [23]. Copyright 2006 WILEY-VCH Verlag GmbH & Co. KGaA, Weinheim.

Chen and Wong et al. reported the synthesis of tetra-oligophenylene-substituted calix[4]arenes by a Pd coupling reaction.[24] Cyano-phenylcalix[4]arene and methyl-biphenyl-calix[4] arene derivatives (Fig. 11.13a) displayed highly ordered smectic supramolecular organization. Tetra-oligophenylene-substituted calix[4]arene cores acted as mesogenic moieties that exhibited liquid crystalline properties. Figure 11.13b shows the proposed structures of the ordered smectic mesophase. The cone-shaped mesogens orient alternatively up and down. The distance between the smectic layers could be tuned by changing the length of the alkyl chains. As the length of the alkyl chains increased, the distance between the smectic layers increased.

(a) (b)

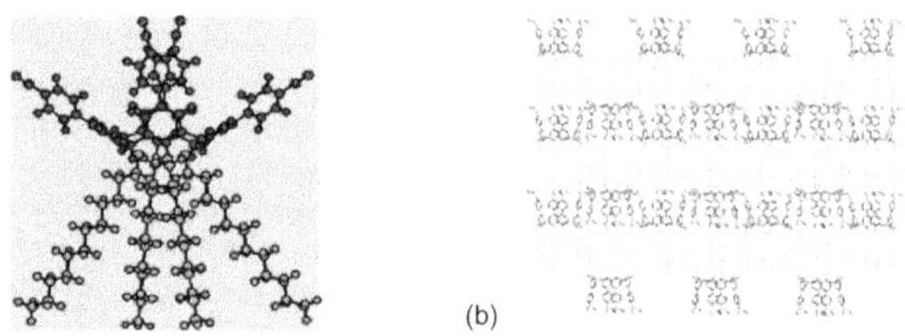

Figure 11.13 (a) MM2-optimized pinched cone conformation of cyano-biphenylcalix[4]arene and (b) schematic representation of the smectic layered molecular arrangement. Reprinted with permission from Ref. [24]. Copyright (2006) American Chemical Society.

11.5 Cucurbuturils

Cucurbit[*n*]urils (CB[*n*], n = 5–10) are pumpkin-shaped macrocyclic compounds.[25–27] Characterization and the host–guest properties of CB[6] were first reported by Mock et al. in 1981.[28] Unfortunately, the low solubility of CB[6] slowed the development of CB[6] chemistry until 2000, when Kim et al. reported the synthesis of CB[*n*] homologs (*n* = 5–11) that showed moderate solubility in water (CB[5]: 420 mM, CB[7]: 370 mM); thus, CB[*n*] chemistry has been developed rapidly since Kim's discovery.[29] Compared with other macrocyclic compounds, such as CDs and calix[*n*]arenes, CB[*n*] have highly symmetrical entrances. Thus, guest molecules can access both faces of CB[*n*]. Moreover, glycoluril units are connected by two methylene bridges, which contributes to the rigid CB[*n*] structure. This rigid structure results in no conformational flexibility, thus CB[*n*] can form host–guest complexes with high affinity and selectivity.

Kim et al. reported the construction of free-standing, single-monomer-thick 2D polymers by covalent self-assembly of hexagonal CB[6] molecules (Fig. 11.14).[30]

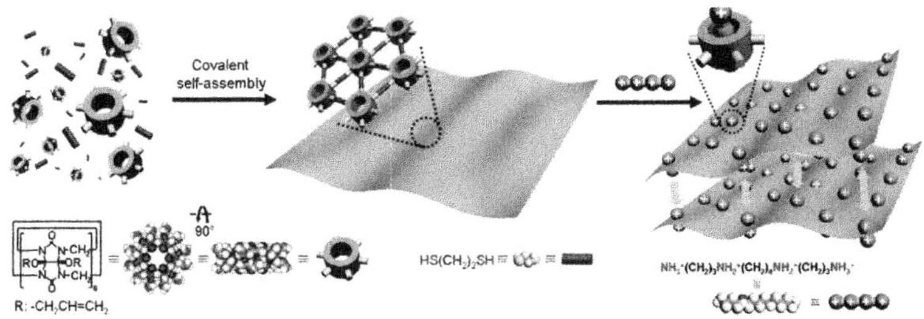

Figure 11.14 Schematic representation of freestanding 2D polymer formation using CB[6] with 12 allyl groups and ethanedithiol through irreversible covalent self-assembly and layer-to-layer separation through the Coulomb repulsion of 2D polymers in which each CB[6] repeating unit is threaded with protonated spermine carrying positive charges on both ends. Reprinted with permission from Ref. [30]. Copyright (2013) American Chemical Society.

To readily synthesize a robust and 2D polymer, they used a thiol–ene click reaction between CB[6] with 12 laterally predisposed reactive alkene groups and 1,2-ethanedithiol (Fig. 11.14, center). Single-monomer-thick layers of the 2D polymer were obtained by the formation of a host–guest complexes between CB[6]s in the 2D polymers and protonated spermine, a thread-like molecule with four positive ammonium groups. Electronic repulsion between the sheets induced the layer-to-layer separation (Fig. 11.14, right). Two-dimensional sheets composed of CB[6]-based polyrotaxanes were constructed by the same group (Fig. 11.15).

First, a pseudorotaxane was produced by threading CB[6] with a diamine axle with two pyridine moieties (Fig. 11.15a).[31] The 2D polyrotaxanes were obtained by reaction of the pseudorotaxane with $Cu(NO_3)_2$ in the presence of oxalate ions. CB[6]-based rotaxanes were incorporated in the 2D sheets, which was confirmed by the X-ray crystal structure of the 2D sheet (Fig. 11.15b–d). Two different cavities existed in the 2D sheets (Fig. 11.15b).

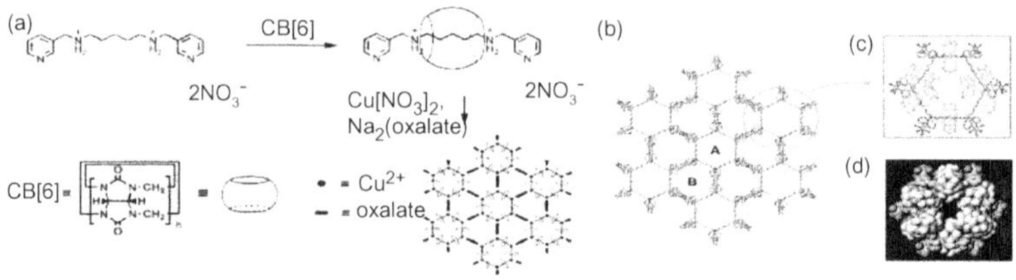

Figure 11.15 (a) Synthesis of 2D polyrotaxanes. The 2D polyrotaxane layer built with a molecular necklace as a secondary building block (b), which is highlighted in a ball-and-stick representation (c) and in a space-filling representation (d). Two different types of cavities A and B exist in the 2D network, as noted. Reproduced with permission from Ref. [31]. C 2001 WILEY-VCH Verlag GmbH, Weinheim.

Cyclic octamer CB[8] exhibits remarkable host–guest properties that are different from those of the smaller CB[6] and CB[7] homologs. CB[8] can capture two guest compounds because its cavity size is larger than that of CB[5–7]s.[32] Zhao, Liun, and Li et al. used CB[8] as a connector to construct 2D supramolecular polymers (Fig. 11.16).[33] They synthesized a rigid 1,3,5-triphenyl benzene carrying three viologen parts. Mixing the triangular molecules and CB[8] in a 2:3 ration resulted in the formation of a 2D supramolecular organic framework with a honeycomb structure. CB[8] also encapsulates two different molecules, electron-donating molecules and electron-accepting molecules, inside the cavity to form a stable 1:1:1 complex. Electron donors do not form a complex with CB[8] in the absence of an electron acceptor. Thus, the major driving force for the ternary 1:1:1 complex was the strong charge transfer (CT) interaction between electron donor and electron acceptor molecules inside the CB[8] cavity. Feng et al. have reported that the ternary complexes can be used to construct 2D polymers (Fig. 11.17).[34]

First, rigid triangle-shaped monomers with three naphthalene donor moieties and a rigid viologen dimer were synthesized. The reaction of the preassembled pseudo[3]rotaxane consisting of two CB[8] rings and the viologen dimer, and a donor monomer in the two-phase system gave a 2D supramolecular monolayer.

Kim et al. discovered that an organic molecular porous material with permanent porosity was obtained by recrystallization of CB[6] from aqueous HCl solution.[35]

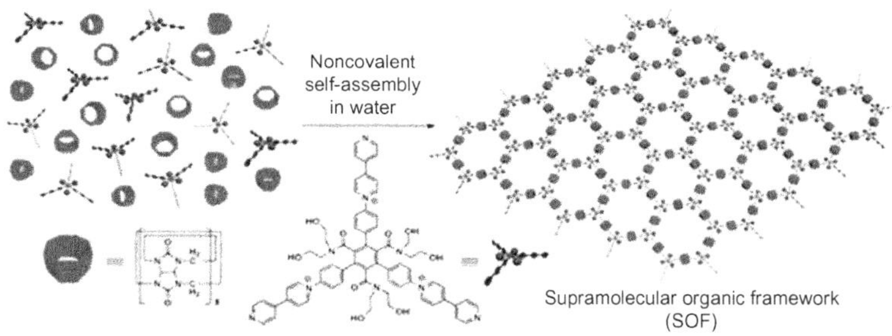

Figure 11.16 Schematic representation of the self-assembly patterns of a rigid 1,3,5-triphenylbenzene carrying three viologen parts with CB[8] in water. Reprinted with permission from Ref. [33]. Copyright (2013) American Chemical Society.

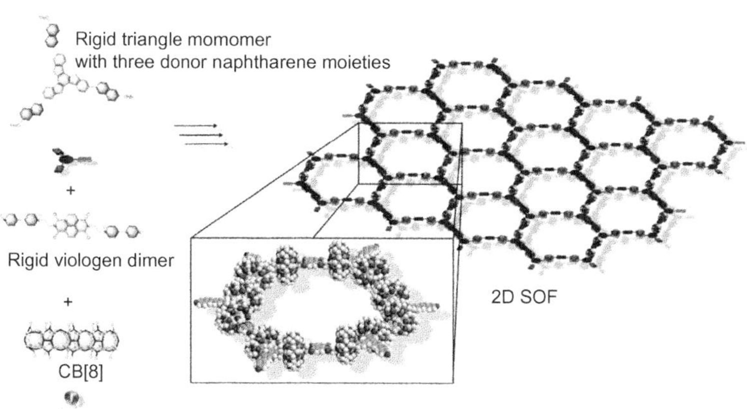

Figure 11.17 Schematic representation of the self-assembly patterns of a rigid 1,3,5-triphenylbenzene carrying three viologen parts with CB[8] in water.

The organic porous material showed highly selective sorption of CO_2 over CO and CH_4. X-ray crystal structure analysis of CO_2 adsorbed on the CB[6]-based organic porous material suggested three independent CO_2 sorption sites (Fig. 11.18a). At sites A (Fig. 11.18b) and B (Fig. 11.18c), CO_2 molecules were located in 1D channels, which were generated by an arrangement of CB[6] molecules. At these sites, the CO_2 molecules interacted with the CB[6]-based organic porous materials by hydrogen bonding and quadrupole–quadrupole interaction in a T-shaped arrangement. Site

C (Fig. 11.18d) was a CB[6] cavity. Two CO_2 molecules were included in the cavity. The CB[6]-based organic porous materials also took up acetylene molecules. The X-ray crystal structure showed that acetylene molecules were adsorbed in the 1D channels. CH/O hydrogen bonding and van der Waals interactions were the main driving forces for the take up of acetylene.

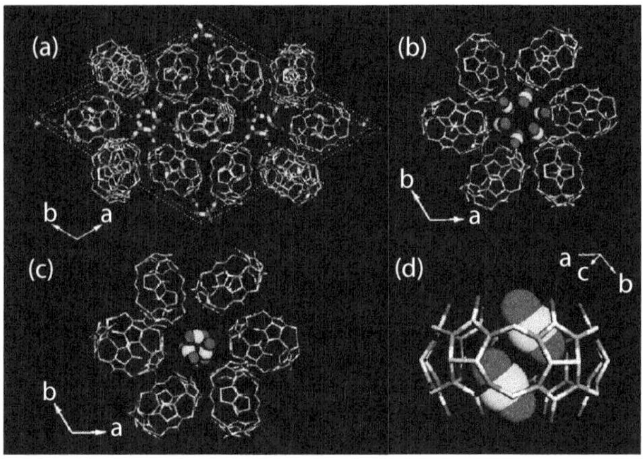

Figure 11.18 (a) X-ray crystal structure of CO_2-adsorbed organic molecular porous material of CB[6], (b) sorption site A, (c) sorption site B, and (d) sorption site C. Reprinted with permission from Ref. [35]. Copyright (2010) American Chemical Society.

11.6 Pillararenes

In 2008, our group reported a new class of pillar-shaped macrocyclic host compounds known as pillar[*n*]arenes.[36] Pillar[*n*]arenes with n = 5–13 have already been produced,[37, 38] but the cyclic pentamers, that is, pillar[5]arenes, and cyclic hexamers, that is, pillar[6]arenes, have been the most widely used because they can be obtained in high yields. Pillar[*n*]arenes are composed of 1,4-dialkoxybenzene units. Their units are connected by methylene bridges at the para-position of the 1,4-dialkoxybenzene units. Therefore, pillar[*n*]arenes have highly symmetrical pillar-shaped structures, which are similar to the highly symmetrical cucurbit[*n*]urils.[39–46] The other interesting aspect of pillar[n]arene chemistry is the versatile functionalization of pillar[n]arenes. The constituent units of pillar[n]arenes are 1,4-dialkoxybenzene derivatives, thus pillar[n]arenes possess $2n$ substituents. For example, pillar[5]arenes and pillar[6]arenes have 10 and 12 alkoxy substituents at both rims, respectively. Deprotection of these alkoxy substituents affords highly reactive pillar[n]arenes bearing OH moieties at both rims. Introducing other reactive groups such as bromo, alkyne, and azido groups on both of the rims of pillar[n]arenes can be also accomplished by various organic synthetic approaches. The presence of different functional groups at the reactive sites significantly affects the physical properties of pillar[n]arenes. Consequently, installing mesogenic moieties at both rims of pillar[n]arenes resulted in liquid crystal pillar[n]arenes. Nierengarten et al. developed thermotropic liquid crystals using a pillar[5]arene carrying 10 mesogenic cyanobiphenyl moieties on both of its rims (Fig. 11.19a).[47] It exhibited an enantiotropic smectic A phase over a wide temperature range from 32°C to 201°C (Fig. 11.19c). In contrast, the unit model bearing two cyanobiphenyl moieties (Fig. 11.19a) only showed a monotropic smectic A phase during the cooling process over a narrow temperature range from 149°C to 125°C. Intermolecular π–π stacking interactions between the neighboring pillar[5]arene cores stabilized the smectic A phase (Fig. 11.19d). Pillar[6]arene bearing 12 cyanobiphenyl moieties also showed a stable smectic A phase over a wide range of temperatures.[48] Wang et al. produced liquid crystals using pillar[5]arenes bearing 10 azobenzene moieties with different spacers (Fig. 11.19e, n = 4 and 10).[49] The pillar[5]arenes showed smectic liquid crystalline phases over a wide temperature range (Fig. 11.19f,g). Furthermore, the distance between the smectic layers can be tuned by the length of the alkyl linkers. The lengths determined by small- and wide-angle X-ray scattering (SAXS/WAXS) analyses were 72.4 and 59.7 Å, respectively (Fig. 11.19h). Recently, Barberá, Deschenaux, and Nierengarten et al. synthesized new liquid crystalline pillar[5]arene derivatives by modification of poly(benzylether) dendrons.[50] The molecules self-assembled into tubular nanostructures within columnar phases. Pillar[5]arenes in the core of the nanotubes captured a guest molecule, 1,6-dicyanobutane, even in the liquid crystal state.

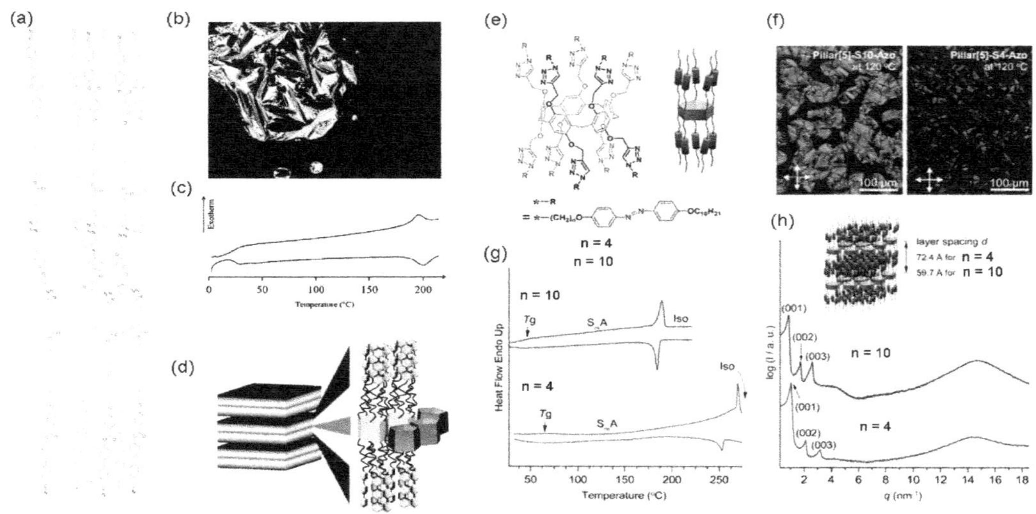

Figure 11.19 Liquid crystal pillar[5]arenes. (a) Pillar[5]arene with mesogenic moieties and the unit model. (b) Polarized optical micrograph of the texture displayed by the pillar[5]arene with mesogenic moieties in the smectic A phase at 183°C. (c) DSC thermogram of the pillar[5]arene with mesogenic moieties in the second heating-cooling cycle. (d) Proposed liquid crystal structure of the pillar[5]arene with mesogenic moieties in the smectic A phase. Reproduced from Ref. [47] with permission from the Royal Society of Chemistry. (e) Chemical structures of the azobenzene-modified pillar[5]arenes with different spacers (n = 4 and 10). (f) Polarized optical micrograph of the azobenzene-modified pillar[5]arenes (n = 4 and n = 10) at 120°C. (g) DSC thermograms of the azobenzene-modified pillar[5]arenes (n = 4 and 10) in the heating-cooling cycle. (h) SAXS/WAXS profiles in their liquid crystalline state (80°C). Reproduced with permission from Ref. [49]. Copyright 2015 WILEY-VCH Verlag GmbH & Co. KGaA, Weinheim.

Pillar[*n*]arenes have highly symmetrical pillar-shaped structures and high free volumes; thus they are good candidates for preparing porous materials for gas and organic vapor adsorption. Yang et al. reported pillar[5]arene crystals as gas adsorption materials. A pillar[5]arene bearing 10 hydroxyl groups exhibited a 3D open network structure, which was stabilized by intermolecular hydrogen-bonding interactions (Fig. 11.20).[51]

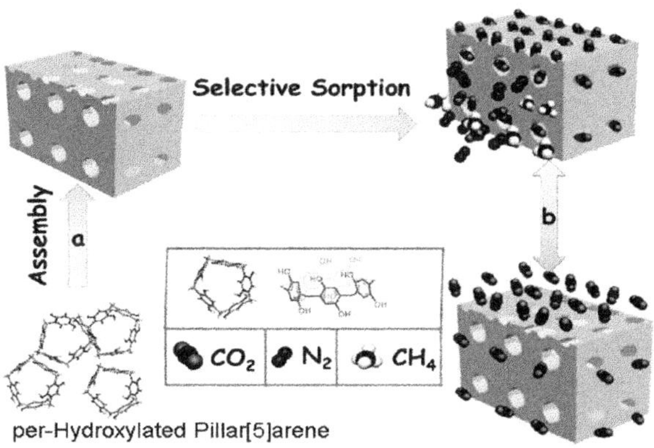

Figure 11.20 Schematic representation of selective gas sorption and desorption by the activated crystals of per-hydroxylated pillar[5]arene. Reproduced with permission from Ref. [31]. Copyright 2014 WILEY-VCH Verlag GmbH & Co. KGaA, Weinheim.

The 3D network structure contained acetone molecules. First, they obtained activated crystals, in which there were no solvents, by drying the crystals at 170°C under reduced pressure. The activation process disturbed the porous structures of the organic crystals in many cases. However, the crystal structure remained unchanged, even after the acetone molecules had been removed because the 3D open network structure was stabilized by intermolecular hydrogen-bonding interactions. The activated crystals absorbed CO_2 with high sorption ability (88 mg g^{-1}, at 25°C, 1 atm). In contrast, CH_4 and N_2 were poorly adsorbed, indicating remarkable CO_2-selectivities over CH_4 (375/1) and N_2 (339/1). The highly selective CO_2 uptake was mainly because of the porous nanostructure of the crystals and the dipole interactions between CO_2 and the hydroxyl groups of the pillar[5]arene.

Our group reported the gas and organic vapor uptake properties of crystalline pillar[6]arene bearing 12 hydroxyl moieties.[52] Hexagonal pillar[6]arene molecules assembled into a two-hexagonal-sheet structure along the *a* and *b* axes, and a 1D channel structure along the *c* axis. Acetone molecules were included in the channels. The 1D channels were retained even after the activation process because the structure was stabilized by multiple hydrogen-bonding interactions in the same way as the pillar[5]arene with 10 hydroxyl groups. This crystalline material not only took up various gases such as CO_2 (3.3 Å), N_2 (3.7 Å), and n-butane (4.3 Å), but also took up organic vapors such as n-hexane (4.9 Å) and cyclohexane (6.7 Å). The broad adsorption characteristics of this crystalline material resulted from the permanent 1D pores with the same cavity size as pillar[6]arene (ca. 6.7 Å), which was larger in size than the gases and organic vapors described above. The pore size of the 1D channels was determined to be 4.10 Å using a molecular probe technique.

We also constructed 2D hexagonal sheets based on the hexagonal-shaped structure of pillar[6]arene (Fig. 11.21a).[53] Generally, molecules with highly symmetrical triangular, tetragonal, and hexagonal molecules formed densely packed structures by self-assembly. Therefore, our group investigated the 2D supramolecular polymerization of hexagonal pillar[6]arene molecules to form 2D hexagonal sheets with microporous pores. 2D supramolecular polymerization can be induced by the oxidation of the six hydroquinone units of a pillar[6]arene to benzoquinone because the CT complexation process between the hydroquinone and benzoquinone units of pillar[6]arenes results in the intermolecular connection of pillar[6]arene molecules (Fig. 11.21b). The highly ordered structure of the 2D sheets formed regular fiber structures. The 2D sheets also had pores (4.04 Å), resulting from the original pore size of the pillar[6]arene crystals (4.10 Å). The 2D sheets were converted to carbons by the calcination of the 2D sheets at 900°C for 2 h under an inert atmosphere. Interestingly, even after the calcination of the 2D sheets, the fibrous structures were completely retained. Furthermore, the fibrous carbon had pores (4.09 Å) that were similar in size to the original starting pillar[6]arene assembly (Fig. 11.21c). This method is a new synthetic approach to create porous carbon with controlled pores at the angstrom level.

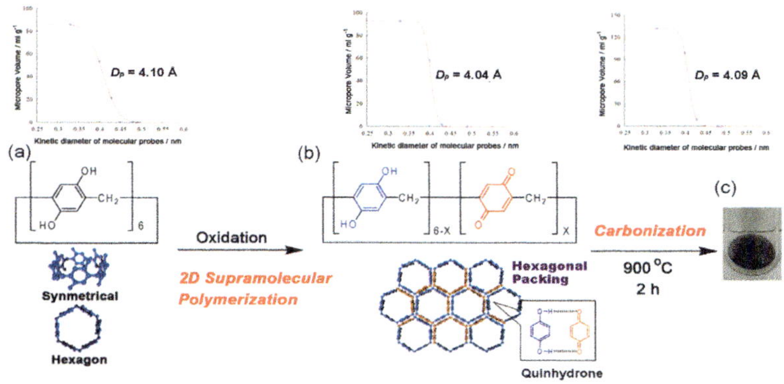

Figure 11.21 (a) Hexagonal structure of pillar[6]arene, (b) 2D supramolecular polymerization by the oxidation of the hydroquinone units, and (c) porous carbon, which was prepared by the carbonization of the CT complex. Reproduced with permission from Ref. [53]. Copyright 2015 WILEY-VCH Verlag GmbH & Co. KGaA, Weinheim.

Our group reported unexpected alkane uptake behavior using activated crystals of pillar[5]arene with 10 ethoxy moieties.[54] In this case, the activation process of the crystals changed the crystal structure from 1D channels to a different crystal structure because there are no strong interactions to stabilize the 1D channels. Consequently, the crystals formed a soft porous material. Therefore, during the adsorption process, crystals without solvent molecules could quantitatively take up linear n-alkanes by making structural changes (Fig. 11.22). The uptake of the *n*-alkanes took place quickly at certain pressures. Furthermore, during the desorption process, the adsorbed linear *n*-alkanes were not released even under reduced pressure because they were stabilized by multiple CH/π interactions in the pillar[5]arene cavity. The pillar[5]arene cavity fits the linear alkanes, but not the branched and cyclic alkanes. Therefore, the cyclic and branched alkanes were not absorbed.

Recently, our group constructed pillar[5]arene thin films using a layer-by-layer method (Fig. 11.23).[55] We synthesized two ionic pillar[5]arene derivatives, pillar[5]arene with 10 cationic and anionic moieties. Pillar[5]arene-based multilayer films were constructed by the layer-by-layer method with alternating layers of cationic and anionic pillar[5]arenes. The films have active pores from the pillar[5]arene cavity, thus allowing shape-selective uptake of dinitrobenzene isomers: the film takes up *para*-dinitrobenzene, but not *ortho*- and *meta*-dinitrobenzene.

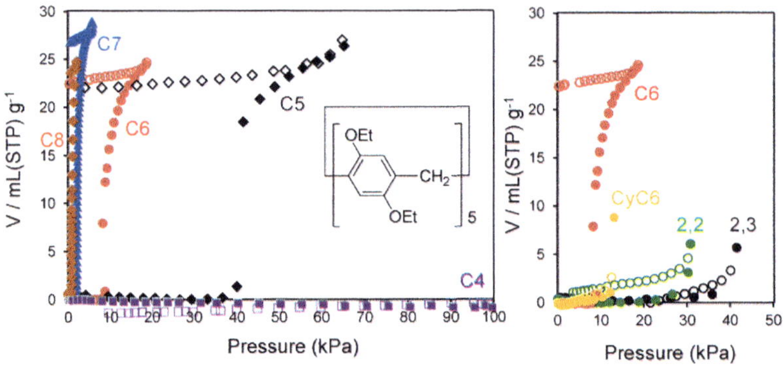

Figure 11.22 Sorption isotherms of activated crystals of pillar[5]arene with 10 ethoxy groups toward the gases and vapors of several linear alkanes, including *n*-butane (C4, purple squares), *n*-pentane (C5, black diamonds), *n*-hexane (C6, red circles), *n*-heptane (C7, blue triangles), and *n*-octane (C8, brown diamonds), as well as several branched alkanes, including 2,2-dimethylbutane (2,2, green circles), 2,3-dimethylbutane (2,3, black circles), and cyclohexane (CyC6, yellow circles) at 25°C. Solid symbols = adsorption; open symbols = desorption. Reproduced with permission from Ref. [54]. Copyright 2015 WILEY-VCH Verlag GmbH & Co. KGaA, Weinheim.

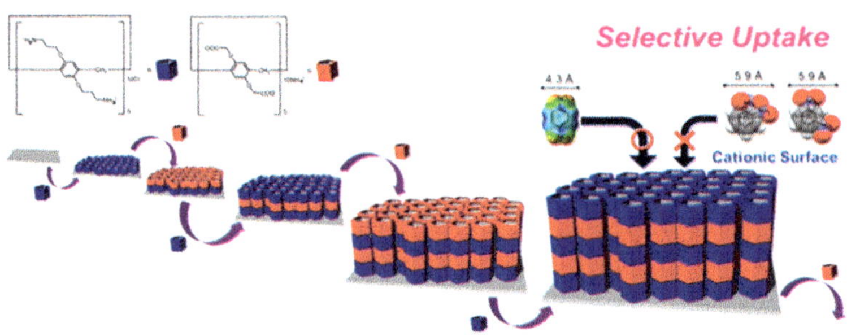

Figure 11.23 Layer-by-layer assembly by the consecutive adsorption of cationic and anionic pillar[5]arenes. The multilayered film showed size-selective molecular recognition properties. Reprinted with permission from Ref. [55]. Copyright (2015) American Chemical Society.

11.7 Other Macrocyclic Compounds

Ordered assembled structures can also be constructed using other macrocyclic compounds. Disk-shaped macrocyclic compounds are often used as building blocks to construct discotic assembled structures because disk-shaped structures tend to assemble into 1D structures.[56] In other words, the structural change from planar disk to non-planar structures induces a phase transition. Norikane et al. reported a phase transition from a liquid crystal to an isotropic phase by UV light irradiation, and from an isotropic to a liquid crystal phase by visible light irradiation or heating (Fig. 11.24).[57]

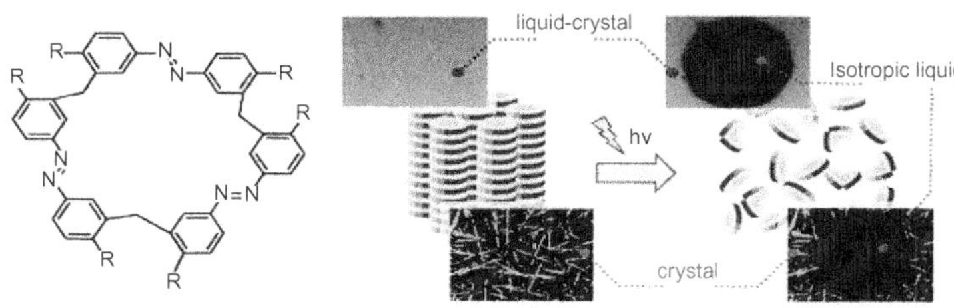

Figure 11.24 Chemical structure of macrocycles consisting of azobenzene units. Polarizing optical photomicrograph changes of the azobenzene macrocycle in liquid crystalline and crystal by irradiation of UV light. Reproduced from Ref. [57] with permission from the Royal Society of Chemistry.

They synthesized macrocycles consisting of azobenzene units. All trans isomers displayed a columnar phase. The UV-irradiated region became dark under a crossed polarizer configuration, indicating a phase transition from the liquid crystal to an isotropic phase by UV irradiation. Furthermore, a crystal film of the macrocyclic compounds in the *trans* form rapidly melts by UV light irradiation owing to the phase transition from the crystal to the isotropic phase.

Chen and Wang et al. produced liquid crystalline macrocyclic azacalix[4]pyridine (Fig. 11.25).[58]

The macrocycle had saddle-shaped cores in the absence of a zinc ion. Thus, the macrocyclic compounds exhibited a columnar phase with helical and tilted molecular packing. Coordination of zinc salts with the macrocyclic compound led to a conformational change from saddle-shaped to flattened structures, which induced a change in the liquid crystal properties.

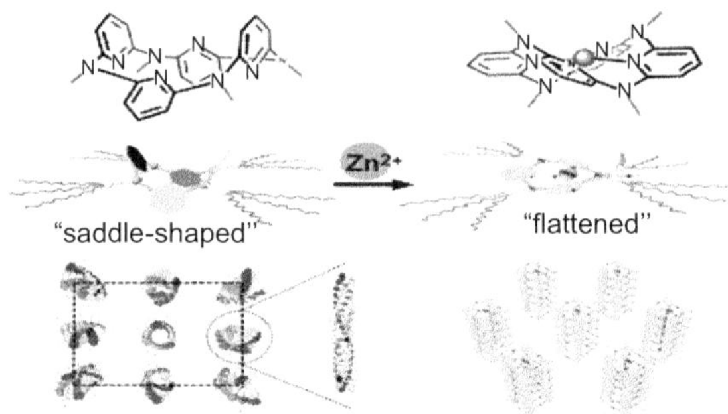

Figure 11.25 Schematic diagram of conformational variation of a azacalix[4] pyridine core by coordinating with a zinc ion. Change in the slant helical column in the orthorhombic lattice (C_r) to Col_h lattice upon addition of Zn^{2+}. Reproduced from Ref. [58] with permission from the Royal Society of Chemistry.

Shimizu et al. prepared a macrocyclic compounds consisting of C-shaped spacers and two urea groups (Fig. 11.26).[59] The macrocycles self-assembled though intermolecular hydrogen bonding between the urea molecules to afford pillar-shaped columnar channels with an inner diameter of ca. 9–10 Å. They used the columnar channels as porous materials for performing organic reactions. The photoreaction of 2-cyclohexenone usually affords a photodimer with four conformers. Photodimerization of 2-cyclohexenone in the columnar channels yielded a head-to-tail dimer, whereas the head-to-head dimer was the major product in the absence of the columnar channels.

Itoh and Aida et al. reported columnar assemblies of liquid crystalline peptidic macrocycles (Fig. 11.27).[60]

They formed conformationally robust bowl-shaped structures owing to internal hydrogen bonding. They exhibited hexagonal columnar liquid crystals in a wide temperature range, including room temperature. The columnar assembly is formed by intermolecular hydrogen-bonding interactions involving the exocyclic amide units. These liquid crystal columns underwent large-area unidirectional orientation upon application of an electric field. The liquid crystal film showed birefringence. However, the area under the electronic field became nonbirefringent, indicating orientation of the liquid crystal columns on the substrate. Even after switching off the electronic field, the unidirectional columnar orientation was maintained for a half year.

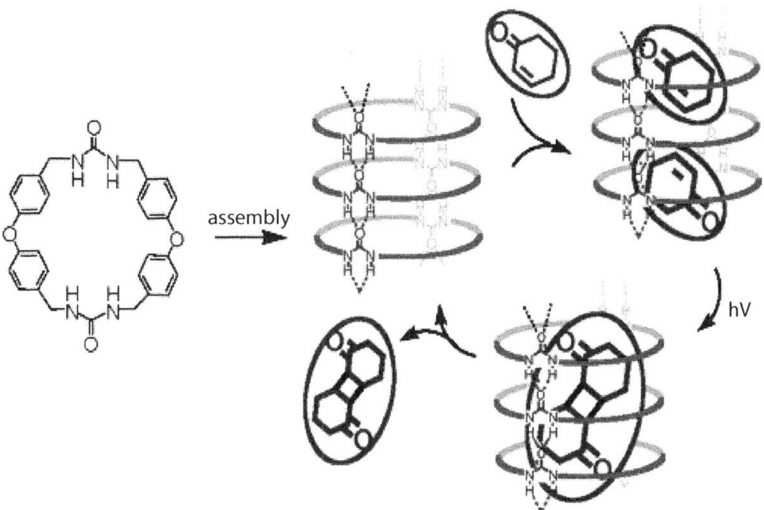

Figure 11.26 Bisurea macrocycle self-assembles into tubular structures that can bind 2-cyclohexenone molecules. Irradiation of the enone in the confined environment selectively yields the head-to-tail photodimer. Reprinted with permission from Ref. [58]. Copyright (2006) American Chemical Society.

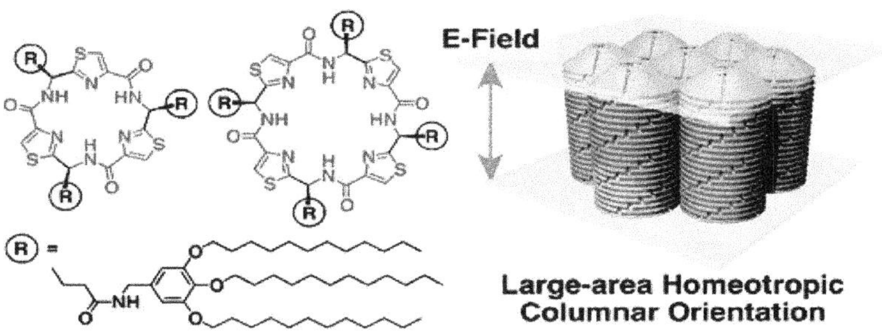

Figure 11.27 Molecular structures of liquid crystalline macrocyclic oligopeptide derivatives self-assemble columnarly with a hexagonal geometry allied by electronic field. Reprinted with permission from Ref. [60]. Copyright (2011) American Chemical Society.

Tashiro and Aida et al. studied the self-assembly of planer triple fused metalloporphyrin dimers containing either hydrophobic paraffinic side chains, hydrophilic triethylene glycol chains, or amphiphilic dimers containing both hydrophobic paraffinic and hydrophilic triethylene glycol chains (Fig. 11.28).[61]

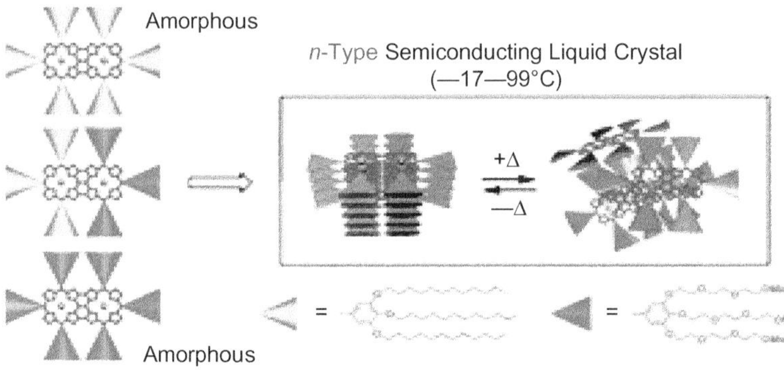

Figure 11.28 A triply fused copper porphyrin dimer with hydrophobic and hydrophilic wedges self-assembled into a columnar liquid crystalline mesophase but gives rise to an amorphous solid when modified with only hydrophobic or hydrophilic wedges. Reprinted with permission from Ref. [61]. Copyright (2008) American Chemical Society.

11.8 Conclusions and Outlook

Macrocyclic compounds are useful building blocks for the construction of a large variety of novel assembled nanometer-size structures because they exhibit versatile functionality, host–guest properties, and highly symmetrical and uniform structures. To align the macrocyclic molecules, mesogenic moieties can be introduced using simple and straightforward methods. The other approach to construct nanometer-size structures is by using highly symmetrical macrocyclic compounds. The assembly of hexagonal molecules to obtain 2D hexagonal sheets is a good approach based on geometric design. Janus structures consisting of hydrophobic and hydrophilic substituents are also an effective approach for the assembly of molecules at the nanometer scale. We hope that the current chapter will stimulate research efforts in that direction.

References

1. Lehn, J. M. *Supramolecular Chemistry*, Wiley: Weinheim (1997).
2. Crini, G. *Chem. Rev.*, **114**, 10940–10975 (2014).
3. Harata, K. *Chem. Rev.*, **98**, 1803–1828 (1998).
4. Kamitori, S., Matsuzaka, O., Kondo, S., Muraoka, S., Okuyama, K., Noguchi, K., Okada, M., Harada, A. *Macromolecules*, **33**, 1500–1502 (2000).
5. Rusa, C. C., Bullions, T. A., Fox, J., Porbeni, F. E., Wang, X., Tonelli, A. E. *Langmuir*, **18**, 10016–10023 (2002).
6. Marui, Y., Kida, T., Akashi, M. *Chem. Mater.*, **22**, 282–284 (2010).
7. Kida, T., Sato, S., Yoshida, H., Teragaki, A., Akashi, M. *Chem. Commun.*, **50**, 14245–14248 (2014).
8. Kida, T., Marui, Y., Miyawaki, K., Kato, E., Akashi, M. *Chem. Commun.*, 3889–3891 (2009).
9. Chen, L., Hu, T.-H., Xie, H.-L., Zhang, H.-L. *J. Polym. Sci., Part A: Polym. Chem.*, **48**, 2838–2845 (2010).
10. Harada, A., Hashidzume, A., Yamaguchi, H., Takashima, Y. *Chem. Rev.*, **109**, 5974–6023 (2009).
11. Harada, A., Li, J., Kamachi, M. *Nature*, **356**, 325–327 (1992).
12. Kidowaki, M., Nakajima, T., Araki, J., Inomata, A., Ishibashi, H., Ito, K. *Macromolecules*, **40**, 6859–6862 (2007).
13. Terao, J., Tsuda, S., Tanaka, Y., Okoshi, K., Fujihara, T., Tsuji, Y., Kambe, N. *J. Am. Chem. Soc.*, **131**, 16004–16005 (2009).
14. Pedersen, C. J. *J. Am. Chem. Soc.*, **89**, 7017–7036 (1967).
15. Steinke, N., Frey, W., Baro, A., Laschat, S., Drees, C., Nimtz, M., Hägele, C., Giesselmann, F. *Chem. -Eur. J.*, **12**, 1026–1035 (2006).
16. Suhan, N. D., Loeb, S. J., Eichhorn, S. H. *J. Am. Chem. Soc.*, **135**, 400–408 (2013).
17. Vukotic, V. N., O'Keefe, C. A., Zhu, K., Harris, K. J., To, C., Schurko, R. W., Loeb, S. J. *J. Am. Chem. Soc.*, **137**, 9643–9651 (2015).
18. Cazacu, A., Tong, C., van der Lee, A., Fyles, T. M., Barboiu, M. *J. Am. Chem. Soc.*, **128**, 9541–9548 (2006).
19. Gutsche, C. D. *Calixarenes*, Royal Society of Chemistry: Cambridge (1989).
20. Fang, X., Guo, H., Yang, F., Wu, Y. *Tetrahedron Lett.*, **56**, 6128–6131 (2015).

21. Guo, H., Yang, F., Liu, W., Lai, J. *Tetrahedron Lett.*, **56**, 866–870 (2015).
22. Yang, F., Xu, B., Guo, H., Xie, J. *Tetrahedron Lett.*, **53**, 1598–1602 (2012).
23. Piermattei, A., Giesbers, M., Marcelis, A. T. M., Mendes, E., Picken, S. J., Crego-Calama, M., Reinhoudt, D. N. *Angew. Chem. Int. Ed.*, **45**, 7543–7546 (2006).
24. Lo, P. K., Chen, D., Meng, Q., Wong, M. S. *Chem. Mater.*, **18**, 3924–3930 (2006).
25. Lagona, J., Mukhopadhyay, P., Chakrabarti, S., Isaacs, L. *Angew. Chem. Int. Ed.*, **44**, 4844–4870 (2005).
26. Lee, J. W., Samal, S., Selvapalam, N., Kim, H.-J., Kim, K. *Acc. Chem. Res.*, **36**, 621–630 (2003).
27. Kaifer, A. E. *Acc. Chem. Res.*, **47**, 2160–2167 (2014).
28. Freeman, W. A., Mock, W. L., Shih, N. Y. *J. Am. Chem. Soc.*, **103**, 7367–7368 (1981).
29. Kim, J., Jung, I.-S., Kim, S.-Y., Lee, E., Kang, J.-K., Sakamoto, S., Yamaguchi, K., Kim, K. *J. Am. Chem. Soc.*, **122**, 540–541 (2000).
30. Baek, K., Yun, G., Kim, Y., Kim, D., Hota, R., Hwang, I., Xu, D., Ko, Y. H., Gu, G. H., Suh, J. H., Park, C. G., Sung, B. J., Kim, K. *J. Am. Chem. Soc.*, **135**, 6523–6528 (2013).
31. Lee, E., Kim, J., Heo, J., Whang, D., Kim, K. *Angew. Chem. Int. Ed.*, **40**, 399–402 (2001).
32. Ko, Y. H., Kim, E., Hwang, I., Kim, K. *Chem. Commun.*, 1305–1315 (2007).
33. Zhang, K.-D., Tian, J., Hanifi, D., Zhang, Y., Sue, A. C.-H., Zhou, T.-Y., Zhang, L., Zhao, X., Liu, Y., Li, Z.-T. *J. Am. Chem. Soc.*, **135**, 17913–17918 (2013).
34. Pfeffermann, M., Dong, R., Graf, R., Zajaczkowski, W., Gorelik, T., Pisula, W., Narita, A., Müllen, K., Feng, X. *J. Am. Chem. Soc.*, **137**, 14525–14532 (2015).
35. Kim, H., Kim, Y., Yoon, M., Lim, S., Park, S. M., Seo, G., Kim, K. *J. Am. Chem. Soc.*, **132**, 12200–12202 (2010).
36. Ogoshi, T., Kanai, S., Fujinami, S., Yamagishi, T. A., Nakamoto, Y. *J. Am. Chem. Soc.*, **130**, 5022–5023 (2008).
37. Ogoshi, T., Ueshima, N., Sakakibara, F., Yamagishi, T., Haino, T. *Org. Lett.*, **16**, 2896–2899 (2014).
38. Hu, X. B., Chen, Z. X., Chen, L., Zhang, L., Hou, J. L., Li, Z. T. *Chem. Commun.*, **48**, 10999–11001 (2012).
39. Ogoshi, T. *J. Incl. Phenom. Macrocycl. Chem.*, **72**, 247–262 (2012).
40. Ogoshi, T., Yamagishi, T. *Eur. J. Org. Chem.*, 2961–2975 (2013).

41. Ogoshi, T., Yamagishi, T. *Bull. Chem. Soc. Jpn.*, **86**, 312–332 (2013).

42. Ogoshi, T., Yamagishi, T. *Chem. Commun.*, **50**, 4776–4787 (2014).

43. Ogoshi, T. *Pillararene*, Royal Society of Chemistry: Cambridge (2016).

44. Xue, M., Yang, Y., Chi, X. D., Zhang, Z. B., Huang, F. *Acc. Chem. Res.*, **45**, 1294–1308 (2012).

45. Strutt, N. L., Zhang, H. C., Schneebeli, S. T., Stoddart, J. F. *Acc. Chem. Res.*, **47**, 2631–2642 (2014).

46. Li, C. *Chem. Commun.*, **50**, 12420–12433 (2014).

47. Nierengarten, I., Guerra, S., Holler, M., Nierengarten, J. F., Deschenaux, R. *Chem. Commun.*, **48**, 8072–8074 (2012).

48. Nierengarten, I., Guerra, S., Holler, M., Karmazin-Brelot, L., Barbera, J., Deschenaux, R., Nierengarten, J. F. *Eur. J. Org. Chem.*, 3675–3684 (2013).

49. Pan, S., Ni, M. F., Mu, B., Li, Q., Hu, X. Y., Lin, C., Chen, D. Z., Wang, L. Y. *Adv. Funct. Mater.*, **25**, 3571–3580 (2015).

50. Nierengarten, I., Guerra, S., Ben Aziza, H., Holler, M., Abidi, R., Barberá, J., Deschenaux, R., Nierengarten, J. F. *Chem. Eur. J.*, **22**, 6185–6189 (2016).

51. Tan, L. L., Li, H. W., Tao, Y. C., Zhang, S. X. A., Wang, B., Yang, Y. W. *Adv. Mater.*, **26**, 7027–7031 (2014).

52. Ogoshi, T., Sueto, R., Yoshikoshi, K., Yamagishi, T. *Chem. Commun.*, **50**, 15209–15211 (2014).

53. Ogoshi, T., Yoshikoshi, K., Sueto, R., Nishihara, H., Yamagishi, T. *Angew. Chem. Int. Ed.*, **54**, 6466–6469 (2015).

54. Ogoshi, T., Sueto, R., Yoshikoshi, K., Sakata, Y., Akine, S., Yamagishi, T. *Angew. Chem. Int. Ed.*, **54**, 9849–9852 (2015).

55. Ogoshi, T., Takashima, S., Yamagishi, T. *J. Am. Chem. Soc.*, **137**, 10962–10964 (2015).

56. Wöhrle, T., Wurzbach, I., Kirres, J., Kostidou, A., Kapernaum, N., Litterscheidt, J., Haenle, J. C., Staffeld, P., Baro, A., Giesselmann, F., Laschat, S. *Chem. Rev.*, **116**, 1139–1241 (2016).

57. Norikane, Y., Hirai, Y., Yoshida, M. *Chem. Commun.*, **47**, 1770–1772 (2011).

58. Fa, S.-X., Chen, X.-F., Yang, S., Wang, D.-X., Zhao, L., Chen, E.-Q., Wang, M.-X. *Chem. Commun.*, **51**, 5112–5115 (2015).

59. Yang, J., Dewal, M. B., Shimizu, L. S. *J. Am. Chem. Soc.*, **128**, 8122–8123 (2006).

60. Sato, K., Itoh, Y., Aida, T. *J. Am. Chem. Soc.*, **133**, 13767–13769 (2011).
61. Sakurai, T., Shi, K., Sato, H., Tashiro, K., Osuka, A., Saeki, A., Seki, S., Tagawa, S., Sasaki, S., Masunaga, H., Osaka, K., Takata, M., Aida, T. *J. Am. Chem. Soc.*, **130**, 13812–13813 (2008).

Biography

Tomoki Ogoshi received his BS (2000), MS (2002), and PhD degrees (2005) from Kyoto University, Japan, under the supervision of Prof.

Yoshiki Chujo. He received JSPS research fellowships for young scientists (2002–2005). He was a JSPS postdoctoral research fellow (2005–2006) in the Graduate School of Science at Osaka University in the group of Prof. Akira Harada. In 2006, he moved to the Graduate School of Natural Science and Technology at Kanazawa University as assistant professor (2006–2010). He was associate professor (2010–2015) and was promoted to professor at the same university at 2015. He has also worked as JST-PRESTO researcher: PRESTO program "Hyper-Nano-Space Design toward Innovative Functionality" since October 2013 under the research supervisor Prof. Kazuyuki Kuroda (Waseda University). He has received the Young Scientist Award in the 88th Spring Meeting of the Chemical Society of Japan (2008), the 23th Selected Lectures by Young Chemists in the 89th Spring Meeting of the Chemical Society of Japan (2009), the Award for Encouragement of Research in Polymer Science (Society of Polymer Science, Japan) (2010), the Young Scientist Award Young Scientist Lecture Award in the 56nd Polymer Research Symposium (2010), the HGCS Japan Award Excellence 2010 (2011), the Chemical Society of Japan Award for Young Chemists (2012), the Cram Lehn Pedersen Prize in Supramolecular Chemistry (Royal Society of Chemistry) (2013), the Commendation for Science and Technology by the Minister of Education, Culture, Sports, Science and Technology (2014), and the 2014 SPSJ Award for the Outstanding Paper in Polymer Journal sponsored by ZEON (2015). His research interests include organic synthesis, and supramolecular and hybrid materials.

Takahiro Kakuta is assistant professor (2016–) at the Graduate School of Natural Science and Technology, Kanazawa University. He

obtained his MS degree of engineering (2011) from the Graduate School of Applied Chemistry and Chemical Engineering, Kogakuin University, and his PhD degree of science (2014) from the Graduate School of Science, Osaka University, under the supervision of Prof. Akira Harada. He was a researcher (2014) at Eindhoven University of Technology in the group of Prof. E. W. Meijer, the Netherlands, and a postdoctoral fellow (2014–2016) at the Graduate School of Engineering, Kyoto University, in the group of Prof. Yoshiki Chujo, Japan. He has received the RSC Poster Award in the 8th International Symposium on Macrocyclic and Supramolecular Chemistry (2013) and the CSJ Student Presentation Award (2014) from the Chemical Society of Japan. His current interests include synthesis polymer, and supramolecular and functional materials.

Tada-Aki Yamagishi received his MS degree (1985) in industrial chemistry from Kanazawa University and PhD degree (1989) in

polymer chemistry from Kyoto University. He was assistant professor (1988–1997) at the Faculty of Engineering, Kanazawa University, and a postdoctoral fellow at Nice-Sophia-Antipolis University in France (1992–1993), was associate professor from 1997 to 2009, and was promoted to professor at the Graduate School of Natural Science and Technology, Kanazawa University, in 2009. He received the Japan Thermosetting Plastics Industry Association Award in 2008. His current interests include synthesis and physical chemistry for highly functional and performance polymers, designed supramolecular chemistry, and hybrid materials.

Chapter 12

Precision Spatial Arrangement of Chromophores Using Cyclic Aromatic Triamide Scaffolds

Koji Takagi[a] **and Ryohei Yamakado**[a,b]
[a]*Life Science and Applied Chemistry, Graduate School of Engineering, Nagoya Institute of Technology, Gokiso, Showa, Nagoya, Aichi 466-8555, Japan*
[b]*College of Pharmaceutical Sciences, Ritsumeikan University, 1-1-1, Nozihigashi, Kusatsu, Shiga 525–8577, Japan*
takagi.koji@nitech.ac.jp

The controlled arrangement of chromophores in space is of much importance for understanding the potential utility as optoelectronic materials. In this chapter, we will firstly review the background of related research fields and clarify the motivation to use cyclic aromatic triamide scaffolds. In the following sections, we will show our recent successful examples where the achiral and chiral arrangement of chromophores are furnished.

Conjugated Objects: Developments, Synthesis, and Applications
Edited by Atsushi Nagai and Koji Takagi

ISBN 978-981-4774-03-1 (Hardcover), 978-1-315-22962-1 (eBook)
www.panstanford.com

12.1 Introduction

12.1.1 Three-Dimensional Arrangement of π-Conjugated Molecules

π-Conjugated molecules play an important role in various optoelectronic devices, including light-emitting diodes (LEDs),[1–2] field-effect transistors (FETs)[3–4], and solar cells (SCs)[5–7]. π-Conjugated molecules normally have a highly planar structure leading to self-aggregation, such as face-to-face π–π stacking and edge-to-face herringbone packing. The large overlap of π-electrons between neighboring molecules is crucial to enhance the charge carrier mobility of semiconducting materials. On the other hand, the dense packing of π-conjugated molecules often decreases or quenches the light emission from compounds, and this is a shortcoming for applying to luminescent devices. Accordingly, the 3D arrangement of π-conjugated molecules should be precisely designed in response to their application fields (Fig. 12.1). The aggregation pattern between π-conjugated molecules has been studied intensively for over three quarters of a century. In 1937, Jelley et al. discovered that cyanine dyes in concentrated aqueous solution exhibit a large red shift of their absorption band as compared to that in diluted solution. This phenomenon was named the Jelley (*J*-) or Scheibe aggregate in honor of their discovery[8–10]. The H-aggregate (H for hypsochromic) is another aggregation pattern of dyes, which leads to a blue shift of their absorption band and a fluorescence quenching[11]. Thus the measurement of absorption spectra is a powerful tool to investigate the aggregation pattern of π-conjugated molecules. To achieve the controlled arrangement of π-conjugated molecules, there are two possible strategies: (i) the chemical modification of π-conjugated molecules' periphery and (ii) the utilization of well-defined tethering units. As a typical example of the first approach, Kobayashi et al. introduced a thiomethyl group at the 6- and 13-positions of pentacene[12]. Due to the S–S and S–π interactions, the modified pentacene adopts the cofacial packing, although unmodified pentacene has the herringbone packing. In this method, the 3D structure of π-conjugated molecules relies on the noncovalent bond interaction and is difficult to be predicted. On the other hand, the second approach based on the covalent bond would be much straightforward because the 3D structure of π-conjugated molecules can be determined by the shape of template molecules.

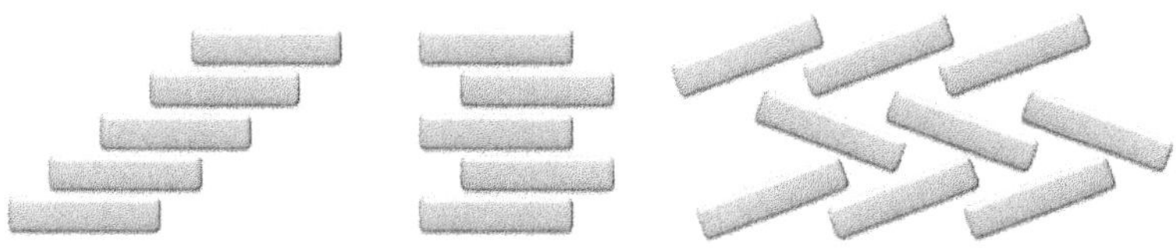

Figure 12.1 Schematic representation of 3D arrangement of π-conjugated molecules.

12.1.2 Utility of Cyclic Oligomers

Cyclic oligomers, including calixarene, cyclophane, pillararene, and cyclodextrin, have unique cavities and have been attracting much attention in the field of host–guest chemistry(Fig. 12.2). On the other hand, the definite conformations of these cyclic oligomers are useful as the template unit for the 3D arrangement of various functional groups attached to the edge of the molecule.

Since cyclophane compounds such as dithia[3.3]metacyclophane and [2.2]paracyclophane[13] prefer the face-to-face conformation of two benzene rings, these compounds can be used as the template for the 3D arrangement of π-conjugated molecules. In particular, [2.2]paracyclophane forces two benzene rings in close proximity each other (ca. 3 Å), which enables the through-space π-electron interaction between appended π-conjugated molecules. Tsuge et al. achieved the spatial arrangement of oligothiophene chromophores using dithia[3.3]metacyclophane[14,15] (Fig. 12.3). The equilibrium between two conformers was studied by the variable-temperature (VT) nuclear magnetic resonance (NMR) spectra, and the compound adopted the *syn*-conformer in the crystalline state. As a result of the close arrangement of π-conjugated systems, an excimer emission was observed even in the diluted solution. Through-space π-conjugated polymers containing the [2.2]paracyclophane unit were prepared by Morisaki and Chujo et al. through the palladium-catalyzed cross-coupling reaction[16–19] (Fig. 12.4). From the measurement of ultraviolet-visible (UV-Vis) absorption and fluorescence emission spectra, these polymers showed excellent charge and energy transfer properties. The high quantum yields were obtained despite the π-stacked structure.

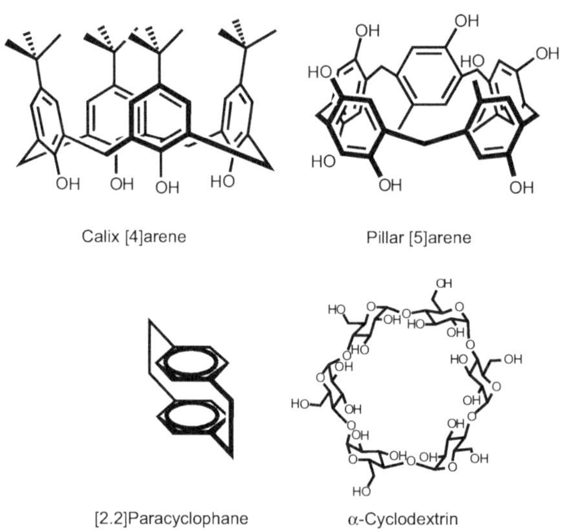

Figure 12.2 Chemical structure of calix[4]arene, pillar[5]arene, [2.2]paracyclophane, and α-cyclodextrin.

Calix[4]arene is a cyclic tetramer made up of alternating phenol and methylene units[20]. The rotation of phenol moiety allows various conformations. Among them, the cone isomer would be suitable for the template for the 3D arrangement of π-conjugated molecules, because the substituents on the upper rim are positioned at the same side[21,22]. Wong et al. introduced oligothiophene chromophores into calix[4]arene[23] (Fig. 12.5). The close proximity of tetra-oligothiophenes lead to a blue shift in the absorption and a red shift in the emission spectra. The fluorescence quantum yield was decreased as compared to that of the model compound corresponding to the repeating unit. These phenomena can be interpreted as a result of the intramolecular chromophore interaction. Würthner et al. prepared the zigzag array of three different perylenebisimide chromophores using the calix[4]arene scaffold and demonstrated sequential energy and electron transfer cascades[24] (Fig. 12.6). The calix[4]arene plays an important role for providing defined distance and angle between chromophores to inhibit the fluorescence quenching and energy loss.

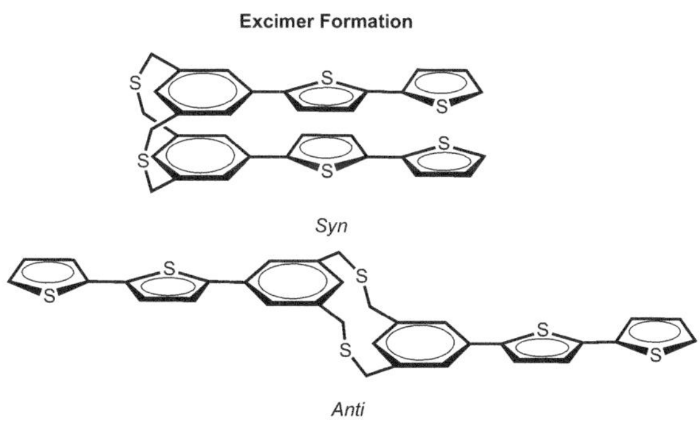

Figure 12.3 Chemical structure of *syn*- and *anti*-isomers of dithia[3.3]metacyclophane bearing oligothiophene chromophores.

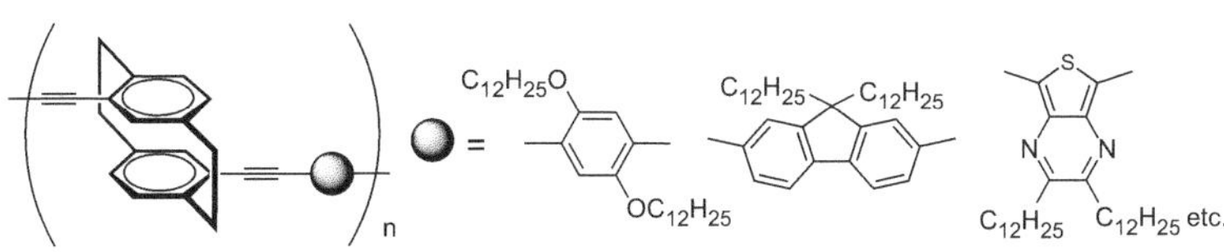

Figure 12.4 Chemical structure of [2.2]paracyclophane-containing through-space conjugated polymers.

As shown in these examples, cyclic oligomers would be suitable templates for the 3D arrangement of π-conjugated moieties, and the conformation and optoelectronic properties of π-conjugated moieties depend on the character of the template. Accordingly, the precision arrangement of π-conjugated molecules using a novel cyclic oligomer is quite interesting to investigate the photophysical and electronic properties of π-conjugated molecules.

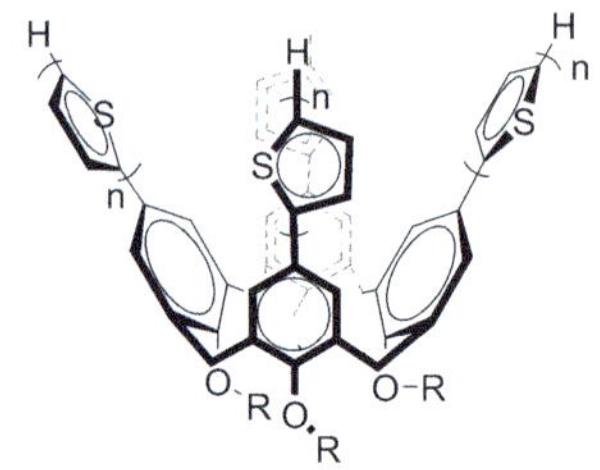

Figure 12.5 Chemical structure of calix[4]arene bearing oligothiophene chromophores.

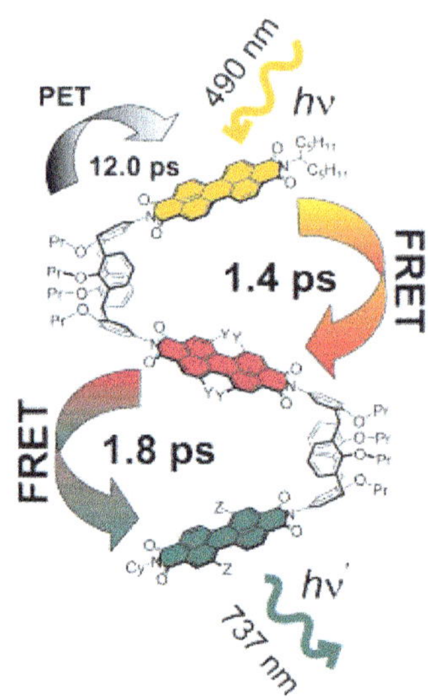

Figure 12.6 Sequence-controlled perylene bisimides array using calix[4]arene scaffold. Reprinted with the permission from Ref. [24]. Copyright (2008) American Chemical Society.

12.1.3 Stereochemistry of Aromatic Tertiary Amide and Characteristic of Cyclic Aromatic Triamide

The crystal structure of *N*-methylbenzanilide was initially revealed by Shudo et al. in 1989[25–28]. *N*-methylbenzanilide exists in the *cis* form in the crystal state, although *N*-unsubstituted benzanilide exists in the *trans* form, both in the crystal and solution states[29, 30].

The *cis* preference of *N*-methylbenzanilide in solution was also confirmed by ^{1}H NMR spectroscopy. Moreover, Shudo et al. showed the steric influence of the *ortho*-substituent on the conformation (Table 12.1). *N*-methylbenzanilide derivatives mostly exist in the *cis* forms in the crystal state. In solution, the *trans/cis* ratio was increased with increasing the number of methyl substituents. The stepwise synthesis of *meta*-cyclic amide trimer (calix[3]amide) was first reported by Stoddart et al. in 1982[31,32] (Scheme 12.1). In 1996, Shudo et al. developed the one-step formation of calix[3]amide[33]. Due to the *cis* preference of the *N*-alkylbenzanilide skeleton (vide supra), calix[3]amide can be efficiently synthesized by the condensation reaction of 3-methylaminobenzoic acid with tetrachlorosilane or dichlorotriphenyl phosphorane[34–36]. On the other hand, Yokozawa et al. reported the synthesis of the *para*-cyclic amide trimer with the large inner cavity from monomers bearing 4-alkylamino and 4'-methoxycarbonyl groups using lithium bis(trimethylsilyl)amide (LiHMDS)[37].

Calix[3]amide has two conformers, in which the *syn*-conformer has three benzene rings in the same orientation relative to the amide bond and the *anti*-conformer has one benzene ring turning in other direction.

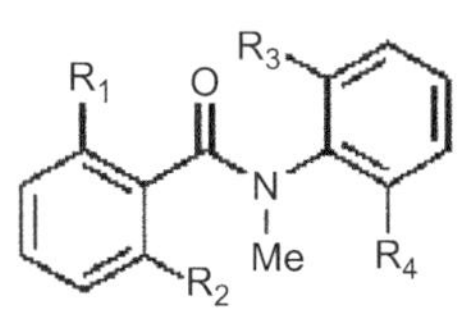

Table 12.1 *Trans/cis* ratio of *N*-methylbenzanilide derivatives **1–7**

Compound	R_1	R_2	R_3	R_4	*Trans/cis*
1	H	H	H	H	1 : 99
2	Me	H	H	H	10 : 90
3	Me	Me	H	H	33 : 67
4	H	H	Me	H	6 : 94
5	H	H	Me	Me	8 : 92
6	Me	H	Me	H	22 : 78
7	Me	Me	Me	Me	100 : 0

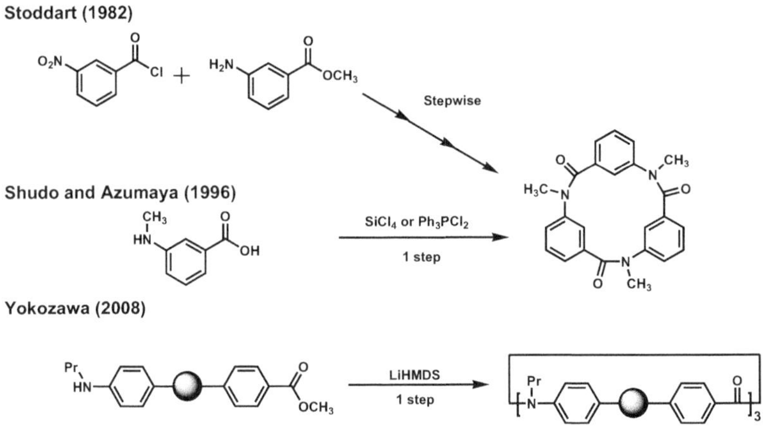

Scheme 12.1 Synthetic route to cyclic amide trimers.

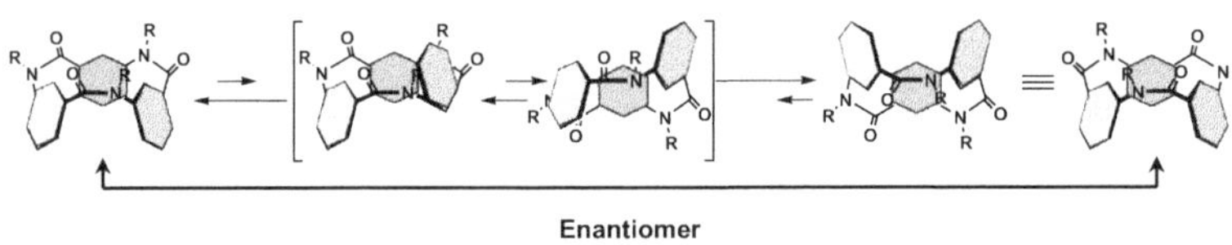

Figure 12.7 Schematic representation of equilibria of calix[3]amide (both ends denote *syn*-isomer and others are *anti*-isomer).

The equilibrium between two conformers was investigated by VT-NMR spectra[33], and the *syn*-conformer generally prefers to the *anti*-conformer (Fig. 12.7). The influence of the substituent at the 5-position on the energy barrier between the *syn*- and *anti*-conformers in solution was discussed[38] (Fig. 12.8, **8**–**12**). The *syn*-/*anti*-ratio of each compound was dependent on the temperature. The percentage of the *syn*-conformer increased with increasing the solvent polarity because of the larger dipole moment of the *syn*-conformer. Functionalization of calix[3]amide was also reported. For example, a calix[3]amide bearing the carboxylic acid group was synthesized[39] (Fig. 12.8, **13**). The hydrogen-bonding interaction with chiral amines resulted in the chirality induction on the calix[3] amide skeleton in solution, and the capsule-type homodimer was obtained in the crystalline state.

Interestingly, calix[3]amide in the *syn*-conformation has a planar chirality based on the direction of the amide bond. However, the optical resolution of enantiomers was impossible owing to the easy racemization through the conformational exchange in solution (Fig. 12.7). To suppress the conformational exchange, the spherical aromatic amide compound consisting of four calix[3]amide rings was synthesized, and the optical resolution was successfully achieved by chiral HPLC[40] (Fig. 12.8, **14** and **14'**).

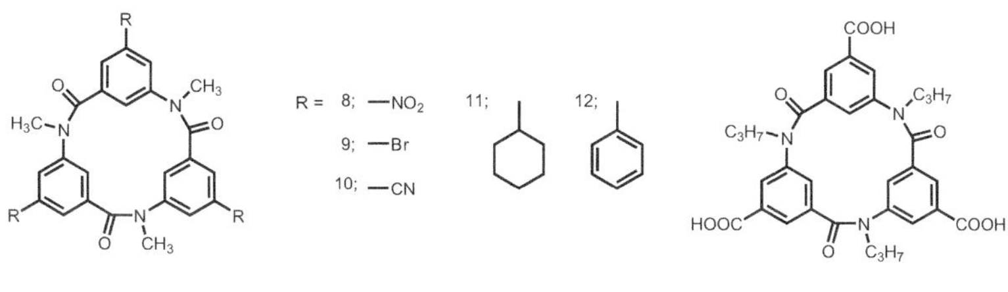

13

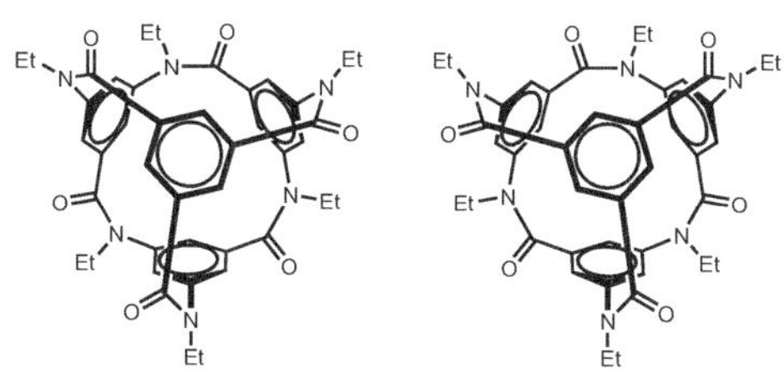

14 14′

Figure 12.8 Chemical structure of calix[3]amide derivatives **8–13** and enantiomeric structure of **14**.

Although the conformation analysis and the host–guest chemistry were investigated, calix[3]amides (cyclic amide trimers) have not been used as the template for the 3D arrangement of π-conjugated molecules yet. The authors focused on three exclusive properties of calix[3]amide: (i) facile introduction of substituents, (ii) conformation character depending on the solvents, and (iii) molecular chirality based on the direction of the amide bond. By virtue of the stereochemistry of the aromatic tertiary amide, it would be possible to carry out the precision spatial arrangement of π-conjugated molecules.

12.2 Achiral Arrangement of Chromophores

12.2.1 Oligothiophene

Oligothiophenes and polythiophenes are important class of π-conjugated organic molecules that have been extensively investigated and explored as the active material for optoelectronic devices. Thus we became interested in investigating the spatial arrangement of oligothiophene chromophores in a controlled fashion by utilizing the calix[3]amide scaffold. We have prepared calix[3]amides carrying bithiophene and terthiophene chromophore in a confined space and studied their conformation in solution. Polymers and model compounds were also prepared for comparison. The absorption/emission properties of these materials were carefully investigated to disclose a potential utility of the calix[3]amide framework[41].

The dropwise addition of lithium bis(trimethylsilyl)amide (LiHMDS) into a dilute tetrahydrofuran (THF) solution of methyl 3-nonylaminobenzoate carrying the bithiophene chromophore at the 5-position followed by the purification by SiO_2 chromatography and preparative GPC yielded a cyclic trimer **BTC3A** (Scheme 12.2). Likewise, a cyclic trimer **TTC3A** carrying the terthiophene chromophore was synthesized. For more detailed investigation into the characteristics of the calix[3]amide scaffold, polymers (**BTPA** and **TTPA**) were also prepared in the presence of phenyl benzoate (initiator) following to the condition of the chain-growth polycondensation reported by Yokozawa et al.[42] Model compounds corresponding to the repeat unit (**BTM** and **TTM**) were also prepared. In $CDCl_3$, the *syn-*/*anti*-conformer ratio of **BTC3A** at 293 K was calculated as 76/24 from the integral ratio in the ^{1}H NMR spectrum, and the population of the *syn*-conformer was decreased to reach 66% at 233 K. The population of the *syn*-conformer at 293 K was increased to reach 86% in $CDCl_3/CD_3OD$ solution (1/1 in volume) in good agreement with the trend of calix[3]amide without the bithiophene chromophore. On the other hand, the value was decreased to 70% at 293 K in THF-d_8 solution. The *syn-*/*anti*-conformer ratio for **TTC3A** at 293 K was 84/26 in $CDCl_3$ solution. Accordingly, the *syn-*/*anti*-conformer ratio of calix[3]amide compounds bearing the oligothiophene unit were found to be influenced by the solvent, temperature, and chromophore.

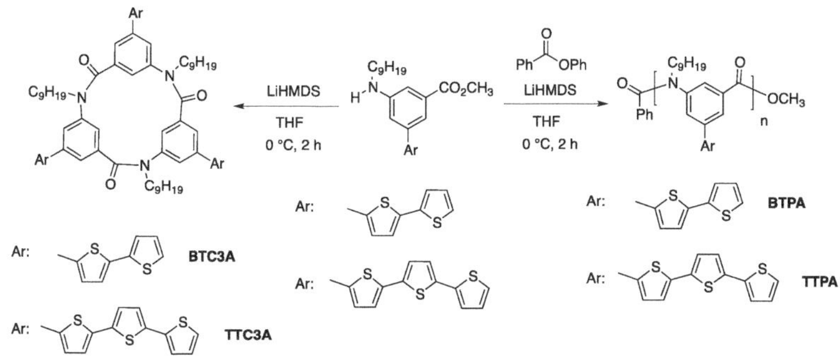

Scheme 12.2 Synthetic route of cyclic trimers (calix[3]amides) and polymers.

The absorption maximum of **BTC3A** (342 nm) shifted to the shorter wavelength region compared with those of **BTPA** (348 nm) and **BTM** (351 nm) (Fig. 12.9A). On the other hand, the emission maxima obviously red-shifted in the order of **BTM** (422 nm), **BTPA** (431 nm), and **BTC3A** (448 nm) (Fig. 12.9B). To gain deeper insights into the large wavelength shift of the emission maximum of **BTC3A**, the UV-Vis and fluorescence spectra of **BTC3A** and **BTM** were collected in various solvents. As a result, the UV-Vis and fluorescence spectra of **BTM** showed a small shift going from a $CHCl_3$ solution to a CH_3OH solution probably because the hydrogen-bonding interaction of CH_3OH with the amide carbonyl group has an influence upon the electronic structure of the bithiophene chromophore.

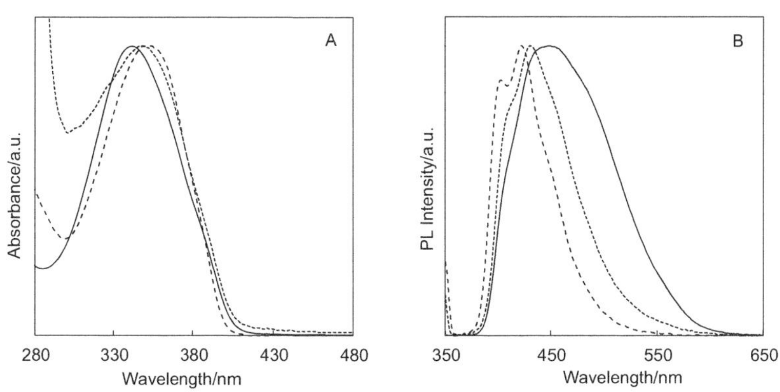

Figure 12.9 (A) UV-Vis and (B) fluorescence spectra of **BTC3A** (solid line), **BTPA** (dotted line), and **BTM** (broken line) in $CHCl_3$ at 298 K (10^{-5} M).

In contrast, **BTC3A** showed a noticeable solvato(fluoro) chromic behavior. The absorption maxima blue-shifted from 342 nm in $CHCl_3$ to 336 nm in CH_3OH, while the emission maxima showed the pronounced and opposite shift from 448 nm in $CHCl_3$ to 481 nm in CH_3OH (Fig. 12.10). The emission maxima demonstrated hypsochromic shifts when the spectra were collected in THF (430 nm). Taking the fact that **BTC3A** preferably adopts the *syn*-conformation in a $CDCl_3/CD_3OD$ solution (vide supra) into account, the self-assembly of bithiophene segments enables the chromophore interaction in a CH_3OH solution.

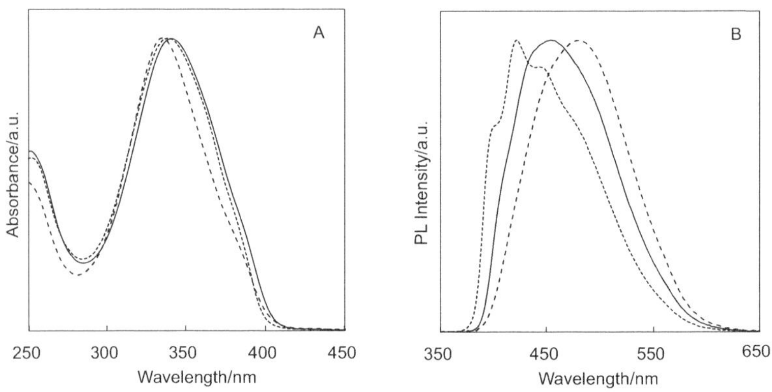

Figure 12.10 (A) UV-Vis and (B) fluorescence spectra of **BTC3A** in THF (dotted line), $CHCl_3$ (solid line), and CH_3OH (broken line) at 298 K (10^{-5} M).

The absorption maximum of **TTC3A** (379 nm) was observed at the shorter-wavelength region compared to those of **TTPA** (386 nm) and **TTM** (390 nm), which is similar to the trend of **BTC3A**. However, the emission maximum of **TTC3A** was detected at 486 nm and the shift width from that of **TTM** was smaller (9 nm) than that of **BTC3A** (24 nm) in spite of the higher population of the *syn*-conformer (vide supra). Since the calix[3]amide framework has a truncated cone structure, the edge-to-edge distance between neighboring terthiophene chromophores in **TTC3A** would be longer than that of bithiophene-carrying **BTC3A**. It is thus considered that the intramolecular interaction between self-assembled chromophores does not effectively work in **TTC3A**. The emission maximum was likewise observed at 486 nm when the fluorescence spectrum was measured in a CH_3OH solution. These results indicate that the calix[3]amide framework would be the intriguing scaffold for the self-assembly of π-conjugated chromophores.

12.2.2 Pyrene

Pyrene, having a high fluorescence quantum yield, shows two characteristic luminescence behaviors[43], namely a monomer emission observed in the diluted solution and an excimer emission derived from the associated dimer in the concentrated solution. The excimer emission can be also detected when pyrenes are forced into the confined space even in the diluted condition. Consequently, pyrene is a useful probe as the DNA fluorescence label to investigate their distance and chain conformation[44]. We will herein describe a novel arrangement of three pyrenes using calix[3]amide as the template to reveal their optical properties[45] (Fig. 12.11).

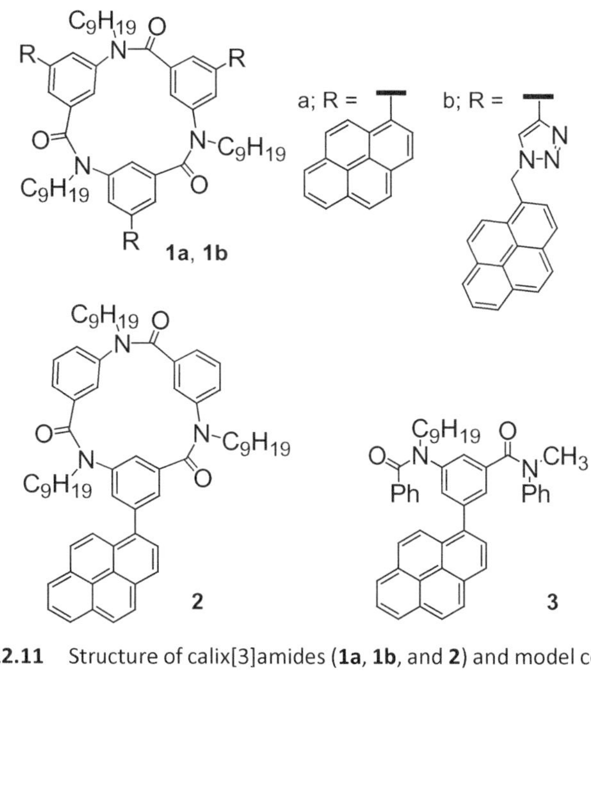

Figure 12.11 Structure of calix[3]amides (**1a**, **1b**, and **2**) and model compound (**3**).

Calix[3]amide bearing three pyrenes (**1a**) was synthesized by the cyclic oligomerization of **6,** as shown in Scheme 12.3. Calix[3]amide (**1b)** having a flexible spacer between pyrene and calix[3]amide was prepared for comparison and the synthetic route is shown in Scheme 12.4. To investigate the optical properties of calix[3]amide in detail, calix[3]amide having one pyrene (**2**) and acyclic model compound (**3**) were prepared. The synthesis of **2** was carried out by the cyclic co-oligomerization of **6** and 3-nonylaminobenzoic acid. The relative arrangement of three pyrenyl groups was determined by the single-crystal X-ray crystallography using calix[3]amide bearing the ethyl substituent on the amide nitrogen (**1a'**) instead of **1a** (Fig. 12.12). **1a'** adopts the *syn*-conformation and three pyrenyl groups are arranged on the same side of the calix[3]amide skeleton. The center-to-center distances between adjacent pyrenyl groups is approximately 6.2 Å and the dihedral angle between three pyrenyl groups is about 61°. The close arrangement of three pyrenyl groups in **1a** was also evident in the ^{1}H NMR spectra measured at low temperature.

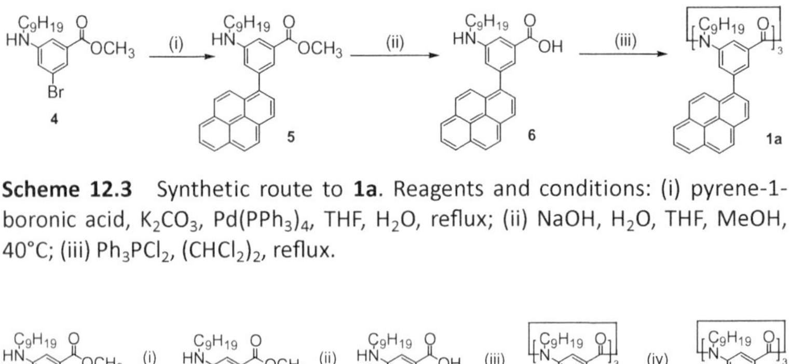

Scheme 12.3 Synthetic route to **1a**. Reagents and conditions: (i) pyrene-1-boronic acid, K_2CO_3, $Pd(PPh_3)_4$, THF, H_2O, reflux; (ii) NaOH, H_2O, THF, MeOH, 40°C; (iii) Ph_3PCl_2, $(CHCl_2)_2$, reflux.

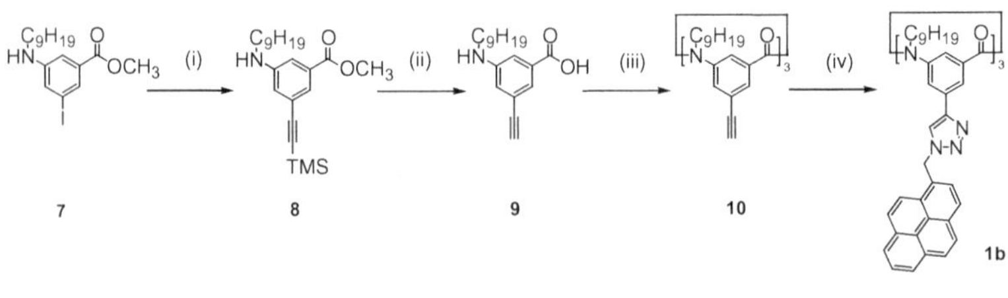

Scheme 12.4 Synthetic route to **1b**. Reagents and conditions: (i) trimethylsilylacetylene, CuI, $Pd(PPh_3)_4$, Et_3N, 50°C; (ii) NaOH, H_2O, THF, MeOH, room temperature; (iii) $SiCl_4$, pyridine, reflux; (iv) 1-azidomethylpyrene, CuI, DMF, 90°C.

The UV-Vis absorption and fluorescence emission spectra were measured in THF solution. As shown in Fig. 12.13A, the absorption maximum wavelengths were similar between **1a**, **2**, and **3**, which indicates that they have comparable effective conjugation length. Hence three pyrenes included in **1a** have negligible interaction in the ground state. The absorption spectrum of **1b** showed the vibronic fine structure characteristic to pyrene (Fig. 12.13B). The fluorescence maximum wavelength of **1a** was red-shifted by 3–5 nm from those of **2** and **3**, and the tailing to the longer wavelength region was observed (Fig. 12.13C). The fluorescence emission from the associated state (excimer emission) was not observed from **1a**. These results indicate that three pyrenes in **1a** have very weak electronic interaction in the excited state because they cannot form the face-to-face stacked structure. In contrast, the fluorescence emission of **1b** appeared as a broad featureless band at the longer wavelength region (476 nm) having the quantum yield of 0.06 (Fig. 12.13D). The flexible methylene spacer reduces the steric hindrance to permit the excimer emission. The fluorescence quantum yield of **3** was quite low (Φ_{fl} = 0.04) resulting from the twisted intramolecular charge transfer (TICT) by the rotation of the amide bonds[46]. On the other hand, the quantum yields of **1a** (Φ_{fl} = 0.36) and **2** (Φ_{fl} = 0.34) were high probably because the cyclic structure suppresses the rotation of the amide bond. It should be noted that the fluorescence quantum yield of **1a** and **2** are almost similar, even though the chromophores are closely located in **1a**. This is also a consequence of the screw-like arrangement of three pyrenes in the calix[3]amide platform.

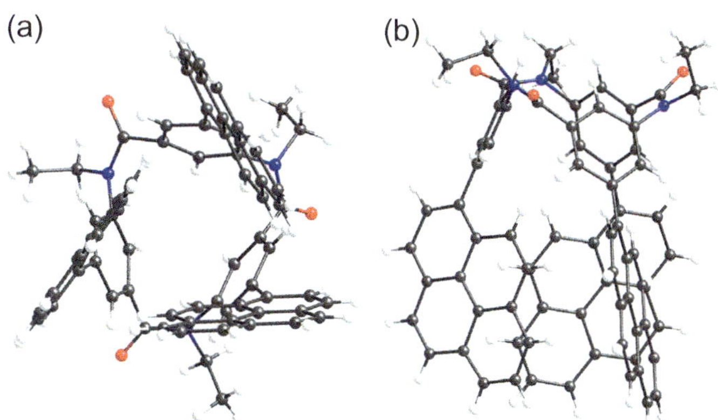

Figure 12.12 Crystal structure of **1a'** depicted as a ball-and-stick model. (a) top view and (b) side view. Included chloroform molecules are omitted for clarity. One of the pyrene groups is disordered to two positions and the occupancies are refined (about 0.60 and 0.40). Disordered atoms of a pyrene rings which have minor occupancy are omitted for clarity.

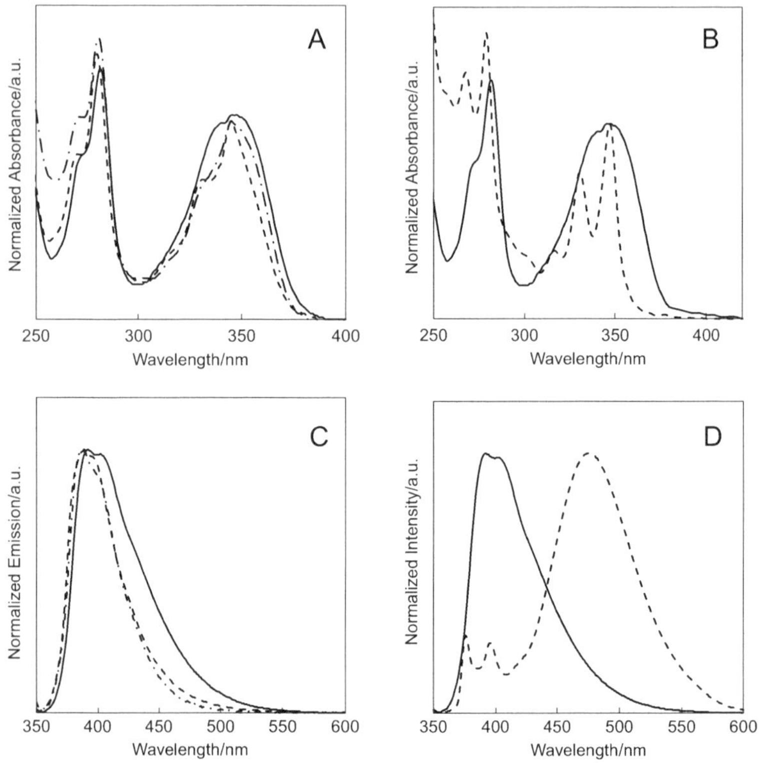

Figure 12.13 Normalized absorption and emission spectra in THF solution ($\lambda_{ex} = \lambda_{abs}$, room temperature). (A and C) **1a** (solid line), **2** (broken line), and **3** (dotted broken line); (B and D) **1a** (solid line) and **1b** (broken line).

12.3 Chiral Arrangement of Chromophores

12.3.1 Bithiophene

Calix[3]amide, which is a cyclic trimer consisting of three benzanilide units, has a planar chirality based on the direction of the amide bond and is therefore considered to be a good candidate as a chiral scaffold (Fig. 12.14A). Although the π-conjugated chromophores were successfully aligned by a single calix[3]amide, as mentioned before, the system did not show any chiroptical properties because of the racemization by rapid flipping of benzene units in solution (Fig. 12.7). We have therefore investigated the helicity induction in three bithiophene units by two chiral calix[3]amide scaffolds (Fig. 12.14B)[47]. To suppress the flipping of benzene units, we designed a tubular molecule (**2**), in which three π-conjugated chromophores were covalently connected by twin calix[3]amides.

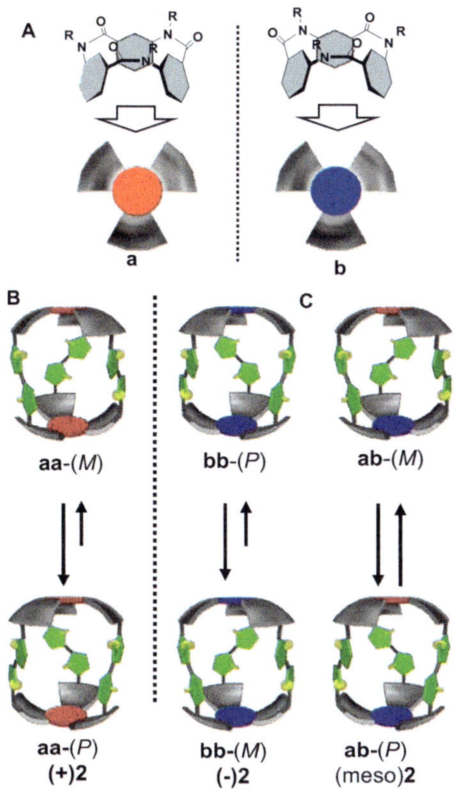

Figure 12.14 Overview and schematic illustrations of the present system with triple-stranded helicity: (A) enantiomers of calix[3]amide (**a** and **b**), (B) equilibria of chiral triple-stranded helical biases of (+)**2** and (−)**2**, and (C) equilibrium of (meso)**2**. (*P*) and (*M*) indicate right- and left-handed helicities of the chromophores, respectively.

The synthetic route to **2** is shown in Scheme 12.5. The cyclic oligomerization was carried out by the dropwise addition of lithium bis(trimethylsilyl)amide (LiHMDS) to a dilute THF solution of **5**. As expected, **2** consisted of three optical isomers in the ratio of 1:2:1 in the chiral high-performance liquid chromatography (HPLC) analysis, generated by the direction of the amide bond of calix[3]amide: one set of enantiomers having the same planar chirality (Fig. 12.14B) and a meso compound that has two different planar chiralities (Fig. 12.14C). Each compound was successfully separated by chiral HPLC.

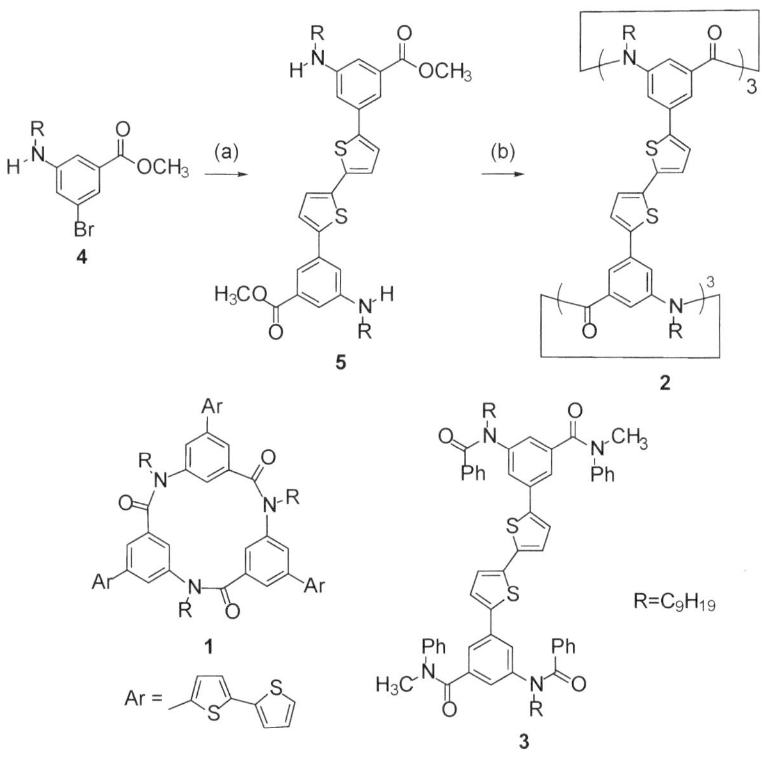

Scheme 12.5 Synthetic route to tubular molecule **2** and chemical structures of **1** and **3**. Reagents and conditions: (a) 5,5′-bis(trimethylstannyl)-2,2′-bithiophene, $Pd(PPh_3)_4$, toluene, reflux; (b) LiHMDS, THF, 0°C.

The specific rotations ($[\alpha]_D^{25}$ ($c = 10^{-3}$ in THF)) of the first and third eluates were −24.8° [(−)**2**] and +27.6° [(+)**2**], respectively. The circular dichroism (CD) spectra of (−)**2** and (+)**2** showed a Cotton effect in the mirror image. In contrast, the second eluate [(meso)**2**] did not show a specific rotation and a Cotton effect. Compared to the ^{1}H NMR spectrum of single calix[3]amide**1**, the spectrum of (−)**2** showed a very simple signal pattern, implying that (−)**2** adopts a well-defined and highly symmetric geometry.

The UV spectrum of (−)**2** in chloroform is shown in Fig. 12.15. The tubular molecule (−)**2** exhibited a large absorption band at 364 nm with a broad shoulder peak at around 420 nm. The observed spectral features can be explained on the basis of Kasha's exciton model[48,49]. As shown in Fig. 12.16, the formation of a cyclic arrangement of the transition dipole moments results in three exciton states as a result of its symmetry; the lower exciton state, which has one nodal plane, is doubly degenerate and optically forbidden as a result of cancellation of the net transition dipole moment, whereas the higher energy state is optically allowed.

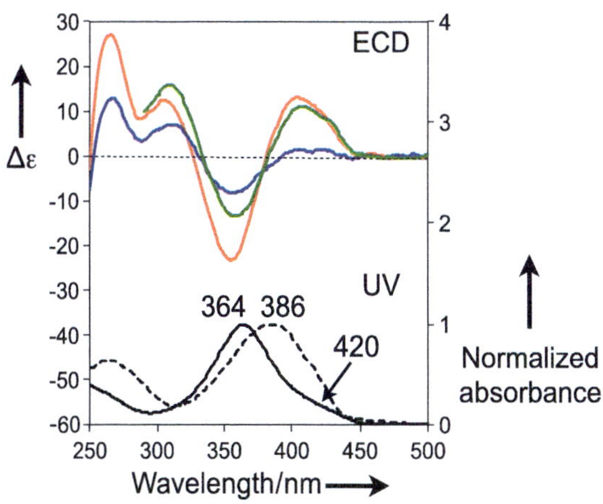

Figure 12.15 UV spectra of **3** (black dot line) and (−)**2** (black solid line) in chloroform, and electronic circular dichroism (ECD) spectra of (−)**2** at room temperature (blue line: in THF; green line: in toluene; red line: in chloroform).

This model reasonably accounts for the experimental observation that the peak position of **3** (386 nm) is in between the peak maximum (364 nm) and the shoulder (ca. 420 nm) of the tubular molecule (−)**2**. The fluorescence spectrum of (−)**2** showed a larger Stokes shift than that of **3**, and the relative quantum yield of (−)**2** (2.4%) was ca. one-sixth smaller than that of **3** (13.4%). Since the internal conversion from the higher exciton state to the lower exciton state is generally much faster than emission, the observed emission of (−)**2** would have occurred from the doubly degenerate forbidden transition state, resulting in weak fluorescence. The ECD of (−)**2** are shown in Fig. 12.15. Strong bisignate ECD signals corresponding to transitions to the exciton states were observed in chloroform and toluene. This clearly indicates that the chiral information of the ordered calix[3]amide has been effectively transferred to the π-conjugated system through the covalent linkage, resulting in a twisted chiral structure of bithiophene chromophores. Although the ECD intensity in THF was considerably weaker than those in chloroform and toluene, we found that the intensity increased significantly with decreasing temperature and completely recovered at room temperature. Since the increase in the ECD intensity in the longer wavelength region was significantly more than that in the shorter wavelength region, the bithiophene chromophores were considered to be aligned in a chiral fashion under low-temperature conditions. These results indicate that (−)**2** can adopt a dynamic chiral structure between two diastereomeric isomers (Fig. 12.14). The preferred chiral structure of the tubular molecule (−)**2** can be determined by considering the excitonic interactions among three transition dipole moments. According to the Rosenfeld equation [$R = Im(\boldsymbol{\mu} \bullet \mathbf{m})$], the ECD sign is readily determined by the relative arrangement of the electric transition dipole moment (**μ**) and the magnetic transition dipole moment (**m**)[50,51]. In the case of the (*M*)-structure (Fig. 12.16), the ECD signal corresponding to the higher exciton state is negative, whereas a positive ECD is predicted for the lower exciton state. As a consequence, it can be concluded that (−)**2** preferentially adopts a left-handed helicity [**aa**-(*M*) or **bb**-(*M*)] in chloroform and toluene at room temperature, and in THF at low temperature. It is noted that a positive bisignate ECD signal was observed, even though (−)**2** has chromophores in the (*M*)-helical arrangement. We successfully determined the absolute configuration using the vibrational circular dichroism (VCD) spectroscopy (Fig. 12.17). The VCD spectra of (−)**2** (blue line) in chloroform showed two dispersion-type signals at 1649 cm^{-1} and 1590 cm^{-1}. The former and latter signals were associated with the C=O stretching of carbonyl groups and C=C stretching of benzene rings, respectively. The mirror-image spectrum was observed for (+)**2** (red line). The theoretical VCD profiles of the two energy-minimized **bb** isomers [**bb**-(*P*) and **bb**-(*M*)] were calculated. Interestingly, the VCD signal pattern of the (*P*)-conformer was the same as that of the (*M*)-conformer, which indicates that the VCD spectra were associated with the planar chirality of calix[3]amide and independent of the twisted chirality of chromophores. The experimental pattern of (−)**2** was consistent with the calculated patterns of **bb**-(*M*) and **bb**-(*P*). The absolute configurations of calix[3]amides of (+)**2** and (−)**2** were, therefore, assigned as **aa** and **bb**, respectively. The VCD and ECD results allowed us to conclude that the planar chiralities of **aa** and **bb** induce, respectively, the *P*- and *M*-helicities in the three bithiophene units in chloroform. In summary, we achieved chiral 3D alignment of π-conjugated chromophores with the triple-stranded helicity. The inherent chirality of calix[3]amide induced a predominantly one-handed helicity to the arrangement of π-conjugated chromophores, without the help of chiral guests or chiral centers.

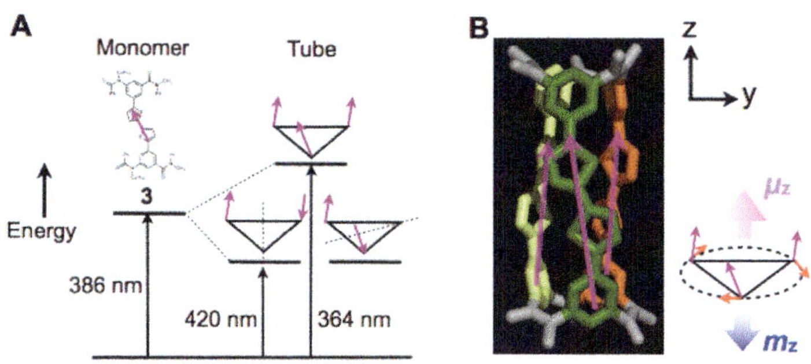

Figure 12.16 (A) Schematic excision band energy diagram for a tube molecule with three light-absorbing units. (B) Optically allowed exciton state of the tube compound with (*M*)-helicity. The pink solid arrows indicate the electric transition dipole moments (**μ**) of each monomer unit. The red solid arrows indicate the *x*, *y* components of **μ**. The pink and blue semitransparent arrows indicate the net electric and magnetic transition dipole moments, respectively.

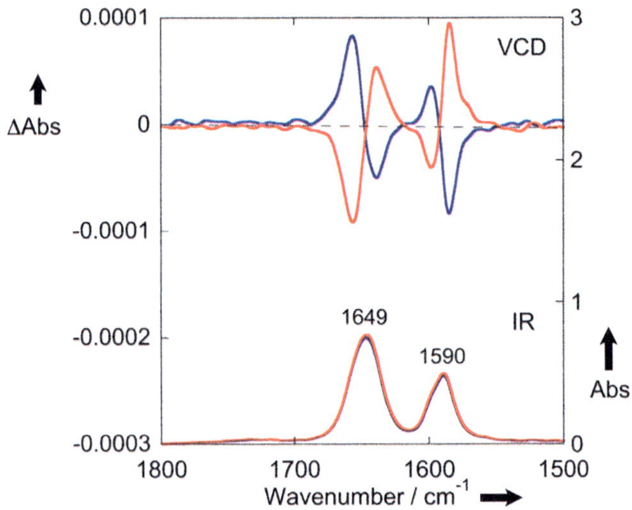

Figure 12.17 IR and VCD spectra of (+)**2** (red line) and (−)**2** (blue line) in chloroform.

12.3.2 Phenylene-ethynylene

The design of an intertwining helical structure is a challenging issue for the construction of well-defined 3D architectures, as exemplified by naturally occurring DNA. One possible strategy for obtaining such helical structures relies on the noncovalent self-assembly including the hydrogen bonding and metal coordination. In contrast, there are not many reports to attain the helical structure using the covalent bond approach despite the robustness over the external environmental change. In particular, the helical structure of π-conjugated chromophores is restricted to a few examples[52–54], probably owing to the rigidity or reactivity of the building unit. We herein describe the synthesis and chiroptical properties of a new cyclic tri(benzamide) carrying bis(phenylethynyl)benzene chromophores (**BPEB_C3A**) arranged in a triple-stranded helical conformation[55].

The synthetic route to **BPEB_C3A** is shown in Scheme 12.6. After the trial and error, **BPEB_C3A** was successfully obtained by the direct cyclic condensation of 2,5-dibromo-4-hexylaminobenzoic acid **1**, followed by the Sonogashira–Hagihara coupling reaction of **DiBr_C3A** with phenylacetylene.

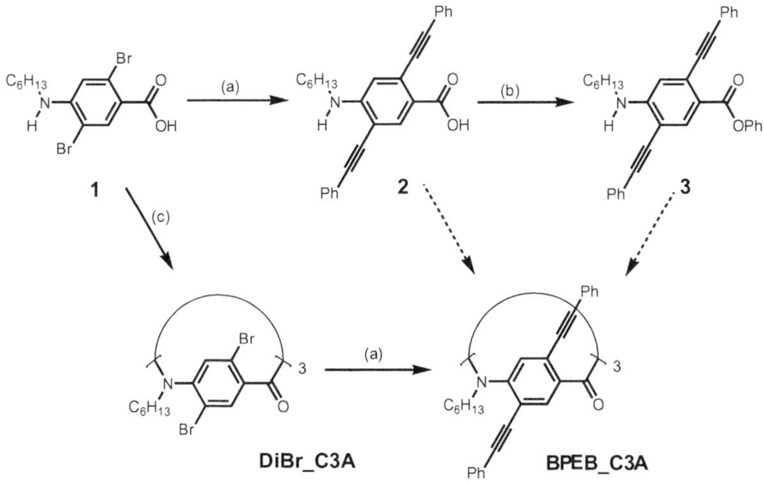

Scheme 12.6 Preparation of cyclic tri(benzamide) **BPEB_C3A**. (a) Phenylacetylene, $Cu_2(OAc)_4$, $Pd(PPh_3)_2Cl_2$, TEA, (b) $SOCl_2$, phenol, and (c) $SiCl_4$, pyridine.

In the first step, the microwave irradiation facilitated the cyclic trimerization with minimum side reactions. The identity of products were confirmed by the spectroscopic data and MALDI-TOF MS. The ^{1}H NMR spectra of **DiBr_C3A** are shown in Fig. 12.18. In the higher-magnetic-field region, the methylene proton signals on the amide nitrogen were clearly split in two positions. This signal separation would be a scribed to the diastereotopic environment of the methylene proton due to the axial chirality of **DiBr_C3A**. If benzene rings rotate around the amide bond, **DiBr_C3A** would exist in the equilibrium between four diastereomers: (*S*,*S*,*S*), (*R*,*R*,*R*), (*R*,*S*,*S*), and (*R*,*R*,*S*). In the lower magnetic field region, however, only two sharp aromatic proton signals assignable to a and b were observed, implying the highly symmetrical structure of **DiBr_C3A**, namely (*S*,*S*,*S*) and (*R*,*R*,*R*) isomers. In addition, the solid-state structure of **DiBr_C3A** was investigated by the single-crystal X-ray analysis, showing a racemic crystal including (*S*,*S*,*S*) and (*R*,*R*,*R*) isomers (Fig. 12.19, space group = $C2/c$, $Z = 8$). Three benzene rings are standing in the direction perpendicular to the amide bond and, hence, the C-O bond length (ca. 1.22 Å) corresponds to the typical C=O bond length (ca. 1.23 Å). Two enantiomers are alternatively stacked to form the columnar structure along the *c* axis. The enantiomers having the different chirality in the neighboring columns are associated by the two noncovalent C=O···Br (halogen-bonding) interactions[56–58] with the length of ca. 2.98 Å, which contributes to the formation of the crystal network structure.

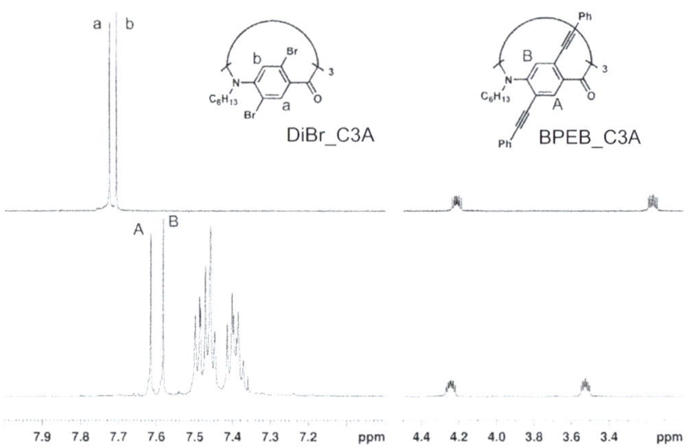

Figure 12.18 ^{1}H NMR spectra (600 MHz, CD_2Cl_2, rt) of (top) **DiBr_C3A** and (bottom) **BPEB_C3A**.

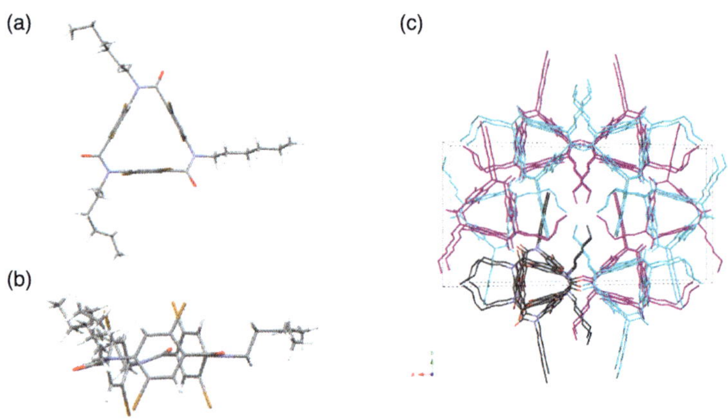

Figure 12.19 Crystal structure of **DiBr_C3A**. (a) Top view, (b) side view, and (c) packing structure along the *c* axis. Hydrogen atoms are omitted for clarify in (c). Magenta- and cyan-colored molecules are enantiomers of each other.

Thus, the cyclization of **1** was found to proceed in a diastereoselective manner (Fig. 12.20). The methylene proton signals on the amide nitrogen and the aromatic proton signals of **BPEB_C3A** were likewise observed as two sharp signals (Fig. 12.18), indicating that the partial rotation of the benzene ring of **DiBr_C3A** (epimerization) did not take place during the Sonogashira–Hagihara coupling reaction. Therefore, **BPEB_C3A** would be expected to contain two optically active enantiomers as a racemic mixture.

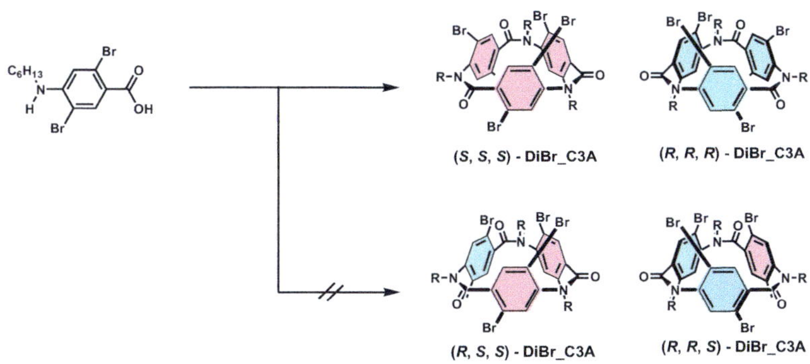

Figure 12.20 Diastereoselective cyclization to give **DiBr_C3A**.

The optical resolution of racemic **BPEB_C3A** was carried out by the chiral HPLC on ChiralPAK IA using 2-propanol/*n*-hexane (20:80) as the mobile phase. The conformation of **BPEB_C3A** was investigated by optimizing the structure with the density functional theory (DFT) calculation at the B3LYP/6-31G(d) level of theory to figure out that bundled bis(phenylethynyl)benzene units form the triple-stranded helical conformation.

To investigate the chiral helical topologies in solution, the CD spectra of isolated chiral **BPEB_C3A** was studied, and the mirroring Cotton effect was observed in THF (Fig. 12.21). The CD spectra of the first fraction showed a positive Cotton effect at 353 nm and 339 nm followed by a negative Cotton effect at 302 nm. The CD anisotropy factors ($\Delta\varepsilon/\varepsilon$) at these wavelengths were found to be $10^3\ g_{abs} = 5.4$, 3.0, and 2.4, respectively. We also found that the CD intensity around 353 nm was increased with lowering temperature. This result may indicate that the rotation around the carbon–carbon triple bond and/or the swinging motion of bis(phenylethynyl)benzene units are suppressed at the low temperature. The UV-Vis absorption and fluorescence emission spectra of **BPEB_C3A** in a de-aerated THF solution and solid state were obtained. The emission spectrum in solution exhibited a broad peak at 440–650 nm in addition to the emission around 340–400 nm having the vibronic structure. The broad emission at the longer wavelength region can be explained by the intramolecular interaction between three bis(phenylethynyl) benzene units and **BPEB_C3A** had a low fluorescence quantum yield (<1%). These results suggest that cyclic aromatic triamide is a reliable scaffold to realize the triple-stranded helical structure of bis(phenylethynyl)benzene units.

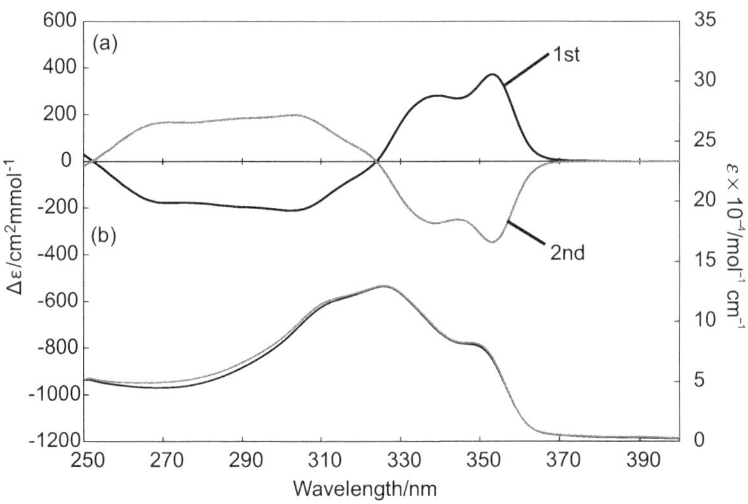

Figure 12.21 (a) CD and (b) UV-Vis spectra of the first and second fractions (3.7×10^{-6} M) in THF at room temperature.

12.4 Conclusions

This chapter deals with the precision spatial arrangement of π-conjugated chromophores based on the characteristic of cyclic aromatic triamide. The introduction surveys the importance of the controlled 3D arrangement of chromophores, the potential utility of cyclic oligomers with well-defined architectures as the template unit, and the stereochemistry of aromatic tertiary amide along with the characteristic of cyclic aromatic triamide. The systematic investigation on the conformation of molecules and the optical properties of π-conjugated moieties gave valuable information for the achiral and chiral arrangement of various chromophores. The authors hope that the study described in this chapter will be widely referred and matured in the future for the construction of novel optoelectronic materials.

References

1. Tang, C. W., Vanslyke, S. A. *Appl. Phys. Lett.*, **51**, 913–915 (1987).
2. Burroughes, J. H., Bradley, D. D. C., Brown, A. R., Marks, R. N., Mackay, K., Friend, R. H., Burns, P. L., Holmes, A. B. *Nature*, **347**, 539–541 (1990).
3. De Silva, A. P., Gunaratne, H. Q. N., Gunnlaugsson, T., Huxley, A. J. M., McCoy, C. P., Rademacher, J. T., Rice, T. E. *Chem. Rev.*, **97**, 1515–1566 (1997).
4. Chabincy, M. L., Salleo, A. *Chem. Mater.*, **16**, 4509–4521 (2004).
5. Tang, C. W. *Appl. Phys. Lett.*, **48**, 183–185 (1986).
6. Hoppe, H., Sariciftci, N. S. *J. Mater. Res.*, **19**, 1924–1945 (2004).
7. Günes, S., Neugebauer, H., Sariciftci, N. S. *Chem. Rev.*, **107**, 1324–1338 (2007).
8. Jelley, E. E. *Nature*, **138**, 1009–1010 (1936).
9. Jelley, E. E. *Nature*, **139**, 631 (1937).
10. Scheibe, G. *Angew. Chem.*, **50**, 51–52 (1937).
11. Brooker, L. G. S., White, F. L., Heseltine, D. W., Keyes, G. H., Dent S. G., VanLare E. J. *J. Photogr. Sci.*, **1**, 173–183 (1953).
12. Kobayashi, K., Shimaoka, R., Kawahata, M., Yamanaka, M., Yamaguchi, K. *Org. Lett.*, **8**, 2385–2388 (2006).
13. Brown, C. J., Farthing, A. C. *Nature*, **164**, 915–916 (1949).
14. Tsuge, A., Hara, T., Motiguchi, T., Yamaji, M. *Chem. Lett.*, **37**, 870–871 (2008).
15. Tsuge, A., Hara, T., Motiguchi, T. *Tetrahedron Lett.*, **50**, 4509–4511 (2009).
16. Morisaki, Y., Chujo, Y. *Angew. Chem. Int. Ed.*, **45**, 6430–6437 (2006).
17. Morisaki, Y., Chujo, Y. *Prog. Polym. Sci.*, **33**, 346–364 (2008).
18. Hopf, H. *Angew. Chem. Int. Ed.*, **47**, 9808-9812 (2008).

19. Morisaki, Y., Chujo, Y. *Bull. Chem. Soc. Jpn.*, **82**, 1070–1082 (2009).
20. Ikeda, A., Shinkai, S. *Chem. Rev.*, **97**, 1713–1734 (1997).
21. Larsen, M., Krebs, F. C., Jørgensen, M., Harrit, N. *J. Org. Chem.*, **63**, 4420–4424 (1998).
22. Alemán, A., Zanuy, D., Casanovas, J. *J. Org. Chem.*, **71**, 6952–6957 (2006).
23. Sun, X. H., Chan, C. S., Wong, M. S., Wong, W. Y. *Tetrahedron*, **62**, 7846–7853 (2006).
24. Hippius, C., van Stokkum, I. H. M., Gsänger, M., Groeneveld, M. M., Williams, R. M., Würthner, F. *J. Phys. Chem. C*, **112**, 2476–2486 (2008).
25. Kagechika, H., Himi, T., Kawauchi, E., Shudo, K. *J. Med. Chem.*, **32**, 2292–2296 (1989).
26. Itai, A., Toriumi, Y., Tomioka, N., Kagechika, H., Azumaya, I., Shudo, K. *Tetrahedron Lett.*, **30**, 6177–6180 (1989).
27. Toriumi Y., Kasuya, A., Itai, A. *J. Org. Chem.*, **55**, 259–263 (1990).
28. Azumaya, I., Kagechika, H., Yamaguchi, K., Shudo, K. *Tetrahedron*, **51**, 5277–5290 (1995).
29. The terms "*cis*" and "*trans*" are conventionally used to show the relative position of aromatic groups connected to the amide unit.
30. Kashino, S., Ito, K., Haisa, M. *Bull. Chem. Soc. Jpn.*, **52**, 365–369 (1979).
31. Elhadi, F. E., Ollis, W. D., Stoddart, J. F., Williams, D. J., Woode, K. A. *Tetrahedron Lett.*, **21**, 4215–4218 (1980).
32. Elhadi, F. E., Ollis, W. D., Stoddart, J. F. *J. Chem. Soc. Perkin Trans. I*, 1727–1732 (1982).
33. Azumaya, I., Kagechika, H., Yamaguchi, K., Shudo, K. *Tetrahedron Lett.*, **37**, 5003–5006 (1996).
34. Azumaya, I., Okamoto, T., Imabeppu, F., Takayanagi, H. *Tetrahedron*, **59**, 2325–2331 (2003).
35. Imabeppu, F., Katagiri, K., Masu, H., Kato, T., Tominaga, M., Therrien, B., Takayanagi, H., Kaji, E., Yamaguchi, K., Kagechika, H., Azumaya, I. *Tetrahedron Lett.*, **47**, 413–416 (2006).
36. Katagiri, K., Sawano, K., Okada, M., Yoshiyasu, S., Shiroyama, R., Ikejima, N., Masu, H., Kato, T., Tominaga, M., Azumaya, I. *J. Mol. Struct.*, **891**, 346–350 (2008).
37. Yokoyama, A., Maruyama, K., Tagami, K., Masu, H., Katagiri, K., Azumaya, I., Yokozawa, T. *Org. Lett.*, **10**, 3207–3210 (2008).

38. Kakuta H., Azumaya, I, Masu, H., Matsumura, M, Yamaguchi, K., Kagechika, H., Tanatani, A. *Tetrahedron*, **66**, 8254–8260 (2010).

39. Fujimoto, N., Matsumura, M., Azumaya, I., Nishiyama, S., Masu, H., Kagechika, H., Tanatani, A. *Chem. Commun.*, **48**, 4809–4811 (2012).

40. Masu, H., Katagiri, K., Kato, T., Kagechika, H., Tominaga, M., Azumaya, I. *J. Org. Chem.*, **73**, 5143–5146 (2008).

41. Takagi, K., Sugimoto, S., Yamakado, R., Nobuke, K. *J. Org. Chem.*, **76**, 2471–2478 (2011).

42. Ohishi, T., Sugi, R., Yokoyama, A., Yokozawa, T. *J. Polym. Sci., Part A: Polym. Chem.*, **44**, 4990–5003 (2006).

43. Winnik, F. M. *Chem. Rev.*, **93**, 587–614 (1993).

44. Isied, S. S., Ogawa, M. Y., Wishart, J. F. *Chem. Rev.*, **92**, 381–394 (1992).

45. Yamakado, R., Matsuoka, S., Suzuki, M., Takagi, K., Katagiri, K., Azumaya, I. *Tetrahedron*, **68**, 9836–9841 (2013).

46. Azumaya, I., Kagechika, H., Fujiwara, Y., Itoh, M., Yamaguchi, K., Shudo, K. *J. Am Chem Soc.*, **113**, 2833–2838 (1991).

47. Yamakado, R., Mikami, K., Takagi, K., Azumaya, I., Sugimoto, S., Matsuoka, S., Suzuki, M., Katagiri, K., Uchiyama, M., Muranaka, A. *Chem. Eur. J.*, **19**, 11853–11857 (2013).

48. Kasha, M., Rawls, H. R., El-Bayoumi, M. A. *Pure Appl. Chem.*, **11**, 371–392 (1965).

49. Kasha, M. *Spectroscopy of the Excited State* (ed. Bartolo, B.), Plenum, New York, 337–363 (1976).

50. Rodger, A., Nordén, B. *Circular Dichroism and Linear Dichroism*, Oxford University Press, Oxford (1997).

51. Kobayashi, N., Muranaka, A., Mack, J. *Circular Dichroism and Magnetic Circular Dichroism Spectroscopy for Organic Chemists*, RSC Publishing, Cambridge (2012).

52. Tominaga, M., Masu, H., Katagiri, K., Kato, T., Azumaya, I. *Org. Lett.*, **7**, 3785–3787 (2005).

53. Tominaga, M., Masu, H., Yamaguchi, K., Azumaya, I. *Chem. Lett.*, **35**, 718–719 (2006).

54. Shomura, R., Higashibayashi, S., Sakurai, H., Matsushita, Y., Sato, A., Higuchi, M. *Tetrahedron Lett.*, **53**, 783–785 (2012).

55. Yamakado, R., Matsuoka, S., Suzuki, M., Takeuchi, D., Masu, H., Azumaya, I., Takagi, K. *Chem. Commun.*, **51**, 5710–5713 (2015).

56. Lommerse, J. P. M., Stone, A. J., Taylor, R., Allen, F. H. *J. Am. Chem. Soc.*, **118**, 3108–3116 (1996).
57. Muniappan, S., Lipstman, S., Goldberg, I. *Chem. Commun.*, 1777–1779 (2008).
58. Pigge, F. C., Vangala, V. R., Kapadia, P. P., Swenson, D. C., Rath, N. P. *Chem. Commun.*, 4726–4728 (2008).

Biography

Dr. Koji Takagi is associate professor at Nagoya Institute of Technology, Japan. He received his PhD in 1998 from Tokyo Institute of Technology under the supervision of Prof. Takeshi Endo. He developed a novel living coordination polymerization of allene derivatives by the π-allylnickel catalyst and earned many experimental skills, not only for the polymerization, but also for the organic synthesis, with the kind guidance of Prof. Ikuyoshi Tomita. He started his academic career at Nagoya Institute of Technology. He launched several research themes, including the controlled radical polymerization by utilizing chalcogen compounds, the synthesis of hyperbranched polymers based on the triazine chemistry, and the development of π-conjugated oligomeric and polymeric materials. The recent topic of his research group can be divided into (i) the precision spatial arrangement of chromophores using well-defined scaffolds, (ii) the controlled synthesis of two-dimensional polythiophene derivatives, and (iii) the manipulation of electron donor–acceptor interaction in fused π-conjugated imidazolium compounds.

Dr. Ryohei Yamakado was born in 1986 in Nagoya (Aichi), Japan. He received his PhD in 2014 from Nagoya Institute of Technology under the guidance of Prof. Koji Takagi and Prof. Masato Suzuki, focusing on the programmable assembly of π-conjugated systems utilizing the stereochemistry of aromatic tertiary amide units. During the spring of 2013, as a visiting fellow, he worked in the group of Prof. Kenneth B. Wagener, Department of Chemistry, the University of Florida, USA. In 2014, he joined the group of Prof. Hiromitsu Maeda, Ritsumeikan University, as a postdoctoral fellow, focusing on the preparation of novel πelectronic ionic species.

Chapter 13

Supramolecular Assemblies of π-Electronic Charged Species

Yohei Haketa, Ryohei Yamakado, and Hiromitsu Maeda
Department of Applied Chemistry, College of Life Sciences, Ritsumeikan University, Kusatsu 525-8577, Japan
maedahir@ph.ritsumei.ac.jp

Well-designed π-electronic charged species form assembled structures through electrostatic and other weak intermolecular interactions. This chapter summarizes the formation of nanoscale morphologies and the resulting fascinating bulk materials such as thermotropic liquid crystals based on the π-electronic charged species.

13.1 Introduction

13.1.1 Molecular Assemblies and Soft Materials

The formation of hierarchically well-defined supramolecular assemblies is currently of great interest in chemistry, biology,

Conjugated Objects: Developments, Synthesis, and Applications
Edited by Atsushi Nagai and Koji Takagi

ISBN 978-981-4774-03-1 (Hardcover), 978-1-315-22962-1 (eBook)
www.panstanford.com

and materials sciences[1]. Because of the limitations in top-down processes for fabricating nanoscale fine materials that have precise arrangements of constituting molecules, the development of rational bottom-up processes is in high demand. Over the past decades, the principles of molecular assembly have been studied using molecules with diverse shapes and electronic states by studying their noncovalent intermolecular interactions such as hydrogen-bonding, metal coordination, van der Waals, π–π, dipole–dipole, and electrostatic interactions. Suitable combinations of appropriate molecules through these noncovalent interactions enable the fabrication of structurally ordered states in both molecular and macroscopic levels. The alignment of molecules via specific interactions forms supramolecular assemblies with various nanoscale architectures (spheres, fibers, rings, vesicles, and tubes) and the resulting soft materials[2] such as supramolecular gels, which are formed by trapping solvent molecules in the 3D networks of nanoscale objects[3]. Liquid crystals are soft materials with molecularly ordered assembled structures, which can be potentially used for transportations of electric charges, ionic species, and molecules[4].

As the scaffolds for the supramolecular assemblies in soft materials, π-electronic molecules play an important role because of their planar geometries and rigid core units that are suitable for the formation of 1D stacking structures through π–π interactions[5]. The periphery of π-conjugated molecules can be modified to induce additional interactions to stabilize the stacking assemblies. Moreover, the electronic states of assemblies can be tuned by the introduction of electron-donating and withdrawing substituents to the π-units. Diverse π-electronic molecules as the scaffolds of the assemblies have been developed. Some of the representative π-electronic molecules include oligophenylenes[6], phenylenevinylenes[7], porphyrins/phthalocyanines[8], perylenebisimides (PBI)[9], and hexa-*peri*-hexabenzocoronenes (HBC)[10]. For example, Müllen et al. reported alkyl-substituted HBC **1** (Fig. 13.1a, i), which afforded fibers in a toluene solution (10^{-4} M) resulting from the 1D π–π stacking of core HBC and van der Waals interactions of alkyl chains (Fig. 13.1a, ii)[11]. HBC **1** formed hexagonally packed liquid crystalline materials exhibiting high charge carrier mobilities[12]. Aida et al. reported the formation of nanotubes comprising amphiphilic HBC **2** and **3** bearing two dodecyl chains on one side and two triethylene glycol (TEG) chains on the other side (Fig. 13.1b, i)[13]. The stabilization of amphiphilic HBC **2** in THF (1 mg mL^{-1}) by heating at 50°C followed by cooling to 30°C afforded nanotubes (Fig. 13.1b, ii). Scanning electron microscopy (SEM) and transmission electron microscopy (TEM) showed tubular morphologies with a uniform diameter of 20 nm, a wall thickness of 3 nm, and an internal diameter of 14 nm. The detailed structural analysis was conducted by synchrotron X-ray diffraction (XRD) techniques[14]. The 2D wide-angle X-ray scattering (WAXS) image of the bundles of aligned nanotubes consisting of HBC **3** clearly showed that the central axes of the helical HBC stacking columns are tilted at 45° relative to the longer axis of the tube (Fig. 13.1b, iii). The hydrophobic alkyl chains are interdigitated, thus arranging hydrophilic TEG chains at the surface of nanotubes. The recent advances of analyzing techniques for nanostructures provide a more clear view of surface information and arrangement of molecules.

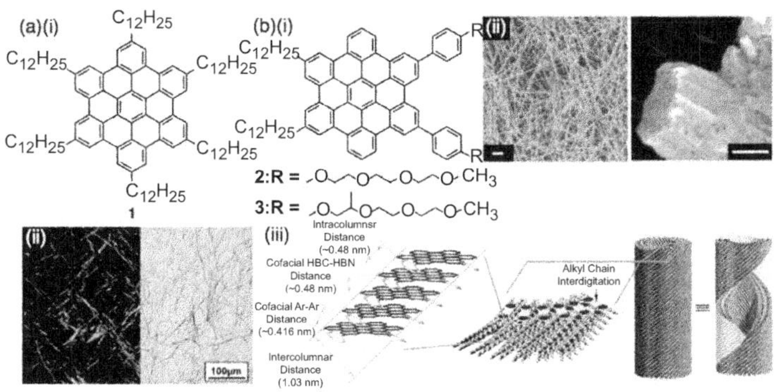

Figure 13.1 (a) (i) Molecular structure of HBC **1** and (ii) polarized optical microscopy (POM) (left) and optical (right) images of the fiber of **1** by drop-casting of a toluene solution (10^{-4} M) (reprinted with permission from Ref.[11]. Copyright (2005) American Chemical Society). (b) (i) Molecular structures of amphiphilic HBC **2** and **3**, (ii) SEM images of the fiber of **2** prepared by drop-casting of a THF suspension (1 mg mL^{-1}) (redrawn from Ref. [13]. Reprinted with permission from AAAS), and (iii) schematic representations of the hierarchical assembling mode of HBC **3** (reprinted with permission from Ref. [14]. Copyright (2008) American Chemical Society).

13.2 Assemblies Comprising Ionic Species

Another important approach for molecular arrangements is the use of electrostatic interactions between oppositely and identically charged species as the building blocks of supramolecular assemblies. This type of assemblies is known as ionic self-assemblies (ISA), which were defined by Faul and Antonietti, and have attracted much attention in recent years especially in the field of material sciences on, for example, surface morphologies, organic–inorganic hybrid materials, and liquid crystals[15]. Diverse building subunits of ISA such as organic charged species accompanied with inorganic and organic counterions are known. With respect to π-electronic species as the building subunits of various assemblies, several strategies can be considered for fabrication of ISA (Fig. 13.2). Four representative ion-pairing modes based on π-electronic charged species are as follows: (I) pairs of electronically neutral π-electronic systems containing charged part(s) and the corresponding (small) counterion(s) (Fig. 13.2a, i,ii), (II) pairs of electronically neutral π-electronic systems containing oppositely charged part(s) (Fig. 13.2a, iii), (III) pairs of charged π-electronic systems and the corresponding (small) counter ion(s) (Fig. 13.2b, i,ii), and (IV) pairs of charged π-electronic systems (Fig. 13.2b, iii).

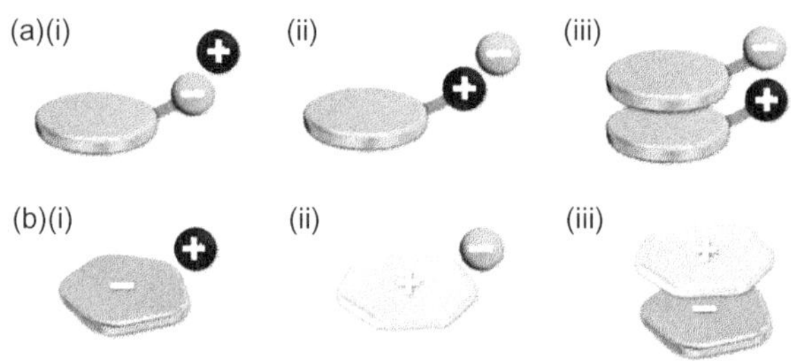

Figure 13.2 Four representative modes of ion-pairing π-electronic species: (a) (i, ii) pairs of electronically neutral π-electronic systems possessing charged part(s) and the corresponding (small) counterion(s) and (iii) pairs of electronically neutral π-electronic systems possessing oppositely charged part(s) and (b) (i, ii) pairs of charged π-electronic systems and the corresponding (small) counterion(s) and (iii) pairs of charged π-electronic systems.

In particular, appropriately designed π-electronic planar charged species (Fig. 13.2b, iii) provide diverse assembling modes by controlling the interactions between them, resulting in the ordered arrangements of ionic components, that is, ion-pairing dimension-controlled assemblies. A *charge-by-charge assembly* is defined as an assembling mode comprising alternately stacking positively and negatively charged species, whereas a *charge-segregated assembly* results from the stacking of identically charged species by overcoming the electrostatic repulsions (Fig. 13.3). Partial contributions of charge-by-charge and charge-segregated assemblies provide intermediate assembling modes. The charge-by-charge mode is effective for constructing columnar assemblies, because the attractive forces for opposite charges support the π–π stacking interactions. In contrast, a charge-segregated assembly can be achieved via the interactions of electron-deficient charged species.

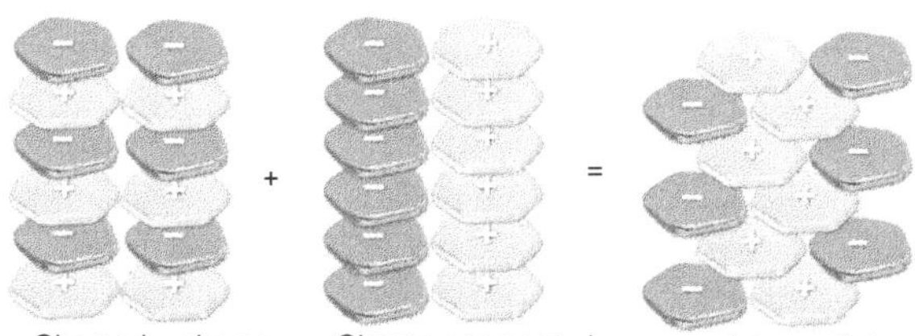

Figure 13.3 Conceptual diagram of the assembling modes comprising planar charged species.

In this chapter, various π-electronic charged species are summarized as the building blocks of ion-pairing supramolecular assemblies, focusing on the recent progresses in their arrangements in nanoscale morphologies and liquid crystals. Furthermore, ion-pairing assemblies based on anion-responsive π-electronic charged species reported by our group are also described later.

13.2.1 Nanostructures Based on Self-Assembling π-Electronic Charged Species

13.2.1.1 Formation of nanofibers

To date, diverse nanofibers have been reported as the assemblies of electronically neutral π-electronic molecules; they may be promising as effective energy-transporting and charge-transporting materials[16]. Planar π-electronic molecules with appropriate substituents for hydrogen-bonding and van der Waals interactions form 1D stacking structures under appropriate external conditions. The charged parts attached to π-electronic units show hydrophilic properties that affect the assembling modes. For example, Müllen et al. reported an HBC-based amphiphile **4** bearing dodecyl chains on one side and alkylimidazolium cations on the other side of the HBC core (Fig. 13.4a, i)[17]. Interestingly, **4** showed solvent-dependent morphologies. The addition of EtOH into a DMSO solution (1 mg mL^{-1}) afforded uniformly dispersed nanofibers with a diameter of ca. 6 nm, as observed by transmission electron microscopy (TEM), which is approximately two times larger than the lateral size (Fig. 13.4a, ii). Moreover, an aqueous solution of **4** after heating afforded nanofibers with diameters of 200–300 nm, as observed by scanning electron microscopy (SEM) (Fig. 13.4a, iii). A precipitate was obtained by the slow diffusion of hexane into an aqueous solution of **4**; it showed uniform fibers with diameters of 10–15 nm and lengths of up to several micrometers (Fig. 13.4a, iv). In the assembled structures in the fibers, the hydrophobic alkyl chains were directed toward the inner parts of the fibers, whereas the hydrophilic polar imidazolium cations were located at the outer surface of the fibers. Similar molecular arrangements were observed in the nanofibers obtained from aqueous solutions (10^{-4} M) comprising amphiphilic sulfonate-pendant bzimpy Pt^{II} complexes **5–8** (Fig. 13.4b, i)[18]. Methyl **5** and methoxy **6** derivatives formed sheet-like morphologies (Fig. 13.4b, ii,iii), whereas the derivatives bearing longer alkyl chains afforded nanofibers. The structures of the nanofibers depended on the alkyl chain lengths such as the diameters of 5.5 and 10.0 nm for **7** and **8**, respectively (Fig. 13.4b, iv,v), as observed by TEM. The diameters of the fibers were twice of the fully stretched molecules, where the hydrophilic sulfonate moieties are arranged at the surface by burying the hydrophobic alkyl chains in the inner parts of the aggregates (Fig. 13.4b, vi,vii)[19].

Similar to the aggregation of amphiphilic molecules in aqueous solutions, as described above, π-electronic molecules exhibit reverse micellar molecular assemblies in hydrocarbon solvents. Baumgartner et al. reported a π-extended phosphonium salt **9** bearing long alkyl chains (Fig. 13.5a, i); fluorescent nanofibers of **9** were obtained from hexane (5 mM), as confirmed by confocal fluorescence microscopy (Fig. 13.5a, ii)[20]. The fibers also formed organogels at this concentration and provided large aggregates on mica, as observed by SEM. Although the powder X-ray diffraction (XRD) of the xerogels showed no clear peaks to identify the molecular arrangement in the nanofibers, their molecular organization may involve the aggregation of the ionic conjugated head groups through ionic interactions inside the fibers along with the arrangement of the hydrophobic alkyl chains towards the surface (Fig. 13.5a, iii).

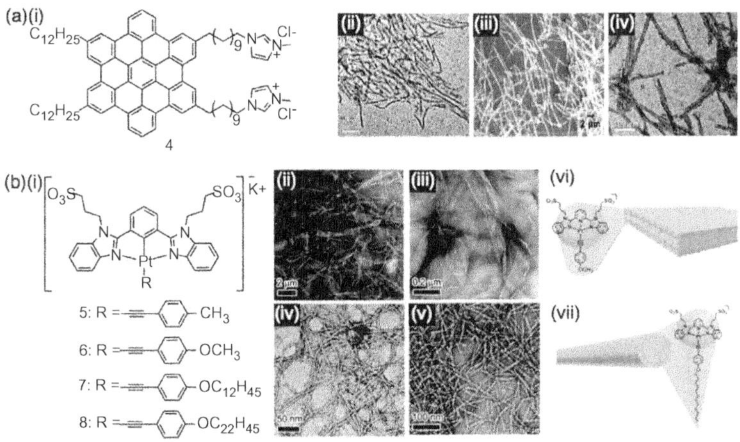

Figure 13.4 (a) (i) Amphiphilic HBC **4**, (ii) TEM image of the nanofibers of **4** obtained from a DMSO–ethanol solution, (iii) SEM image of the nanofibers of **4** obtained from an aqueous solution, and (iv) TEM image of the nanofibers of **4** from an aqueous solution in a hexane-saturated chamber (redrawn from Ref. [17] with permission from the Royal Society of Chemistry). (b) (i) Amphiphilic Pt^{II} complexes **5–8** possessing different alkyl substituents, (ii, iii) TEM images (negatively stained) of **5** and **6** from aqueous solutions, (iv, v) TEM images (negatively stained) of **7** and **8** from aqueous solutions, (vi) an assembling mode in sheet structures, (vii) an assembling mode in fiber structures (redrawn from Ref. [18b] with permission from the Royal Society of Chemistry).

The utilization of π-electronic charged aromatics as the building blocks of assemblies is particularly attractive, because the stacking of π-electronic systems can be tuned by the counterions. Müllen et al. reported the formation of nanoscale uniform belt-like morphologies using 9-phenyl-benzo[1,2]quinolizino[3,4,5,6-*fed*] phenanthridinylium (PQP^+) cations **10–14** (Fig. 13.5b, i)[21]. The widths and lengths of the belts were 200–800 nm and 0.8–8 μm, respectively, with a thickness of 30 nm, as confirmed by SEM, TEM, and atomic force microscopy (AFM) (Fig. 13.5b, ii,iii). The WAXS of the nanobelts showed the reflections corresponding to (100), (200), and (300), indicating lamellar arrangements with a π-plane stacking of ca. 0.35 nm. The *d*-spacing of (100) corresponds to the length of the two fully interdigitated layers of PQP^+ cation, indicating that the dodecane sulfonate generated the lamellar bilayer with a staggered cofacial stacking of the aromatic cores, stabilizing the structures (Fig. 13.5b, iv). The ordered arrangement of π-electronic charged species was responsible for their ionic conductivity[21]; thus the nanobelt of **14** obtained as a mixture with $LiClO_4$ showed anion conductivity of $\sigma = 2.4 \pm 0.6 \times 10^{-7}$ S cm^{-1}, which is comparable to those of other supramolecular ion conductors.

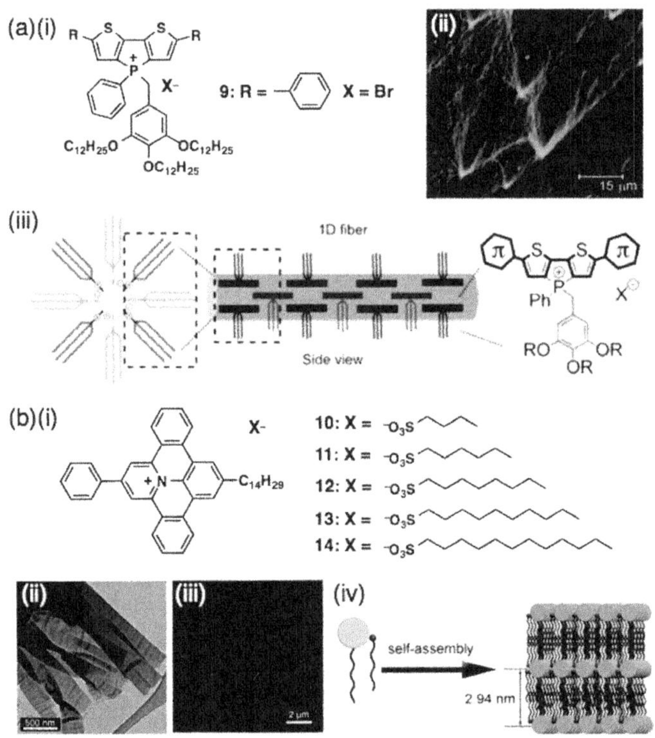

Figure 13.5 (a) (i) Phosphonium salt **9**, (ii) confocal fluorescence microscopy of **9**, and (iii) assembling mode of phosphonium salt **9** (redrawn from Ref. [20]. Copyright 2012 Wiley). (b) (i) PQP salts with various alkyl-substituted sulfonates **10–14**, (ii) TEMs and (iii) SEM of **14** obtained by the drop-casting of a MeOH solution, and (iv) assembling mode of **14** (redrawn with permission from Ref. [21]. Copyright 2011 WILEY-VCH Verlag GmbH & Co. KGaA, Weinheim).

13.2.1.2 Formation of nanotubes

The tubular nanostructures of π-electronic charged species are relatively uncommon compared to those of electronically neutral molecules. This can be partially attributed to the sizes and properties of counterions, which often hinder the formation of tubular structures. Shelnutt et al. reported the formation of tubular nanostructures using a mixture of different porphyrins: *meso*-tetra(*p*-sulfonatophenyl)porphyrin **15** and *meso*-tetra(*p*-pyridyl) porphyrin Sn^{IV} complex **16** (Fig. 13.6a, i)[22]. By mixing porphyrins **15** and **16** in water (pH 2) and storing in dark for 72 h, uniform nanotubes of 50–70 nm in the outer diameter and 20 nm in wall thickness were observed by TEM (Fig. 13.6a, ii).

Nanotubes were also obtained from a π-electronic core unit with amphiphilic ionic moieties and hydrophobic alkyl chains by forming bilayers, where the ionic moieties and alkyl chains are located at the outer and inner surfaces, respectively. For example, π-electronic amphiphilic cyanine **17** provided J-aggregates in an aqueous solution, forming nanotubes. The inner and outer surfaces of the nanotubes are covered with hydrophilic sulfoxide ionic parts (Fig. 13.6b, i), and inside of the nanotubes is packed with hydrophobic alkyl chains as shown by AFM (Fig. 13.6b, ii) and cryogenic TEM (cryo-TEM)[23]. The sample grids were immersed into liquid ethane for vitrification, and the TEM of the sample was measured at a low temperature (–178°C). The nanotubes showed a uniform diameter of 13 ± 5 nm[24].

As a better reason for investigating π-electronic charged species for assemblies, their arrangement can be tuned by counterions. The modification of π-electronic cations is more feasible in organic synthesis compared to π-electronic anions because they are more reactive because of the presence of excess electrons. Anionic species have diverse sizes and electronic states as well as different geometries, as shown by the spherical Cl^- and Br^-, linear N_3^- and SCN^-, trigonal planar CO_3^{2-} and NO_3^-, tetrahedral PO_4^{3-} and SO_4^{2-}, and octahedral $Fe(CN)_6^{4-}$ and $Co(CN)_6^{3-}$. Some examples of assemblies obtained from π-electronic species controlled by anions have been reported. For example, the anion-dependent morphological changes of PQP^+ derivatives as the Cl^- **18** and BF_4^- **19** salts have been reported (Fig. 13.6c, i)[25]. The drop casting of a methanol solution of **18** and rapid drying of the solvent afforded nanoribbons with a width of 80 nm

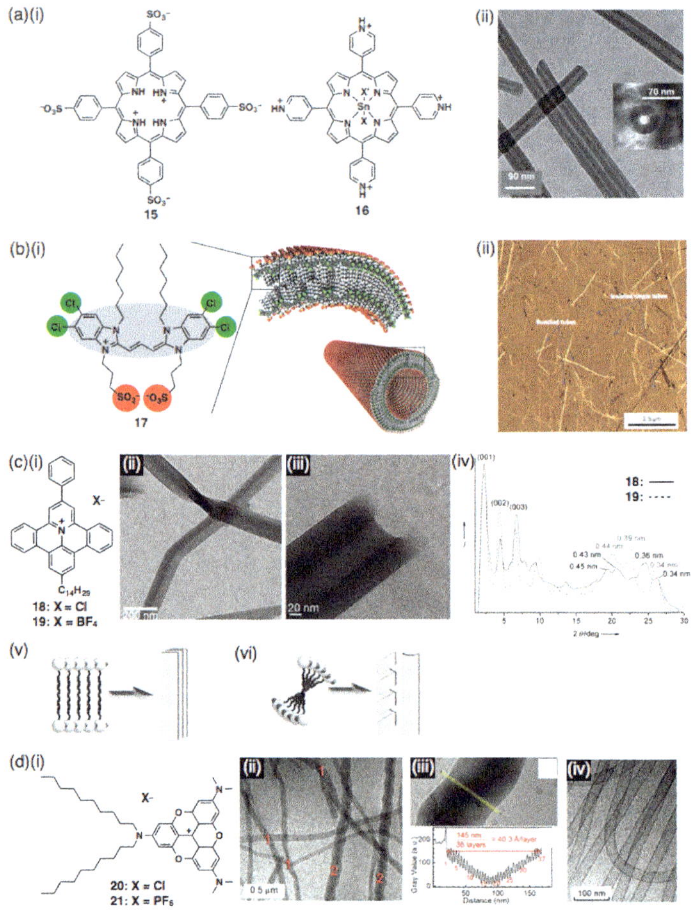

Figure 13.6 (a) (i) Ionic porphyrins **15** and **16** and (ii) TEM of the nanotubes obtained from an aqueous solution (pH 2) consisting of both **15** and **16** (reprinted with permission from Ref. [22]. Copyright (2004) American Chemical Society). (b) (i) Amphiphilic cyanine **17** and the packing structure of the nanotube and (ii) AFM image of the nanotube (redrawn from Ref. [23a]. Copyright 2009, Nature Publishing Group). (c) (i) PQP^+ salts **18** and **19**, (ii) nanofibers of **18** in MeOH (10^{-1} M), (iii) nanotubes of **19** in MeOH (10^{-1} M), and (iv) XRD of the dried powders of **18** and **19** and the assembling modes of (v) **18** and (vi) **19** (redrawn with permission from Ref. [25]. Copyright 2007 WILEY-VCH Verlag GmbH & Co. KGaA, Weinheim). (d) (i) Trioxatriangulenium salts **20** and **21**, (ii) TEM image of **21** in 10% $MeCN/H_2O$, where labels 1 and 2 refer nanoribbons and nanorods, respectively, (iii) cross-sectional analysis of the gray scale along the yellow line in a nanorod, and (iv) TEM image of **20** as nanotubes in 10% $MeCN/H_2O$ (redrawn with permission from Ref. [26]. Copyright 2014 WILEY-VCH Verlag GmbH & Co. KGaA, Weinheim).

and lengths ranging from 0.5–2 μm, as shown by SEM and TEM (Fig. 13.6c, ii). Salt **19** showed the formation of nanotubes with a wall thickness of 40–60 nm, inner diameters of 20–50 nm, and lengths of ca. 5 μm (Fig. 13.6c, iii). The XRD of **18** and **19** in the small-angle region showed a similar lamellar stacking for both the assemblies (Fig. 13.6c, iv). However, XRD in the wide-angle region shows the differences in the stacking modes: The cationic PQP^+ head group alternately stacked with Cl^- in **18** to afford a face-to-face stacking, whereas the larger BF_4^- anion in **19** caused a slipped face-to-face orientation. Consequently, the slipped stacking of PQP^+ rotates along the axis of stacking structures, forming higher 1D aggregates.

Another unique example of nanotubes that depend on counteranions was reported by Laursen et al., such as the assemblies of a decyl-substituted trioxatriangulenium cation ($TOTA^+$) as Cl^- **20** and PF_6^- **21** salts (Fig. 13.6d, i)[26]. A 10% MeCN/aqueous solution of **21** formed nanoribbons and nanorods (Fig. 13.6d, ii) with widths of 60–100 nm and 150 nm, respectively, as shown by cryo-TEM. The cross-sectional TEM analysis (Fig. 13.6d, iii) and WAXS showed a layer thickness of 40 Å, corresponding to approximately twice the sum of the core unit and decyl chains. Furthermore, a peak at 3.5 Å in the WAXS clearly indicated that the core cationic π-unit cofacially stacked, providing a columnar alignment. Nanotubes were formed when the counteranion was changed from PF_6^- to Cl^- under the same aggregation conditions. The nanotubes showed a uniform width of 29 ± 2 nm and wall thickness of ca. 4 nm (Fig. 13.6d, iv). As a plausible mechanism, the solvation of anions and the resulting electrostatic interactions between the ions are important. Thus, as the solvation by water molecules is much larger for Cl^- in **20** compared to PF_6^- in **21**, less solvated **21** induced the hydrophobic surface condition and enabled a reasonable stacking of bilayers to form nanoribbons; in contrast, the significantly water-solvated Cl^- in **20** stabilized the bilayer surface, thus affording nanotubes. The control of surface solvation by tuning counteranions showed the advantage of the use of π-electronic charged species for fabricating nanostructures.

Charged π-electronic molecules form 1D nanostructures such as fibers and nanotubes, depending on the balance of hydrophilicity (ionic parts) and hydrophobicity (alkyl or other nonionic parts) based on π-electronic planes. Furthermore, the assemblies can be tuned by counterions by controlling the size and solvation.

13.3 Ionic Liquid Crystals Based on π-Electronic Charged Species

As described above, π-electronic charged species with appropriate substituents for intermolecular interactions form various self-assembled nanoscale objects. The basic strategy for the assemblies of ionic species is applicable for the formation of liquid crystalline states, known as ionic liquid crystals[27]. However, π-electronic charged species are often highly crystalline because of electrostatic interactions, and interactions between building blocks should be controlled for achieving stable mesophases. Therefore, flexible aliphatic alkyl chains are often used owing to efficient van der Waals interactions that help to assemble molecules into various organized states. A very simple yet efficient method for preparing charged species is the attachment of ionic moieties, such as for cations, ammonium, phosphonium, imidazolium, pyridinium, guanidiniums, and amidiniums, and for anions, sulfonate and carboxylate. For example, Kato et al. reported the preparation of fan-shaped imidazolium salts **22** and **23**,[28] ammonium salts **24**–**26**,[29] and phosphonium salts **27**–**29** (Fig. 13.7a–c)[30] bearing long alkoxybenzyl chains. Notably, imidazolium salts **22** and **23** formed only hexagonal columnar (Col_h) mesophases, whereas ammonium salts **24**–**26** and phosphonium salts **27**–**29** formed bicontinuous cubic (Cub_{bi}) and Col_h structures depending on the alkyl chain lengths. It is important to balance the sizes and geometries of nonionic (alkyl chains) and ionic parts for tuning the mesophase structures. Furthermore, these ionic liquid crystals exhibited ion-conductive properties[31].

Aida et al. developed imidazolium cation-appended triphenylene derivatives **30**–**32** by introducing cationic moieties to π-electronic core units (Fig. 13.7d, ii)[32]. A detailed analysis using a synchrotron radiation XRD at SPring-8 showed that **30**–**32** formed Col_h structures and Cub_{bi} assemblies (Fig. 13.7d)[33]. The temperature range of mesophases can be tuned by changing the lengths of alkyl chains and counteranions, as observed in tetradecyl derivative **32** exhibiting a Cub_{bi} mesophase at 24°C–221°C. The XRD analysis also showed a peak with a *d*-spacing of 3.5 Å corresponding to the π-stacking distance of triphenylene columns in the Cub_{bi} mesophase. The Cub_{bi} and Col_h mesophases exhibited comparable charge carrier conductivity, as measured by the flash-photolysis time-resolved microwave conductivity (FP-TRMC) method.

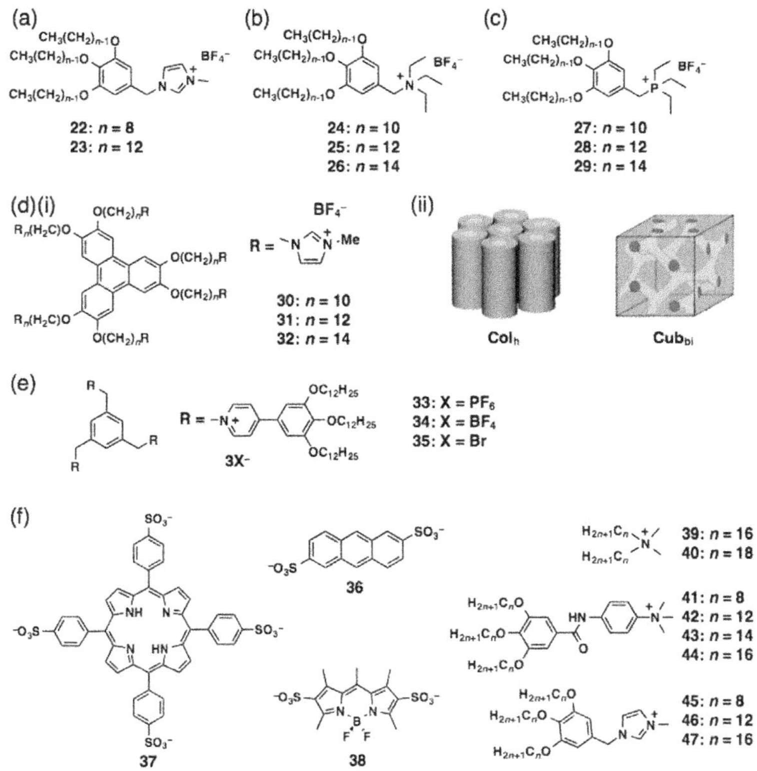

Figure 13.7 (a) Imidazolium-based ion pairs **22** and **23**, (b) ammonium-based ion pairs **24–26**, (c) phosphonium-based ion pairs **27–29**, (d) (i) imidazolium-appended triphenylene ion pairs **30–32**, (ii) schematic models of Col_h and Cub_{bi} mesophases (reprinted with permission from Ref. [33a]. Copyright (2009) American Chemical Society), (e) pyridinium tripodal ion pairs **33–35**, and (f) various ion pairs based on π-electronic sulfonates **36–38** and ammonium and imidazolium countercations **39–47**.

The pyridinium derivatives also showed ionic liquid crystals with columnar structures. Tridodecyloxy-substituted phenyl pyridinium tripodal derivatives **33–35** exhibited rectangular columnar (Col_r) and Col_h mesophase structures over wide temperature ranges[34]. Interestingly, **35** as a Br^- salt showed a micellar cubic (Cub) mesophase, where each micelle was assumed to contain ca. 16 molecules. Compared to PF_6^- and BF_4^-, smaller-size counteranions such as Br^- constructed more tightly packed ionic pyridinium cores, forming a Cub phase rather than columnar structures.

Both cations and anions can be used as the π-electronic species in the building blocks of mesophase structures. As ion pairs[35], Ziessel et al. prepared sulfonate-modified negatively charged anthracene **36**,[35a] porphyrin **37**,[35b] and borondipyrromethene (BODIPY) **38**,[35b–d] containing alkylammonium cations **39–44** and imidazolium cations **45–47** with long alkoxy chains as the counterparts. In these ion-pairing systems, anthracene–ammonium ion pairs $\mathbf{36{\cdot}39_2}$–$\mathbf{44_2}$ showed no mesophases, whereas anthracene–imidazolium ion pairs $\mathbf{36{\cdot}45_2}$–$\mathbf{47_2}$ showed mesophases as confirmed by polarizing optical microscopy (POM) and differential scanning calorimetry (DSC). XRD showed the exact molecular arrangements, indicating Col_h structures similar to other ion pairs such as porphyrin–ammonium $\mathbf{37{\cdot}41_4}$ and $\mathbf{37{\cdot}42_4}$, BODIPY–ammonium $\mathbf{38{\cdot}42_2}$–$\mathbf{44_2}$, and BODIPY–imidazolium $\mathbf{38{\cdot}45_2}$–$\mathbf{47_2}$.

Highly organized structures based on the stacking of π-electronic species can be achieved by incorporating the charged part inside the aromatic cores rather than at the peripheries of the aromatic cores. However, very few examples of highly organized structures using π-electronic ionic species as building blocks of liquid crystals are known due to difficulties in their synthesis, as shown by the PQP^+ cations[21,25,36]. Counteranions have been modified by introducing alkyl chains, for example, alkyl-substituted sulfonates (Fig. 13.8, i). Indeed, ion pair **49** showed a typical birefringent POM texture (Fig. 13.8, ii), whereas ion pair **48** exhibited a needle-like crystalline state. Although only one phase transition of **49** was observed at 210°C, WAXS showed the formation of a discotic columnar structure with a clear peak at 3.5 Å, indicating the formation of π–π stacking. Furthermore, a peak at 7.0 Å, twice the distance of a regular π–π stacking, probably corresponds to the staggered arrangement of PQP^+ in the solid state (Fig. 13.8, iii).

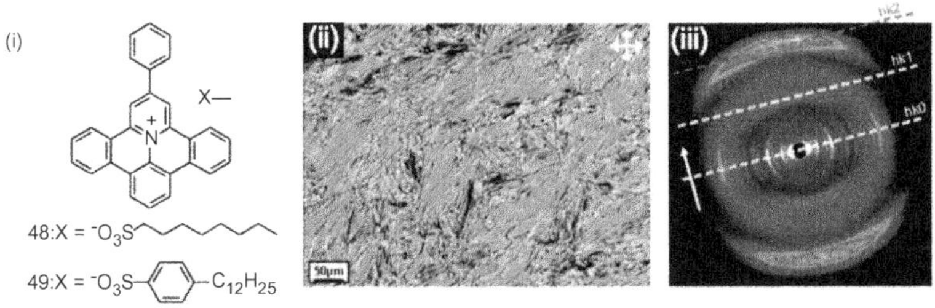

Figure 13.8 (i) Ion pairs of PQP cation with dodecyl-substituted sulfonates **48** and **49**, (ii) 2D WAXS of **48** at 30°C, and (iii) POM of **48**. Reprinted with permission from Ref. [36]. Copyright (2009) American Chemical Society.

13.4 Ion-Based Supramolecular Assemblies Based on π-Electronic Anionic Species

As mentioned above, many ion-based assemblies are based on π-electronic species attached to ionic (cationic and anionic) moieties and those with a positively charged core. The use of anionic species as the building blocks of dimension-controlled assemblies is significantly limited because it is very difficult to prepare π-electronic anions because of the presence of excess electrons that easily undergo electrophilic attack. Although carbanions have been investigated as the components of hydrocarbon salts and ionic charge transfer (CT) complexes[37,38], very few examples of ion-pairing assemblies have been reported in the solid state[39]. Therefore, the preparation of planar anions can be achieved by the anion binding of electronically neutral π-electronic molecules (Fig. 13.9a).

Planar receptor–anion complexes have been widely reported in the solid state. Flood et al. (Fig. 13.9b, i)[40] and Craig et al. (Fig. 13.9b, ii)[41] independently reported the synthesis of triazole-based anion receptors using click chemistry. The crystal structure of **51**·Cl^- as a tetrabutylammonium (TBA) salt exhibited a relatively planar structure and a charge-by-charge assembly with alternately stacking **51**·Cl^- and TBA^+ units (Fig. 13.9b, iii). Thus far, most of the self-assemblies of receptor–anion complexes have been investigated in the solid state or 2D surfaces[42].

Since 2005, we have reported a series of dipyrrolyldiketone difluoroboron (BF_2) complexes **52–60** (representative compounds, Fig. 13.10) as anion-responsive π-electronic molecules with the potential to form π-electronic anionic species[43–49]. As the two pyrrole rings are not located at the preorganized positions for anion binding, they are inverted upon the addition of anions in the solution state (Fig. 13.10a). An efficient anion binding by two pyrrole NH and one bridging CH moieties was observed by ^{1}H NMR, where the signals corresponding to the receptors and receptor–anion complexes were independently observed due to slow anion-binding processes.

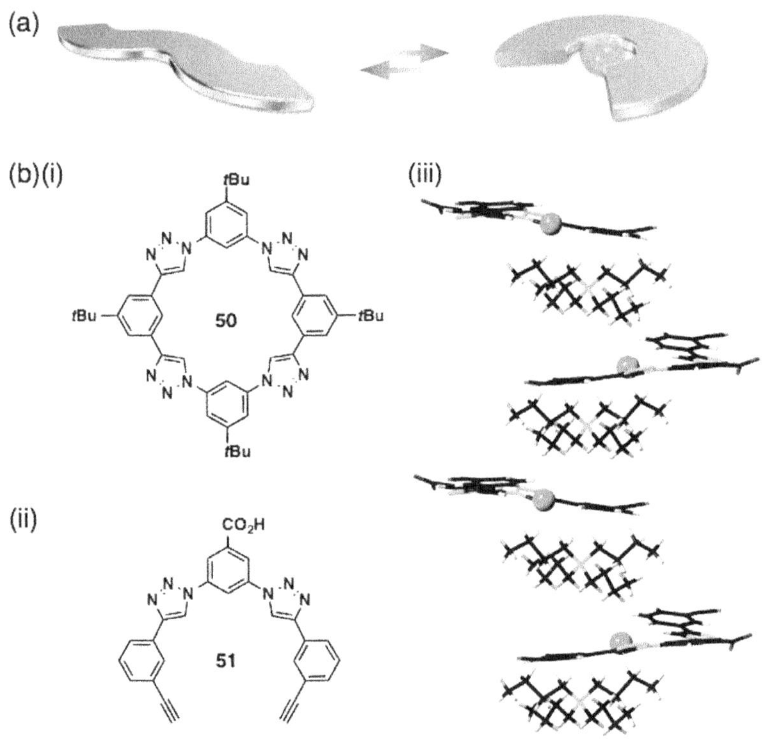

Figure 13.9 (a) Schematic representation of the formation of a receptor–anion complex as a planar anion and (b) (i) triazole-based macrocycle **50**, (ii) triazole-based acyclic anion receptor **51**, and (iii) solid-state packing diagram of **51**·Cl⁻ (redrawn from a cif file: CCDC-663329).

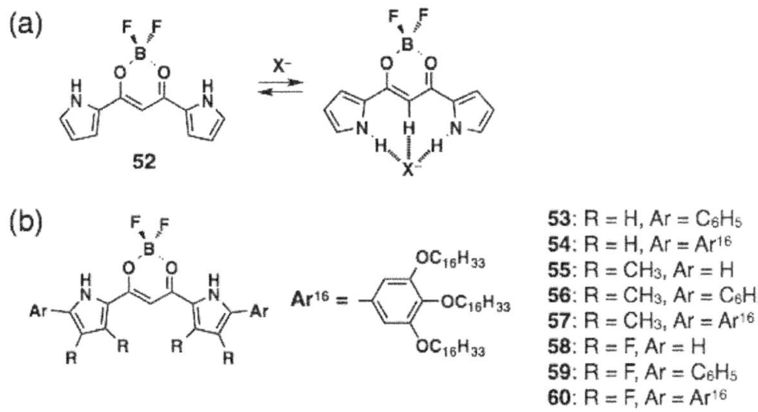

Figure 13.10 (a) Pyrrole-based anion receptor **52** and the anion-binding mode and (b) representative anion receptor derivatives **53–60**.

As various modifications at the periphery of anion-responsive π-electronic molecules (Fig. 13.10b) help to control their anion-binding geometries, electronic states, solubility, and assembling modes; the details of the recent progress are summarized in other reviews[44]. In some representative examples, α-aryl groups have been used for introducing various substituents to the core unit for further applications[45], as shown by alkoxy-substituted receptors forming anion-responsive supramolecular gels[45a] and liquid crystals[45d]. The pyrrole β-positions can be modified[46,47] for achieving selective reactions at the α-positions such as the iodination and coupling reactions of aryl moieties to obtain various functionalities for various electronic properties[47b], substituent-dependent assemblies[47f], metal-bridged oligomers for fascinating binding modes[47e], covalently linked oligomers for high anion-binding affinity[47d], and ion-pairing chirality induction exhibiting circularly polarized luminescence (CPL)[47g]. Furthermore, the boron unit of receptor molecules can also be modified[48]. For example, catechol–boron complexes formed ditopic receptors by connecting π-conjugated backbone and receptor units[48a], and 1,1'-bi-2-naphthol (BINOL)-boron-modified receptors exhibited the anion-responsive enhancement of CPL due to the distorted π-planes[48c].

13.4.1 Assemblies of Receptor–Anion Complexes in the Solid State

In the solid state, receptor–anion complexes exhibited various assembling modes in combination with countercations depending on the geometries and substituents of receptors and cations. *α*-Aryl-substituted receptors **53**, **56**, and **59** formed planar [1+1]-type Cl^- complexes **53**·Cl^- (Fig. 13.11a), **56**·Cl^- (Fig. 13.11b), and **59**·Cl^- (Fig. 13.11c) with tetrapropylammonium (TPA^+) for **53**·Cl^- and **56**·Cl^- and TBA^+ for **59**·Cl^- as countercations. The receptor–anion complexes and cations formed charge-by-charge assemblies by the one-by-one stacking of receptor–Cl^- complexes and cations for **53**·Cl^- and **56**·Cl^-, whereas **59**·Cl^- showed a two-by-two stacking mode. When planar 4,8,12-tripropyl-4,8,12-triazatriangulenium $(TATA^{C3})^+$ cation[50] was used instead of bulky tetraalkylammonium cations, the ion pair **53**·Cl^--$(TATA^{C3})^+$ formed a charge-by-charge stacking assembly with a smaller distance of 6.85 and 7.29 Å between **53**·Cl^- units for $(TATA^{C3})^+$ and TPA^+ salts, respectively (Fig. 13.11d). Furthermore, both *β*-methyl **56**·Cl^--$(TATA^{C3})^+$ (Fig. 13.11e) and *β*-fluorinated **59**·Cl^--$(TATA^{C3})^+$ (Fig. 13.11f) also formed charge-by-charge assemblies, whereas the larger Cl···Cl distances along the columns were 12.43 and 10.57 Å, respectively. The smaller overlapping of receptor–Cl^- complexes and $(TATA^{C3})^+$ in **56**·Cl^--$(TATA^{C3})^+$ and **59**·Cl^--$(TATA^{C3})^+$ indicates that the *β*-substituents of the receptor molecules may interfere with the formation of stable charge-by-charge columnar structures.

13.4.2 Ion-Based Supramolecular Gels

Dimension-controlled assembled structures can be formed by the modification of receptors with aliphatic moieties. Anion receptor **54** with suitable alkyl chain lengths gelated hydrocarbon solvents such as octane (Fig. 13.12a, i)[45a]. The octane gel exhibited characteristic electro-optical properties, as observed by UV/vis and fluorescence luminescence (FL) measurements, in sharp contrast to the monomeric state in the solution state. Supramolecular organogels were formed via noncovalent interactions between the π-electronic moieties and their substituents, as shown by AFM, SEM, and XRD analyses (Fig. 13.12a, i). The addition of Cl^- as a TBA^+ salt to the gel fabricated the gradual transition of the gel to the solution state (Fig. 13.12a, ii). The gel decomposition can be attributed to the formation of soluble ion pair **54**·Cl^--TBA^+ in nonpolar solvents (Fig. 13.12a, iii). In contrast, the ion pair **60**·Cl^--TBA^+ was insoluble in octane, indicating that the receptor–anion complex formed another state rather than the decomposition of the gel state.

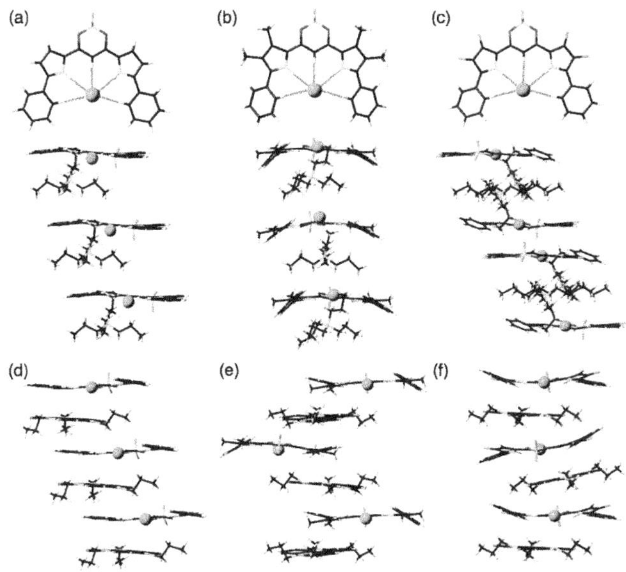

Figure 13.11 Top- and side-packing views from single-crystal X-ray analysis of (a) **53**·Cl^--TPA^+, (b) **56**·Cl^--TPA^+, (c) **59**·Cl^--TBA^+, (d) **53**·Cl^--$(TATA^{C3})^+$, (e) **56**·Cl^--$(TATA^{C3})^+$, and (f) **59**·Cl^--$(TATA^{C3})^+$ (redrawn from cif files: CCDC-646480, 781980, 78198, 745780, 853090, and 853091, respectively).

The use of planar cations afforded soft materials based on planar charged species that are different from the states comprising bulky cations. In contrast to the formation of the solution state of **54** and TBACl in octane, **54**·Cl^--$(TATA^{C3})^+$ afforded an opaque octane gel with a gel-to-solution transition temperature of 8°C, higher than that of **54**, whereas **54**·Cl^--$(TATA^{C16})^+$ afforded precipitates, indicating that appropriate planar cations such as $(TATA^{C3})^+$ were suitable for the formation of gel states (Fig. 13.12b)[49a]. The optical microscopy (OM), AFM, and SEM analyses of the xerogel of **54**·Cl^--$(TATA^{C3})^+$

showed the formation of entangled fibrils with a diameter and length of 0.1–2 μm and >100 μm, respectively (Fig. 13.12b). The detailed structure of the fibrous xerogel of **54**·Cl^--$(TATA^{C3})^+$ was investigated by synchrotron XRD (BL40B2 at SPring-8), exhibiting a Col_h arrangement with a = 4.25 nm and c = 0.73 nm based on a trimeric assembly (Z = 3 for ρ = 1) (Fig. 13.12c). The c value of 0.73 nm derived from the repeating stacking distance of the two planar components, **54**·Cl^- and $(TATA^{C3})^+$ in this case, indicates the formation of a charge-by-charge assembly. In contrast to **54**·Cl^--$(TATA^{C3})^+$, the precipitates of **54**·Cl^--$(TATA^{C16})^+$ showed the formation of a discotic lamellar structure with an intercolumnar distance of ca. 5 nm, indicating that longer alkyl chains at the periphery of a $(TATA^{C3})^+$ cation interfere with the formation of an ordered structure such as a Col_h phase.

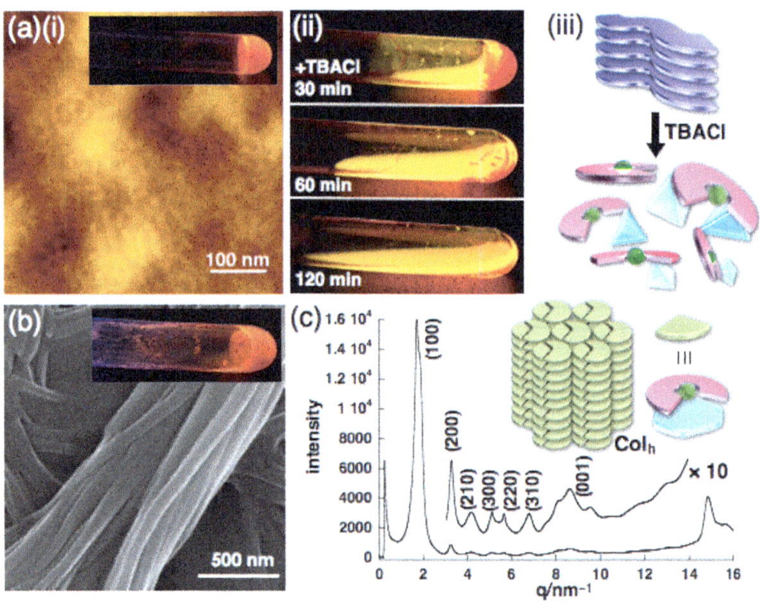

Figure 13.12 (a) (i) AFM image in the tapping mode of **54** from the octane gel (10 mg mL^{-1}) (inset: photograph of the gel under UV_{365nm} light), (ii) transition of the octane gel of **54** at 20°C upon the addition of Cl^- as a TBA^+ salt in solid form (under $UV_{365\ nm}$ light), and (iii) possible anion-responsive process of the gel-to-solution transition. (b) SEM image at 20°C (inset: photograph of the gel under $UV_{365\ nm}$ light) and (c) XRD pattern at 25°C and a proposed assembled model of the xerogel of **54**·Cl^--$(TATA^{C3})^+$ from octane.

13.4.3 Ion-Based Liquid Crystals Based on Planar Charged Species

Receptors bearing aliphatic alkyl chains form assemblies as Col_h-based mesophases through π–π stacking and van der Waals interactions[45d]. Semifluoroalkyl-substituted receptors also form highly ordered assemblies as Col_h mesophases through interactions between alkyl units and those between fluoroalkyl units, resulting in high charge carrier mobilities[45h]. Based on the charge-by-charge columnar assembly, as observed by the fibril morphology, the formation of ion-based liquid crystals was attempted. In fact, the POM analysis of the xerogel of **54**·Cl^--$(TATA^{C3})^+$ exhibited a focal conic texture in the mesophase, and the transition temperature was also obtained by DSC analysis (Fig. 13.13a, i). The synchrotron XRD pattern shows relatively sharp peaks corresponding to the Col_h phase with a = 4.64 nm and c = 0.73 nm based on a tetrameric assembly (Z = 3.58 for ρ = 1) (Fig. 13.13a, ii). The c value of 0.73 nm corresponds to the stacking distance of **54**·Cl^- and $(TATA^{C3})^+$, indicating the formation of charge-by-charge assemblies, as observed in the crystal structure of **53**·Cl^--$(TATA^{C3})^+$ (6.85 Å).

The stacking assemblies even in mesophases can be modulated by the modification of the pyrrole β-positions in the receptors. In fact, the ion pairs of β-methyl-substituted **57**·Cl^--$(TATA^{C3})^+$ and β-fluorine-substituted **60**·Cl^--$(TATA^{C3})^+$ afforded a similar Col_h mesophase of a = 4.99 nm, c = 0.37 nm, and Z = 2.05 for ρ = 1 and that of a = 4.92 nm, c = 0.37 nm, and Z = 1.98 for ρ = 1, respectively (Fig. 13.13b, ii)[49e]. In this case, the characteristic sharp XRD peaks of c = 0.37 nm were comparable to regular π–π stacking distances, in contrast to the peak at ca. 0.73 nm in the charge-by-charge assembly of **54**·Cl^--$(TATA^{C3})^+$. These observations strongly indicate the contributions of charge-segregated assemblies where the local stacking structures of identically charged planes are formed, resulting in columnar structures providing the Col_h mesophases. The charge-segregated structures have resulted from the distorted aryl rings owing to the β-substituents as observed in the single-crystal X-ray structures (Fig. 13.11e,f). Furthermore, the time-of-flight (TOF) electrical conductivity measurements of ion-based materials **54**·Cl^--$(TATA^{C3})^+$, **57**·Cl^--$(TATA^{C3})^+$, and **60**·Cl^--$(TATA^{C3})^+$ indicated ambipolar charge carrier transporting behaviors with well-balanced values at high mobilities (10^{-2}–10^{-3} $cm^2\ V^{-1}\ s^{-1}$) for both holes and electrons without special purification procedures.

Because of van der Waals interactions, mesophases were observed even for the receptor–anion complexes with bulky cations, causing gel–sol transitions in octane. Thus, the Col_h mesophase based on a charge-by-charge assembly was observed for **54**·Cl^--TBA^+ (Fig. 13.13c)[49c]. In this case, the XRD analysis showed a broad diffraction identified as the repeating distance of the charge-by-charge stacking of **54**·Cl^- and TBA^+, indicating a more weakly stacking structure compared to those comprising planar $(TATA^{C3})^+$ cations. Furthermore, a combination of **54**·Cl^- and diverse tetra-alkylammonium cations afforded Col_h structures with very low mesophase temperatures due to the less efficient stacking of bulky cations.

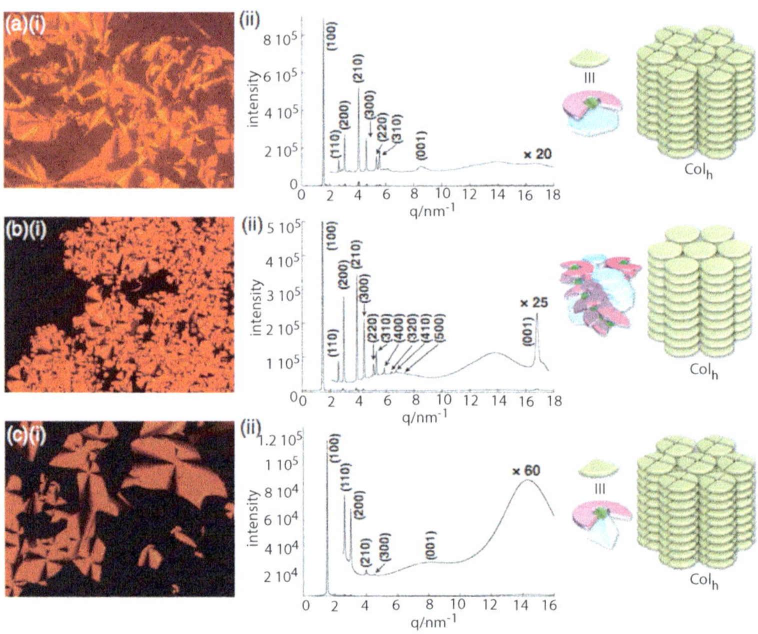

Figure 13.13 (i) POM images and (ii) XRD patterns and proposed assembled models of (a) **54**·Cl^--$(TATA^{C3})^+$ at (i) 70°C and (ii) 101°C, (b) **57**·Cl^--$(TATA^{C3})^+$ at (i) 150°C and (ii) 101°C, and (c) **54**·Cl^--TBA^+ at (i, ii) 62°C upon cooling from Iso.

Another interesting example of the preparation of a countercation is the anion complex of π-electronic dicationic receptors. In fact, phenylene- and pyrimidine-bridged bis(imidazolium) dications **61** and **62** formed monocationic receptor–Cl^- complexes containing a free Cl^- (Fig. 13.14, i)[49h]. The free Cl^- was subsequently captured by an electronically neutral anion receptor **54**, forming a negatively charged planar receptor–Cl^- complex. For example, ion pair **54**·Cl^--**62**$^{2+}$·Cl^- exhibited a broken-fan-shaped POM texture as the mesophase (Fig. 13.14, ii). The synchrotron XRD of **54**·Cl^--**62**$^{2+}$·Cl^- at 115°C as in the mesophase showed a Col_h structure with a = 4.34 nm and a broad and weak peak at ca. 0.75 nm. This was derived from the repeating stacking ion pairs with the contribution of a charge-by-charge assembly (Fig. 13.14, iii). A Col_h structure was also observed at 20°C of the first cooling with a lattice parameter of a = 4.73 nm along with a more clear peak at 0.75 nm. Therefore, the average number of ion pairs in an assembled unit (Z) was estimated as ca. 3 for ρ = 1. The similarity in the sizes and shapes of two ionic components is an important factor for forming stable assemblies containing multiple charged components. Furthermore, the FP-TRMC technique showed the 1D charge carrier transporting property of **54**·Cl^--**62**$^{2+}$·Cl^- with a mobility of 0.05 $cm^2\,V^{-1}\,s^{-1}$.

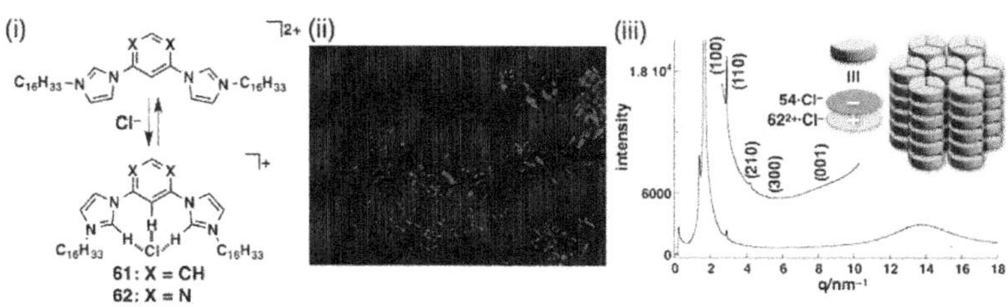

Figure 13.14 (i) Positively charged anion receptors **61** and **62** and the formation of cationic receptor–Cl^- complexes, (ii) POM texture, and (iii) synchrotron XRD pattern of **54**·Cl^--**62**$^{2+}$·Cl^- at 115°C upon cooling from Iso.

Various assemblies can be obtained by the introduction of aliphatic chains to the building subunits as shown by the above examples. Another strategy is available for the combination of building subunits: the receptor without long alkyl chains can be combined with the anion accompanying cations bearing long alkyl chains. For example, benzyltrialkylammoniums **63**$^+$ and **64**$^+$ and benzylpyridinium **65**$^+$ as the chloride salts (Fig. 13.15a, i) were combined with β-unsubstituted **52** and **53** and β-fluorinated **58** and **59**, resulting in the construction of ion-based assemblies as the

Col_h and Col_r mesophases (Fig. 13.15a, ii,iii)[49f]. In addition to the modification of cations, anions can also be modified for benzoate derivatives by introducing alkyl chains. A combination of receptors **52** and **53** with alkoxy-substituted gallic carboxylates **66**$^-$–**68**$^-$ (Fig. 13.15b, i) as the TBA^+ salts formed lamellar mesophases supported by charge-by-charge assemblies (Fig. 13.15b, ii,iii). These ion-pairing materials showed fairly high charge carrier transporting properties, as shown by 0.05 ± 0.01 and 0.22 ± 0.03 cm^2 V^{-1} s^{-1} for **53**·Cl^--**64**$^+$ and **58**·Cl^--**63**$^+$, respectively, whereas **53**·**66**$^-$-TBA^+ exhibited a mobility of 0.04 cm^2 V^{-1} s^{-1} at 70°C.

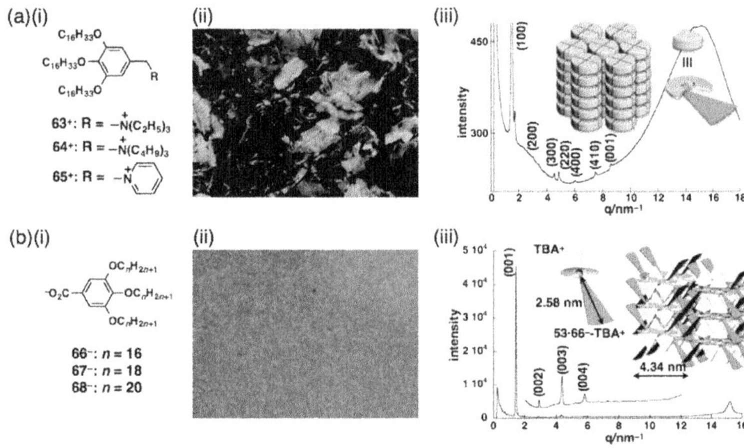

Figure 13.15 (a) (i) Modified cations **63**$^+$–**65**$^+$, (ii) POM texture of **53**·Cl^--**64**$^+$ at 125°C (second heating), and (iii) synchrotron XRD pattern of **53**·Cl^--**64**$^+$ at 90°C (second heating); (b) (i) modified anions **66**$^-$–**68**$^-$, (ii) POM texture of **53**·**66**$^-$-TBA^+ at 80.5°C upon cooling from Iso, and (iii) synchrotron XRD pattern of **53**·**66**$^-$-TBA^+ as the solid state at room temperature.

13.5 Summary

π-Electronic charged species exhibited fascinating nanoscale morphologies and liquid crystalline states based on the characteristic assembling modes because of electrostatic and various weak intermolecular interactions. In this decade, diverse ionic species, including planar receptor–ion complexes, have been used as the components of ion-pairing assemblies. Furthermore, π-electronic ion-pairing materials showed promising properties such as ion conductivities and charge carrier transport behaviors. Further design and synthesis of π-electronic ionic species and ion-responsive π-electronic molecules[51–56] would provide fascinating ion-pairing materials and devices, as the combination of oppositely charged species is also an important factor.

References

1. (a) Tsoucaris, G., ed. *Current Challenges on Large Supramolecular Assemblies, NATO Science Series*, Kluwer (1999); (b) Würthner, F., ed. *Supramolecular Dye Chemistry, Topics in Current Chemistry*, Springer, **258**, 1–324 (2005); (c) Atwood, J. L., Steed, J. W., eds. *Organic Nanostructures*, Wiley (2007); (d) Gale, P. A., Steed, J. W., eds. *Supramolecular Chemistry: From Molecules to Nanomaterials*, John Wiley & Sons (2012).
2. Hamley, I. W. *Introduction to Soft Matter: Polymers, Colloids, Amphiphiles and Liquid Crystals*, John Wiley & Sons (2000).
3. (a) Fages, F., ed. *Low Molecular Mass Gelators, Topics in Current Chemistry*, Springer, **256**, 1–283 (2005); (b) Babu, S. S., Praveen, V. K., Ajayaghosh, A. *Chem. Rev.*, **114**, 1973–2129 (2014).
4. (a) Kato, T., Mizoshita, N., Kishimoto, K. *Angew. Chem. Int. Ed.*, **45**, 38–68 (2006); (b) Kato, T., ed. *Liquid Crystalline Functional Assemblies and Their Supramolecular Structures, Structure and Bonding*, Springer, **128**, 1–237 (2008); (c) Kato, T., Yasuda, T., Kamikawa, Y., Yoshio, M. *Chem. Commun.*, 729–739 (2009); (d) Kaafarani, B. R. *Chem. Mater.*, **23**, 378–396 (2011).
5. Hoeben, F. J. M., Jonkheijm, P., Meijer, E. W., Schenning, A. P. H. J. *Chem. Rev.*, **105**, 1491–1546 (2005).
6. Ryu, J.-H., Hong, D.-J., Lee, M. *Chem. Commun.*, 1043–1054 (2008).
7. Ajayaghosh, A., Praveen, V. K. *Acc. Chem. Res.*, **40**, 644–656 (2007).
8. Yoshimoto, S., Kobayashi, N., in *Functional Phthalocyanine Molecular Materials*, Jiang, J., ed., Springer, **135**, 137–167 (2010).
9. Würthner, F., Saha-Möller, C. R., Fimmel, B., Ogi, S., Leowanawat, P., Schmidt, D. *Chem. Rev.*, **116**, 962–1052 (2016).
10. Wu, J., Pisula, W., Müllen, K. *Chem. Rev.*, **107**, 718–747 (2007).
11. Kastler, M., Pisula, W., Wasserfallen, D., Pakula, T., Müllen, K. *J. Am. Chem. Soc.*, **127**, 4286–4296 (2005).

12. Sergeyev, S., Pisula, W., Geerts, Y. H. *Chem. Soc. Rev.*, **36**, 1902–1929 (2007).
13. Hill, J. P., Jin, W., Kosaka, A., Fukushima, T., Ichihara, H., Shimomura, T., Ito, K., Hashizume, T., Ishii, N., Aida, T. *Science*, **304**, 1481–1483 (2004).
14. Jin, W., Yamamoto, Y., Fukushima, T., Ishii, N., Kim, J., Kato, K., Takata, M., Aida, T. *J. Am. Chem. Soc.*, **130**, 9434–9440 (2008).
15. (a) Faul, C. F. J., Antonietti, M. *Adv. Mater.*, **15**, 673–683 (2003); (b) Faul, C. F. J. *Acc. Chem. Res.*, **47**, 3428–3438 (2014).
16. Kim, F. S., Ren, G., Jenekhe, S. A. *Chem. Mater.*, **23**, 682–732 (2011).
17. Hamaoui, B. E., Zhi, L., Pisula, W., Kolb, U., Wu, J., Müllen, K. *Chem. Commun.*, 2384–2386 (2007).
18. (a) Po, C., Tam, A. Y. Y., Wong, K. M. C., Yam, V. W. W. *J. Am. Chem. Soc.*, **133**, 12136–12143 (2011); (b) Po, C., Tam, A. Y. Y., Yam, V. W. W. *Chem. Sci.*, **5**, 2688–2695 (2014).
19. Zana, R., ed. *Dynamics of Surfactant Self-Assemblies: Micelles, Microemulsions, Vesicles, and Lyotropic Phases, Surfactant Science Series*, CRC Press, **125**, 1–536 (2005).
20. Ren, Y., Kan, W. H., Thangadurai, V., Baumgartner, T. *Angew. Chem. Int. Ed.*, **51**, 3964–3968 (2012).
21. Wu, D., Liu, R., Pisula, W., Feng, X., Müllen, K. *Angew. Chem. Int. Ed.*, **50**, 2791–2794 (2011).
22. Wang, Z., Medforth, C. J., Shelnutt, J. A. *J. Am. Chem. Soc.*, **126**, 15954–15955 (2004).
23. (a) Eisele, D. M., Knoester, J., Kirstein, S., Rabe, J. P., Bout, D. A. V. *Nat. Nanotechnol.*, **4**, 658–663 (2009); (b) Eisele, D. M., Cone, C. W., Bloemsma, E. A., Vlaming, S. M., van der Kwaak, C. G. F., Silbey, R. J., Bawendi, M. G., Knoester, J., Rabe, J. P., Vanden Bout, D. A. *Nat. Chem.*, **4**, 655–662 (2012).
24. von Berlepsch, H., Kirstein, S., Hania, R., Pugžlys A., Böttcher, C. *J. Phys. Chem. B*, **111**, 1701–1711 (2007).
25. Wu, D., Zhi, L., Bodwell, G. J., Cui, G., Tsao, N., Müllen, K. *Angew. Chem. Int. Ed.*, **46**, 5417–5420 (2007).
26. Shi, D., Schwall, C., Sfintes, G., Thyrhaug, E., Hammershøj, P., Cárdenas, M., Simonsen, J. B., Laursen, B. W. *Chem. Eur. J.*, **20**, 6853–6856 (2014).
27. (a) Binnemans, K. *Chem. Rev.*, **105**, 4148–4202 (2005); (b) Axenov, K. V., Laschat, S. *Materials*, **4**, 206–259 (2011).
28. Yoshio, M., Mukai, T., Ohno, H., Kato, T. *J. Am. Chem. Soc.*, **126**, 994–995 (2004).

29. Ichikawa, T., Yoshio, M., Hamasaki, A., Mukai, T., Ohno, H., Kato, T. *J. Am. Chem. Soc.*, **129**, 10662–10663 (2007).

30. Ichikawa, T., Yoshio, M., Hamasaki, A., Taguchi, S., Liu, F., Zeng, X. Ungar, G., Ohno, H., Kato, T. *J. Am. Chem. Soc.*, **134**, 2634–2643 (2012).

31. Shimura, H., Yoshio, M., Hoshino, K., Mukai, T., Ohno, H., Kato, T. *J. Am. Chem. Soc.*, **130**, 1759–1765 (2008).

32. Motoyanagi, J., Fukushima, T., Aida, T. *Chem. Commun.*, 101–103 (2005).

33. (a) Alam, M. A., Motoyanagi, J., Yamamoto, Y., Fukushima, T., Kim, J., Kato, K., Takata, M., Saeki, A., Seki, S., Tagawa, S., Aida, T. *J. Am. Chem. Soc.*, **131**, 17722–17723 (2009); (b) Lee, J. J., Yamaguchi, A., Alam, M. A., Yamamoto, Y., Fukushima, T., Kato, K., Takata, M., Fujita, N., Aida, T. *Angew. Chem. Int. Ed.*, **51**, 8490–8494 (2012).

34. Tanabe, K., Yasuda, T., Kato, T. *Chem. Lett.*, **37**, 1208–1209 (2008).

35. (a) Olivier, J.-H., Camerel, F., Barberá, J., Retailleau, P., Ziessel, R. *Chem. Eur. J.*, **15**, 8163–8174 (2009); (b) Olivier, J.-H., Camerel, F., Ulrich, G., Barberá, J., Ziessel, R. *Chem. Eur. J.*, **16**, 7134–7142 (2010); (c) Olivier, J.-H., Barberá, J., Bahaidarah, E., Harriman, A., Ziessel, R. *J. Am. Chem. Soc.*, **134**, 6100–6103 (2012); (d) Camerel, F., Ulrich, G., Barberá, J., Ziessel, R. *Chem. Eur. J.*, **13**, 2189–2200 (2007).

36. Wu, D., Pisula, W., Enkelmann, V., Feng, X., Müllen, K. *J. Am. Chem. Soc.*, **131**, 9620–9621 (2009).

37. (a) Kuhn, R., Rewicki, D. *Angew. Chem. Int. Ed. Engl.*, **6**, 635–636 (1967); (b) Okamoto, K., Kitagawa, T., Takeuchi, K., Komatsu, K., Kinoshita, T., Aonuma, S., Nagai, M., Miyabo, A. *J. Org. Chem.*, **55**, 996–1002 (1990).

38. Le Goff, E., LaCount, R. B. *J. Am. Chem. Soc.*, **85**, 1354–1355 (1963).

39. (a) Bruce, M. I., Humphrey, P. A., Skelton, B. W., White, A. H. *Aust. J. Chem.*, **37**, 2441–2446 (1984); (b) Bruce, M. I., Humphrey, P. A., Skelton, B. W., White, A. H. *Aust. J. Chem.*, **39**, 165–169 (1986); (c) Richardson, C., Reed, C. A. *Chem. Commun.*, 706–707 (2004); (d) Less, R. J., McPartlin, M., Rawson, J. M., Wood, P. T., Wright, D. S. *Chem. Eur. J.*, **16**, 13723–13728 (2010).

40. Li, Y., Flood, A. H. *Angew. Chem. Int. Ed.*, **47**, 2649–2652 (2008).

41. Juwarker, H., Lenhardt, J. M., Pham, D. M., Craig, S. L. *Angew. Chem. Int. Ed.*, **47**, 3740–3743 (2008).

42. (a) Maeda, H., Shirai, T., Uemura, S. *Chem. Commun.*, **49**, 5310–5312 (2013); (b) Hirsch, B. E., McDonald, K. P., Qiao, B., Flood, A. H., Tait, S. L. *ACS Nano*, **8**, 10858–10869 (2014).

43. (a) Maeda, H., Kusunose, Y. *Chem. Eur. J.*, **11**, 5661–5666 (2005).

44. (a) Dong, B., Maeda, H. *Chem. Commun.*, **49**, 4085–4099 (2013); (b) Maeda, H. *Bull. Chem. Soc. Jpn.*, **86**, 1359–1399 (2013); (c) Haketa, Y., Maeda, H. *Chem. Commun.*, **53**, in press (DOI:10.1039/c6cc10255g) (2017).

45. (a) Maeda, H., Haketa, Y., Nakanishi, T. *J. Am. Chem. Soc.*, **129**, 13661–13674 (2007); (b) Maeda, H., Eifuku, N. *Chem. Lett.*, **38**, 208–209 (2009); (c) Maeda, H., Ito, Y., Haketa, Y., Eifuku, N., Lee, E., Lee, M., Hashishin, T., Kaneko, K. *Chem. Eur. J.*, **15**, 3706–3719 (2009); (d) Maeda, H., Terashima, Y., Haketa, Y., Asano, A., Honsho, Y., Seki, S., Shimizu, M., Mukai, H., Ohta, K. *Chem. Commun.*, **46**, 4559–4561 (2010); (e) Maeda, H., Eifuku, N., Haketa, Y., Lee, E., Lee, M. *Phys. Chem. Chem. Phys.*, **13**, 3843–3850 (2011); (f) Maeda, H., Terashima, Y. *Chem. Commun.*, **47**, 7620–7622 (2011); (g) Maeda, H., Kinoshita, K., Naritani, K., Bando, Y. *Chem. Commun.*, **47**, 8241–8243 (2011); (h) Terashima, Y., Sakurai, T., Bando, Y., Seki, S., Maeda, H. *Chem. Mater.*, **25**, 2656–2662 (2013); (i) Haketa, Y., Katayama, D., Fukunaga, S., Bando, Y., Sakurai, T., Matsuda, W., Seki, S., Maeda, H. *Chem. Asian J.*, **11**, 2025–2029 (2016).

46. (a) Maeda, H., Ito, Y. *Inorg. Chem.*, **45**, 8205–8210 (2006); (b) Maeda, H., Kusunose, Y., Mihashi, Y., Mizoguchi, T. *J. Org. Chem.*, **72**, 2612–2616 (2007); (c) Maeda, H., Haketa, Y., Bando, Y., Sakamoto, S. *Synth. Met.*, **159**, 792–796 (2009); (d) Maeda, H., Bando, Y., Haketa, Y., Honsho, Y., Seki, S., Nakajima, H., Tohnai, N. *Chem. Eur. J.*, **16**, 10994–11002 (2010); (e) Bando, Y., Sakamoto, S., Yamada, I., Haketa, Y., Maeda, H. *Chem. Commun.*, **48**, 2301–2303 (2012); (f) Bando, Y., Sakurai, T., Seki, S., Maeda, H. *Chem. Asian J.*, **8**, 2088–2095 (2013).

47. (a) Maeda, H., Haketa, Y. *Org. Biomol. Chem.*, **6**, 3091–3095 (2008); (b) Maeda, H., Mihashi, Y., Haketa, Y. *Org. Lett.*, **10**, 3179–3182 (2008); (c) Maeda, H., Fujii, R., Haketa, Y. *Eur. J. Org. Chem.*, 1469–1482 (2010); (d) Haketa, Y., Maeda, H. *Chem. Eur. J.*, **17**, 1485–1492 (2011); (e) Maeda, H., Kitaguchi, K., Haketa, Y. *Chem. Commun.*, **47**, 9342–9344 (2011); (f) Haketa, Y., Sakamoto, S., Chigusa, K., Nakanishi, T., Maeda, H. *J. Org. Chem.*, **76**, 5177–5184 (2011); (g) Haketa, Y., Bando, Y., Takaishi, K., Uchiyama, M., Muranaka, A., Naito, M., Shibaguchi, H., Kawai, T., Maeda, H. *Angew. Chem. Int. Ed.*, **51**, 7967–7971 (2012); (h) Maeda, H., Shirai, T., Bando, Y., Takaishi, K., Uchiyama, M., Muranaka, A., Kawai, T., Naito, M. *Org. Lett.*, **15**, 6006–6009 (2013).

48. (a) Maeda, H., Fujii, Y., Mihashi, Y. *Chem. Commun.*, 4285–4287 (2008); (b) Maeda, H., Takayama, M., Kobayashi, K., Shinmori, H. *Org. Biomol. Chem.*, **8**, 4308–4315 (2010); (c) Maeda, H., Bando, Y., Shimomura, K., Yamada, I., Naito, M., Nobusawa, K., Tsumatori, H., Kawai, T. *J. Am.

Chem. Soc., **133**, 9266–9269 (2011); (d) Terashima, Y., Takayama, M., Isozaki, K., Maeda, H. *Chem. Commun.*, **49**, 2506–2508 (2013).

49. (a) Haketa, Y., Sasaki, S., Ohta, N., Masunaga, H., Ogawa, H., Mizuno, N., Araoka, F., Takezoe, H., Maeda, H. *Angew. Chem. Int. Ed.*, **49**, 10079–10083 (2010); (b) Maeda, H., Naritani, K., Honsho, Y., Seki, S. *J. Am. Chem. Soc.*, **133**, 8896–8899 (2011); (c) Dong, B., Terashima, Y., Haketa, Y., Maeda, H. *Chem. Eur. J.*, **18**, 3460–3463 (2012); (d) Haketa, Y., Takayama, M., Maeda, H. *Org. Biomol. Chem.*, **10**, 2603–2606 (2012); (e) Haketa, Y., Honsho, Y., Seki, S., Maeda, H. *Chem. Eur. J.*, **18**, 7016–7020 (2012); (f) Dong, B., Sakurai, T., Honsho, Y., Seki, S., Maeda, H. *J. Am. Chem. Soc.*, **135**, 1284–1287 (2013); (g) Maeda, H., Hane, W., Bando, Y., Terashima, Y., Haketa, Y., Shibaguchi, H., Kawai, T., Naito, M., Takaishi, K., Uchiyama, M., Muranaka, A. *Chem. Eur. J.*, **19**, 16263–16271 (2013); (h) Dong, B., Sakurai, T., Bando, Y., Seki, S., Takaishi, K., Uchiyama, M., Muranaka, A., Maeda, H. *J. Am. Chem. Soc.*, **135**, 14797–14805 (2013); (i) Sekiya, R., Tsutsui, Y., Choi, W., Sakurai, T., Seki, S., Bando, Y., Maeda, H. *Chem. Commun.*, **50**, 10615–10618 (2014).

50. (a) Laursen, B. W., Krebs, F. C. *Angew. Chem. Int. Ed.*, **39**, 3432–3434 (2000); (b) Laursen, B. W., Krebs, F. C. *Chem. Eur. J.*, **7**, 1773–1783 (2001). See also ref. 49a for the preparation of $(TATA^{C3})^{+}\cdot Cl^{-}$.

51. (a) Yamakado, R., Sakurai, T., Matsuda, W., Seki, S., Yasuda, N., Akine, S., Maeda, H. *Chem. Eur. J.*, **22**, 626–638 (2016); (b) Yamakado, R., Maeda, H. *J. Photochem. Photobiol. A*, **331**, 215–223 (2016); (c) Yamakado, R., Sato, R., Shigeta, Y., Maeda, H. *J. Org. Chem.*, **81**, 8530–8536 (2016); (d) Yamakado, R., Ashida, Y., Sato, R., Shigeta, Y., Yasuda, N., Maeda, H. *Chem. Eur. J.*, **23**, in press (DOI:10.1002/chem.201605765) (2017).

52. (a) Maeda, H., Fukui, A., Yamakado, R., Yasuda, N. *Chem. Commun.*, **51**, 17572–17575 (2015); (b) Sasano, Y., Yasuda, N., Maeda, H. *Chem. Asian J.*, **11**, 3423–3429 (2016).

53. Bando, Y., Haketa, Y., Sakurai, T., Matsuda, W., Seki, S., Takaya, H., Maeda, H. *Chem. Eur. J.*, **22**, 7843–7850 (2016).

54. (a) Maeda, H., Ito, Y., Kusunose, Y., Nakanishi, T. *Chem. Commun.*, 1136–1138 (2007); (b) Maeda, H., Chigusa, K., Sakurai, T., Ohta, K., Uemura, S., Seki, S. *Chem. Eur. J.*, **19**, 9224–9233 (2013); (c) Maeda, H., Chigusa, K., Yamakado, R., Sakurai, T., Seki, S. *Chem. Eur. J.*, **21**, 9520–9527 (2015).

55. (a) Nakamura, K., Yasuda, N., Maeda, H. *Chem. Commun.*, **52**, 7157–7160 (2016); (b) Haketa, Y., Takasago, R., Maeda, H. *Chem. Commun.*, **52**, 7364–7367 (2016); (c) Haketa, Y., Tamura, Y., Yasuda, N., Maeda, H. *Org. Biomol. Chem.*, **14**, 8035–8038 (2016).

56. Naito, W., Yasuda, N., Morimoto, T., Shigeta, Y., Takaya, H., Hisaki, I., Maeda, H. *Org. Lett.*, **18**, 3006–3009 (2016).

Biography

Yohei Haketa (center): Yohei Haketa was born in 1984 in Ueda (Nagano), Japan. He received his PhD in 2011 from Ritsumeikan University, under the guidance of Prof. Hiromitsu Maeda, focusing on the assemblies of π-conjugated anion receptors. He was selected as a research fellow of the Japan Society for the Promotion of Science (JSPS) in 2009–2012. In 2012–2015, he worked in Asahi Kasei Corporation as a researcher. In 2015, he joined the group of Prof. Hiromitsu Maeda at Ritsumeikan University as a postdoctoral fellow and started an academic career as assistant professor from April 2017, focusing on the dimension-controlled assemblies of π-electronic ionic species.

Ryohei Yamakado (left): Ryohei Yamakado was born in 1986 in Nagoya (Aichi), Japan. He received his PhD in 2014 from Nagoya Institute of Technology, under the guidance of Prof. Koji Takagi and Prof. Masato Suzuki, focusing on the programmable assembly of π-conjugated systems utilizing the stereochemistry of aromatic tertiary amide units. During the spring of 2013, as a visiting fellow, he worked in the group of Prof. Kenneth B. Wagener, Department of Chemistry, the University of Florida, USA. In 2014, he joined the group of Prof. Hiromitsu Maeda, Ritsumeikan University, as a postdoctoral fellow, focusing on the preparation of novel π-electronic ionic species.

Hiromitsu Maeda (right): Hiromitsu Maeda was born in 1976 in Kadoma (Osaka), Japan. He received his PhD in 2004 from Kyoto University, under the guidance of Prof. Hiroyuki Furuta (Kyushu University) and Prof. Atsuhiro Osuka. He was selected as a research fellow of the Japan Society for the Promotion of Science (JSPS) in

2001–2004. During the autumn of 2001, he worked as a visiting fellow in the group of Prof. Jonathan L. Sessler, the University of Texas at Austin, USA. In 2004, he started an academic career as a lecturer in the Department of Bioscience and Biotechnology, Faculty of Science and Engineering, Ritsumeikan University, and in 2005 he became an associate professor. In 2008, he was transferred to the College of Pharmaceutical Sciences of the same university, wherein he was promoted to a professor in 2014. In April 2016, he moved (practically came back) to Department of Applied Chemistry, College of Life Sciences, Ritsumeikan University. In 2006–2008, he was also a visiting associate professor in the Institute for Molecular Science, and in 2007–2011, he was a researcher of PRESTO (Structure Control and Function), JST.

Chapter 14

Theory Study and Present CT Trend of Organic Charge Transfer Complexes

Atsushi Nagai[a,b] and Daniel J. Siegwart[b]
[a]*Department of Materials Molecular Science, National Institute for Natural Science, Institute for Molecular Science (IMS), Higashiyama, Okazaki, Aichi 447-8787, Japan*
[b]*Department of Biochemistry, University of Texas Southwestern (UTSW) Medical Center, 6001 Forest Park Road, ND2-504E, Dallas, TX 75390-8807, USA*
nagai@ims.ac.jp

Novel charge transfer (CT) complexes based on organic molecules with π-conjugated ring structures gain high attention because they offer the possibility to tailor the electronic structure of organic electronic devices. In this chapter, how to prepare and observe organic CT complexes and studies and applications of the use of organic CT interactions between donors and acceptors are reviewed in detail.

14.1 Introduction

Approaches to charge transfer (CT) complexes lend themselves to many interesting tests. Novel CT compounds based on organic

Conjugated Objects: Developments, Synthesis, and Applications
Edited by Atsushi Nagai and Koji Takagi

ISBN 978-981-4774-03-1 (Hardcover), 978-1-315-22962-1 (eBook)
www.panstanford.com

molecules with π-conjugated ring objects attract high attention because they offer the possibility to tailor the electronic structure of organic electronic devices.[1] The wave functions of electronic π-systems of the donor and acceptor constituents can be sensitively tuned via a change of the change density by adding specific ligand groups at the periphery of the molecule. Characteristic of such compounds is a highly correlated electronic ground state that can show superconductivity, spin density waves, or charge density waves.[2] Thus, this class of compounds is highly interesting not only because of potential applications but also for fundamental research. Growing crystals in a solvent, however, is prohibitive for the standard techniques of electronic structure investigations because of the unavolidable surface contaminations.

Typically, the donor and acceptor molecules are mixed, although no chemical reaction occurs; an absorption band, not present in the spectra of either of the pure compound, may appear in UV/vis region. On the other hand, the formation of a CT complex enables us visually to change its characteristic color. For example, solutions of iodine in benzene have a very strong absorption band at 297 nm. This absorption band has been interpreted as the result of a molecule of benzene acting as an electron donor and interesting strong with the acceptor, iodine. In this particular case, the electronic excited state can be viewed as a pair $C_6H_6^+$ and I_2^- produced by a π-electron transfer from benzene to iodine.

Of course, this phenomenon is not general and special energetic conditions have to be fulfilled for this type of interaction to occur between a pair of the molecules. Using quantum mechanical language, for strong CT interaction, the donor and acceptor molecular orbitals must be very close in energy and these must overlap a large orbital; in other words, the geometry of the two partners must allow for such an overlap. Typically, these conditions are fulfilled when high-energy orbitals from a donor interest with a low-energy empty orbital from an acceptor.

The experimental quantity easily obtained is the frequency of the CT absorption band, V_{CT}. One finds that V_{CT} is correlated to electron affinity of the accepotor molecule. E_A and to inonization potential of donor molecule, I_b. Furthermore, one looks at a series of molecular complexes, that is, varying I_b, for which the acceptor is same, that is, E_A content. V_{CT} and ΔG^0 show each a simple dependence on I_b.

14.2 Theory of the Charge Transfer Complex

The main points are presented here.[3] However, a complete grasp of the derivation is not essential in interpreting the results which can be understood with semi-intuitive arguments. Mulliken's theory attempts to deal with the formation of molecular complexes both inorganic or organic; in general, such species would be the result of interaction between a Lewis acid A (electron acceptor) and a Lewis base (electron donor). The formation of the weak complex AB,

$$A + B \rightarrow AB \tag{14.1}$$

is due to attractive dipole and polarization forces between the two molecules. Note that, in principle, it is possible that higher-order complexes be formed (more than two moleules), but only 1:1 complexes will be considered here. Also, it will be assumed that all the valance electrons are paired up in A and B and therefore not available to form a covalent bond between A and B.

Let ψ (A, B) be the wave function for the molecular structure AB arising from the weak dipolarization interactions. The nature of partners in the formation of the complexes (electron donor/acceptor) give rise to an ionic contribution in the bonding which can be viewed as A^{-}–B^{+} ; Let $\psi(A^{-} B^{+})$ be the wave function for this resonance structure*. Using the above two resonace structure as a basis set, the wave function corresponding to the ground state of the complex is expressed as

$$|\psi_G\rangle = a|\ \psi(A, B)\rangle + b|\ \psi(A^{-} B^{+})\rangle \tag{14.2}$$

Using standard quantum mechanical formalism, the ground state energy E_G is written as

$$E_G = \langle\psi_G|\ H\ |\psi_G\rangle \tag{14.3}$$

where H is the exact Hamiltonian operator including nuclei and electrons for both molecules for the complex and with the extra condition that ψ_G be normalized, that is,

$$\langle\psi_G|\ H\ |\psi_G\rangle = 1 = a^2 + b^2 + 2abS \tag{14.4}$$

where S, the overlap integral, is

$$S = \langle\psi(A, B)\ |\ \psi(A^{-} B^{+})\rangle \tag{14.5}$$

A simple substitution of Eqs. 14.2 and 14.4 into Eq. 14.3 leads to one expression for E_G:

$$E_G = \frac{a^2E_0 + 2abH_{01} + b^2 E_1}{a^2 + 2abS + b^2} \tag{14.6}$$

where E_0 and E_1 are the energies corresponding to the pure states AB and A^-–B^+ respectively and H_{01} the mixing energy:

$$E_0 = <\psi(A, B) \mid H \mid \psi(A, B)> \tag{14.7}$$

$$E_1 = <\psi(A^- B^+) \mid H \mid \psi(A^- B^+)> \tag{14.8}$$

$$H_{01} = <\psi(A, B) \mid H \mid \psi(A^-B^+)> \tag{14.9}$$

Now the aim is to express E_G and the parameters a and b as a function of these three energies, which in principle can be easily calculated since they correspond to the ideal with respect to variables a and b. Equation 14.6 must satisfy the conditions

$$\delta E_G/\delta a = \delta E_G/\delta_b = 0 \tag{14.10}$$

which results in having to solve the following equations:

$$a(E_0 - E_G) + b(H_{01} - E_GS) = 0 \tag{14.11a}$$

$$a(H_{01} - E_GS) + b(E_1 - E_G) = 0 \tag{14.11b}$$

These two simultaneous linear equations can have solutions (other than the trivial solution $a = b = 0$) only if the associated determinant is zero, that is,

$$(E_0 - E_G)(E_1 - E_G) - (H_{01} - E_GS)^2 = 0 \tag{14.12}$$

At this stage, one should notice that the resonance form AB is the major contribution to the ground state of the molecular complex, which implies that $\psi_G \approx \psi(A, B)$ (i.e., $|a| >> |b|$ in Eq. 14.2) and $E_G \approx E_0$. Then, by substituting E_0 for E_G (except in $E_0 - E_G$) one obtains an approximate expression for E_G:

$$E_G \approx E_0 - (H_{01} - E_0S)^2/(E_1 - E_0) \tag{14.13}$$

and from Eq. 14.11, a relationship between a and b:

$$b/a \approx -(H_{01} - E_0S)/(E_1 - E_0) \tag{14.14}$$

Now for the electronic excited state whose energy E^* will be approximately E_1 and which represents mostly the dative resonance from A^- – B^+, using the same basis set as before, the corresponding wave function is written as

$$|\psi^*> = a^* \mid \psi(A, B)> + b^* \mid \psi(A^- B^+)> \tag{14.15}$$

where $\psi^* \approx \psi(A^- B^+)$ (or $|a^*| << |b^*|$). The same derivation as for the ground state can be applied to the excited state and one gets readily

$$E^* \approx E_1 + (H_{01} - E_1S)^2/(E_1 - E_0) \quad (14.16)$$

and

$$a^*/b^* \approx (H_{01} - E_1S)/(E_1 - E_0) \quad (14.17)$$

Comparing Eqs. 14.14 and 14.17, and providing that the value of the overlap S is small, the following is sure:

$$b/a \approx -a^*/b^* \quad (14.18)$$

This last equation verifies that the assumptions made lead to internally consistent results; the ground state wave function has mostly a $\psi(A, B)$ character, while in the excited state, the dominant term is $\psi(A^- B^+)$. In this context, it is justified to regard the transion from the ground state to the excited state as a transfer of an electron from B to A, resulting in a CT spectrum. The energy of the transition is given by (using cm^{-1} units)

$$\nu_{CT} = 1/hc\,(E^* - E_G) \approx 1/hc\,(E_1 - E_0 + \delta) \quad (14.19)$$

where the small correction term δ can be explicitly written (see Eqs. 14.13 and 14.16):

$$\delta = \frac{(H_{01} - E_1S)^2 + (H_{01} - E_0S)^2}{E_1 - E_0} \quad (14.20)$$

To introduce into Eq. 14.19 a molecular parameter readily available, it is convenient to break down the CT process into the following elementary steps:

AB	$\rightarrow$	A + B	$-E_0$	Dissociation of complex
$A + e^-$	$\rightarrow$	A^-	$-E_A$	Electron affinity of the Lewis acid
B	$\rightarrow$	$B^+ + e^-$	I_b	Ionization potential of the Lewis base
$A^- + B^+$	$\rightarrow$	$A^- B^+$	$-e^2/r_{AB} + C'_{AB}$	Coulomb interaction + other
A, B	$\rightarrow$	$A^- B^+$	$(E_1 - E_0)$	Charge transfer energy

Therefore

$$(E_1 - E_0) \approx I_b - E_A - e^2/r_{AB} + C_{AB} \quad (14.21)$$

where $C_{AB} = C'_{AB} - E_0$ represents the difference between all additional energy terms in the dative and no-bond resonance forms. Equations 14.19 and 14.21 allow us to relate the experimentally observable transition energy to the ionization potential of the electron donor

$$\nu_{CT} \approx 1/hc\,(I_b - P) \tag{14.22}$$

where $P = E_A + e^2 / r_{AB} - C_{AB}$ is nearly constant for a series of complexes having a common acceptor since E_A is constant, while the other two terms are not expected to change significantly, providing there are no unusual steric effects in the series.

For condensed systems like the ones under consideration, the energy terms can be identified with enthapy terms, which in turn, very linearly with standard free energy for the present set of complexes.[5] Going back to Eqs. 14.13 and 14.21 and identifying ΔG_0 with E_G one relates ΔG_0 to I_b according to

$$\Delta G_0 \approx M - Q/(I_p - P) \tag{14.23}$$

where M and Q are constants. The right-hand side of Eq. 14.23 represents a hyperbola which over a limited region is practically linear.

Finally, the standard free energy is related to the equilibrium content of complex (an experimentally accessible parameter) through the expression

$$\Delta G^0 = -RT\ln K \tag{14.24}$$

14.3 Practical Considerations of Charge Transfer Complexes

The energy of the CT band is given by Eq. 14.22 due to complex formation. The exact location of this band will depend obviously on I_b and E_A; in particular, if the electron acceptor has a low electron affinity while the electron donor has high ionization potential, the CT band may well be located into the UV region, where it would interfere with other absorption bands making it impossible to observe. For the series of complexes used in the present experiment, the CT band shows up in the visible and is easily identified to measure the corresponding absorbance with a spectrophotometer. At the desired wavelength, Beer–Lambert's law relates the concentration to the absorbance (or optical density, OD) at wavelength λ according to

$$\mathrm{OD}\lambda = \varepsilon_\lambda c l \tag{14.25}$$

where ε is the molar extinction coefficient at the particular wavelenth, c the concetration, and l the path length of light through the absorbing medium. For the present solution, at the maximum absorption in the CT band, Eq. 14.26 takes the form

$$OD_{max}/l = \varepsilon_A [A] + \varepsilon_B [B] + \varepsilon_{AB} [AB] \quad (14.26)$$

Let $[A]_0$ and $[B]_0$ be the known nominal concentrations of A and B when the mixture is prepared. One must have

$$[A]_0 = [A] + [AB] \quad (14.27)$$

and

$$[B]_0 = [B] + [AB] \quad (14.28)$$

In addtion, species A and B and the CT complex AB are in equilibrium with the constant

$$K = [AB] / [A][B] \quad (14.29)$$

in which concentrations are taken to be equal to activities.

Of all these quantities, OD_{max}, the absorbance of the CT complex solution, is measured experimentally and ε_A and ε_B can be obtained from the fact that in general one of the components is in large excess—say, B—such that $[B]_0 >> [A]_0$ and therefore $[B] \approx [B]_0$. Equation 14.25 can be solved for [AB] using this last assumption:

$$[AB] = K[A][B] \approx k ([A]_0 - [AB]) [B]_0 \quad (14.30)$$

and after rearrangement

$$[AB] \approx K[A]_0 [B]_0 / (1 + K[B]_0) \quad (14.31)$$

Substitution of Eq. 14.31 into Eq. 14.26, neglecting ε_A and ε_B and rearrangment leads to

$$OD_{max} / l[A]_0 = \varepsilon_{AB}K[B]_0 / (1 + K[B]_0) \quad (14.32)$$

Taking the reciprocal of both sides of Eq. 14.32, one gets an equation of the form

$$U = s / [B]_0 + i \quad (14.33a)$$

where $U = l[A]_0/OD_{max}$, with $s = 1/\varepsilon_{AB}K$ and $i = 1/\varepsilon_{AB}$. (14.33b)

Equation 14.33a shows that a plot of U versus $1 / [B]_0$ should be linear, where the slope is s and the interrept is i. Then from the definitions of s and i, the equilibrium constant and the molar extinction coefficient of the complex can be obtained:

$$K = i / s \text{ and } \varepsilon_{AB} = 1/i \quad (14.34)$$

14.4 Molecular Pair Enables Forming Organic CT Complexes

14.4.1 Account for Donor and Acceptor Chromophores

The present donor and acceptor chromophores were introduced in the context of their propensity to form CT complexes, enabling constructing supramolecular assemblies.[4] Recently, the large number of molecules reported so far makes it practically impossible to provide a comprehensive account on their structures and properties in this chapter. This section has chosen a few selected examples of representative π-donors (**1–12**) and π-acceptors (**13–24**) (Fig. 14.1). It is important to note that this list does not include either lone pair (n) donors, such as aliphatic amines, phosphines, and alcohols, or σ acceptors, such as I_2, Br_2, and ICN, in order of the concern with π-donors and π-acceptors.[5] Hexamethylbenzene (**1**), tetramethylbenzne (**2**), *N,N*-dimethylaniline (**3**), and many other structurally related benzene derivatives substituted with electron-donating groups were identified as donor units early on. Likewise 1,3,5-trinitrobenzene (**13**), tetracyanobenzene (**14**), and structurally similar molecules with electron-withdrawing groups act as acceptor counterparts of the formation of CT complexes. 2,3-dichloro-5,6-dicyano π-benzoquinone (**15**), tetrafluoro-*p*-benzoquinone (**16**), and other related compounds (quinone derivatives) have been also examined as acceptor units of CT interactions with a benzene ring containing donors.[5] Aromatic derivatives with larger π-conjugative surfaces, such as napthalene (**4**), dialkoxy-napthalene (**5**), anthracene (**6**), pyrene (**7**), carbazole (**8**), and triphenylene (**10**), can act as donor components in the formation of CT complexes with acceptors such as 2,4,7-trinitrofluorenone (**16**), pyromellitic diimide (**20**), naphthalene-diimide (**21**), mellitic triimide (**22**), and viologen (**18**) derivatives. Different donor–acceptor pairs have found widespread interest in the construction of a diverse range of supramolecular assemblies. Tetrathiafulvalene (TTA) (**9**) and its derivatives[6,7] show an important class of donors that can form efficient CT complexes with 7,7,8,8-tetracyano-*p*-quinodimethane (TCNQ, **17**), undergoing metallic behavior[8], and also with **18**, **20**, and **21** for the formation of interlocked compounds such as rotaxanes and catenanes.[9–12] A few recent examples suggest that C_3-symmetric donors such as hexakisalkoxytriphenlene (**11**) and truxene derivatives (**12**) can make a highly stable donor–acceptor pair with the complementary acceptors hexaazatriphenylene (**23**)[13] and truxenone derivatives (**24**)[14], respectively, owing to their almost identical shape and size.

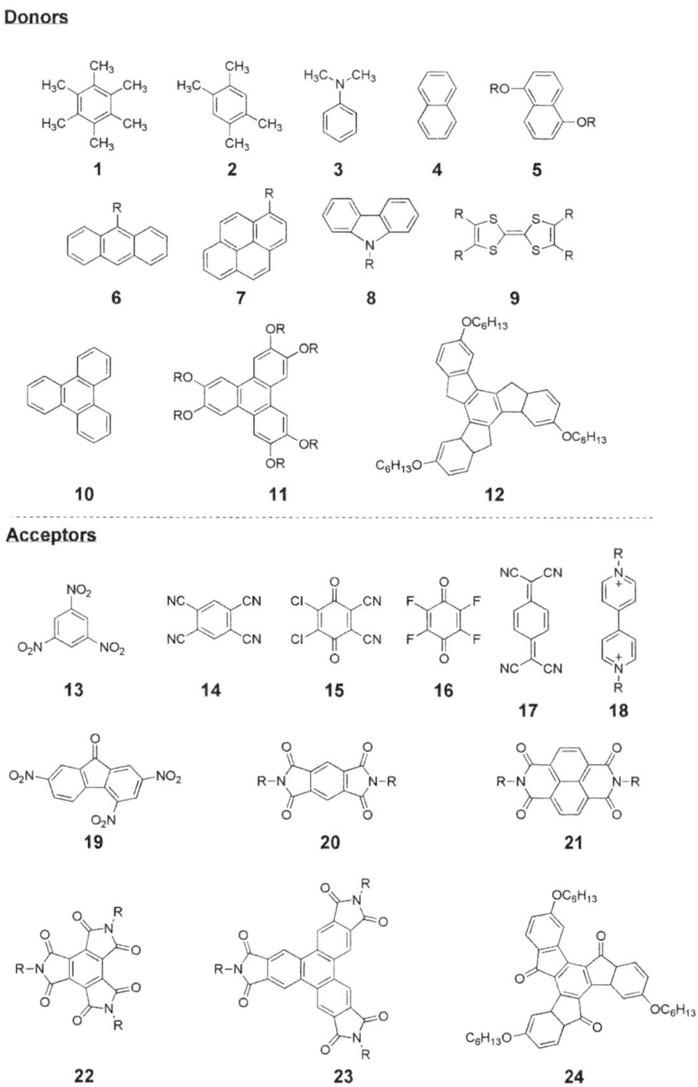

Figure 14.1 Structures of representative π-donors and π-acceptors.

14.4.2 Search for CT Complexes

In accordance with molecular orbitals, the CT band results in an electronic transition from the highest occupied molecular orbital (HOMO) of the donor to the lowest unoccupied molecular orbital (LUMO) of the acceptor. Formation of a CT complex can be readily appeared as a new single or multiple low-energy absorption bands, while the absorption peaks of the individual donor and acceptor components significantly remain. Increasing the concentration of one of the components often allows to visualize the CT band.[13,14] Therefore, CT complexes also exhibit a new emission band having a mirror-image relationship with the CT absorption spectra. They are also routinely accounted by nuclear magnetic resonance (NMR) techniques,[15] in which face-to-face alternate donor–acceptor stacking results in up-field shifts of the aromatic protons because of the shielding effect. In addition, 2D NMR spectroscopic tools have been used in the recent past to defect spatial communication among the protons from the alternately stacked donor and acceptor units. Recently, electronic structures of donor–acceptor assemblies have been examined by ultraviolet photoelectron spectroscopy.[16]

14.5 CT Trend Research

14.5.1 Development of CT Mechanochromism

Recently, organic complexes with electron donor (D) and electron acceptor (A) moieties have emerged as promising candidates for advanced mechanochromic materials due to their intramolecular CT capability. This endows the materials with charge ordering, spin density waves (SDWs), spin-Peierls ground state, and superconductivity.[17] Aromatic molecules, such as coronene, are typical examples of π-electron systems able to donate electrons, depending on their ionization potential I. As a typical example, 7,7,8,8-tetracyanoquinodimethane (TCNQ) can be easily reduced to form the open-shell electron radical anion $TCNQ^-$ when it comes into contact with electron donors. Recently, a three-component system comprising two different benzenoid hydrocarbons together with one molecule of TCNQ was reported.[18] The pyrene–TCNQ complex was synthesized with a molecular ratio of 1:1 in toluene under refluxing conditions. Notably, the compound crystallized in a monoclinic system, with a space group of $P2_1/n$ with half of the molecules of pyrene and TCNQ as the asymmetric part of the unit cell. The pyrene moiety is tilted with respect to the TCNQ along the stacks, with an angle between mean planes of 6° (Fig. 14.2). If this principle could be introduced into an oligomer ore polymer chain backbone, a force-sensitive CT complex as the chemical unit (mechanophore) would be linked into the polymer and oligomer backbones. Mechanical loading would then drive conversion of the mechanophore to a new chemical species. Therefore, we expect the possibility of completely alternating stacking for the CT complex between pyrene and TCNQ using mechanical grinding. This is to say, the mechanochromism[19] phenomenon will be observed, which is characterized by changes in color (absorption) and/or luminescence as a result of mechanical grinding or pressing of a sample in the solid state. Accordingly, the mechanical grinding of pyrene–TCNQ complexes in the solid state would lead to a much higher increase of high CT complex absorption and the color of the grinded solid will change. To create the system described above utilized the simple structure of squaramide, which is typically synthesized via the stepwise reaction of commercially available squaric acid with amines and aromatic amines.[20] The hydrogen bond donor–acceptor properties of squaramide have attracted interest for various applications in supramolecular chemistry, for example, as catalysts[21], molecular sensors,[22] and in molecular recognition as receptors for charged guests.[23] Moreover, there is a growing number of bioactive squaramides involved in cancerous,[24] inflammatory,[25] and infectious[26] diseases, among others.

Herein, Nagai et al. have reported the synthesis of π-conjugated squaric acid diamide-type oligomers composed of 1,3- or 1,6-diaminopyrenes and squaric acid with TCNQ (molecular ratio 1:1:1), oligo(sq-alt-1,3ampy) and oligo(sq-alt-1,6py). The obtained oligomers show bright orange colors in the solid state at room temperature. Upon mechanical grinding, oligo(sq-alt-1,6py) changes from orange to a metallic deep-green color. In contrast, oligo(sq-alt-1,3py) does not change color upon mechanical grinding, because of the inferior plane angle compared to mechanochromic oligo(sq-alt-1,6py).

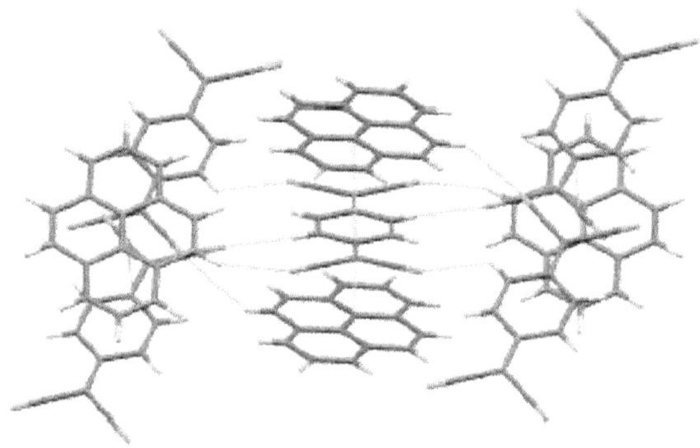

Figure 14.2 Short interaction in a crystal of pyrene-TCNQ (1:1).

Initially, the pyrene-conjugated squaramide oligomers for the ligand oligomer of the CT complex were synthesized under solvothermal conditions through the condensation reaction of squaric acid and 1,3- or 1,6-diaminopyrene in toluene:*n*-butanol (3:1 by volume) at 120°C for 2 days (Scheme 14.1). The resulting precipitates were collected by filtration, washed with methanol and acetone, and dried at 100°C under vacuum to provide oligo(sq-*alt*-1,3py) and oligo(sq-*alt*-1,6py) as bright oranges in 65% and 72%, respectively. When combination of *n*-butanol with other aromatic solvents, such as mesitylene and benzene, were used, the resulting solid showed lower yields and crystallinity.

The Fourier transform infrared (FTIR) spectra of oligo(sq-*alt*-1,3py) and oligo(sq-*alt*-1,6py) exhibited three vibration bands at 1503, 1523, and 1590 cm^{-1}, characteristic squamide of C=O bonds. Elemental analysis of oligo(sq-*alt*-1,3py) and oligo(sq-*alt*-1,6py) revealed that the H, C, and N contents for oligo(sq-*alt*-1,3py) were 3.34%, 69.51%, and 12.44%, respectively, and the H, C, and N contents for oligo(sq-*alt*-1,6py) were 3.22%, 69.01%, and 13.2%, respectively, which are close to the calculated values of 3.34%, 68.24%, and 13.26% expected for an infinite squaric acid 1,3- or 1,6-diamidepyrene-conjugated oligomers.

The UV-Vis absorption spectroscopy was used to evaluate the electornic properties of oligo(sq-*alt*-1,3py) and oligo(sq-*alt*-1,6py) relative to 1,3-diamiopyerene and 1,6-diaminopyrene (Fig. 14.3a).

Both oligo(sq-*alt*-1,3py) and oligo(sq-*alt*-1,6py) (red curves) exhibited respective strong absorption bands at 527 and 530 nm, which were red-shifted by over 107 and 87 nm relative to those of 1,3-diamiopyerene (black dashed curve) and 1,6-diaminopyrene (black curve), respectively. The significant bathochromic shifts and extremely broad absorption bands of oligo(sq-*alt*-1,3py) band oligo(sq-*alt*-1,6py) suggest delocalization of the π-electrons in the squaramide-pyrene oligomer backbone. Moreover, a slightly larger red shift of oligo(sq-*alt*-1,6py) than that of oligo(sq-*alt*-1,3py) means that a squaraine unit connected at the pyrene-1,6 position shows higher conjugation than that at the pyrene-1,3 position, that is, the connection of the 1,6-position in the pyrene unit leads to extend larger π-conjugation.

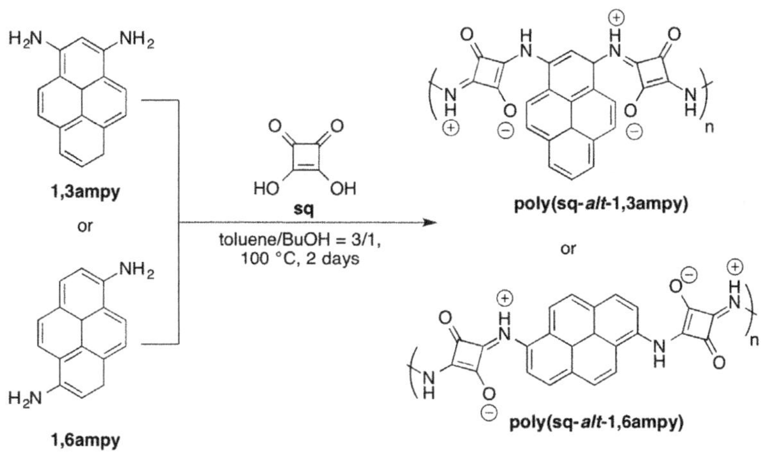

Scheme 14.1 Synthesis of poly(sq-*alt*-Xampy); X = 1,3: poly(sq-*alt*-1,3-ampy) and X = 1,6: poly(sq-*alt*-1,6ampy).

The powder X-ray diffraction (PXRD) measurements of oligo(sq-*alt*-1,3py) and oligo(sq-*alt*-1,6py) are shown in Fig. 14.3b. The PXRD pattern of oligo(sq-*alt*-1,3py) has considerably broaden weak peaks, which is a characteristic amorphous structure containing crystal objects. However, the XRD pattern of oligo(sq-*alt*-1,6py) displayed a strong, sharp peak at 7.839° (11.2693 Å) and four weak peaks at 8.961° (9.8609 Å), 13.538° (6.5351 Å), 24.444° (3.6386 Å), and 25.920° (3.4347 Å).

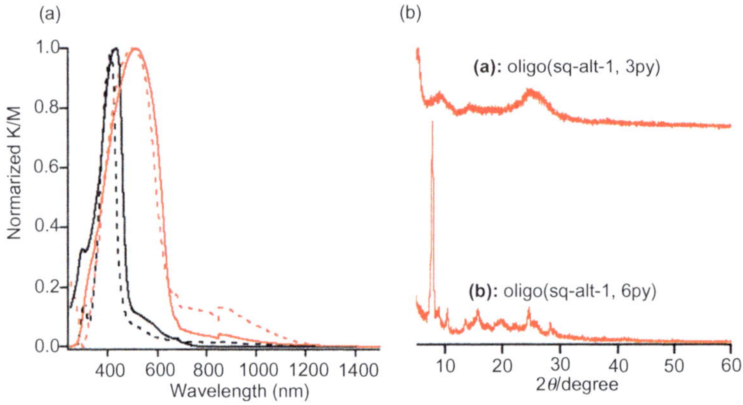

Figure 14.3 (a) Absorption spectra of 1,3-diaminopyrene (black dashed curve), 1,6-diaminopyrene (black solid curve), oligo(sq-*alt*-1,3py) (red dashed curve), and oligo(sq-*alt*-1,6py) (red solid curve).

To provide a more effective understanding for the structure and photophysical behavior of oligo(sq-*alt*-1,3py) and oligo(sq-*alt*-1,6py), we employed a theoretical calculation of model compounds using density functional theory (DFT) and time-dependent DFT (TD-DFT) at the B3LYP/6-31G(d)/B3LYP/6-311G(d) level. The 1,3-connected model compound has a torsional structure in the 1,3-position of pyrene. In the contrary, the 1,6-connected model compound shows high planarity, indicating that oligo(sq-*alt*-1,6py) possesses both higher crystallinity and larger extent of π-conjugation than oligo(sq-*alt*-1,3py) (Fig. 14.4). The LUMO is spread out down to the C=O and C–O$^-$ groups of the squaraine moiety in the 1,6-connected model compound, although the orbital is not spread on their groups of the 1,3-connected model compound. These result indicates that larger π-conjugation of oligo(sq-*alt*-1,6py) is extended than that of oligo(sq-*alt*-1,3py).

Next, we conducted the condensation reaction of squaric acid and 1,3- or 1,6-diaminopyrenes in the presence of TCNQ under same conditions (Scheme 14.2) to afford the TCNQ-coordinated oligomers, oligo(sq-*alt*-1,3py)-TCNQ and oligo(sq-*alt*-1,6py)-TCNQ. The obtained solids were afforded in 54% and 78% yields and were similar in color to those of oligo(sq-*alt*-1,3py) and oligo(sq-*alt*-

1,6py), that is, orange color. The structures of the obtained oligomers were determined by FTIR spectroscopy. Although oligo(sq-*alt*-1,3py)-TCNQ does not have the C=N band in the oligomer backbone, oligo(sq-*alt*-1,6py)-TCNQ showed a vibration band at 1789 cm^{-1}, assignable to TCNQ of a C=N bond. Furthermore, elemental analysis of oligo(sq-*alt*-1,6py)-TCNQ revealed that the H, C, and N contents were 3.91%, 71.84%, and 11.45%, respectively, which are close to the calculated values of 3.91%, 69.84%, and 16.71% expected for an infinite CT-coordinated squaric acid 1,6-diamidepyrene conjugated oligomer. However, elemental analysis of the 1,3- conjugated oligomer was poor agreement with that of calculated values.

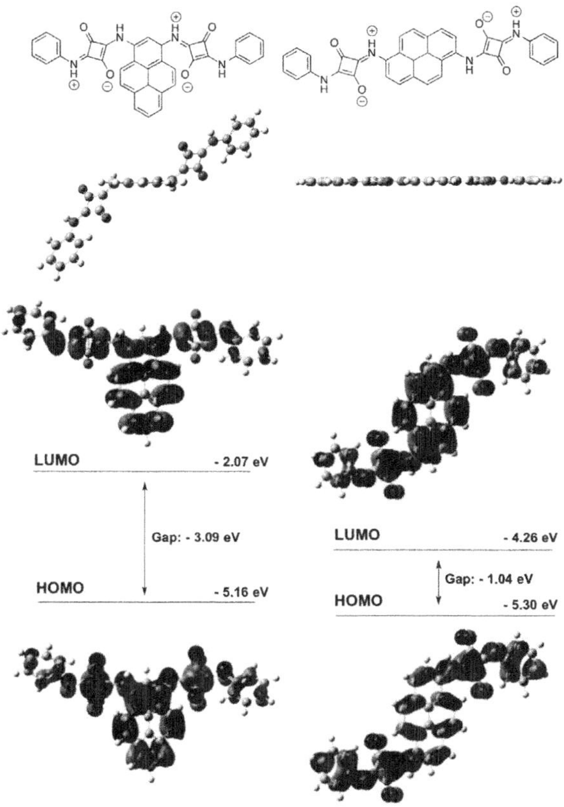

Figure 14.4 Molecular orbital diagrams for HOMO and LUMO of model compounds (B3LYP/6-311G(d)// B3LYP/6-311G(d)).

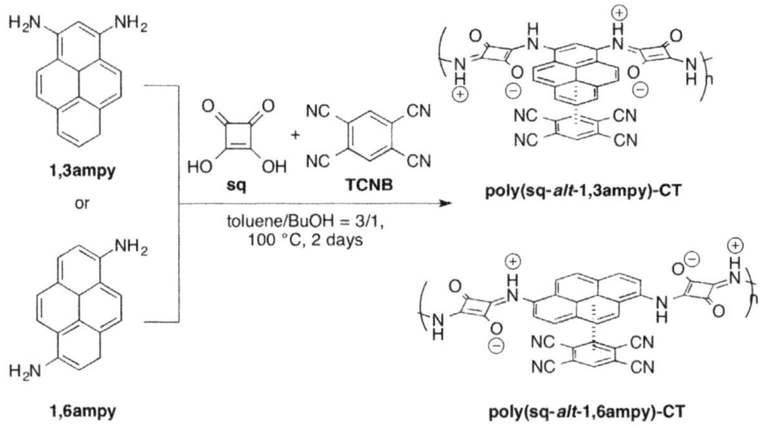

Scheme 14.2 Condensation of squaric acid and X-diaminopyrenes with TCNQ; X = 1,3: oligo(sq-*alt*-1,3py)-TCNQ and X = 1,6: oligo(sq-*alt*-1,6py)-TCNQ.

The CT complex formation in oligo(sq-*alt*-1,3py)-TCNQ and oligo(sq-*alt*-1,6py)-TCNQ was also confirmed by UV-Vis absorption spectroscopy (Fig. 14.5a). Although oligo(sq-*alt*-1,3py)-TCNQ shows no CT band due to the torsion structure of the squaraine moiety from the DFT calculation, the CT band of oligo(sq-*alt*-1,6py)-TCNQ was observed at 933 nm. According to XRD analysis, a strong peak of oligo(sq-*alt*-1,6py) at 7.839° disappeared. New four sharp peaks of oligo(sq-*alt*-1,6py)-TCNQ appeared at 2.409° (36.6422 Å), 4.858° (18.1754 Å), and 5.322° (16.5902 Å), probably causing the formations of CT complex and hydrogen-bonding interactions of sqaramide, and the weak peaks at 23.735° (3.7457 Å), and 26.321° (3.3833 Å) were assigned to π–π stacking interactions, differing from the XRD pattern of oligo(sq-*alt*-1,6py) (Fig. 14.5b). These data indicated that the condensation reaction of squaric acid and 1,6-diaminopyrene with TCNQ forms the oligo(sq-*alt*-1,6py) bearing the CT complex structure.

The unique features of a oligo(sq-*al*t-1,6py)-TCNQ inspired us to explore the absorption response gained upon treating with mechanical force stimuli. We anticipated that interlayer interaction (hydrogen-bonding interactions of squaramide and π–π interactions) in the structure should change upon external force stimuli as grinding, causing the increased CT complex intensity of absorption of a oligo(sq-*al*t-1,6py)-TCNQ.

When the crystalline oligomer bearing a weak CT complex was treated with strong grinding in an agate mortar with a pestle, the color of oligo(sq-*al*t-1,6py)-TCNQ changed from orange to metallic deep green (Fig. 14.5b,c). From the measurement of the UV-Vis spectra of oligo(sq-*al*t-1,6py)-TCNQ before and after grinding, when the pyrene absorption in oligo(sq-*al*t-1,6py)-TCNQ at 510 nm was decreased, the intensity of the CT complex absorption at 949 nm in oligo(sq-*al*t-1,6py)-TCNQ was increased exceedingly. The intensity of grinding oligo(sq-*al*t-1,6py)-TCNQ in the XRD pattern was increased in the area of π–π stacking interaction at 23.735°–28.644°. These results indicated that crystalline oligo(sq-*al*t-1,6py)-TCNQ has an efficiently mechanochromic phenomenon. From these results, we tried to employ another acceptor, tetracyanobenzene (TCNB), which led to the same mechanochromic phenomenon by mechanical grinding.

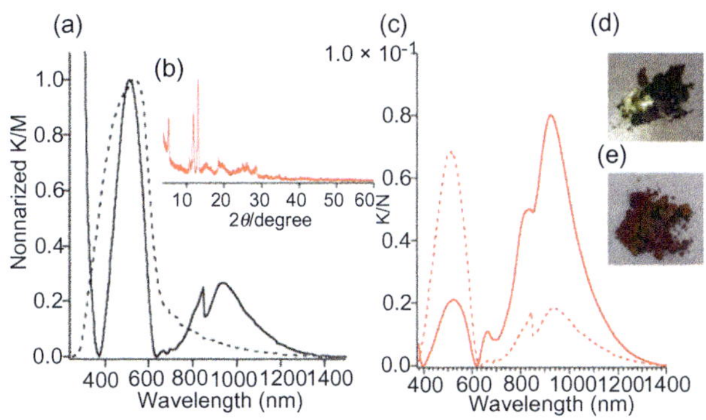

Figure 14.5 (a) Absorption spectra of oligo(sq-*alt*-1,3py)-TCNQ (black dashed curve) and oligo(sq-*alt*-1,6py)-TCNQ (black solid curve), (b) PXRD profiles of oligo(sq-*al*t-1,6py)-TCNQ, (c) grinding oligo(sq-*alt*-1,6py)-TCNQ (red solid curve) and gentle ground oligo(sq-*alt*-1,6py)-TCNQ (red dashed curve), (d) grinding state, and (e) ground state.

This section introduced the synthesis of π-conjugated squaric acid diamide-type oligomers composed of 1,3- or 1,6-diaminopyrenes and squaric acid with TCNQ (molecular ratio 1:1:1), oligo(sq-alt-1,3ampy) and oligo(sq-alt-1,6py). The obtained oligomers exhibited a bright orange color in the solid state at room temperature. The CT complex absorption color was strikingly changed from orange to metallic deep green upon mechanical grinding and could be recovered by sonication in toluene. Furthermore, this system was applicable for the use of other acceptors such as TCNB. This research reported herein is the first example of mechanochromic materials prepared by the introduction of CT complexes into π-conjugated oligomers (Fig. 14.6).[27]

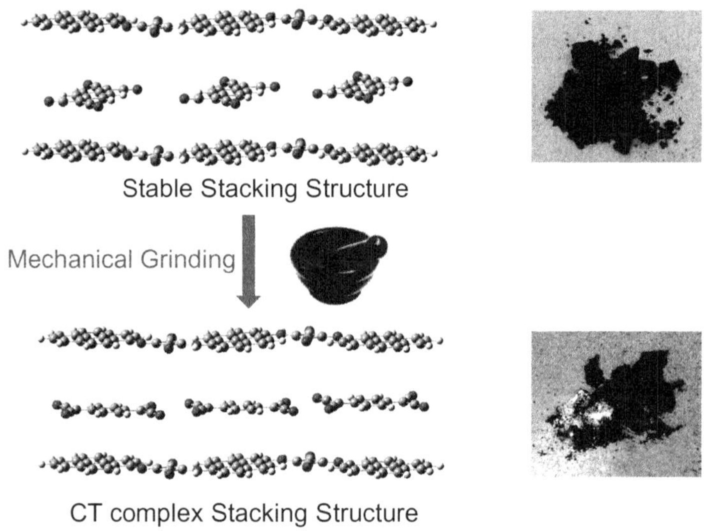

Figure 14.6 Mechanochromism of π-conjugated squaric acid diamide-type oligomers composed of 1,6-diaminopyrenes and squaric acid with TCNQ by mechanical grinding.

14.5.2 Biomedical Imaging Application of CT-Active Biomaterial

In this section, Nagai and Siegwart et al. have provided evidence that a semicrystalline template composed of cellulose chemically bearing CT complexes consisting of pyrene donors and TCNQ, TCNB (1,2,4,5-tetracyanobenzene), and TCNE (tetracyanoethylene) acceptors emit multicolored fluorescence in solution and in biological settings.[28] Noncovalent organic complexes with electron donor (D) and electron acceptor (A) moieties have emerged as promising candidates for advanced materials with diverse applications (including mechanochromic materials) due to their intermolecular CT capability. Pyrene–TCNQ complexes have been shown to crystalize in a monoclinic system with a spacing group of $P2_1/n$ with half of the molecules of pyrene and TCNQ as asymmetric parts of the unit cell, in which the pyrene moiety is tilted with respect to TCNQ along stacks and at an angle between mean planes of 6°.[30] As mentioned before, the first example of mechanochromic conjugated CT materials using pyrene as a donor and TCNQ or TCNB as two acceptors into π-conjugated oligomer backbones has been demonstrated.[27] The conjugated materials showed an interesting color change from bright orange to metallic deep green upon mechanical grinding in the solid state at room temperature, which was reversible by sonication in toluene. However, this property was lost in solution. To overcome this limitation, we envisioned ordering the CT complexes into a templated design on a polymer backbone to promote enhanced tumor retention for imaging applications. It is known that the modular structure of CT complexes coupled with environmental dependence on CT properties allows the control over crystal structure and optical properties by alternating CT interactions between electron donors and acceptors. Because small-molecule CT complex particles have shown utility in biomedical imaging,[29] we were inspired to design CT complexes based on supramolecular self-assembly of natural polymers, which would enable enhanced tumor retention for in vivo imaging applications.

Bioinspired materials with unique and intriguing functionalities, such as the self-cleaning ability of plants in nature, have attracted much attention for the rational design of next-generation materials.[31,32] Polymers can be used as a template to construct ordered bioinspired structures such as helical objects because of their facile processability, low cost, and well-defined architecture.[31–34] Cellulose is one of the most abundant biopolymers and has been converted into functional fibers, hydrogels, films, and microspheres.[35–40] That is, utilization of cellulose as a semicrystalline template awakens our interest in not only the active CT complexes under higher-order surroundings but also applications for biomedical imaging.

This section introduces newly the preparation of CT complex-connecting cellulose. The blue and red emission of CT complexes appended to a cellulose template was induced by using nanoprecipitation, in which a dimethyl sulfoxide (DMSO) solution of CT-cellulose was precipitated drop wise into a large amount of water under stirring to form nanoparticles (NPs) with an aggregation-induced emission (AIE) effect. Furthermore, due to the biocompatible nature of the CT emission-active material, the cellulose-CT NPs were nontoxic to cells in vitro and could be used for biological imaging.

Pyrene-connected methyl cellulose (Py-cellulose) as a ligand for CT complex formation was prepared from the condensation reaction of 1-pyrenebutanoyl chloride and 2,5-di-*O*-methyl cellulose (HO-cellulose-OMe; M_n = 27,186, M_w/M_n = 3.12), which were synthesized according to previous reports[41] at room temperature. After the reaction was allowed to proceed overnight, the precipitates were collected using a centrifuge and washed with excess ethanol and acetone to provide the corresponding Py-cellulose (M_n = 39,443, M_w/M_n = 3.30) in 60% yield (Scheme 14.3a). The transfer from hydroxyl to pyrenebutyl groups was confirmed using IR spectroscopy where a peak at 1737cm^{-1}, ascribed to the carbonyl group that forms during the esterification reaction, can be clearly observed, and the size-exclusion chromatography (SEC) analysis also showed a successful monomodal peak before and after the esterification reaction. In ^{1}H NMR spectra, the broad peaks assignable to the aromatic pyrene group in Py-cellulose appear at 7.49–8.37 ppm after the esterification reaction, although HO-cellulose-OMe does not have any peaks in this area.

Next, we conducted the CT-coating reaction of Py-cellulose with three acceptors (TCNQ, TCNE, or TCNB) in the mixed solvent of toluene/DMAc (9/1 = v/v, conc. = 0.2 or 0.02 mol/L) under solvothermal conditions (Scheme 14.3b). After 48 h, the precipitates were collected using a centrifuge and washed with excess ethanol three times and dried under vacuum at 50°C to afford the expected CT complex–coated cellulose in 49%–51% (0.02 g/L) and 64%–70% (0.2 g/L) yields. The different yields of the CT-coated celluloses were a result of solvent dependency. As shown in Scheme 14.3b, performing a CT-coating reaction at high concentration (0.2 g/L) allows an intermolecular CT reaction, resulting in the physical crosslinking reaction between Py-cellulose strands, as evidenced by noticeably reduced solubility in DMSO. On the contrary, the CT reaction in the mixed solvent (0.02g/L) showed completely homogeneous deep-color solutions such as deep green (TCNQ), pale yellow (TCNE), and orange (TCNB), implying that the CT-coating reaction of Py-cellulose using only TCNQ progressed efficiently (Fig. 14.7) to form linear polymers.

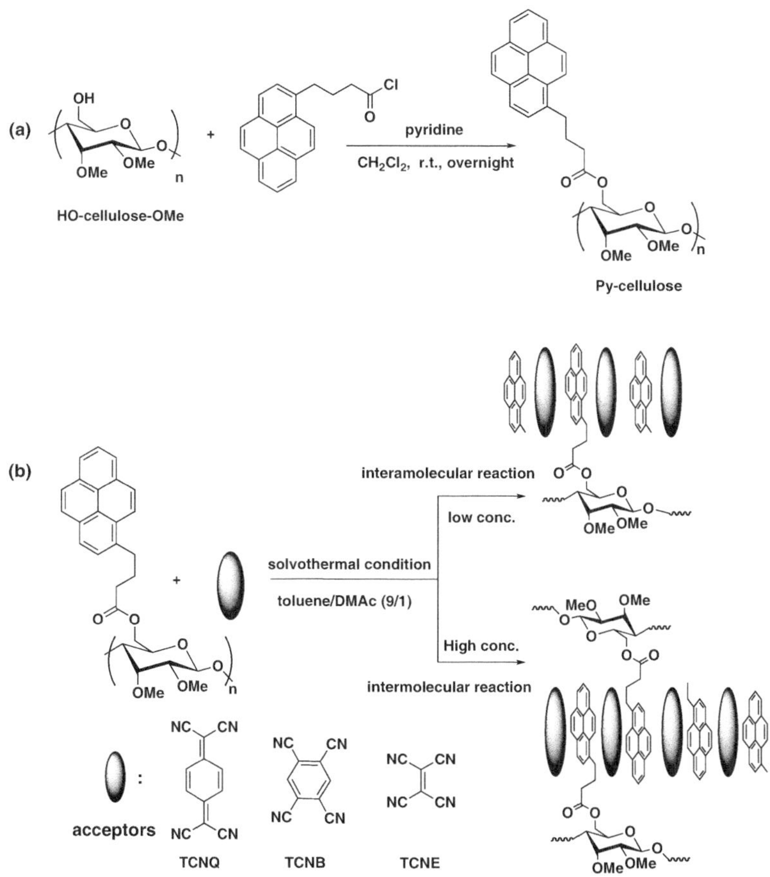

Scheme 14.3 (a) Synthesis of pyrene-connected methyl cellulose (Py-cellulose) and (b) the practicable reaction of Py-cellulose and acceptors such as TCNQ, TCNB, and TCNE.

Infrared (IR) spectroscopy analysis of the solid CT complexes confirmed the CT complex formation, as indicated by the shift of the CN stretching frequency in comparison to the analogous frequency in the neutral acceptor monomers.[40] Linear TCNQ-CT exhibited a CN stretching frequency of 2218 cm^{-1}, assigned to the formation of CT complexes, as compared to TCNQ monomer (2230 cm^{-1}). Meanwhile, linear TCNE-CT showed a very weak broad peak around 2220 cm^{-1}, which indicates poor complex formation, likely due to lower electron affinity. The IR spectra of crosslinked TCNQ-CT showed bimodal CN stretching at 2217 cm^{-1} and 2183 cm^{-1}, while interestingly at the higher concentration, TCNE-CT showed a broad band at 2216 cm^{-1}, indicating that reaction at high concentration enables both CT coating and crosslinking (intermolecular and intramolecular reactions between pyrene and acceptors). While the TCNB monomer showed a defined CN stretching frequency at 2245 cm^{-1}, neither linear nor crosslinked TCNB-CT had any peaks in this region, which can be attributed to higher steric hindrance as compared to TCNQ and TCNE. Thermal degradation behavior of all CT complexes was examined by thermogravimetric analysis (TGA) (Fig. 14.8a and Table 1 in Fig. 14.8b). Evidence of the crosslinking was apparent when comparing the TGA curve of the CT complexes when formed at high or low concentration. For TCNQ-CT-cellulose, the final weight percentage (T = 500°C) for the high-concentration complex was 44.03%, while the low concentration complex was 32.49%, which confirms the crosslinked conformation of the CT complex. The same trend was observed for TCNE-CT-cellulose (wt%final linear 28.15%, crosslinked 37.49%). Meanwhile, for TCNB, the final weight percentage of the linear form was 31.45%, while the crosslinked form was 28.05%. This suggests that no crosslinking occurred, likely due to the steric hindrance of this acceptor.

Figure 14.7 Photographs of linear TCNQ-, TCNE-, and TCNB-CT-cellulose solids.

The UV-Vis absorptions of Py-cellulose, TCNQ-, TCNE-, and TCNB-CT-linear cellulose relative to monomeric TCNQ, TCNE, and TCNB, respectively, were examined in DMSO solution (Fig. 14.9). The acceptors exhibited absorption maxima at 400 (TCNQ), 300 (TCNE), and 317 (TCNB) nm, and only TCNQ showed two broad bands between 651 and 889 nm. Vibration absorption maxima at 345 nm of TCNQ-, TCNB-, and TCNQ-CT-celluloses red-shifted from that of Py-cellulose at 341 nm (Fig. 14.9a). TCNQ-CT-cellulose showed broad bands in the red to near-IR region, which overlap with monomeric TCNQ. Consequently, the CT-cellulose has no typical CT bands at high wavelength in DMSO. These data show that slight red shifts of pyrene moieties on the cellulose indicate weak CT π-stacking interaction that was influenced by the highly ordered cellulose-based fiber. Interestingly, upon sonication of the DMSO solution of TCNQ-CT, the peak at 345 nm reverts back to that 341 nm, and the high-wavelength peaks disappear, suggesting disruption of the weak π–π stacking interactions (Fig. 14.9b).

(a)

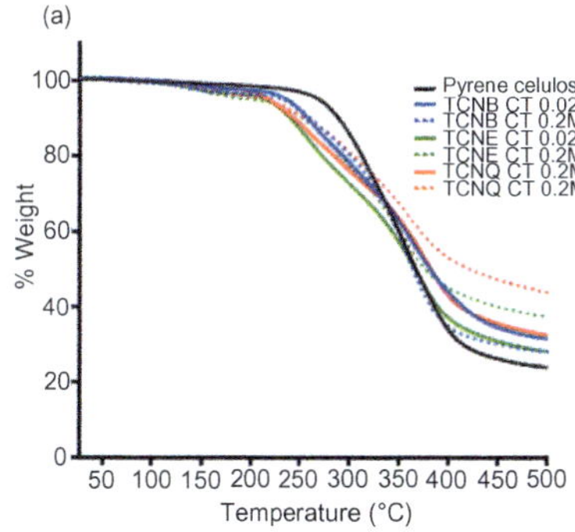

(b)

Table 1.

Sample	$T_{5\%wt\ loss}$ (°C)	$T_{10\%wt\ loss}$ (°C)	%wt final (500 °C)
Py-cellulose	270.67	290.17	24.00
TCNB	281.83	298.33	4.09
TCNB CT 0.02M	237.33	259.33	31.45
TCNB CT 0.02M	232.67	262.17	28.05
TCNB	151.17	161.50	1.36
TCNB CT 0.02M	215.67	241.67	28.15
TCNB CT 0.02M	196.83	254.50	37.49
TCNQ	284.83	297.50	11.49
TCNB CT 0.02M	219.00	245.33	32.49
TCNB CT 0.02M	217.00	256.67	44.03

Figure 14.8 (a) TGA measurements of TCNQ-CT-cellulose (red solid, 0.02 g/L; red dotted 0.2 g/L), TCNE-CT-cellulose (green solid, 0.02 g/L; green dotted 0.2 g/L), and TCNB-CT-cellulose (blue solid, 0.02 g/L; blue dotted 0.2 g/L) and (b) table data of acceptors and Py-cellulose and CT complexes.

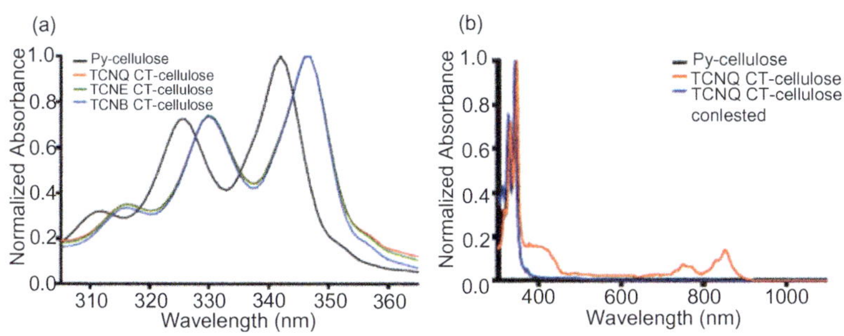

Figure 14.9 (a) Absorbance spectra of CT complexes dissolved in DMSO and (b) the effect of sonication on DMSO solutions for TCNQ-CT-cellulose.

To generate improved CT-cellulose absorbance and emission properties, we employed the AIE approach to increase aggregation interactions between donors and acceptors by forming NPs of the CT complexes. TCNQ-CT-cellulose NPs were formed by nanoprecipitation to improve the ordering form in the cellulose-bearing pyrene and TCNQ pair.[42] For this purpose, 100 μL TCNQ-CT-cellulose (c = 1 g/L in DMSO) was added drop-wise to deionized water (900 μL) under stirring. The size of the resulting NPs was 36 ± 4 nm as measured by dynamic light scattering (DLS). Figure 14.10 shows UV-Vis and fluorescence spectra of TCNQ-CT-cellulose NPs in aqueous solution as compared to Py-DOMC NPs. Because NPs formed from TCNE-CT and TCNB-CT did not show any new absorbance peaks, these were not further analyzed. Figure 14.10a shows the UV spectra of TCNQ-CT-cellulose NPs, Py-DOMC NPs, and a TCNQ suspension in aqueous solution (0.02 g/L). TCNQ-CT-cellulose NPs exhibit two distinct broad absorbance bands at 485 and 620 nm, which can be differentiated from the bands of TCNQ monomer. These new broad peaks are due to formation of the CT complex.[27] The semicrystalline cellulose template helps in order to maintain a regular CT stacking structure between pyrene and TCNQ. Next, photoluminescence (PL) experiments of TCNQ-CT-cellulose NPs compared to Py-DOMC NPs were carried out in aqueous solution containing a small amount of DMSO (0.01 or 0.02 g/L). A decrease of the excimer emission peak of the pyrene moiety at 450 nm results in the CT reformation by nanoprecipitation, suggesting that the aggregation state inhibits a high-order array of the CT complexes in the NP. However, the active CT NP formed between TCNQ and pyrene moieties in the cellulose emits blue light at 376–540 nm. Interestingly, TCNQ-CT-cellulose NPs showed efficient red emission at 602 nm upon the excitation of 485 nm (CT first broad-band maximum). In addition, excitation at 620 nm results in a small emission peak at 777 nm (Fig. 14.10b).

Because we are interested in the potential for these cellulose-based CT NPs to be used in biomedical imaging, we next measured the cytotoxicity of the NPs to cells in vitro. TCNQ-CT-cellulose NPs were incubated with HeLa cells at different concentrations and cell viability was quantified using the MTT assay. TCNQ-CT-cellulose NPs were nontoxic to cells at the concentrations tested (Fig. 14.11).

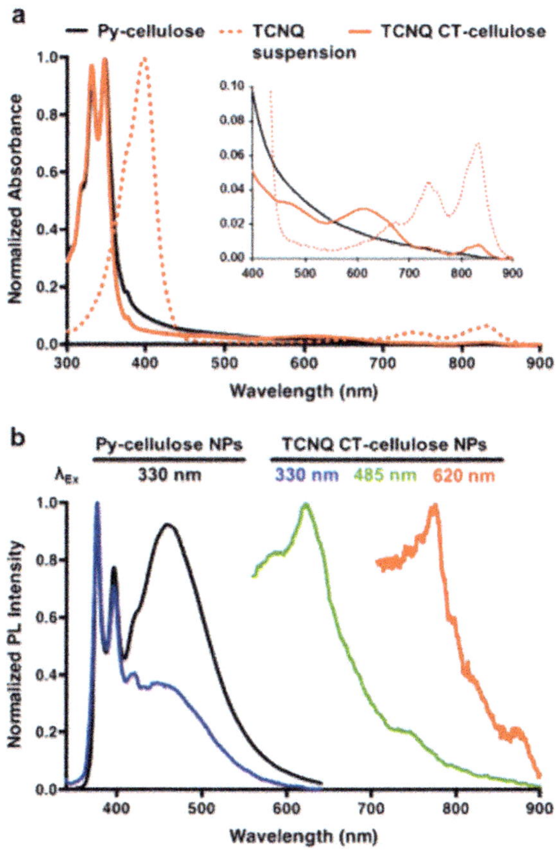

Figure 14.10 (a) UV-Vis absorption of TCNQ-CT-cellulose NPs, Py-cellulose NPs, and a TCNQ suspension and (b) photoluminescence of TCNQ-CT-cellulose NPs in aqueous solution (0.02 g/L).

Next, we explored the potential for the CT mechanism to be preserved in biological settings in vitro and in vivo. TCNQ-CT-cellulose NPs were stabilized by coating with Tween 20 to reduce NP aggregation. They were added to growing HeLa cells and then imaged after 3 h of incubation in a proof-of-principle experiment. Figure 14.12a shows effective cell uptake of TCNQ-CT-cellulose NPs into HeLa cells. The complexes were imaged using the excitation of the pyrene moiety (λ_{Ex} 405 nm) and near the absorbance of the CT complexes (λ_{Ex} 639 nm). Excitation of the pyrene moiety resulted in strong intracellular signal for both TCNQ-CT and Py-cellulose NPs. Red emission was only observable for the TCNQ-CT NPs, suggesting successful CT. Importantly, the CT mechanism was preserved in serum-containing media.

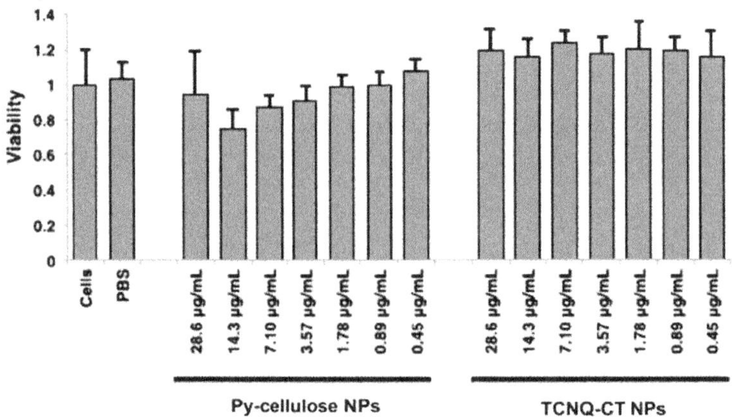

Figure 14.11 CellTiter-Glo assay to determine cytotoxicity of the TCNQ-CT NPs and Py-DOMC NPs. Formulations were prepared by nanoprecipitation in water at 0.1 g/L final concentration and dialyzed against 1X PBS for 4 h. The nanoparticles (80 µL) were added to HeLa cells in a 96-well plate in 200 µL growth medium and incubated for 36 h prior to reading the assay. Values are normalized to untreated cells and plotted +/– standard deviation (N = 4 or 5).

Finally, we injected TCNQ-CT-cellulose NPs into mice bearing RPMI-7951 melanoma xenograft tumors to test whether CT could be used as an imaging principle in vivo. After 4 h, tumors were harvested, embedded, frozen, and sectioned. Confocal microscopy was used to image the tumor sections (Fig. 14.12b). The results show that TCNQ-CT-cellulose NPs likely aggregated after injection but that the CT mechanism was preserved. Similar to the in vitro results, red emission was observable for the TCNQ-CT NPs. This is suggestive of successful CT in vivo after delivery of TCNQ-CT NPs. It is worth noting that the emission was particularly strong in areas of aggregation. Therefore, these TCNQ-CT-cellulose NPs that utilize AIE for effective CT represent a significant advancement for the use of CT complexes for biomedical imaging.

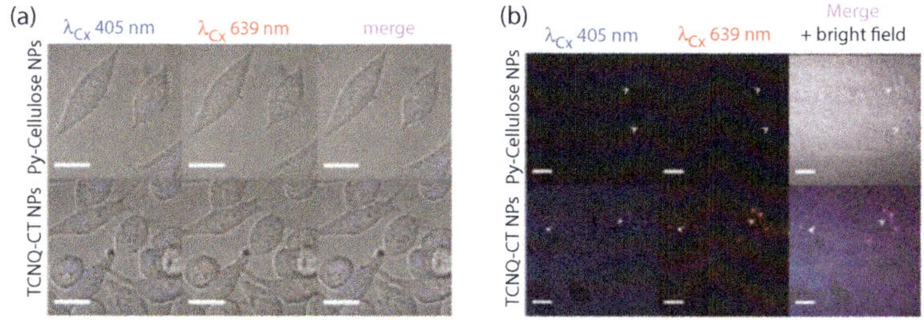

Figure 14.12 (a) Cellular uptake of TCNQ-CT-cellulose NPs into HeLa cells in vitro. TCNQ-CT-cellulose NPs show red emission after excitation at 639 nm, while Py-DOMC NPs do not. Scale bar = 20 μm. (b) Imaging of TCNQ-CT-cellulose NPs in vivo to mice bearing RPMI-7951 melanoma xenograft tumors. Arrows indicate regions of concentrated NPs. Scale bar = 20 μm.

In conclusion of this section, we report the synthesis of new CT complexes based on ordered cellulose biomaterials. Appending CT complexes onto a cellulose backbone enabled control of the orientation and environment of the donor and acceptor molecules of the CT complexes, thus arranging the pyrene–TCNQ interactions to allow detectable absorption and emission properties. This led to unique CT emission properties in the red to near-IR regions. This paper demonstrates that by using an AIE approach through NP formation, the CT mechanism can occur in biologically relevant conditions in vitro and in vivo, suggesting future utility of this mechanism for a variety of biomedical imaging applications (Fig. 14.13).

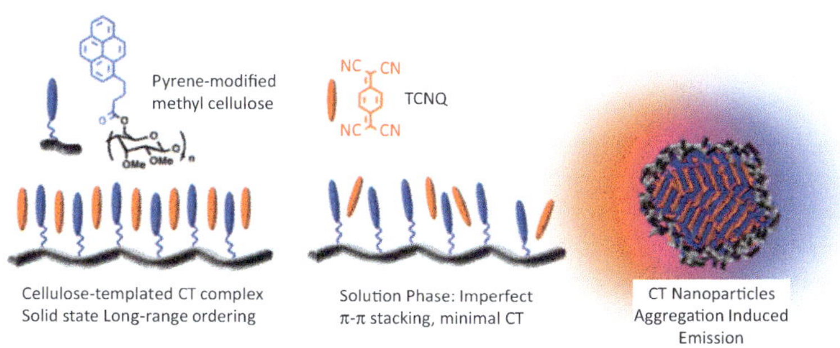

Figure 14.13 CT complex nanoparticles based on a semicrystalline cellulose template.

14.6 Conclusions

In this chapter, we focused and reviewed CT complex interaction as one of weak noncovalent bonds for application not only in electronic fields but also in biofields. This interaction of small molecule has inspired chemists and students to explore new concepts such as CT mechanisms and new materials of several fields such as electronic, superamolecular, and biological materials. Therefore, we hope that chemists and students over the world use this textbook as one of the textbooks on how to consider methodologies and concepts of CT interaction, extending from small-molecule results to macromolecules application.

References

1. Wu, J., Pisula, W., Müllen, K. *Chem. Rev.*, **107**, 718 (2007).
2. Williams, J. M., Ferraro, J. R., Thorn, R. J., Carlson, K. D., Geiser, U., Wang, H. H., Kini, A. M., Wangbo, M.-H. *Organic Supercondutors* (including Fullerens), Prentice Hall: Englewood Cliffs (1992).
3. Mulliken, R. S. *J. Am. Chem. Soc.*, **74**, 811–824 (1952).
4. McGlynn, S. P. *Chem. Rev.*, **58**, 1113–1156 (1958).
5. Foster, R. *Organic Charge-Transfer Complexes*, Academic Press: London (1969).
6. Nielsen, M. B., Lomholt, C., Becher, J. *Chem. Soc. Rev.*, **29**, 153–164 (2000).
7. Sato, G., Yoshida, Y. *The Chemical Record*, **11**, 124 (2011).
8. Ferraris, J., Cowan, D. O., Walatka, V. V., Perlstein, J. H. *J. Am. Chem. Soc.*, **95**, 948 (1973).
9. Amabilino, D. B., Stoddart, J. F. *Chem. Rev.*, **95**, 2725 (1995).
10. Balzani, V., Gómez-López, M., Stoddart, J. F. *Acc. Chem. Res.*, **31**, 405–414 (1998).
11. Ma, X., Tian, H. *Chem. Soc. Rev.*, **39**, 70 (2010).
12. van Dongen, S. F. M., Cantekin, S., Elemans, J. A. A. W., Rowan, A. E., Notle, R. J. M. *Chem. Soc. Rev.*, **43**, 99 (2014).
13. Klivansky, L. M., Hanifi, D., Koshkakarayan, G., Holycross, D. R., Gorski, E. K., Wu, Q., Chai, M., Liu, Y. *Chem. Sci.*, **3**, 2009 (2012).
14. Wang, J.-Y., Yan, J., Ding, L., Ma, Y., Pei, J. *Adv. Funct.*, **19**, 1746–1752 (2009).

15. Zych, A. J., Iverson, B. L. *J. Am. Chem. Soc.*, **122**, 8898 (2000).

16. Medjanik, K., Kutuyakhov, D., Nepijko, S. A., Schönhense, G., Naghavi, S., Alijani, V., Felser, C., Koch, N., Rieger, R., Baumgartend, M., Müllen, K. *Phys. Chem. Chem. Phys.*, **12**, 7184 (2010).

17. Jerome, D. *Chem. Rev.*, **104**, 5565 (2004).

18. Dobrowolski, M. A., Garbarino, G., Mezouar, M., Ciesielski, A., Cyranski, M. K. *CrystEngComm*, **16**, 415 (2014).

19. (a) Sagara Y., Kato, T. *Angew. Chem. Int. Ed.*, **47**, 5175 (2004); (b) Sagara, Y., Kato, T. *Nat. Chem.*, **1**, 605 (2009); (c) Assefa, Z. Omary, M. A., McBurnett, B. G., Mohamed, A. A., Patterson, H. H., Staples, R. J., Facker, J. P. *Inorg. Chem.*, **41**, 6274 (2002).

20. Schmidt, A. H. *Synthesis*, 961 (1980); Charton, J., Charruault, L., Deprez-Poulain, R. Deprez, B. *Comb. Chem. High Throughput Screening*, **11**, 294 (2008).

21. (a) Aleman, J., Parra, A., Jiang, H., Jørgensen, K. A. *Chem. Eur. J.*, **17**, 6890 (2011); (b) Busschaert, N., Kirby, I. L., Young, S., Coles, S. J., Horton, P. N., Light, M. E., Gale, P. A. *Angew. Chem. Int. Ed.*, **51**, 4426 (2012); (c) Xu, D. Q., Wang, Y. F., Zhang, W., Luo, S. P., Zhong, A. G., Xia, A. B., Xu, Z. Y. *Chem. Eur. J.*, **16**, 4177 (2010).

22. (a) Delgado-Pinar, E., Rotger, C., Costa, A., Pina, M. N., Jimenez, H. R., Alarcon, J., Garcıa-Espana, E. *Chem. Commun.*, **48**, 2609 (2012); (b) Lim, N. C., Pavlova, S. V. Bruckner, C. *Inorg. Chem.*, **48**, 1173 (2008).

23. (a) Soberats, B., Martınez, L., Sanna, E., Sampedro, A., Rotger, C., Costa, A. *Chem. Eur. J.*, **18**, 7533 (2012); (b) Ramalingam, V., Domaradzki, M. E., Jang, S. Muthyala, R. S. *Org. Lett.*, **10**, 3315 (2008).

24. Villalonga, P., Fernandez de Mattos, S., Ramis, G., Obrador-Hevia, A., Sampedro, A., Rotger, C., Costa, A. *ChemMedChem*, **7**, 1472 (2012).

25. (a) Brand, S., de Candole, B. C., Brown, J. A. *Org. Lett.*, **5**, 2343 (2003); (b) Tsujishita, H., Gerdes, C. *Bioorg. Med. Chem. Lett.*, **17**, 6151 (2007).

26. (a) De Lamo Marin, S., Xu, Y., Meijler, M. M., Janda, K. D. *Bioorg. Med. Chem. Lett.*, **17**, 1549 (2007); (b) Gloria, P., Gut, J., Goncalves, L. M., Rosenthal, P. J., Moreira, R., Santos, M. M. M. *Bioorg. Med. Chem.*, **19**, 7542 (2011).

27. Nagai, A., Okabe, Y. *Chem. Commun.*, **50**, 10052 (2014).

28. Nagai, A., Miller, J. B., Du, J., Kos, P., Stefan, M. C., Siegwart, D. J. *Chem. Commun.*, **51**, 11868 (2015).

29. Dobrowolski, M. A., Garbarino, G., Mezouar, M., Ciesielski, A., Cyranski, M. K. *CrystEngComm*, **16**, 415 (2014).

30. Roach, P., Shirtcliffe, N. J., Newton, M. I. *Soft Matter*, **4**, 224 (2008).
31. Sun, T., Feng, L., Gao, X., Jiang, L. *Acc. Chem. Res.*, **38**, 644 (2005).
32. Bleek, K., Taubert, A. *Acta Biomater.*, **9**, 6283 (2013).
33. Bonderer, L. J., Studart, A. R., Gauckler, L. J. *Science*, **319**, 1069 (2003).
34. Ma, M., Guo, L., Anderson, D. G., Langer, R. *Science*, **339**, 186 (2013).
35. Thomas, A., Goettmann, F., Antonietti, M. *Chem. Mater.*, **20**, 738 (2008).
36. Cai, J., Zhang, L., Zhou, J., Qi, H., Chen, H., Kondo, T., Chen, X., Chu, B. *Adv. Mater.*, **19**, 821 (2007).
37. He, M., Zhao, Y., Duan, J., Wang, Z., Chen, Y., Zhang, L. *ACS Appl. Mater. Interfaces*, **6**, 1872 (2014).
38. Qi, H., Chang, C., Zhang, L. *Green Chem.*, **11**, 177 (2009).
39. Luo, X., Liu, S., Zhou, J., Zhang, L. *J. Mater. Chem.*, **19**, 3538 (2009).
40. Chi, X., Besnard, C., Thorsmølle, V. K., Butko, V. Y., Taylor, A. J., Siegrist, T., Ramirez, A. P., *Chem. Mater.*, **16**, 5751 (2004).
41. Hornig, S., Heinze, T., Becer, C. R., Schubert, U. S. *J. Mater. Chem.*, **19**, 3838 (2009).
42. Dillon, R. J., Bardeen, C. J. *J. Phys. Chem. A*, **116**, 5145 (2012).

Biography

These personal experiences of Atsushi Nagai cover activities during his PhD study program, including how to consider precisely polymerization mechanisms and design freely and broadly the polymerizable monomer structures, under the guidance of Prof. Takeshi Endo at Yamagata University (2005), Japan, who shared with Nagai his love of fundamental science. The activities pursued during his postdoctoral fellowship and as assistant professor with Prof. Yoshiki Chujo at Kyoto University, Japan, have also given him a new idea that can be used in the classroom, including fundamental scientific discoveries that can translate into actual functional materials such as conjugated polymers and boron materials from academic research. In 2010, he cotaught a lot of international students and postdoctors belonging to the Graduate University for Advanced Studies (Chemical, Optical Chemical, and Polymer Synthetic Project) at the Institute for Molecular Science (IMS), Japan. He was assistant professor working on the project design and concept. While at Yamagata University during 2002–2005, he served as a teaching assistant and recitation instructor of the advanced placement chemistry course, including both classroom lectures and laboratory supervision. He has a lot of experience giving lectures at conferences and group meetings using multimedia and utilizes those skills to effectively communicate the subject matter. In 2014, he joined the University of Texas Southwestern Medical Center as visiting assistant professor. In addition to any basic course in general chemistry, organic chemistry, polymer chemistry, and optical chemistry, Nagai has designed and taught about next-generation materials, for example, bioelectronic polymers, electronic porous materials, and new precisely optical nature-providing polymers. Since 2016, Nagai is researching the developments of novel CT complex hydrogels and teaching polymer chemistry. Moreover, he actively speaks on research on new next-generation polymer materials all over the world.

Dr. Daniel J. Siegwart is assistant professor at the University of Texas Southwestern Medical Center in Dallas, TX, U.SA. USA, in the Simmons

Comprehensive Cancer Center and the Department of Biochemistry. He obtained his PhD at Carnegie Mellon University (CMU) under the guidance of Prof. Krzysztof Matyjaszewski, who developed atom transfer radical polymerization (ATRP), one of the most powerful controlled/living radical polymerization (CRP) techniques. Dr. Siegwart received the Joseph A. Solomon Memorial Fellowship in Chemistry at CMU and was also awarded a National Science Foundation (NSF) East Asia and Pacific Summer Institutes (EAPSI) Fellowship to study at the University of Tokyo with Prof. Kazunori Kataoka in 2006. To apply his background in polymer chemistry to translational medical applications, he conducted National Institutes of Health (NIH) NRSA-sponsored postdoctoral research with Prof. Robert Langer at MIT and focused on combinatorial, high-throughput methods in material discovery. There, he directed a project reporting the first large library of >1500 structurally defined core-shell nanoparticles that identified key chemical functionalities for designing polymers for siRNA delivery. At UT Southwestern, the Siegwart Research Group's long-term goals are to (i) develop improved materials for effective siRNA and miRNA delivery to tumors; (ii) discover theranostic probes to image and treat cancer through nucleic acid recognition, including probes that utilize the charge transfer (CT) mechanism; and (iii) globally understand how the physical and chemical properties of materials affect interactions with biological systems in vitro and in vivo in the context of improving cancer therapies. They The group aspires to utilize chemistry and engineering to make a beneficial impact on human health through improved cancer therapies.

Chapter 15

Construction of a Conjugation System with Heteroatoms in Polymer Main Chains

Kazuo Tanaka and Yoshiki Chujo
Department of Polymer Chemistry, Graduate School of Engineering, Kyoto University, Katsura, Nishikyo-ku, Kyoto 615-8510, Japan
kazuo123@chujo.synchem.kyoto-u.ac.jp; chujo@chujo.synchem.kyoto-u.ac.jp

From the conjugation systems through the polymer main chains, various useful characteristics such as light absorption property, emission, and charge carrier ability can be observed. These properties are practically applicable for the fabrication of modern organic optical and/or electronic devices such as light-emitting diodes, laser amplification, and photovoltaic cells. In addition, by introducing heteroatoms into the main-chain conjugation, further unique functions are often obtained. The specific functions originated from the heteroatoms are feasible not only to improve the device efficiencies but also to bring a breakthrough in the device development. In this chapter, the studies on heteroatom-containing conjugated polymers are explained. Especially, the strategies for

---

Conjugated Objects: Developments, Synthesis, and Applications
Edited by Atsushi Nagai and Koji Takagi

ISBN 978-981-4774-03-1 (Hardcover), 978-1-315-22962-1 (eBook)
www.panstanford.com

the extension of the conjugation via the heteroatoms through the polymer main chains and their unique optical properties are explained. Initially, the introduction of organoboron complexes, which are a class of optically functionalized substances into polymer main chains, is described. As an application of the properties of organoboron polymers, we will introduce the recent hot topics on the design and synthesis of conjugated polymers having aggregation-induced emission properties, which are promising to be a key phenomenon for obtaining bright emissive solid materials. Next, as other examples of the introduction of heavier atoms, the construction of the conjugation systems involving gallium and silicon elements is described. The synthetic strategy and the electronic structures are demonstrated. Finally, the general synthetic protocol of these polymers is also mentioned.

15.1 Introduction

The development of conjugated polymers is a topic with high relevance. Unique optical, electric, and spintronic characteristics have been discovered from the conjugated polymers. Based on these properties various kinds of modern optical and electric devices were fabricated. Indeed, for several decades, advanced devices and materials such as paper displays, colorful solar cells, and field-effect transistors were practically realized. The conjugated polymers have flexibility in the molecular design for receiving intended functions. According to the preprogrammed design, these functions can be tuned. For example, bright emissions can be obtained in the desired wavelength regions from the donor–acceptor-type alternating copolymers by selecting the combination of the component monomer units.[1] In addition, we can readily add material properties such as solubility, film formability, and durability into the materials by modulating side chains. Therefore, not only the fundamental studies for clarifying the mechanisms on these functions but also the application researches for constructing various types of practical opto- and/or electronic devices vigorously proceed.

By incorporating heteroatoms or organometallic complexes into the conjugation system, peculiar functions originated from each element are introduced into the polymers. The synergetic effects with a heteroatom and conjugation system often express the significant characteristics which are unable to be obtained only with C, H, O, and N elements. Moreover, instable species or semistable transition states can be allowed to exist in the conjugated polymers, leading to the expression of superior properties which transcend the intrinsic characteristics of the components. Thus, we can say that the heteroatom-containing polymers are attractive platform for realizing high-performance materials based on the molecular design. We have recently proposed a new concept of "element block" consisting of various groups of heteroatoms.[2] From this viewpoint, we pay attention to the heteroatom- and organometallic complex-containing conjugated polymers as a representative example of element blocks. Initially, the series of organoboron polymers and their optical properties are introduced. Especially, the enhanced emissions by the condensation are described as one of the possible applications for the optical materials. Next, the construction of the main-chain-type conjugated polymers containing gallium and hypervalent silicon are presented. The strategy for receiving conjugated polymers involving unusual states is explained.

15.2 Conjugation through Boron Elements

Boron belongs to Group 13 and has one less electron than carbon. By connecting the conjugation system with the tricoordinated boron atom, the conjugation can be extended via boron due to the vacant p orbital even in the polymers.[3] Various intriguing characteristics were obtained, such as low-lying lowest unoccupied molecular orbitals (LUMOs), ease of reduction, and a bathochromic shift of the absorption and emission spectra. These low-lying LUMO levels are advantageous for the design of organic light-emitting devices.[4] Because of strong electron affinity of tricoordinated boron, the conjugation systems can show the strong affinity toward Lewis bases. Based on this fact, chemical sensors were constructed for anion sensing.[5] As another instance, even rare electronic states such as antiaromaticity were realized in the borafluorene strucutre with the tricoordinate state of boron.[6] However, because of the intrinsic instability of the tricoordinate state of boron, the synthesis and material usages are still challenging. In contrast to the tricoordinate boron, the organoboron complexes with the tetracoordinate boron generally show higher stability. So far, various types of luminescent materials involving the tetracoordinate boron have been reported. For example, boron dipyrromethene (BODIPY) derivatives are well known to have superior optical properties such as large light absorption and emission ability, sharp spectra, and high photostability and are used for a wide variety of applications in both material science and biotechnology.[7]

15.3 Optical Properties of Organoboron Polymers

Initially, optical properties of conjugated polymers containing organoboron complexes are introduced. A wide variety of the complexes containing tetracoordinate boron have been reported, and unique optical properties were observed. In this section, the preparation of the emissive polymers and the control of their optical properties are illustrated. The light in the near-infrared (NIR) region (750 nm to 1000 nm) is versatile not only in optoelectric devices but also in biotechnical tools because of relatively higher permeability through various media.[8] Thus, the development of the NIR-emissive polymers is still strongly required. In particular, the NIR-emissive materials with high quantum yields and sharp emission spectra represented as full-width half-maxima (FWHM) have been desired to obtain pure color. The expansion of the conjugated system is a feasible strategy for obtaining red-shifted emission spectra originated from the narrow bandgap between the frontier orbitals. The main-chain-type polymers composed of aromatic ring-fused BODIPY (boron di(iso)indomethene, BODIN, Scheme 15.1) possessing bright emission in the red light region were prepared.[9] The series of conjugated copolymers having alternating comonomers were obtained using a diiodo-substituted BODIN monomer via the metal-catalyzed cross-coupling reaction. The bathochromic shifts in the absorption and emission bands were observed compared to those of monomers. These data indicate the extension of π-conjugation through the polymer main chains. Notably, it was observed that the synthesized polymers emitted in the range from the deep-red to the NIR region with the emission spectral maxima at around 691–720 nm with high quantum yields (Φ_F = 33%–49%, Table 15.1). Moreover, high photostabilities and thermal stabilities of these polymers were observed.

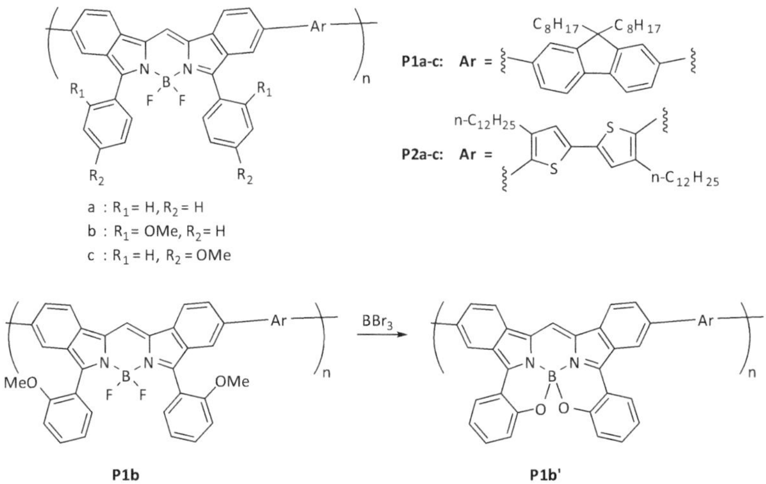

Scheme 15.1 Chemical structures of BODIN-containing polymers and polymer reactions to form intramolecular crosslinking.

Table 15.1 Photoluminescence properties of **4a–d** and **5a–c** in the chloroform solution

Compound	$\lambda_{PL,\,max}$ (nm)	$\Delta\lambda_{1/2}$ (nm)	$\Delta\lambda_{1/2}$ (cm^{-1})	Φ_{PL}
P1a	697	32	660	0.79
P1b	686	38	830	0.67
P1c	710	33	660	0.68
P1b’	758	24	420	0.37
P2a	701	45	910	0.40
P2b	692	45	940	0.51
P2c	714	40	780	0.42

Next, the polymer reaction was carried out for obtaining the red-shifted emission with a very small FWHM based on the above polymers containing BODIN. Burgess et al. have previously reported that by the treatment with tribromoborane, the demethylation and subsequently the intramolecular crosslinking through the intermediary of the boron atom occur in 3,5-*o*-methoxyphenyl-substituted BODIPY (Scheme 15.1).[10] Due to the improvement of the planar conformation at the dipyrromethene nucleus, a dramatic red shift and emission enhancement originated from extended conjugation through BODIPY were observed. On the basis of this result, we applied for constructing a robust conjugation system through the polymer main chain with the BODIN-containing polymers.

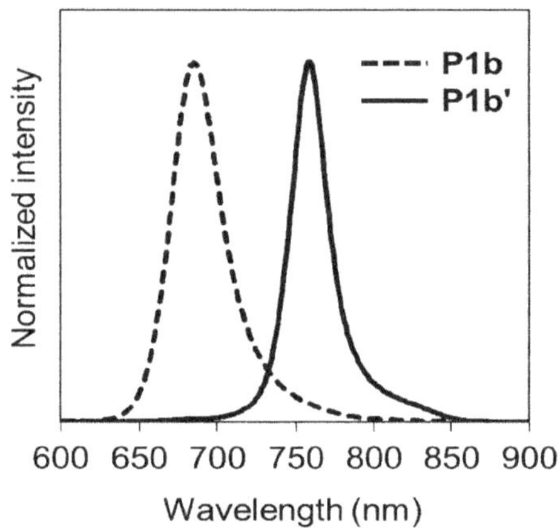

Figure 15.1 Emission spectra of **P1b** before and after BBr_3 treatment. Reprinted with permission from Ref. [9]. Copyright 2013 Wiley-VCH Verlag GmbH & Co. KGaA, Weinheim.

The quantitative demethylation and following intramolecular crosslinking in the presence of tribromoborane in dichloromethane were accomplished with **P1b**. The same reaction was applied to polymer **P1b'**, and the reaction progress was monitored. The significant peak shifts and the disappearance of the methyl groups were observed in ^{1}H and ^{11}B nuclear magnetic resonance (NMR) spectra of the polymer **P1b'**. These data indicate that the formation of the intramolecular crosslinking was confirmed. Furthermore, the similar size-exclusion chromatograms were obtained from **P1b** and **P1b'**. These results suggest the polymer main chains should be maintained during the reactions. According to these characterization data, we summarized that the polymer structure should be quantitatively transformed via efficient intramolecular crosslinking through the intermediary of the boron atom. As shown in Fig. 15.1, the resulting polymer **P1b'** after this treatment presented both drastically larger red-shifted and sharper emission spectrum (λ_{PL} = 758 nm, $\Delta\lambda_{1/2}$ = 420 cm^{-1}) in the NIR region than that of the parent polymer **P1b** with deep-red emission (λ_{PL} = 686 nm, $\Delta\lambda_{1/2}$ = 830 cm^{-1}). These optical properties are feasible for the preparation of the highly emissive materials with pure color.

15.4 Emissive Solid Materials with Aggregation-Induced Emission Properties

In general, the brilliant emissions of conjugated polymers disappear by condensation. This fact means that optical properties should be lost in the solid or film states. These behaviors are called concentration quenching or aggregation-caused quenching (ACQ). On the contrary, it has been reported that some of compounds showed stronger emission in the aggregation states. Tang et al. firstly showed unique behaviors with tetraphenyl-substituted silole compounds in 2001.[11] Only when the compounds were suspended in the poor solvent, the bright emission can be observed. This phenomenon is called aggregation-induced emission (AIE), and the application for highly emissive solid materials is expected to overcome ACQ. In particular, since ACQ directly contributes to the decreases of the device efficiency in the electroluminescent (EL) devices, the development of new AIE-active molecules is still of great interest in the development of modern electric devices. In this section, we explain the recent progress of the molecular design and the development of AIE-active conjugated polymers based on organoboron complexes.

15.5 Development of Boron Ketoiminate Polymers

Boron diketonates (Fig. 15.2) having a tetracoordinate boron atom are one of the classical organoboron complexes. Because of a wide acceptability of the β-diketonate ligand, various metal complexes including boron can be readily constructed. Boron diketonates show high stability toward oxygen and moisture. Only under the strong alkaline condition or in the presence of an excess amount of fluoride anion, the degradation would occur. The most of boron diketonates have the difluoride forms and have high planarity. Boron diketonates have been used as a building block for constructing nanoarchitectures. After the complexation, the significant optical properties can be obtained from boron diketonates. For example, 1,3-diphenyl-2,3-propanedione which is the ligand moiety of boron diketonate mainly focused in this review has slight emission in the visible region. After introduction of difluoroboron, the bright emission can be observed from the resulting boron diketonate complex (boron difluoride dibenzoylmethane) in the blue region. By the substituent effect on the both sides of aryl groups, the optical properties can be tuned. Since the earlier reports on the optical properties, versatile functions have been still now reported. For instance, the regulation of optical properties by physical stimuli was accomplished with the boron diketonate derivatives.[12–14] Boron diketonates commonly have weak electron-accepting ability. Therefore, Lewis bases can often form the interaction with boron diketonate complexes. By connecting with electron-donating units, the intermolecular charge transfer state should be formed, resulting in the strong emissive properties. Moreover, by conjugation with the other functional units and by introducing into the conjugation systems, interesting luminescence properties were presented. However, similarly to other fluorescent molecules, optical properties would be lost in the solid state via the critical ACQ effect. We successfully transformed from the ACQ molecule to the AIE-active one based on boron diketonate derivative. We explain the chemistry of the AIE-active organoboron complexes.

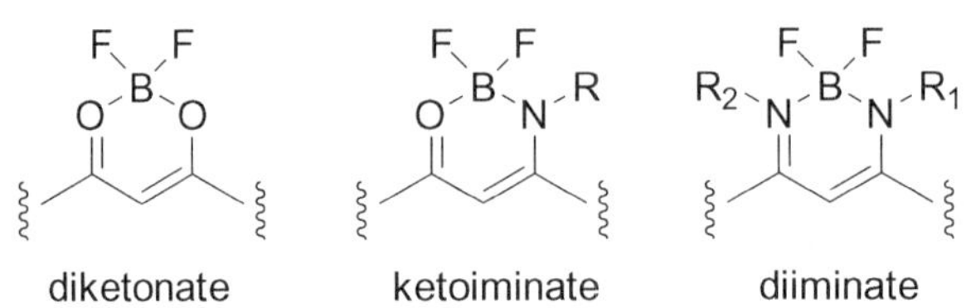

Figure 15.2 Chemical structures of boron complexes.

According to the proposed mechanism in the previous reports, the introduction of bulky groups into the molecules is effective for receiving the AIE property.[11] In the solution state, the excitation energy can be consumed by the intramolecular tumbling of these substituents, resulting in the annihilation of the emission. On the other hand, these molecular motions would be suppressed in the aggregation state. Furthermore, the stacking and intermolecular interaction can be also disturbed by these bulky groups in the condensed state. As a result, the ACQ can be efficiently avoided, leading to the solid-emissive property. To realize the disturbance of the molecular interaction, we designed the boron ketoiminate structure (Fig. 15.3a).[15] By replacing one of oxygen atoms to nitrogen, the functional group can be incorporated. In addition, we expected to introduce the distortion in the molecules by employing the relatively weaker B–N bond than the B–O bond. The series of boron ketoiminates were synthesized and evaluated the physical properties, including the AIE behavior.

Figure 15.3b shows the emission spectra of ketoiminates and diketonate. Boron diketonate exhibited strong blue emission in tetrahydrofuran (THF) (5.0×10^{-5} M, $\Phi_{PL} = 0.91$), and ACQ occurred in the aggregation state ($\Phi_{PL} = 0.36$). On the other hand, ketoiminates showed very weak emissions in the THF solutions (5.0×10^{-5} M, $\Phi_{PL} \leq 0.01$). Notably, the aggregation states of ketoiminates in the mixed solvent of THF/H_2O (1/9 v/v) showed large increases in their emission intensities ($\Phi_{PL} = 0.30$–0.76). These results clearly indicate that ketoiminates are AIE-active molecules. The six-membered ring including B–N bond should play an important role to develop the AIE properties.

The AIE behaviors of boron ketoiminates can be significantly supported by the crystal packing structures. From single-crystal X-ray analyses, it was revealed that a molecular structure of diketonate has high planarity, while ketoiminates involved asymmetric structures and especially distortion in the complex originated from relatively-longer bond lengths of B–N.[16] These data mean that the boron-chelating rings on ketoiminates should be more flexible than that on diketonate. Therefore, the energy consumption should occur in the solution state. As a result, the slight emission was observed from the solutions containing ketoiminates. On the other hand, these molecular tumbling should be suppressed in the aggregation state. Moreover, because of the steric hindrance of the substituent at the nitrogen atom, the intermolecular interaction and stacking could be disturbed, resulting in the inhibition of the ACQ. Thus, the AIE behavior can be obtained. As we mention later, from the following researches with boron diiminates, the similar data were obtained from X-ray crystal analyses as well as from the optical measurements. Thereby, the above mechanism is the most likely explanation on the AIE behavior of boron ketoiminates and diiminates.

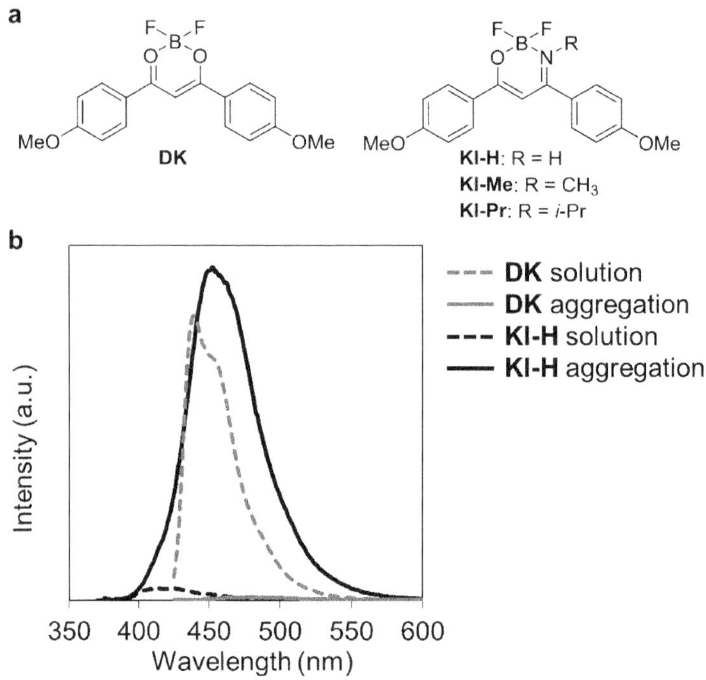

Figure 15.3 Chemical structures (a) and emission spectra (b) of boron complexes. The aggregation was induced by adding water (99 vol%) to the THF solutions. Reprinted with permission from Ref. [15]. Copyright 2013 Wiley-VCH Verlag GmbH & Co. KGaA, Weinheim.

Next, based on boron ketoiminate, the main-chain-type conjugated polymers were synthesized for obtaining the polymeric AIE-active materials.[17] The series of boron ketoiminates shown in Fig. 15.4 were designed. Ketoimine derivatives can form three possible tautomeric structures: ketoimine, enolimine, and enaminoketone forms (Scheme 15.2). It is known that the enaminoketone form has the highest thermal stability of the three tautomers.[18] By the boron complexation, the structures of the ketoiminates would be transformed to be a similar electronic structure to the enolimine form.[19,20] Thereby, it can be said that the tautomerization of ketoimine can be induced by the introduction of boron. Furthermore, it is presumed that the boron-induced tautomerization could change the electronic interaction mediated through the ketoimine moiety. Finally, the boron complexation at the ketoimine skeleton induces the extension of the conjugated system in the polymer main chain. Thus, we have a large possibility to obtain AIE-active materials with tunability to the optical properties based on the boron ketoiminate materials.

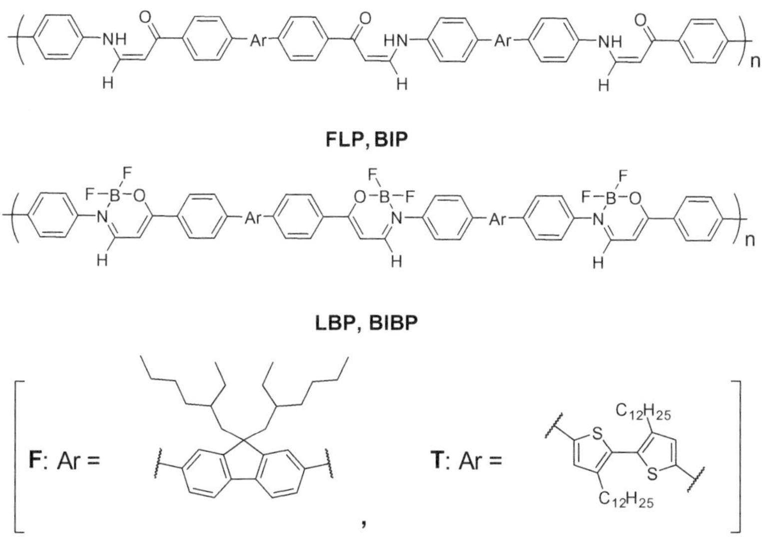

Figure 15.4 Chemical structures of ketoiminate polymers.

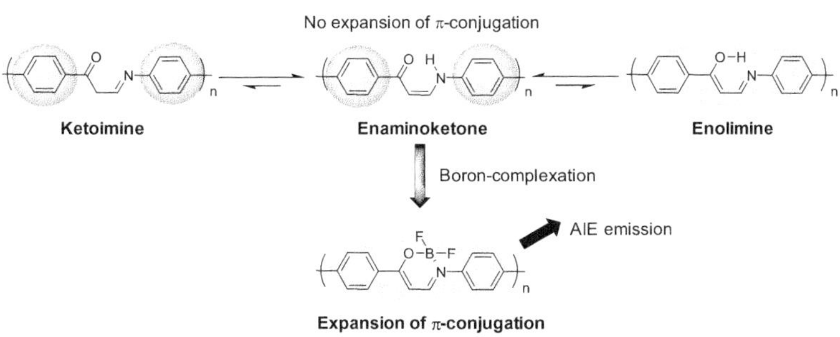

Scheme 15.2 Tautomerization of ketoimines and immobilization at the enaminoketone form by boron complexation.

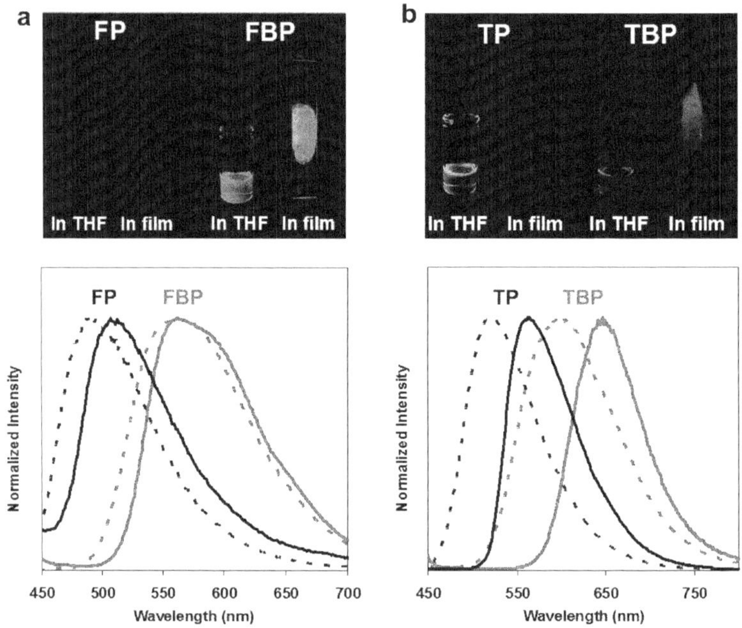

Figure 15.5 Photographs and emission spectra of (a) fluorene-based polymers and (b) bithiophene-based polymers in THF and solid state under UV irradiation. Reprinted with permission from Ref. [17]. Copyright (2014) American Chemical Society.

From the emission spectra, it was found that the boron ketoiminate-containing polymers showed much higher quantum yields in the solid state (**FLBP**: Φ_{PL} = 0.13, **BIBP**: Φ_{PL} = 0.06) than those in THF (**FLBP**: Φ_{PL} = 0.10, **BIBP**: Φ_{PL} = 0.04), as shown in Fig. 15.5. These results clearly indicate that the synthesized polymers should be AIE-active polymers. On the other hand, without boron complexation, the typical ACQ behaviors were obtained from the polymers. In THF, relatively higher quantum yields were observed (**FLP**: Φ_{PL} = 0.01, **BIP**: Φ_{PL} = 0.09), while the solid sample showed the decreases of the quantum yields (**FLP**: Φ_{PL} < 0.01, **BIP**: Φ_{PL} = 0.02). This fact means that the boron complexation to the ketoimine units is responsible for the expression of the AIE properties. It should be mentioned that by changing the comonomer unit, the color of AIE was altered. This result represents that optical properties can be tuned by the molecular design. To demonstrate the AIE behaviors of the boron ketoiminate polymers, the dependency of the emission properties on the solvent compositions was examined in the THF/H_2O solvent mixture system. The enhancement of the emission intensities was observed by increasing the water content ($H_2O \geq 90$ vol%). The aggregation of the polymers should cause the emission enhancement. These data can be summarized that we accomplished to obtain the AIE-active polymers with the color tunability. The introduction of the boron can induce the structural torsion in the conjugated systems. As a result, the ACQ effect might be avoided in the film state.

15.6 Crystallization-Induced Emission from Boron Diiminates

By replacing one of the oxygen atoms with nitrogen in diketonate, the AIE properties were obtained. Next, we investigated the properties of boron diiminates.[21] Compared to boron diketonates and boron ketoiminates, there is room in boron diiminates for functionalization deriving from their two nitrogen atoms which can bond various substituents (Fig. 15.2). Since the regulation of intermolecular interactions in solid state is very important to show strong emission, boron diiminates with high functionality have a high potential as AIE-active materials.

The optical properties of diiminates were investigated. Similarly to ketoiminates, diiminates also presented the AIE properties. The emission spectra showed slight luminescence in THF ($\Phi < 0.01$), whereas crystallization led to large increases in their emission intensities. These results clearly indicate that the synthesized boron diiminates are AIE-active molecules. Next, the modulation of the optical properties was evaluated in the solid states. Two types of solid samples were prepared in the amorphous and crystalline states. In the emission spectra, strong emission of diiminate **DI** was observed from the crystalline samples (Fig. 15.6). On the other hand, the amorphous samples presented weaker emission. It should be emphasized that the crystallization induced approximately 15-fold larger emission efficiencies from the amorphous state ($\Phi_{PL,crystal} = 0.59$, $\Phi_{PL,amorphous} = 0.04$) with a diiminate derivative. From these results, it is clearly shown that boron diiminates are the crystallization-induced emission (CIE)-active units. Moreover, the colors of the CIE can be modulated from blue to red by the substituents.

DI

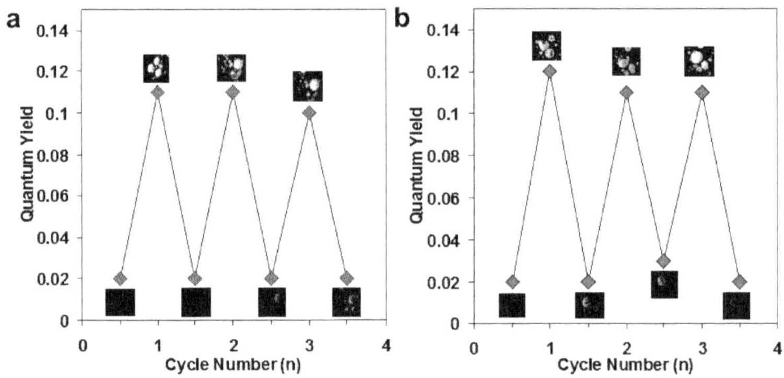

Figure 15.6 Repeated switching of the emission of **DI** between amorphous and crystalline states by (a) fuming–heating and (b) heating–cooling cycles. Reprinted with permission from Ref. [21]. Copyright 2014 Wiley-VCH Verlag GmbH & Co. KGaA, Weinheim.

The diiminate-containing polymers also showed the AIE properties.[22] By changing the comonomer unit, the optical properties such as color can be tuned. Based on these polymers, the dynamic modulation of the optical properties was performed by acid and base sensing experiments (Scheme 15.3). By exposing to the trifluoroacetic acid (TFA) and triethylamine (NEt_3) vapors, the change in the emission properties was monitored with the film samples. Accordingly, the AIE properties of boron diiminates-containing polymers can be dynamically changed. When the film was placed in a small container saturated with TFA vapors, the fumed film showed a blue-shifted emission spectrum than that of the parent sample. Emission color changed from red to yellow. Furthermore, the optical properties of the TFA-exposed film were converted back into the parent state again by exposing to NEt_3 vapor. These data clearly show that the optical properties can be tuned reversibly. The strong donor–acceptor interaction between the amino group and the boron-chelating ring could be suppressed by the protonation of the amino group. As a result, the color was changed. Boron diiminate-containing polymers are promised to be applicable for various kinds of film-type optical sensors.

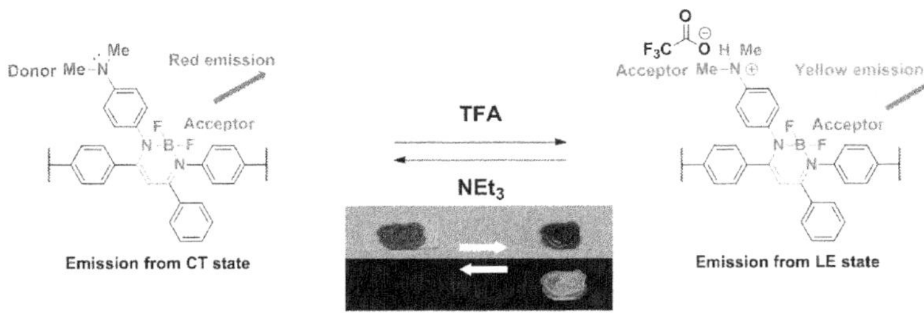

Scheme 15.3 The pH-responsive behavior of the diiminate polymer and a photograph of pH sensing. Reproduced with permission from Ref. [22]. Copyright (2014) American Chemical Society.

To extend this idea about the application of the AIE-active conjugated polymers as the film-type optical sensors, another system was constructed. The films with the AIE-active copolymers composed of fluorenes and the sulfide-substituted boron diiminates were prepared, and the changes in the optical properties by the oxidation were evaluated (Scheme 15.4).[23] Hydrogen peroxide (H_2O_2) is classified as reactive oxygen species and is endogenously produced in energy metabolism. The excessive H_2O_2 induces cell damage, leading to many pathological problems such as diabetes, cardiovascular diseases, cancer, etc. Thereby, the development of the facile detection techniques is strongly required. By soaking the film samples of the polymer in the solution containing H_2O_2, the emission intensity from the film gradually increased during the incubation. The quantum yields were correspondingly enhanced up to three times larger by oxidation at the sulfide groups. This would be attributed to the facilitating charge transfer-emissive fluorescence between the fluorene units and the boron diiminates having the enhanced electron-withdrawing ability by replacing from the sulfide to the sulfoxide groups. These data indicate that the synthesized polymer can work as a film-type optical sensor with turn-on fluorescence. Hence, it is proposed that the AIE-active polymeric materials should be a suitable platform as a bioprobe for longitudinal monitoring the biorelated reactions.

Scheme 15.4 Schematic illustration for H_2O_2 detection with the film-type sensor having oxidation-induced AIE property. Reproduced from Ref. [23] with permission from the Royal Society of Chemistry.

15.7 Conjugated Polymers Containing Gallium Atoms

The boron atom belongs to Group 13, and except for the above-mentioned, versatile properties such as opto- and/or electric materials have been found from the organoboron-containing conjugated molecules and polymers.[24–30] By replacing the heavier atoms, the expression of further functions is expected. In contrast to organoboron complexes, there are limited numbers of reports on electronic properties of organogallium compounds.[31,32] Because of the instability to air and moisture, the synthesis of the organogallium compounds is challenging. Thereby, the role of gallium atoms in the conjugation and the contribution to the optical properties is still unclear. Recently, we have reported on synthesis of air- and moisture-stable gallafluorenes by employing the protection group to form tetracoordinate gallium atom (Fig. 15.7).[33] Because of high durability, a series of optical properties can be monitored. However, the electronic interaction between gallium atom and the conjugation system was not achieved yet. With this background, we started developing stable conjugated polymers including tetracoordinate gallium atoms and aimed to reveal the conjugation with gallium atoms experimentally.[34] We designed and synthesized gallium-containing polymers with 2,4-di-*tert*-butyl-6-[(dimethylamino) methyl]phenyl (Mamx) groups. By the coordination with the amine, the improvement of the stability and the immobilization of the conformation around the gallium atoms were expected.[35] The polymerization products were obtained via conventional palladium-catalyzed cross-coupling reactions with di-halogenated monomers. From the ^{1}H NMR spectra, the polymers provided similar peaks with the corresponding monomers. From these data, we concluded that the coupling polymerization progressed successfully without cleavage of bonds of gallium and carbon atoms. The synthesized polymers showed high stability and solubility in the common organic solvents enough for the optical measurements.

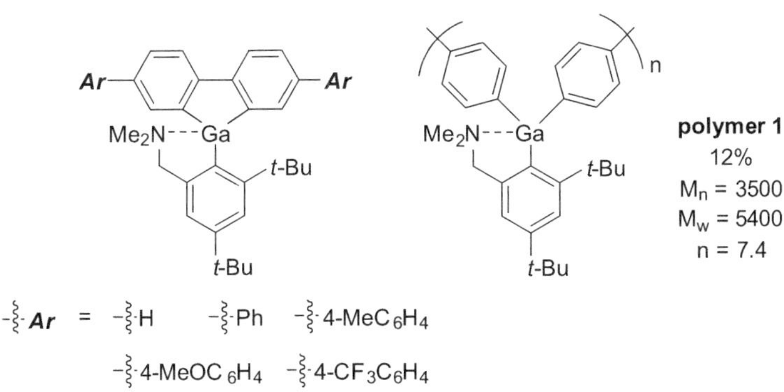

Figure 15.7 Chemical structures of tetracoordinate gallium compounds.

The UV-Vis absorption properties of biphenyl, model **3**, and polymer **1** were compared in the chloroform solutions to examine the electronic structures at the ground state (Fig. 15.8). Model **3** showed the absorption peak at 263 nm and the onset of a peak at 294 nm. The top and onset of the absorption peak of polymer **1** were observed in the red-shifted regions compared to those of model **3** (275 nm and 306 nm), respectively. Both compounds showed larger bathochromic shifts in the absorption spectra by 13 nm and 25 nm compared to that of biphenyl. Since polymer **1** showed the red-shifted absorption peak relative to model **3**, it is clearly indicated that each biphenyl unit in the polymer should electronically interact through the gallium atoms. In other words, the electronic conjugation is obtained at the ground state. Emission properties of organogallium compounds including polymers are under investigation.

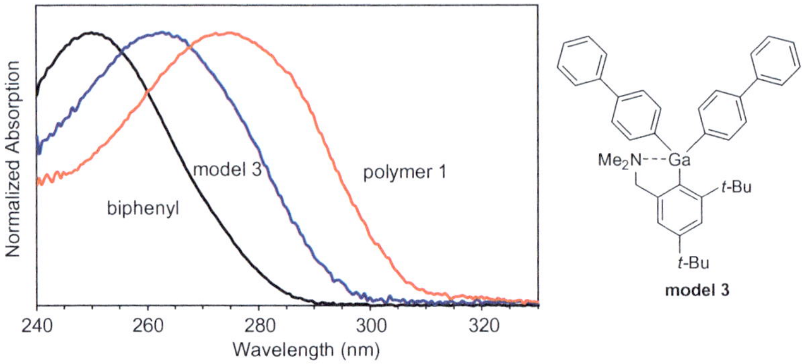

Figure 15.8 UV-Vis absorption spectra of biphenyl, model **3**, and polymer **1** in chloroform. Reprinted from Ref. [34] with permission from the Royal Society of Chemistry.

15.8 Conjugated Polymers Containing Hypervalent Silicon Atoms

Organosilicon compounds such as siloles have been also paid attention for the application of optoelectronic devices.[36–39] Because of the conjugation effect between σ* (Si–C) and π* (C–C) orbitals, the low-lying energy levels of their LUMO are induced. Thereby, organosilicon compounds are expected to be low-bandgap materials which are promised to improve the conversion efficiency of optoelectronic devices.[40] Indeed, the increases of open-circuit voltage in solar cells were reported.[41] Moreover, the conjugated polymers involving the pentacoordinate states of silicon were synthesized by the dative bond formation with nitrogen atoms in the ligands.[42–47] Their specific electronic structures involving three-center four-electron bonds as well as intrinsic properties of silicon were examined. However, due to the weak dative bonds, there is still much room to establish robust synthetic methodology for pentacoordinate organosilicon complexes and explore unique functions originated from the pentacoordinate states of silicon.

By employing the benzoquinoline ligand, the series of pentacoordinate organosilicon complexes can be obtained (Fig. 15.9).[48–50] Moreover, the conformation changes around the silicon center were observed by changing the substituents on the pentacoordinate silicon atoms. As a result, the tuning of the emission was accomplished. At this stage, there were several reports on the substituent effects on the optical properties of the complexes. On the other hand, electronic interaction via the conjugation systems has not been confirmed yet. Therefore, we next aimed to incorporate the pentacoordinate silicon complexes into the polymers and extend the conjugation system through the main chains. Especially, we expected to observe the drastic changes of optical properties by the substituents via the conjugation system through the polymer main chains including pentacoordinate silicon. In this section, our recent works on the synthesis and comonomer-dependent optical properties of the conjugated polymers composed of the pentacoordinate silicon complexes are explained.

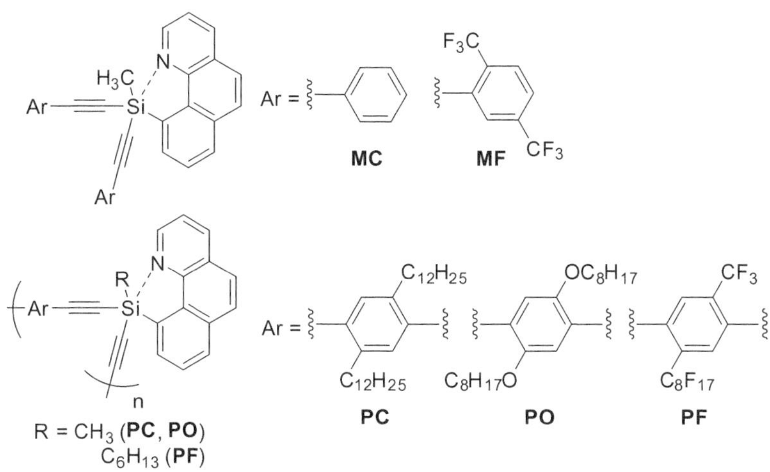

Figure 15.9 Chemical structures of pentacoordinate silicon complexes.

The polymers were prepared via Sonogashira–Hagihara cross-coupling reactions with various kinds of the comonomers.[50] The synthesis and characterization were performed with the conventional methods. Initially, the optical properties of the synthesized polymers are demonstrated from UV-Vis absorption measurements. Figure 15.10a shows the spectra from the THF solutions. The model compounds **MC** and **MF** provided slight absorption bands over 300 nm assigned as the prohibited transition between frontier orbitals of the benzo[*h*]quinoline ligand.[51] On the other hand, polymers

PO and **PF** showed the absorption bands with the peaks at 400 nm and 360 nm, respectively. Intrinsic absorptions observed from 250 nm to 300 nm assigned as the π–π* transition of phenylene-ethynylene were hardly shifted with the decreases of the molar extinction coefficients. These data indicate that the π-conjugation should be formed through the polymer main chains, leading to the appearance of new absorption bands in the longer-wavelength region than those of intrinsic absorption bands. In addition, because of the comonomer-dependent peak shifts, it is implied that the new absorption bands might be induced by the electronic interactions between the pentacoordinate silicon complex and the fluoroalkyl- or alkoxy-substituted benzenes.

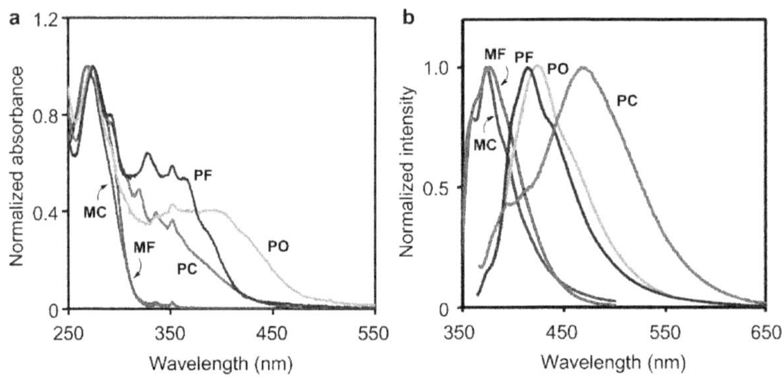

Figure 15.10 UV-Vis (a) and emission (b) spectra of the THF solutions containing pentacoordinate silicon compounds. Reprinted from Ref. [50] with permission from the Royal Society of Chemistry.

Emission properties of the pentacoordinate silicon complexes including the polymers were measured in the same THF solutions (Fig. 15.10b). Models **MC** and **MF** presented weak emissions with peaks around 370 nm (**MC**: Φ_F = 0.05, **MF**: Φ_F = 0.08). In contrast, emission bands with stronger intensity were obtained from the polymers in the longer-wavelength region. Polymer **PC** showed the stronger emission with the peak at 474 nm than those of monomers (Φ_F = 0.10). This result also suggests the extension of the π-conjugation through the pentacoordinate silicon. Moreover, the emission peaks can be modulated by the electron-withdrawing or electron-donating group at the comonomers: The emission band was obtained from polymer **PO** in the relatively shorter-wavelength region ($\lambda_{em,max}$ = 463 nm, Φ_F = 0.19). The polymer **PF** gave the strongest emission with the peak at 417 nm (Φ_F = 0.24). From these results, it is clearly indicated that the optical properties of the pentacoordinate silicon complex can be tuned by altering the type of comonomers through the conjugation system. According to the mechanism studies, the intramolecular charge transfer through the conjugation systems at the excited state should occur in the polymers including the pentacoordinate silicon. This fact means that the elongation of the conjugation system over the pentacoordinate silicon should be achieved in the main chains.

15.9 General Synthesis of Heteroatom-Containing Conjugated Polymers

Similar synthetic reactions for common conjugated polymers are applicable for the preparation of the conjugated polymers presented here. If the enough stability can be obtained from the complexes with the adequate functional groups, Suzuki–Miyaura, Sonogashira–Hagihara, Mizorogi–Heck, Negishi, Yamamoto, and Stille cross-coupling reactions can be used for preparing alternating copolymers. If the monomer has enough durability under oxidation, oxidative polymerization can provide the homopolymers. By using the ethynyl linker obtained from Sonogashira–Hagihara coupling, the steric hindrance between the monomer units can be released. Therefore, the expansion of the conjugated systems can be expected. The replacement of benzene to thiophene is also valid design to improve the planarity for extending the conjugation system. The typical protocols for synthesis and characterization are described here to obtain the polymers: For the polymerization, the monomer unit is needed to have dual polymerizable functional groups such as halogen, boronic acid, ethynyl, and vinyl groups. The cross coupling reactions were executed in the presence of the metal catalysts with the comonomers. The monomer consumption and the chain growth can be monitored with the profiles of a size-exclusion chromatography. After finishing the increase of the molecular weight of the product, the reaction solution is poured into the poor solvents (methanol, and so on) to remove the metal species and monomers.

To wash thoroughly, the products were dissolved and reprecipitated several times and if needed treated with the Soxhlet extraction. After removing the solvent, the polymers were often obtained as powders. The number-average molecular weights (M_n) and the molecular weight distributions (M_w/M_n) of the polymers were evaluated by a size-exclusion chromatography with the standard polymers. The chemical structures of the polymeric products were characterized with the series of NMR measurements. From the integration ratios of the peaks in the ^{1}H NMR spectra between the organometallic complex and the comonomer, the formation of the alternating sequence can be confirmed. In particular, when the Sonogashira–Hagihara coupling is used in the polymerization reaction, the defect of the monomer unit is possible via the Gracer-type side reaction which can also generate highly emissive polymers. The mass measurements with the matrix-assisted laser desorption and ionization method are also a valid tool to evaluate the chain length and the end groups. The formation of the complexes can be confirmed by comparing the chemical shifts of the peaks in the multinuclear NMR spectra to those of the model compounds. To investigate the expansion of the conjugation system, we generally prepared the polymers having at least five repeat units for ignoring the undesired effect on the optical properties from the end groups. The synthesized polymers were dissolved into the organic solvents with 10^{-6} M concentration of the monomer unit. The spin-coating or casting onto the substrates provides the film sample. The quantum yields of the emission were determined as an absolute value using the integration sphere.

15.10 Conclusions

The construction of the heteroatom- and organometallic complex-containing polymers are focused in this review. In the case of organoboron complexes, so far, a relatively larger number of fundamentally unique and practically useful functions have been found. On the other hand, there is still much room to explore the robust synthesis and functions of the heteroatom-containing conjugation system with heavier elements at this stage. The conjugated polymers are a valid platform not only for utilizing peculiar functionality of each element but also for expressing superior properties to those of the components. Furthermore, it is relatively easy to obtain practical devices with these polymers because of intrinsic superior material properties of polymers. Therefore, there are huge potential to obtain the unknown functions and materials which have never been seen before.

References

1. Hirose, A., Tanaka, K., Tamashima, K., Chujo, Y. *Tetrahedron Lett.*, **55**, 6477–6481 (2014).
2. Chujo, Y., Tanaka, K. *Bull. Chem. Soc. Jpn.*, **88**, 633–643 (2015).
3. Matsumi, N., Naka, K., Chujo, Y. *J. Am. Chem. Soc.*, **120**, 5112–5113 (1998).
4. Sun, C., Hudson, Z. M., Helander, M. G., Lu, Z.-H., Wang, S. *Organometallics*, **30**, 5552–5555 (2011).
5. Miyata, M., Chujo, Y. *Polym. J.*, **34**, 967–969 (2002).
6. Braunschweig, H., Kupfer, T. *Chem. Commun.*, **47**, 10903–10914 (2011).
7. Ulrich, G., Ziessel, R., Harriman, A. *Angew. Chem. Int. Ed.*, **47**, 1184–1201 (2008).
8. Weissleder, R. *Nat. Biotechnol.*, **19**, 316–317 (2001).
9. Yoshii, R., Nagai, A., Tanaka, K., Chujo, Y. *J. Polym. Sci., Part A: Polym. Chem.*, **51**, 1726–1733 (2013).
10. Kim, H., Burghart, A., Welch, M. B., Reibenspies, J., Burgess, K. *Chem. Commun.*, 1889–1890 (1999).
11. Luo, J., Xie, Z., Lam, J. W. Y., Cheng, L., Chen, H., Qiu, C., Kwok, H. S., Zhan, X., Liu, Y., Zhu, D., Tang, B. Z. *Chem. Commun.*, 1740–1741 (2001).
12. Zhang, G., Palmer, G. M., Dewhirst, M. W., Fraser, C. L. *Nat. Mater.*, **8**, 747–751 (2009).
13. Zhang, G., Singer, J. P., Kooi, S. E., Evans, R. E., Thomas, E. L., Fraser, C. L. *J. Mater. Chem.*, **21**, 8295–8299 (2011).
14. Zhang, G., Lu, J., Sabat, M., Fraser, C. L. *J. Am. Chem. Soc.*, **132**, 2160–2162 (2010).
15. Yoshii, R., Nagai, A., Tanaka, K., Chujo, Y. *Chem. Eur. J.*, **19**, 4506–4512 (2013).
16. Macedo, F. P., Gwengo, C., Lindeman, S. V., Smith, M. D., Gardinier, J. R. *Eur. J. Inorg. Chem.*, 3200–3211 (2008).
17. Yoshii, R., Tanaka, K., Chujo, Y. *Macromolecules*, **47**, 2268–2278 (2014).

18. Da Silva, M. A. V. R., Da Silva, M. D. M. C. R., Paiva, J. P. A., Nogueira, I. M. C. S., Damas, A. M., Barkley, J. V., Harding, M. M., Akello, M. J., Pilcher, C. *J. Chem. Soc., Perkin Trans. II*, 1765–1769 (1993).

19. Itoh, K., Okazaki, K., Fujimoto, M. *Aust. J. Chem.*, **56**, 1209–1214 (2003).

20. Macedo, F. P., Gwengo, C., Lindeman, S. V., Smith, M. D., Gardinier, J. R. *Eur. J. Inorg. Chem.*, 3200–3211 (2008).

21. Yoshii, R., Hirose, A., Tanaka, K., Chujo, Y. *Chem. Eur. J.*, **20**, 8320–8324 (2014).

22. Yoshii, R., Hirose, A., Tanaka, K., Chujo, Y. *J. Am. Chem. Soc.*, **136**, 18131–18139 (2014).

23. Hirose, A., Tanaka, K., Yoshii, R., Chujo, Y. *Polym. Chem.*, **6**, 5590–5595 (2015).

24. Tanaka, K., Chujo, Y. *Macromol. Rapid Commun.*, **33**, 1235–1255 (2012).

25. Tanaka, K., Yamane, H., Yoshii, R., Chujo, Y. *Bioorg. Med. Chem.*, **21**, 2715–2719 (2013).

26. Yeo, H., Tanaka, K., Chujo, Y. *Macromolecules*, **46**, 2599–2605 (2013).

27. Yoshii, R., Yamane, H., Tanaka, K., Chujo, Y. *Macromolecules*, **47**, 3755–3760 (2014).

28. Kajiwara, Y., Nagai, A., Tanaka, K., Chujo, Y. *J. Mater. Chem. C*, **1**, 4437–4444 (2013).

29. Yoshii, R., Yamane, H., Nagai, A., Tanaka, K., Taka, H., Kita, H., Chujo, Y. *Macromolecules*, **47**, 2316–2323 (2014).

30. Yoshii, R., Nagai, A., Tanaka, K., Chujo, Y. *Macromol. Rapid Commun.*, **35**, 1315–1319 (2014).

31. Bagh, B., Gilroy, J. B., Staubitz, A., Müller, J. *J. Am. Chem. Soc.*, **132**, 1794–1795 (2010).

32. Wang, X. M., Fan, R. Q., Qiang, L. S., Li, W. Q., Wang, P., Zhang, H. J., Yang, Y. L. *Chem. Commun.*, **50**, 5023–5026 (2014).

33. Matsumoto, T., Tanaka, K., Chujo, Y. *J. Am. Chem. Soc.*, **135**, 4211–4214 (2013).

34. Matsumoto, T., Onishi, Y., Tanaka, K., Chujo, Y. *Chem. Commun.*, **50**, 15740–15743 (2014).

35. Bagh, B., Schatte, G., Green, J. C., Müller, J. *J. Am. Chem. Soc.*, **134**, 7924–7936 (2012).

36. Yuan, M., Yang, P., Durban, M. M., Luscombe, C. K. *Macromolecules*, **45**, 5934–5940 (2012).

37. Schroeder, B. C., Sshraf, R. S., Thomas, S., White, A. J. P., Biniek, L., Nielsen, C. B., Zhang, W., Huang, Z., Tuladhar, P. S., Watkins, S. E., Anthopoulos, T. D., Durrant, J. R., McCulloch, I. *Chem. Commun.*, **48**, 7699–7701 (2012).

38. Wang, J. Y., Hau, S. K., Yip, H. L., Davies, J. A., Chen, K. S., Zhang, Y., Sun, Y., Jen, A. K. Y. *Chem. Mater.*, **23**, 765–767 (2011).

39. Ashraf, R. S., Chen, Z., Leem, D. S., Bronstein, H., Zhang, W., Schroeder, B., Geerts, Y., Smith, J., Watkins, S., Anthopoulos, T. D., Sirringhaus, H., de Mello, J. C., Heeney, M., McCulloch, I. *Chem. Mater.*, **23**, 768–770 (2011).

40. Yamaguchi, S., Tamao, K. *J. Chem. Soc., Dalton Trans.*, 3693–3702 (1998).

41. Ohshita, J. *Macromol. Chem. Phys.*, **210**, 1360–1370 (2009).

42. Douglas, W. E., Guy, D. M. H., Kar, A. K., Wang, C. *Chem. Commun.*, 2125–2126 (1998).

43. Douglas, W. E., Klapshina, L. G., Kuzhelev, A. S., Peng, W., Semenov, V. V. *J. Mater. Chem.*, **13**, 2809–2813 (2003).

44. Douglas, W. E., Benfield, R. E., Antipov, O. L., Klapshina, L. G., Kuzhelev, A. S., Guy, D. M. H., Jones, R. G., Mustafa, A., Domrachev, G. A. *Phys. Chem. Chem. Phys.*, **2**, 3195–3201 (2000).

45. Antipov, O. L., Domrachev, G. A., Douglas, W. E., Guy, D. M. H., Klapshina, L. G., Koritin, A. I., Kuzhelev, A. S., Semenov, V. V. *Appl. Organometal. Chem.*, **14**, 640–643 (2000).

46. Douglas, W. E., Kuzhelev, A. S., Yurasova, I. V., Antipov, O. L., Klapshina, L. G., Semenov, V. V., Domrachev, G. A., Lopatina, T. I., Guy, D. M. H. *Phys. Chem. Chem. Phys.*, **4**, 109–114 (2002).

47. Bushuk, S. B., Carré, F. H., Guy, D. M. H., Douglas, W. E., Kalvinkovskya, Y. A., Klapshina, L. G., Rubinov, A. N., Stupak, A. P., Bushuk, B. A. *Polyhedron*, **23**, 2615–2623 (2004).

48. Tokoro, Y., Yeo, H., Tanaka, K., Chujo, Y. *Chem. Commun.*, **48**, 8541–8543 (2012).

49. Tokoro, Y., Tanaka, K., Chujo, Y. *Org. Lett.*, **15**, 2366–2369 (2013).

50. Tokoro, Y., Yeo, H., Tanaka, K., Chujo, Y. *Polym. Chem.*, **4**, 5237–5242 (2014).

51. Hamai, S. *Bull. Chem. Soc. Jpn.*, **79**, 1039–1045 (2006).

Biography

Kazuo Tanaka received his PhD in 2004 from Kyoto University, Japan, and worked in Stanford University (USA), Kyoto University, and RIKEN as a postdoctoral fellow. Until this time, he has worked on synthetic chemistry using DNA for developing DNA wires to construct a molecular circuit and fluorescent nucleobases to detect point mutations in DNA. In 2007, he moved to the Department of Polymer Chemistry, Graduate School of Engineering, Kyoto University, and in 2014, he was promoted to lecturer and then associate professor in 2015. His research projects especially focus on the design of new functional materials based on heteroatom-containing polymers and organic–inorganic polymer hybrids.

Yoshiki Chujo completed his PhD at Kyoto University in 1980 and then joined Nagoya University as an assistant professor in 1981. In 1983, he joined the group of J. McGrath at Virginia Polytechnic Institute as a postdoctoral research fellow. He returned to Kyoto University as a lecturer in 1986 and has been professor of polymer chemistry there since 1994. His research interests focus on polymer synthesis, inorganic polymers, and polymeric hybrid materials.

Chapter 16

Nanoscale Carbon Materials: Carbon Nanotubes

Takahiro Maruyama
Department of Applied Chemistry, Meijo University, Shiogamaguchi, Tempaku, Nagoya 468-8502, Japan
takamaru@meijo-u.ac.jp

Nanocarbon materials composed of sp^2-hybridized bonding, as represented by carbon nanotubes (CNTs) and graphene, have fascinating properties due to their mechanical strength and electrical conductivity. In particular, CNTs can have various structures depending on their chirality and several electronic properties that may be useful for various applications in the future. The properties of CNTs are reviewed in this chapter, beginning with the fundamental theory about the bonding of carbon atoms. Then, recent studies regarding single-walled CNTs, especially the growth method using Pt catalysts, are presented as potential techniques for developing applications of CNTs. The remaining issues that must be solved regarding the application of CNTs are mentioned, and the techniques for growing CNTs to approach these issues are reviewed.

Conjugated Objects: Developments, Synthesis, and Applications
Edited by Atsushi Nagai and Koji Takagi

ISBN 978-981-4774-03-1 (Hardcover), 978-1-315-22962-1 (eBook)
www.panstanford.com

16.1 Introduction

Starting from the review of bonding and hybridization orbitals of small hydrocarbon molecules, this chapter then introduces the structure and electronic state of typical nanocarbon materials, such as carbon nanotubes (CNTs) and graphene, and finally deals with the recent topics of synthesis of single-walled carbon nanotubes (SWCNTs). Readers require knowledge of basic quantum chemistry, especially the electron configuration, and the hybridization orbitals of small molecules. Readers also need to understand the basis of molecular orbital theory, that is, the Hückel method. As part of the overview of nanocarbon materials, the readers should know basic band theory to understand the electronic states of carbon atom 2D networks and the density of states of CNTs. In the last part, the recent topic of synthesis of CNTs is presented simply based on elementary knowledge of materials science, although readers would benefit from some prior understanding of analysis techniques such as electron microscopy and photoluminescence (PL) spectroscopy to comprehend the details of CNT properties.

16.2 Hybridization of Orbitals of Carbon Atom

Carbon is the sixth element in the periodic table and has six electrons, two in the 1s state, and four in the valence state occupying the 2s and 2p orbitals (Fig. 16.1). The two electrons occupying the 1s orbitals are tightly bound to the nucleus and are positioned at approximately 285 eV below the vacuum level. These electrons are called core electrons, and there is a small possibility that the 1s orbitals will overlap neighboring atomic orbitals. These core electrons shield the nucleus, reducing the effective positive charge. The four electrons in the 2s and 2p orbitals are distributed in the outer shells of the carbon atoms and can participate in atomic bonding. The energy difference between the four electrons in the 2s and 2p orbitals is small, and is less than the energy gained if a C–C bond is formed. Consequently, when carbon atoms bond with each other, the 2s and 2p orbitals can mix and form sp^n (n = 1, 2, 3)-hybridized orbitals (Fig. 16.2).

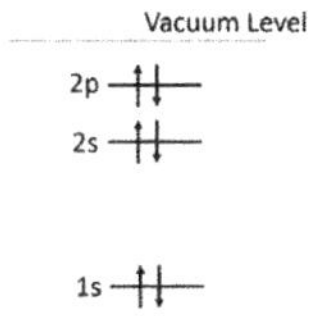

Figure 16.1 Energy levels of one carbon atom.

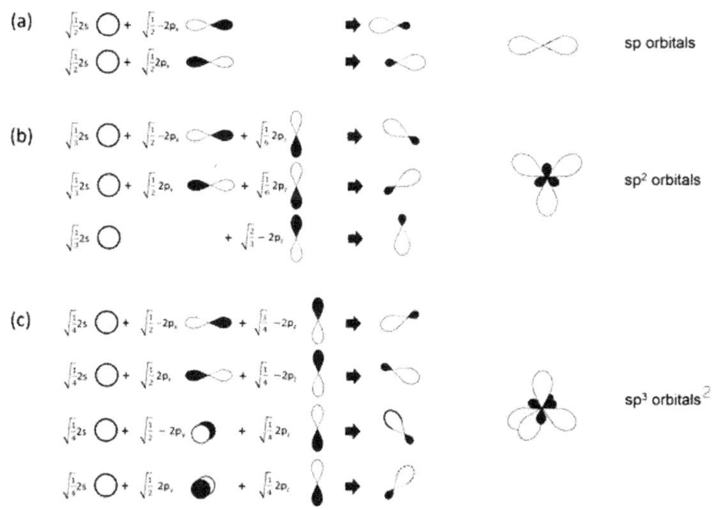

Figure 16.2 Schematics for the sp, sp^2, and sp^3 hybridization and bonding formation.[1]

One-half of the 2s orbital on a carbon atom can be mixed with one-half of the $2p_x$ orbital on a carbon atom to create a degenerate pair of hybrid orbitals. These hybrid orbitals are called sp hybrid orbitals, where the $2p_y$ and $2p_z$ orbitals are unused (Fig. 16.2a). The 2s orbital on carbon can also mix with the $2p_x$ and $2p_z$ orbitals, taking one-third of the 2s orbital successively in each case and one-half of the $2p_x$ and one-sixth of the $2p_z$ in two combinations to create two hybrids. The remaining two-thirds of the $2p_z$ orbital form the third hybrid. These three orbitals are called sp^2 hybrid orbitals (Fig. 16.2b), where the unused $2p_y$ orbital is left perpendicular to the planes, including the three sp^2 orbitals. For the sp^3 hybrids, there are four hybrid orbitals, two combinations formed by the mixing of one-quarter of the 2s orbital with one-half of the $2p_x$ and one-quarter of the $2p_z$ orbital to create one pair of hybrids and two combinations formed by the mixing of one-quarter of the 2s orbital with one-half of the $2p_y$ and one-quarter of the $2p_z$ orbitals to form the other two pairs of hybrids (Fig. 16.2). The resultant sp^n (n = 1, 2, 3)-hybridized orbitals are shown in Fig. 16.2 in the conventional styles.

Most of the C–C and C–H bonds in the organic molecules are formed by these hybridized orbitals. Methane is formed by the 1s orbitals of the four hydrogen atoms and the 2s, $2p_x$, $2p_y$ and $2p_z$ orbitals of one carbon atom. Here, the 1s orbital of a hydrogen atom overlaps an sp^3 hybrid orbital on carbon, leading to a σCH-bonding orbital or a σ*CH antibonding orbital. Four bonds between carbon and hydrogen atoms, in which two electrons are shared in each bond, point toward the corners of a regular tetrahedron. Consequently, methane forms a regular tetrahedron shape and the bond angle between each C–H bond is 109.5° (Fig. 16.3). Ethane has a C–C bond with six C–H bonds. The 1s orbitals in the hydrogen atoms and the 2s, $2p_x$, $2p_y$ and $2p_z$ orbitals in the two carbon atoms result in a set of seven bonding molecular orbitals. In the hybridization picture, the C–C bond results from the bonding overlap of the sp^3-hybridized orbitals on carbon with each other. This bond is largely made up of 2s orbitals on carbon with some $2p_x$ orbitals and is called a σ-bond. The bond length and bond strength of the C–C bond are 154 pm and 377 kJ/mol, respectively. The bond angle between each C–H bond is slightly different from 109.5° (Fig. 16.3). For ethylene, orbitals are formed from the 1s orbitals of the four hydrogen atoms and the 2s, $2p_x$, $2p_y$, and $2p_z$ orbitals of the two carbon atoms. The 1s orbitals on the hydrogen and the 2s, $2p_x$ and $2p_y$ orbitals on the carbon make up the sp^2 bonds, forming σ-bonds between the two carbons. The $2p_z$ orbitals of the carbon atoms, which do not participate in σ-bonding, overlap and form a π-bond. Consequently, both σ-bonds from overlapping sp^2-sp^2 hybrids and π-bonds from overlapping $2p_z$–$2p_z$ form the C=C bond in ethylene, where four electrons are shared between the two carbon atoms. The bond length and bond strength of the C=C bond are 134 pm and 728 kJ/mol, respectively, which indicates the formation of stronger bonds than ethane (Fig. 16.3). Acetylene is a representative molecule composed of sp hybrid orbitals and is formed by bonding between the 1s orbitals of two hydrogen atoms and the 2s, $2p_x$, $2p_y$ and $2p_z$ orbitals of two carbon atoms. In this molecule, the sp hybrid orbitals of two carbon atoms overlap with each other, forming two sp–sp bonds or σ-bonds. In acetylene, the other two p_y and p_z orbitals of each carbon atom overlap, forming two π-bonds. Consequently, the two carbon atoms in acetylene share six electrons; that is, a triple bond is formed. The other sp orbital in the carbon atom forms a σ-bond with a hydrogen atom. Due to the sp–sp bonds, the H–C–C bond angle in acetylene is 180° and the bond length of C–C and its bond strength are 120 pm and 965 kJ/mol (Fig. 16.3), respectively, indicating the strongest bond formation among the three molecules.

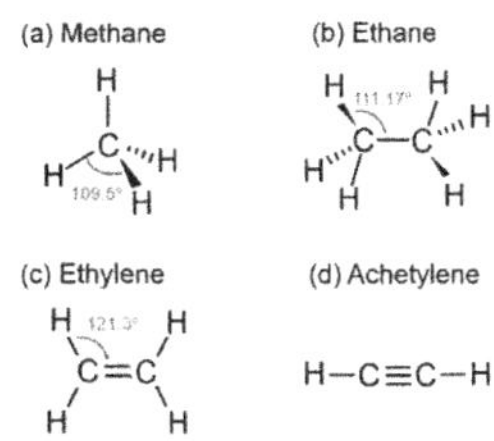

Figure 16.3 Molecular structure with the bond angles and lengths for methane, ethane, ethylene, and acetylene.

16.3 Stabilization by π-Bonding

Here, stabilization by π-bonding is shown using the small molecules ethane and butadiene (Fig. 16.4).[2] In ethane, π-bonding occurs between two carbon atoms, which are accompanied with σ-bonding composed of the sp^2 hybrids. This π-bond is formed by the $2p_z$ orbitals of the carbon atoms. The π-orbital of ethane, ψ_π, can be written as follows:

$$\Psi_\pi = C_1\phi_{2p_{z1}} + C_2\phi_{2p_{z2}} \tag{16.1}$$

where $\phi_{2p_{z1}}$ and $\phi_{2p_{z2}}$ are the $2p_z$ orbitals of the two carbon atoms of ethane (C_1 and C_2 in Fig. 16.4a).

The secular equation for this wave function is represented as follows:

$$\begin{vmatrix} H_{11} - ES_{11} & H_{12} - ES_{12} \\ H_{21} - ES_{21} & H_{22} - ES_{22} \end{vmatrix} = 0 \tag{16.2}$$

Here,

$$H_{ij} = \int \phi_i H \phi_j d\tau \qquad (16.3)$$

and

$$S_{ij} = \int \phi_i \phi_j d\tau \qquad (16.4)$$

where $\phi_i, \phi_j = \phi_{2p_{z1}}$ or $\phi_{2p_{z2}}$ and H is the Hamiltonian of this system.

In ethane, $\phi_{2p_{z1}}$ and $\phi_{2p_{z2}}$ are equivalent; therefore, $H_{11} = H_{22}$ and $H_{12} = H_{21}$. Based on Hückel theory, $S_{ij} = \delta_{ij}$, and Eq. 16.2 can be described by

$$\begin{vmatrix} \alpha - E & \beta \\ \beta & \alpha - E \end{vmatrix} = 0 \qquad (16.5),$$

where $H_{ii} = \alpha$ and $H_{ij} = \beta\ (i \neq j)$. The solution of this secular equation is given by

$$E = \alpha \pm \beta \qquad (16.6)$$

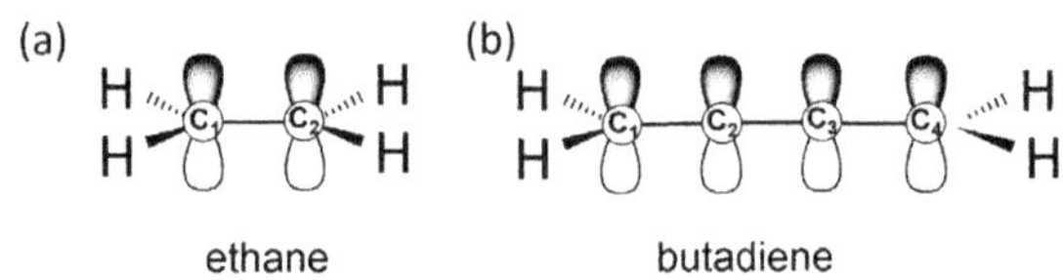

Figure 16.4 Molecular structures of (a) ethane and (b) butadiene.

From experimental data, the value of β has been determined to equal −75 kJ/mol, and the resultant energies $E = \alpha \pm \beta$ correspond to the bonding and antibonding orbitals of ethane. That is, the energy difference in the bonding and antibonding orbitals is $2\beta = -150$ kJ/mol. The total energy of the two π-electrons of ethane is

$$E = 2\alpha \pm 2\beta \qquad (16.7)$$

Butadiene is composed of four carbon atoms, all of which have $2p_z$ orbitals that form π-bonds. In Fig. 16.4b, the three C–C bonds in butadiene are shown as straight, for simplicity, although butadiene has conformers. If a linear alignment is assumed, the π-bonding is described as follows:

$$\Psi_\pi = \sum_{j=1}^{4} C_j \phi_{2p_{zj}} \qquad (16.8)$$

where ϕ_{2pzj} is the jth carbon atom in butadiene. Using a similar procedure based on Hückel theory, the secular equation is described as follows:

$$\begin{vmatrix} \alpha - E & \beta & 0 & 0 \\ \beta & \alpha - E & \beta & 0 \\ 0 & \beta & \alpha - E & \beta \\ 0 & 0 & \beta & \alpha - E \end{vmatrix} = 0 \qquad (16.9)$$

where $H_{ii} = \int \phi_i H \phi_i d\tau = \alpha$ and, $H_{ij} = \int \phi_i H \phi_j d\tau = \beta (j = i \pm 1)$ or $0 (j \neq i, i \pm 1)$. The solution of this secular equation is given by $E = \alpha \pm 0.618\beta$ and $E = \alpha \pm 1.618\beta$, and the electron configuration in the energy levels of butadiene is shown in Fig. 16.5.

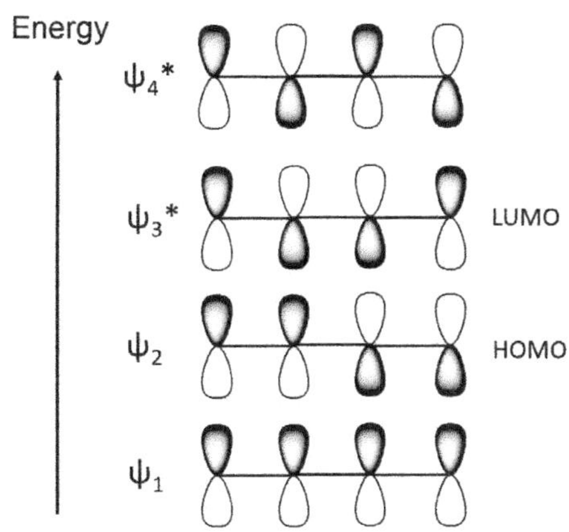

Figure 16.5 Energy levels of the molecular orbitals of butadiene.

The four π-electrons in butadiene occupy the lowest and second-lowest energy levels. Consequently, the total energy of the four π-electrons is

$$E = 4\alpha \pm 4.472\beta. \qquad (16.10)$$

It should be noted that the total energy of butadiene is not twice of that of ethane, but that the former is lower than the latter by 0.472β. This indicates that the delocalization of π-electrons stabilizes the total electron energy of a molecule.

This stabilization of the lowest-energy orbital in the conjugated system is generally seen in the π-molecular orbitals in the longer molecules. As the conjugated system becomes longer, the energy of the lowest orbital becomes lower, stabilizing the molecular structure. In aromatic systems, the energy stabilization of π-orbitals becomes more remarkable. In the case of benzene (Fig. 16.6), where each C–C bond is formed by one sp^2 bond and π-bonds derived from the $2p_z$ orbitals of carbon atoms, the total energy of the π-bonding is 8β, where β is the energy of ethylene. This means that the π-energy of benzene (8β) is lowered by one β unit more than the π-energy for three linearly conjugated π-bonds. Thus, remarkable stabilization occurs in aromatic molecules. This striking property would lead to the robust property of carbon materials composed of sp^2 hybrid bonding, that is, the formation of a carbon hexagon.

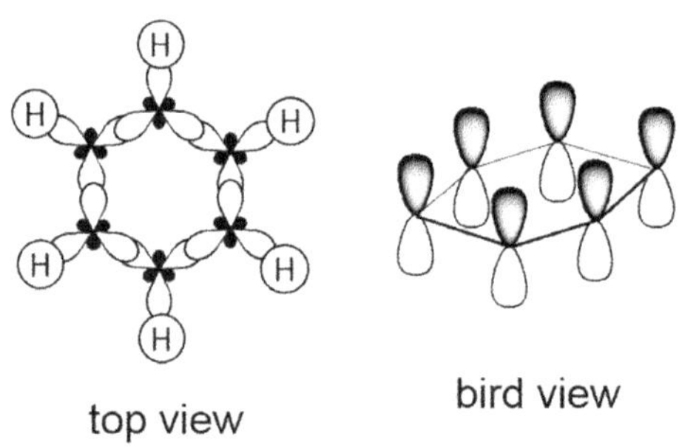

Figure 16.6 π-bands of benzene.

16.4 Electronic Structure of sp^2 Nanocarbons

Carbon atoms can form three types of hybridized bonds: sp, sp^2, and sp^3; therefore, carbon atoms can have various allotropes that use these hybridized bonds. A carbon atom in diamond has four sp^3 orbitals, which form a regular tetrahedron (Fig. 16.7a). In contrast, in graphite, the 2s and two 2p orbitals mix to form three sp^2 orbitals, which have in-plane configurations (Fig. 16.7b). These planar structures are composed of carbon hexagons, that is, a graphene sheet, and are layer-stacked to form graphite structures. Carbyne is a known example of sp bonding.[3] Carbyne is a chain of carbon atoms that is held together by either double or alternating single and triple atomic bonds (Fig. 16.7c). In the case in which carbyne is formed by sequential double bonds, carbyne can be regarded as an sp-hybridized carbon chain. Recently, nanometer-size carbon materials, *nanocarbons*, have attracted a great deal of attention because of their unique structural and electronic properties. As carbon materials on the nanometer scale, nanodiamond is also known, which is a fine diamond crystal 4–5 nm in diameter that can be produced using an explosion technique. However, here we focus on nanocarbon materials that are formed by sp^2-hybridized bonding, that is, graphene and CNTs.[4] As nanocarbon materials composed of sp^2 hybridization bonding, fullerenes, graphene, and CNTs are well known (Fig. 16.8). A fullerene is a molecule composed entirely of carbon in the form of a hollow sphere and a variety of other shapes, depending on the number of carbon atoms. The structure of spherical fullerenes resembles a soccer ball. Fullerenes have been used for medical applications, such as high-performance MRI contrast agents, and are commercially used in cosmetics. The discovery of molecules composed of only carbon atoms in the shape of a soccer ball had an impact, and the Nobel Prize in Chemistry was awarded in 1996 to R. F. Curl, H. W. Kroto, and R. E. Smalley for discovering fullerenes.[5] CNTs were discovered by S. Iijima in 1991.[6] CNTs have been found to have remarkable mechanical strength, thermal conductivity, and electronic properties; Small-diameter SWCNTs have a high Young's modulus and high tensile strength, and the calculated Young's modulus for an individual SWCNT is ~0.64 TPa.[7] The measured room-temperature thermal conductivity for an individual MWCNT is beyond 3000 W/mK,[8] which is larger than that of natural diamond and the basal plane of graphite. In particular, CNTs have enormous carrier mobility (>100,000 cm^2/Vs)[9] and high current density (~10^9 A/cm^2) [10]; therefore, CNTs have been anticipated to play an important role in the next generation of electronics. The Nobel Prize in physics was awarded to A. Geim and K. Novoselov in 2010, who fabricated "real" graphene and demonstrated its exotic electronic properties.[11] Preceding CNT studies strongly supported the study of graphene, especially its characterization. Here, we focus on graphene and CNTs in our overview of the properties of nanocarbons.

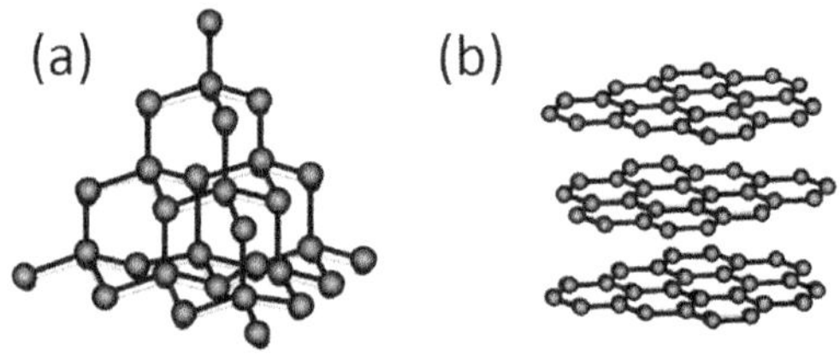

(c) ......–C≡C–C≡C–C≡C–C≡C–C≡C–......

or

......=C=C=C=C=C=C=C=C=C=C=......

Figure 16.7 Crystal structures of (a) diamond, (b) graphite, and (c) carbyne.

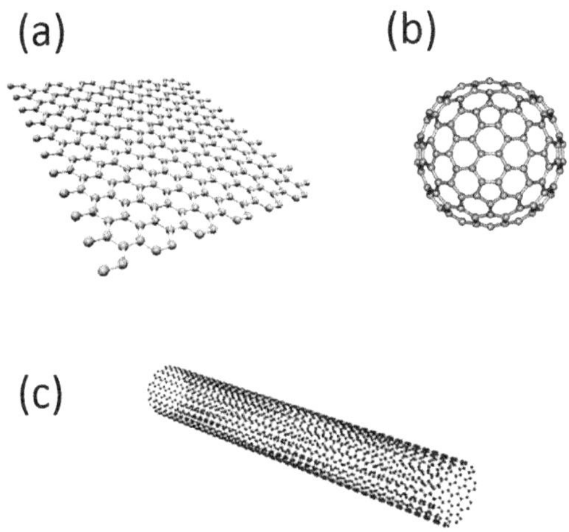

Figure 16.8 Crystal structures of (a) graphene, (b) fullerene, and (c) carbon nanotube.

Graphene is a single-layer sheet composed of a planar honeycomb lattice of sp^2-hybridized carbon atoms (Fig. 16.8a). Due to its unique crystal structure with only one atomic layer and its extraordinary electronic properties, graphene has been studied extensively in the last decade. The unique electronic structure of graphene having a zero bandgap was first reported by P. R. Wallace in 1947.[12] Later, D. P. DiVincernzo and E. J. Mele noted massless behavior based on the

Dirac equation.[13] However, only a few studies have been published in the preceding century because graphene is difficult to fabricate in a real laboratory to clarify its properties. In 2004, A. Geim and K. Novoselov succeeded in the fabrication of a real graphene layer from graphite by using the Scotch tape technique.[14] These authors demonstrated the tremendous electronic properties of graphene, including the anomalous quantum Hall effect. Due to the exceptional properties of graphene, applications in various fields have been pursued, such as field-electron transistors (FETs), transparent conducting electrodes, and energy storage devices, including electrodes for Li ion batteries, supercapacitors, etc.

There are two carbon atoms in a unit cell of graphene, A and B, as shown in Fig. 16.9a, with the unit vectors $\mathbf{a_1}$ and $\mathbf{a_2}$. Here, the real space unit vectors $\mathbf{a_1}$ and $\mathbf{a_2}$ are expressed in Cartesian coordinates as follows:

$$\mathbf{a_1} = \left(\frac{\sqrt{3}}{2}a, \frac{a}{2}\right),\ \mathbf{a_2} = \left(\frac{\sqrt{3}}{2}a, -\frac{a}{2}\right) \tag{16.11}$$

where $a = |\mathbf{a_1}| = |\mathbf{a_2}| = 1.42 \times \sqrt{3} = 0.246\ \text{nm}$ is the lattice constant of monolayer graphene. The unit cell in reciprocal space is shown in Fig. 16.9b and is described by the unit vectors $\mathbf{b_1}$ and $\mathbf{b_2}$ of the reciprocal lattice, which is given as follows:

$$\mathbf{b_1} = \left(\frac{2\pi}{\sqrt{3}a}, \frac{2\pi}{a}\right),\ \mathbf{b_2} = \left(\frac{2\pi}{\sqrt{3}a}, -\frac{2\pi}{a}\right). \tag{16.12}$$

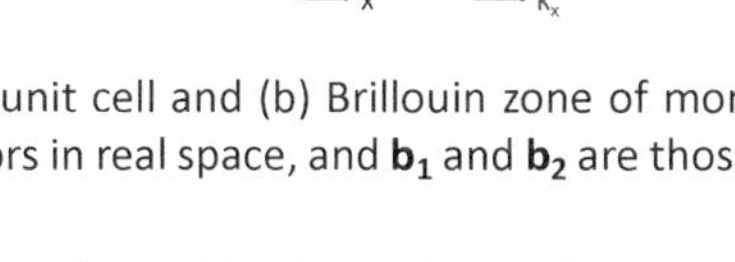

Figure 16.9 (a) The unit cell and (b) Brillouin zone of monolayer graphene. $\mathbf{a_1}$ and $\mathbf{a_2}$ are unit vectors in real space, and $\mathbf{b_1}$ and $\mathbf{b_2}$ are those in reciprocal space.

The unit vectors $\mathbf{b_1}$ and $\mathbf{b_2}$ in reciprocal space are rotated by 30° from the unit vectors $\mathbf{a_1}$ and $\mathbf{a_2}$ in real space.

In monolayer graphene, three of the four valence electrons of a carbon atom form σ-bonds composed of sp^2 hybrid orbitals, and the remaining electron forms π-bonds, which are composed of $2p_z$ orbitals. The energy band structure of monolayer graphene is composed of these σ- and π-bonds, which are referred to as the σ- and π-energy bands. Particularly, the energy band near the Fermi level is formed by π-bands, which determine the electronic properties of graphene. Figure 16.10 shows the electronic energy dispersion relations for the π-bands of monolayer graphene along the high-symmetry axes.[15] The upper half of the energy dispersion curves describes the π*-energy antibonding band, and the lower half is the π-energy band. Because there are two π-electrons per unit cell, these two π-electrons fully occupy the lower π-band. Consequently, the π-band is filled by spin-up and spin-down electrons, while the π*-band is empty. The upper π*-band and the lower π-band are degenerate at the K (K') points through which the Fermi energy passes for an undoped graphene monolayer. It should be noted that the valence and conduction bands come into the K (K') point with a linear $E(k)$ relation and that there is no bandgap at the K (K') point. In addition, it is interesting to note that the linear dispersion is the solution to the following massless Dirac Hamiltonian at K (K'):

$$\mathrm{H} = \hbar \upsilon_F(\sigma \cdot \kappa) \tag{16.13}$$

where $\kappa = -i\nabla$ and σ is the Pauli matrix operating in the space of the electron wave function amplitude on the A and B sublattices of graphene.[16]

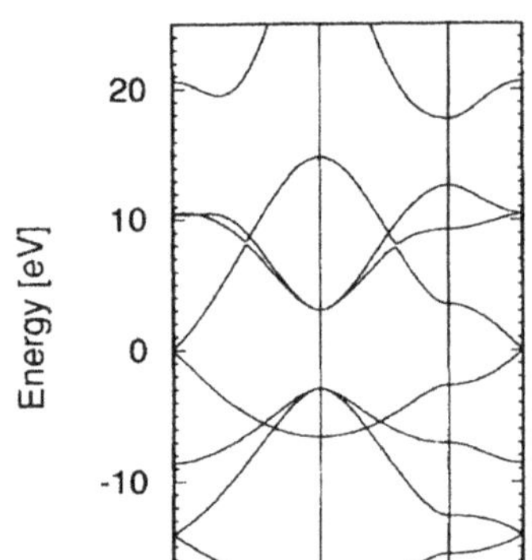

Figure 16.10 Energy dispersion of monolayer graphene. Reprinted (figure) with permission from Ref. [15]. Copyright (1992) by the American Physical Society.

An SWCNT is constructed by rolling up a graphene layer into a seamless cylinder.[16] The nanotube structure is uniquely determined by the chiral vector $\mathbf{C_h}$, which spans the circumference of the cylinder when the graphene layer is rolled up into a tube. The chiral vector can be written in the form $\mathbf{C_h} = n\mathbf{a_1} + m\mathbf{a_2}$, where n and m are coefficients of the real space unit vectors $\mathbf{a_1}$ and $\mathbf{a_2}$ and they are integers. In general, an SWCNT, which is composed of a chiral vector $\mathbf{C_h} = n\mathbf{a_1} + m\mathbf{a_2}$, is called an (n, m) nanotube. The nanotube can also be characterized by its diameter, d_t, and chiral angle, θ, which is the angle between the chiral vector $\mathbf{C_h}$ and a zigzag direction (Fig. 16.11). Both d_t and θ are expressed in terms of the indices n and m by the relations $d_t = a\sqrt{n^2 + nm + m^2}/\pi$ and $\tan\theta = \sqrt{3}m/(2n + m)$, where $a = |\mathbf{a_1}| = |\mathbf{a_2}| = \sqrt{3}a_{c-c} = 0.246\ \text{nm}$ is the lattice constant for the graphene layer and $a_{c-c} = 0.142$ nm is the nearest-neighbor C–C distance. Due to the sixfold symmetry of one graphene layer, all nonequivalent nanotubes can be characterized by the (n, m) pairs of integers, where $0 \le m \le n$. Achiral nanotubes are known as zigzag $(m = 0)$ and armchair $(m = n)$ nanotubes.

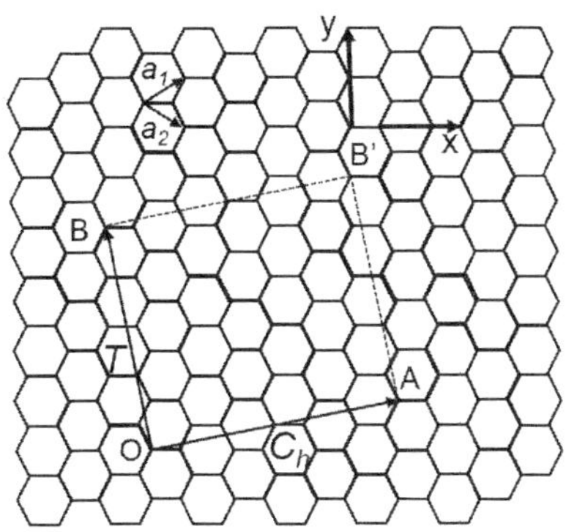

Figure 16.11 An unrolled carbon nanotube projected on the graphene layer. When the graphene layer is rolled up to the nanotube, the chiral vector $\mathbf{C_h}$ turns to the circumference of the nanotube and the translation vector $\mathbf{T}$ is aligned along the cylinder axis.

The electronic structure of an SWCNT can be obtained simply from that of the graphene layer. By using periodic boundary conditions in the circumferential direction denoted by the chiral vector $\mathbf{C_h}$, the wave vector associated with the $\mathbf{C_h}$ direction becomes quantized and the wave vector associated with the direction of the translational vector **T** along the nanotube axis remains continuous for a nanotube of infinite length. By folding the energy dispersion relations of the graphene layer, the energy dispersion of an SWCNT can be obtained. The resultant electronic structure depends on the chiral index (n, m). If $n - m = 3l$ (where l is an integer), the resulting nanotube shows metallic behavior, and if $n - m \neq 3l$, the resulting nanotube shows semiconducting behavior. Thus, the number of semiconducting nanotubes is roughly twice that of the metallic nanotubes. The electronic density of states (DOS) of a metallic SWCNT and a semiconducting one are shown in Fig. 16.12. For both SWCNTs, $1/\sqrt{E}$ -behaviors are observed in the valence and conduction bands, and the van Hove singularities (vHS) appear, which are labeled by $E_i^{(v)}$ and $E_i^{(c)}$ in Fig. 16.12. These DOS are characteristic of a 1D structure. For the metallic SWCNT, there is a small DOS near the Fermi level, while an energy gap is seen for the semiconducting SWCNT. This energy gap depends roughly on $1/d_t$, the reciprocal of the nanotube diameter d_t (Fig. 16.13).[17]

$$E_g \propto \frac{1}{d_t}. \tag{16.15}$$

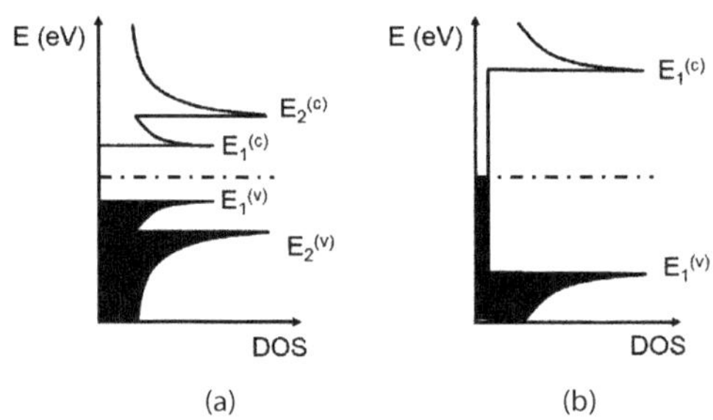

Figure 16.12 Density of states (DOS) of (a) a semiconducting carbon nanotube and (b) a metallic carbon nanotube.

An SWCNT 1 nm in diameter has a bandgap of approximately 0.9 eV, and as the diameter increases, its bandgap becomes narrower. This means that emission and absorption of SWCNTs are strong in the infrared region. For applications in FETs, a larger bandgap is better; that is, SWCNTs with small diameters of less than 1 nm are beneficial.

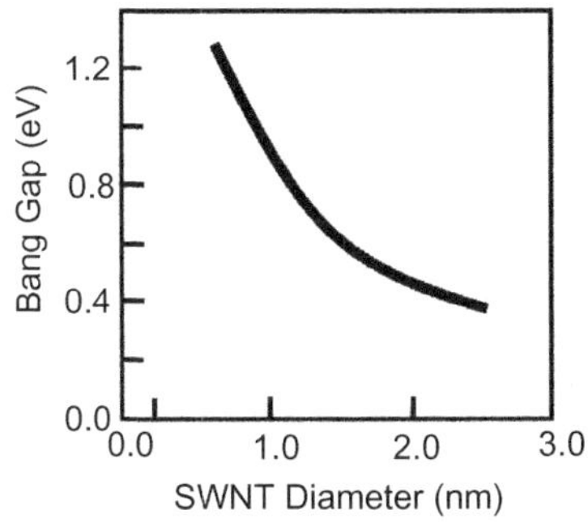

Figure 16.13 Relationship between bandgap and diameters for semiconducting SWCNTs.[17]

16.5 Carbon Nanotube: Classification and Growth Methods

CNTs are members of the nanocarbon family and are constructed by sp^2-hybridized bonding. The graphitic sp^2 bond is 33% stronger than the sp^3 bond of diamond and is one of the hardest substances in nature. Consequently, CNTs have a high tensile strength of nearly 100 GPa,[18] and their elastic modulus is in the terapascal range.[7,19] Therefore, applications to reinforce materials in composites have already been employed in commercial products. In contrast, applications in electronics are still in development, although CNTs have high carrier mobility values and can carry high electrical currents. In this section, classification and growth methods of CNTs are reviewed.

CNTs have cylindrical structures that are formed by rolling up graphene walls (Fig. 16.8c). When the graphene sheet is a monolayer, the CNTs are called SWCNTs (Fig. 16.14a). When more than one graphene sheet constitutes the CNTs, they are called multiwalled carbon nanotubes (MWCNTs) (Fig. 16.14c). In the cases of two graphene sheets, sometimes they are uniquely classified and called double-walled carbon nanotubes (Fig. 16.14b). Consequently, the structure of a CNT is identified using three parameters: chirality, length, and the number of graphene walls. In general, the length of the CNTs is on the order of micrometers, but millimeter-scale

CNTs have been realized by several research groups.[21,22] The diameters of SWCNTs are approximately 1–3 nm, while those of MWCNTs are generally much larger and range from 10 to 100 nm. Therefore, from the standpoint of mechanical strength, MWCNTs have an advantage. In addition, MWCNTs can carry a high-density electrical current, which makes them suitable for applications as interconnections in electronic devices. In contrast, SWCNTs have specified electronic structures, and some of them can work as semiconducting materials, which are appropriate for transistors in electronic devices. Additionally, SWCNTs are smaller and can have larger specific surface areas, which is beneficial for their application as supercapacitors or as electrodes in Li ion batteries.

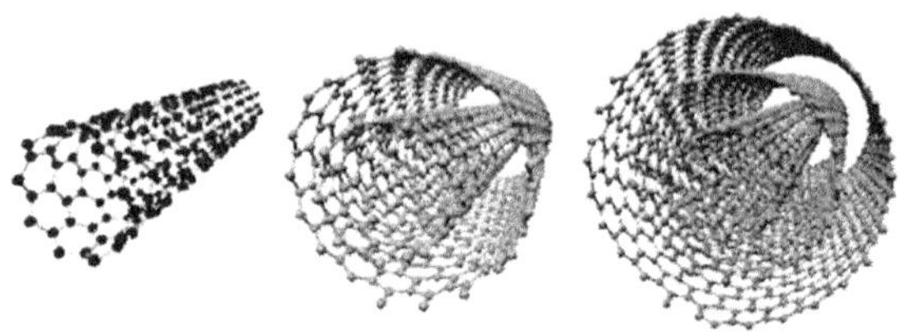

(a) (b) (c)

Figure 16.14 (a) Single-walled carbon nanotube (SWCNT), (b) double-walled carbon nanotube, and (c) multiwalled carbon nanotube (MWCNT). Reprinted from Ref. [20] with courtesy of Thomas Hirschmann (physicsworld.com).

Among the three structural parameters, chirality is the most important because it determines the electronic structure of an SWCNT, as stated in the previous section. In addition, the diameters of SWCNTs are determined by their chirality. Whether an SWCNT is semiconducting or metallic depends on its chirality, and if it is a semiconductor, the bandgap depends on its chirality. Therefore, to realize applications in electronic devices, it is important to obtain SWCNTs with a single chirality. However, selective growth of SWCNTs with unique chirality has not been realized and has been an important issue in the CNT research field. To avoid this problem, purification techniques have been developed to obtain SWCNTs of a single chirality, and highly purified SWCNTs separated into either metallic or semiconducting species have been obtained.[23] Progress in the techniques for realizing selective growth for SWCNTs with unique chirality is still an important issue for avoiding pollution during the purification process, and the process of obtaining SWCNTs with a single chirality will become simple, if it is realized.

To date, three main growth techniques have been employed for CNT growth: arc discharge,[24] laser ablation[25], and chemical vapor deposition (CVD).[26,27] S. Iijima discovered CNTs grown by the arc discharge method.[6] In early studies, it was believed that high temperatures (>1000°C) were necessary for growing CNTs because the melting point of carbon is extremely high (>3500°C); thus, early research proceeded with CNTs grown by arc discharge or laser ablation methods. Later, CNT growth below 1000°C was achieved using the CVD method.[26,27] In addition, CVD growth has potential for large-area growth, which has merit for commercial purposes.[28] In CVD, a CNT is formed from a metal catalyst particle under a feedstock gas flow containing carbon atoms (Fig. 16.15). In general, catalyst particles are usually deposited on substrates to support them, but in some methods, catalysts are supplied in a flow gas, such as organometallic molecules. Various types of hydrocarbon[28–30] (CH_4, C_2H_2, C_2H_4), alcohol[31] (C_2H_5OH), and carbon oxide[32] (CO) gases are used as the feedstock gas. If a feedstock gas excited by plasma is used, it is called plasma-enhanced chemical vapor deposition (PECVD)[33,34], and if a raw feedstock gas is used, it is simply called thermal CVD. For its convenience and low cost, the thermal CVD method is mostly used for CNT growth and is suitable for mass production toward commercial uses.

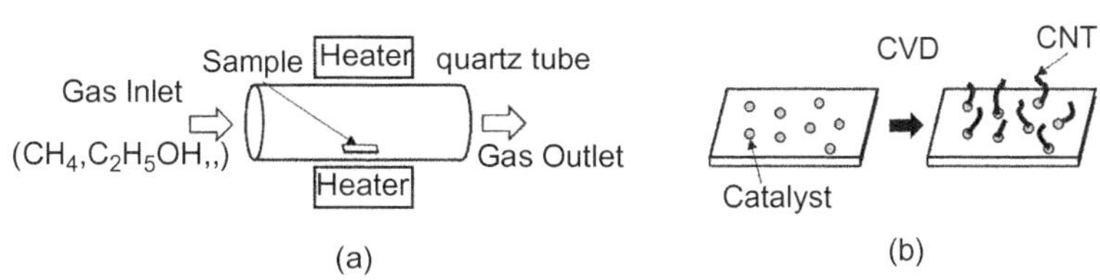

Figure 16.15 (a) Typical CVD system for SWCNT growth and (b) schematic of the growth process of SWCNTs from catalysts by CVD.

In CVD growth, transition metal catalysts, such as Fe, Co, and Ni, have been widely used for both SWCNT and MWCNT growth because these catalysts are not only easily available but also able to produce CNTs efficiently.[21,22,26,28,30,33] Currently, Fe, Co, Ni, and alloys including these metals are used in most CVD processes for

CNT growth. A supergrowth technique that can produce millimeter-scale SWCNTs through the addition of a small amount of water to the feedstock gas also uses Fe as a catalyst and has attained a high-rate growth.[22] However, others have attempted to grow SWCNTs using other catalysts. So far, SWCNT growth methods using noble metal catalysts such as Au, Ag and Cu have been reported[35], and narrowing chirality has been observed for SWCNTs using Au catalysts in a plasma CVD process.[34] Amazingly, it has also been reported that nonmetal nanometer-size particles, such as SiC particles[36] and nanodiamond[37], act as catalysts for SWCNT growth. Although the SWCNT yield from these nonmetal catalysts are not large, these results show the possibility of developing novel catalyst materials for SWCNT growth.

16.6 SWCNT Growth from Platinum Catalysts

16.6.1 Growth Conditions

As mentioned in previous sections, transition metal catalysts have been widely used for SWCNT growth. However, the diameters of SWCNTs grown from these catalysts have generally been distributed between 1 and 3 nm. For applications in electronic devices such as FETs, smaller diameters are suitable because the bandgap of a semiconducting SWCNT is roughly inversely proportional to the diameter. To reduce the diameters of SWCNTs, bimetal catalysts are generally used, such as CoMo[38], CoPt[39], CoW[40], FeCo[41], FeCu[42], and so on. On the other hand, other metal catalysts, such as noble metals, have been developed to achieve SWCNTs with narrow chirality distributions. One group of notable catalysts among them consists of platinum group metals because they are widely used in many practical applications, in addition to the use for jewelry. Pt is used as a catalyst in the catalytic converter in vehicle emission control devices, which catalytically reform straight-run naphthas into higher-octane gasoline and Adams' catalysts for hydrogenation.[43] However, few studies have been reported regarding SWCNT growth using platinum group metal catalysts.[35] In this section, SWCNT growth from Pt catalysts is reviewed[44–47] and compared with SWCNT growth from transition metal catalysts.[48]

One interesting property of Pt catalysts is their high efficiency in the formation of SWCNTs. Figures 16.16a and 16.16b show scanning electron microscopy (SEM) images of SWCNTs grown at 700°C from Pt and Co catalysts on SiO_2/Si substrates using a cold-wall CVD system in a high vacuum with ethanol as the carbon source.[44] The ethanol pressures used were 1×10^{-3} and 1×10^{-1} Pa for Pt and Co, respectively. Web-like SWCNTs are grown from both catalysts, suggesting that the SWCNT yields are similar between the two catalysts. In this CVD system, the ethanol gas is supplied by a nozzle onto the substrate surface, while the ethanol pressure is measured by a vacuum gauge on the inside wall of the CVD chamber; thus, the pressure is not the same as that on the catalysts. However, considering that the former is roughly proportional to the latter, this result indicates that the lower ethanol pressure is more suitable for Pt catalysts during SWCNT growth than Co catalysts. In addition, Pt catalysts act effectively to grow SWCNTs. Transmission electron microscopy (TEM) images show that the SWCNT diameters from Pt catalysts were less than 1.2 nm (Fig. 16.16c).

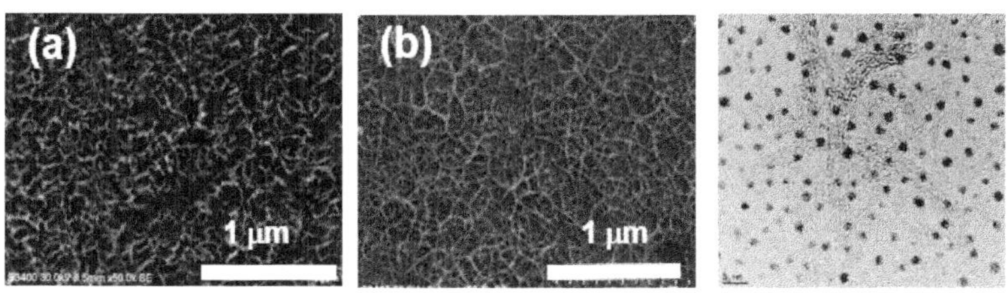

Figure 16.16 SEM images of SWCNTs grown from (a) Pt and (b) Co catalysts. (c) TEM images of SWCNTs grown from Pt catalysts.[45]

Raman spectroscopy is widely used to characterize SWCNTs because it not only determines the crystalline quality from the intensity ratio of the G band to the D band, but also identifies the SWCNT diameters from the Raman shifts of the radical breathing mode (RBM) peaks.[49] In addition, for web-like SWCNTs grown on the Si substrates, the SWCNT yield can be compared using the intensity ratio of the G band peak to the Si derived peak at 520 cm^{-1} (G/Si intensity ratio) in the Raman spectra. Figure 16.17 shows the dependence of the G/Si intensity ratio in Raman spectra for SWCNTs on the ethanol pressures during the growth process. As the SWCNT yield increases, the G/Si intensity ratio becomes strong, indicating an increase in the SWCNT yield. It should be noted that the SWCNT yield from Pt catalysts reaches a maximum at an ethanol pressure of 1×10^{-3} Pa, while the yield from Co catalysts increases with increasing ethanol pressure. This result confirms that Pt is more efficient for obtaining SWCNTs on SiO_2/Si substrates. The SWCNT yield (G/Si intensity ratio) against both growth temperatures and ethanol pressures for SWCNTs grown from Pt is shown in Fig. 16.18. The optimal ethanol pressure to obtain the highest SWCNT yield is 1×10^{-3} Pa at 700°C, but it decreased to 1×10^{-5} Pa between 400°C and 600°C. A decrease in the optimal carbon feedstock gas pressure with the reduction of growth temperature is generally observed in SWCNT growth by CVD[50], potentially due to decreases in the dissociation rate of carbon feedstock molecules on the catalyst particle surface decreasing as the growth temperature was reduced.

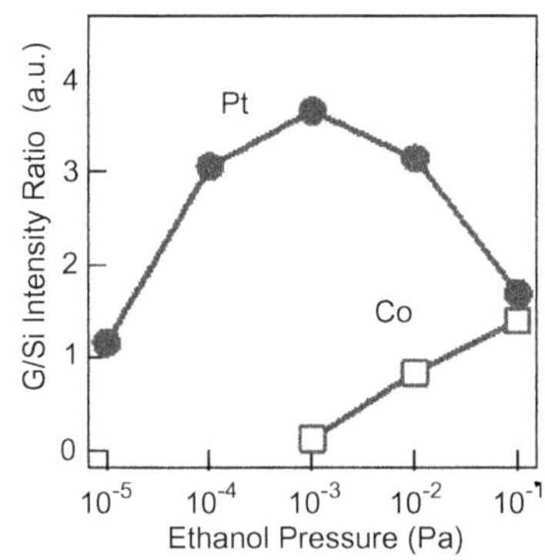

Figure 16.17 Dependence of the G band intensity on the Si-derived peak in the Raman spectra obtained at the ethanol pressure for SWCNTs grown at 700°C from Pt and Co catalysts using the alcohol gas source method.[44]

Another interesting characteristic of Pt catalysts is that small-diameter SWCNTs are likely to grow from Pt catalysts, as suggested by the TEM image in Fig. 16.16. Figure 16.19 shows the diameter distribution of SWCNTs grown from Pt and Co, which are estimated from Raman shifts of RBM peaks using the relationship d (nm) = $248/\omega$ (cm^{-1}), where d is the SWCNT diameter and ω is the Raman shift of its RBM peaks.[49] Because the RBM peak intensity is enhanced under the resonance condition due to the presence of vHS in DOS, the diameter distributions are different depending on the excitation wave numbers. However, irrespective of the excitation wavelength, the diameters of SWCNTs from Pt are smaller than those from Co. The diameters grown from Pt are distributed from 0.7 to 1.4 nm, and most of them are between 0.8 and 1.2 nm, while those from Co are distributed from 0.9 to 1.7 nm. TEM observations support these diameter distributions. Thus, it was concluded that SWCNTs grown from Pt catalysts have smaller diameters. As mentioned in the previous section, the bandgap of an SWCNT is roughly inversely proportional to its diameter. Thus, the small-diameter SWCNTs from Pt will be useful for applications in electronics, such as FETs.

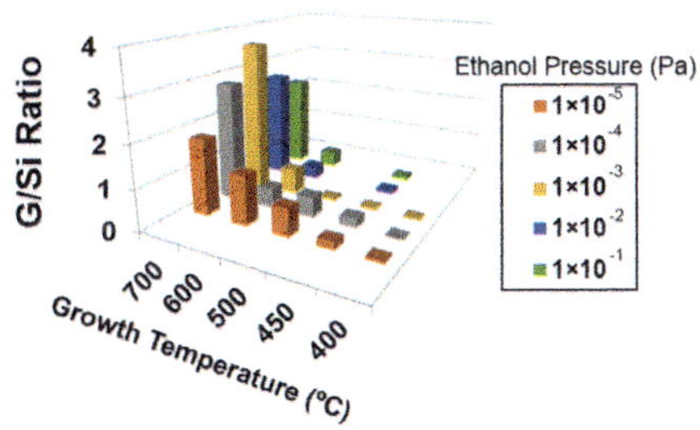

Figure 16.18 (a) Histogram of the intensity ratio of the G band to the Si peak (520 cm^{-1}) in the Raman spectra (G/Si ratio) as a function of growth temperature and ethanol pressure.

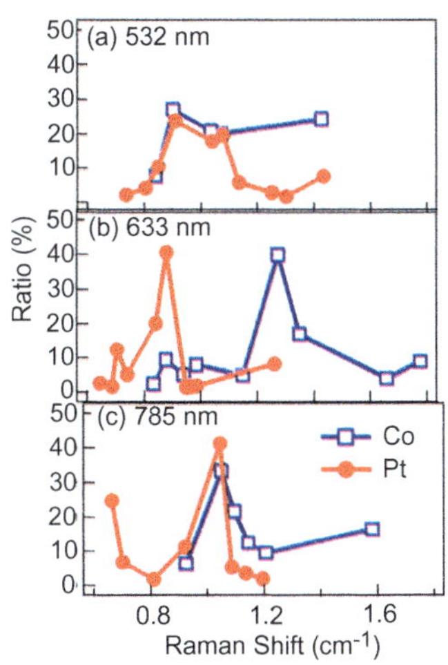

Figure 16.19 Diameter distributions of SWCNTs grown from Co and Pt catalysts estimated from Raman spectra measured by three excitation wavelengths.

16.6.2 Chirality Distribution of SWCNTs from Pt Catalysts

For applications such as transistors, semiconducting SWCNTs are necessary. However, to date, a selective growth technique to obtain semiconducting SWCNTs has not yet been attained. To determine whether an SWCNT is semiconducting or metallic, it is necessary to determine the chirality distribution of the SWCNTs. Here, the chirality distribution of SWCNTs grown from Pt catalysts is shown. Figure 16.20a shows the photoluminescence (PL) spectrum of SWCNTs grown from Pt at 700°C and a chiral map of SWCNTs determined from PL and Raman spectra is shown in Fig. 16.20b. It can be confirmed that most of SWCNTs have diameters of less than 1.0 nm and are near-armchair type, as has been observed for other metal catalysts. Although the chirality distribution is relatively narrow, both metallic and semiconducting SWCNTs are grown from Pt catalysts on SiO_2/Si substrates.

One way to improve the ratio of semiconducting SWCNTs is to use other materials to support catalyst particles because the adhesion energy between the catalyst metal and the supporting layer affects the migration and aggregation of catalyst materials.[51] For example, amorphous Al_2O_3 layers are often used to support metal catalysts because they suppress the catalyst migration, enhancing the SWCNT yield.[22,52] Recently, it was reported that acid treatments of graphene layers allow SWCNTs to be grown directly on them.[53] Figure 16.21 shows SWCNTs grown from Pt catalysts on chemically treated graphene layers, which are exfoliated from a graphite surface. Several RBM peaks are observed in the RBM region of the Raman spectra (Fig. 16.21c), indicating that SWCNTs are grown from Pt on the graphene layers. Raman measurements using lasers with various excitation wavelengths indicate that most of the SWCNTs grown from Pt on graphene layers are semiconductors and that their diameters are less than 1.1 nm (Fig. 16.21b). It is interesting that SWCNTs grow on only the graphene layers, even though Pt particles are well distributed across the surface of the graphite substrate (Fig. 16.21a). Considering that Pt is easy to move on the raw graphite surface, this finding suggests that the damage induced by the acid treatment for the graphene layers on the graphite is larger than that for the other areas on the graphite. Consequently, Pt particles are only well anchored at the defect sites of the graphene layers, leading to SWCNT growth. In the other areas of the graphite surface, the damage is not enough to anchor the Pt catalysts, which could cause aggregation of the Pt catalysts and no SWCNT growth.

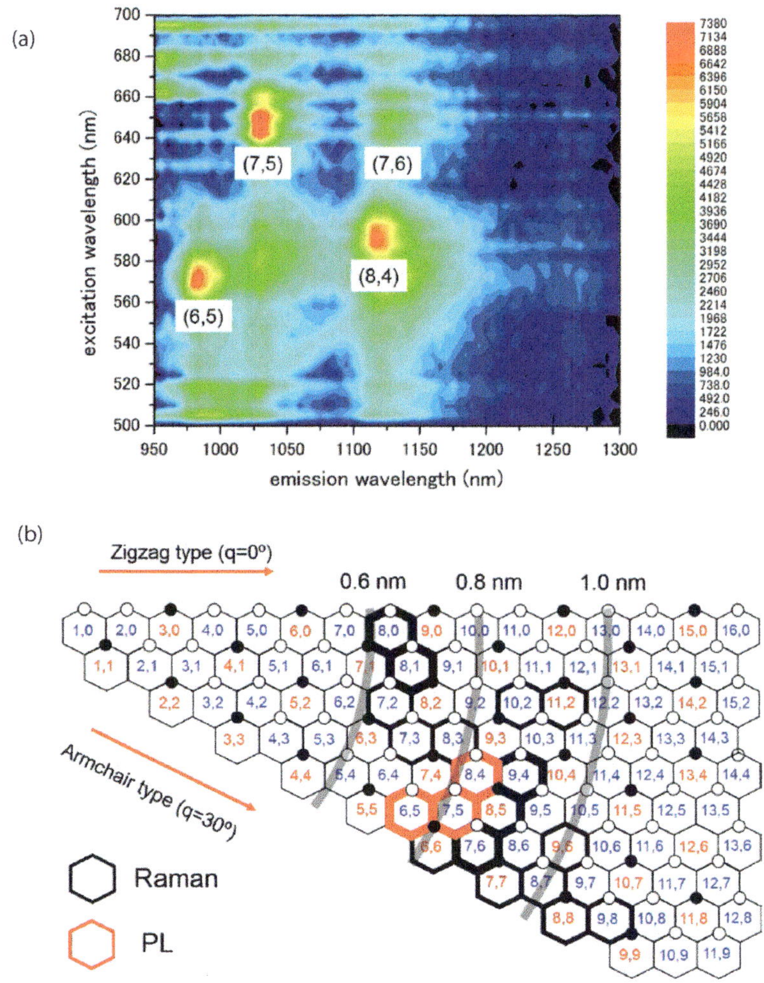

Figure 16.20 PL map and chiral map of SWCNTs grown from Pt catalysts at 700°C.

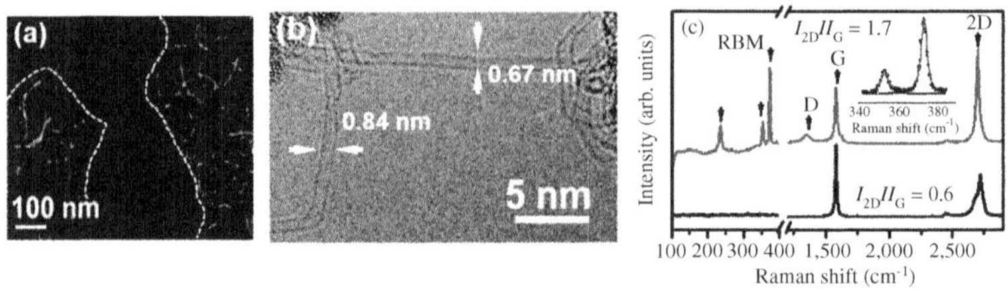

Figure 16.21 (a) SEM image, (b) TEM image, and (c) Raman spectra of SWCNTs grown on exfoliated graphene layers on graphite from Pt catalysts at 700°C.[53]

The relationship between the catalyst particle size and the SWCNT diameter has also been an interesting issue because it is believed that the catalyst particle size determines the SWCNT diameter grown from it. For the transition metal catalysts, the SWCNT diameters are similar to the catalyst particle size in some cases,[54,55] but in other cases, the former is smaller than the latter.[56–58] This result could depend on growth conditions, such as the solubility of carbon atoms in metal catalyst particles at the growth temperature. The diameter distribution of Pt particles after SWCNT growth at 700°C is shown in Fig. 16.22. In this experiment, pulsed arc plasma gun (ULVAC APG-1000) was used for deposition of Pt particles. Most of Pt particle sizes are distributed between 1.0 and 2.6 nm. At this growth condition, most of SWCNT diameters are less than 1.2 nm and are smaller than the Pt particle sizes. This indicates that the diameters of SWCNTs grown from Pt particles are smaller than the particle size. The tendency of SWCNT growth from Pt catalysts would support the growth of small-diameter SWCNTs.

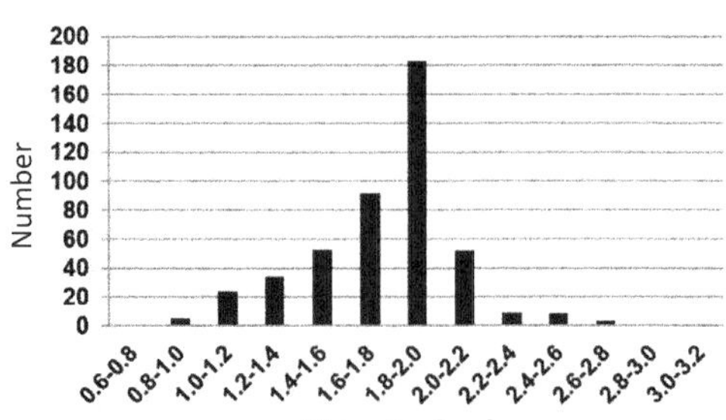

Figure 16.22 Diameter distribution of Pt particle sizes after SWCNT growth at 700°C.

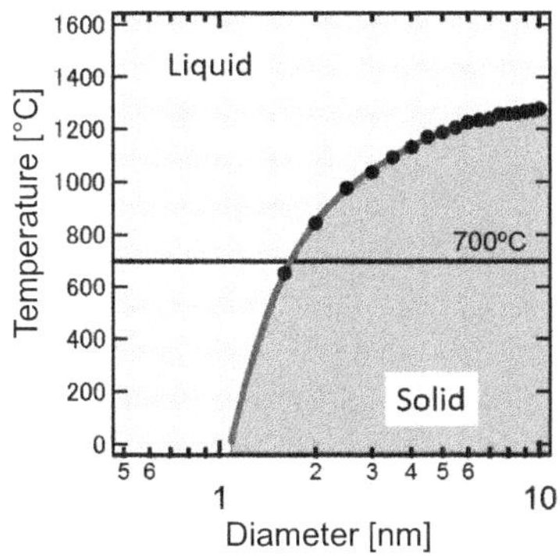

Figure 16.23 Dependence of the melting point of a Pt particle on its diameter.[61]

Whether the catalyst particles are solid or liquid during SWCNT growth has been an important issue in the field of CNT synthesis. So far, the vapor-liquid-solid (VLS) model has been widely adopted to explain the growth process of SWCNTs by CVD.[59] In this model, catalyst particles are in the liquid state at the growth temperature, and SWCNTs are formed by their supersaturation, followed by the segregation of hexagons on the catalyst surface. However, in situ TEM observations at the growth temperature were carried out for SWCNT growth from Fe and Co catalyst particles, which showed that Fe and Co particles change their shapes during SWCNT growth, but lattice fringes were observed on the TEM images of the catalyst particles.[55,60] This indicates that these catalysts are solid, keeping their crystal structures, but the atoms on their surfaces are mobile due to the melting point depression caused by the Gibbs–Thomson effect. Pt has a higher melting point (1772°C) than Fe (1538°C) and Co (1495°C); thus, the possibility that Pt is solid during SWCNT growth is high, although in situ observations of Pt catalysts have not been attempted. Furthermore, a molecular dynamics simulation showed that the melting point of Pt nanoparticles decreases drastically when the particle diameter is below 10 nm (Fig. 16.23) and showed that the melting point of Pt with diameters between 1 and 2 nm is below 800°C.[61] Therefore, Pt particles could also be solid, as seen for transition metal catalysts during SWCNT growth. In situ TEM observations also showed that Fe_3C is formed during the SWCNT growth, suggesting that SWCNTs grow after the supersaturation of Fe particles.[55] On the other hand, the solubility of C in Pt is low, so there is a possibility that both the decomposition of ethanol molecules and the formation of hexagons occur near the surface region of the Pt particles. If this is the case, the growth process from Pt catalysts is very different from that of Fe catalysts.

16.7 Conclusions

In this chapter, properties and growth mechanisms of CNTs are reviewed as representatives of nanocarbon materials composed of sp^2 hybrid orbitals. Next, we focused on SWCNT growth using Pt catalysts and show their specified characteristics relative to SWCNT growth in the presence of transition metal catalysts. Twenty years have passed since CNTs were discovered, and due to endeavors aimed at reducing production costs, applications to composite materials have become more practical. However, the application of CNTs in electronics for practical use has not been attained as initially expected. To realize the selective growth of SWCNTs with a single chirality, further development of growth techniques is still necessary.

References

1. Freming, I. *Molecular Orbitals and Organic Chemical Reactions*, John Wiley & Sons: UK (2009).
2. McQarrie, D. A., Simon, J. D. *Physical Chemistry: A Molecular Approach*, University Science: Sausalito, CA (1997).
3. Smith, P. P., Buseck, P. R. *Science*, **216**, 984–986 (1982).
4. Jorio, A., Dresselhaus, M. S., Dresselhaus, G. *Carbon Nanotubes*, Springer-Verlag: Berlin, Heidelberg (2008).
5. Kroto, H. W., Heath, J. R., Obrien, S. C., Curl, R. F., Smalley, R. E. *Nature*, **318**, 162–163 (1985).
6. Iijima, S. *Nature*, **354**, 56–58 (1991).
7. Gao, G., Çagin, T., Goddard, W. A. *Nanotechnology*, **9**, 184–191 (1998).
8. Kim, P., Shi, L., Majumdar, A., McEuen, P. L. *Phys. Rev. Lett.*, **87**, 215502 (2001).
9. Durkop, T., Getty, S. A., Cobas, E., Fuhrer, M. S. *Nano Lett.*, **4**, 35–39 (2004).

10. Yao, Z., Kane, C. L., Dekker, C. *Phys. Rev. Lett.*, **84**, 2941–2944 (2000).
11. Geim, A. K., Novoselov, K. S. *Nat. Mater.*, **6**, 183–191 (2007).
12. Wallace, P. R. *Phys. Rev.*, **71**, 622–634 (1947).
13. DiVincenzo, D. P., Mele, E. J. *Phys. Rev. B*, **29**, 1685–1694 (1984).
14. Novoselov, K. S., Geim, A. K., Morozov, S. V., Jiang, D., Zhang, Y., Dubonos, S. V., Grigorieva, I. V., Firsov, A. A. *Science*, **306**, 666–669 (2004).
15. Saito, R., Fujita, M., Dresselhaus, G., Dresselhaus, M. S. *Phys. Rev. B*, **46**, 1804–1811 (1992).
16. Saito, R., Dresselhaus, G., Dresselhaus, M. S. *Physical Properties of Carbon Nanotubes*, Imperial College Press: London (1998).
17. Matsuda, Y., Tahir-Kheli, J., Goddard, W. A. *J. Phys. Chem. Lett.*, **1**, 2946–2950 (2010).
18. Yu, M., Lourie, O., Dyer, M. J., Moloni, K., Kelly, T. F., Ruoff, R. S. *Science*, **287**, 637–640 (2000).
19. Treacy, M. M., Ebbesen, T. W., Gibson, J. M. *Nature*, **381**, 678–680 (1996).
20. Dumé, B. physicsworld.com News: Feb. 13 (2013).
21. Li, S., Yu, Z., Rutherglen, C., Burke, P. J. *Nano Lett.*, **4**, 2003–2007 (2004).
22. Hata, K., Futaba, D. N., Mizuno, K., Namai, T., Yumua, M., Iijima, S. *Science*, **306**, 1362–1364 (2004).
23. Liu, H., Tanaka, T., Urabe, Y., Kataura, H. *Nano Lett.*, **13**, 1996–2003 (2013).
24. Ebbeseb, T. W., Ajayan, P. M. *Nature*, **358**, 220–222 (1992).
25. Thess, A., Lee, R., Nikolaev, P., Dai, H. J., Petit, P., Robert, J., Xu, C. H., Lee, Y. H., Kim, S. G., Rinzler, A. G., Colbert, D. T., Scuseria, G. E., Tomanek, D., Fischer, J. E., Smalley, R. E. *Science*, **273**, 483–487 (1996).
26. Yudasaka, M., Kikuchi, R., Matsui, T., Ohki, Y., Yoshimura, S., Ota, E. *Appl. Phys. Lett.*, **67**, 2477–2479 (1995).
27. Dai, H., Rinzler, A. G., Thess, A., Nikolaev, P., Colbert, D. T., Smalley, R. E. *Chem. Phys. Lett.*, **260**, 471–475 (1996).
28. Cassell, A. M., Raymakers, J. A., Kong, J., Dai, H. J. *Phys. Chem. B*, **103**, 6484–6492 (1999).
29. Ci, L., Xie, S., Tang, D., Yan, X., Li, Y., Liu, Z., Zou, X., Zhou, W., Wang, G. *Chem. Phys. Lett.*, **349**, 191–195 (2001).
30. Hafner, J. H., Cheung, C. L., Liever, C. M. *J. Am. Chem. Soc.*, **121**, 9750–9751 (1999).

31. Maruyama, S., Kojima, R., Miyauchi, Y., Chiashi, S., Kohno, M. *Chem. Phys. Lett.*, **360**, 229–234 (2002).

32. Bribujiwsju, M. J., Willis, P. A., Colbert, D. T., Smith, K. A., Smalley, R. E. *J. Vac. Sci. Technol. A*, **19**, 1800 (2001).

33. Qin, L. C., Zhou, D., Krauss, A. R., Gruen, D. M. *Appl. Phys. Lett.*, **72**, 3437–3439 (1998).

34. Ghorannevis, Z., Kato, T., Kaneko, T., Hatakeyama, R. *J. Am. Chem. Soc.*, **132**, 9570–9572 (2010).

35. Takagi, D., Homma, Y., Hibino, H., Suzuki, S., Kobayashi, Y. *Nano Lett.*, **6**, 2642–2645 (2006).

36. Takagi, D., Hibino, H., Suzuki, S., Kobayashi, Y., Homma, Y. *Nano Lett.*, **7**, 2272–2275 (2007).

37. Takagi, D., Kobayashi, Y., Homma, Y. *J. Am. Chem. Soc.*, **131**, 6922–6923 (2009).

38. Hu, M., Murakami, Y., Ogura, M., Maruyama, S., Okubo, T. *J. Catal.*, **225**, 230–239 (2004).

39. Liu, B., Ren, W., Li, S., Liu, C., Cheng, H. M. *Chem. Commun.*, **48**, 2409–2411 (2012).

40. Yang, F., Wang, X., Zhang, D., Yang, J., Luo, D., Xu, Z., Wei, J., Wang, J., Xu, Z., Peng, F., Li, X., Li, R., Li, Y., Li, M., Bai, X., Li, Y. *Nature*, **510**, 522–524 (2014).

41. Liu, B. C., Yu, B., Zhang, M. X. *Chem. Phys. Lett.*, **407**, 232–236 (2005).

42. He, M., Chernov, A. I., Fedotov, P. V., Obraztsova, E. D., Sanio, J., Rikkinen, E., Jiang, H., Zhu, Z., Tian, Y., Kauppinen, E. I., Niemela, M., Krause, A. O. *J. Am. Chem. Soc.*, **132**, 13994–13993 (2010).

43. Voorheers, V., Adams, R. *J. Am. Chem. Soc.*, **44**, 1397–1405 (1922).

44. Maruyama, T., Mizutani, Y., Naritsuka, S., Iijima, S. *Mater. Express*, **1**, 267–272 (2011).

45. Mizutani, Y., Fukuoka, N., Naritsuka, S., Maruyama, T., Iijima, S. *Diamond Relat. Mater.*, **26**, 78–82 (2012).

46. Fukuoka, N., Mizutani, Y., Naritsuka, S., Maruyama, T., Iijima, S. *Jpn. J. Appl. Phys.*, **51**, 06FD23 (2012).

47. Kondo, H., Fukuoka, N., Ghosh, R., Naritsuka, S., Maruyama, T., Iijima, S. *Jpn. J. Appl. Phys.*, **52**, 06GD02 (2013).

48. Tanioku, K., Maruyama, T., Naritsuka, S. *Diamond Relat. Mater.*, **17**, 589–593 (2008).

49. Jorio, A., Dresselhaus, M., Saito, R., Dresselhaus, G. F. *Raman Spectroscopy in Graphene Related Systems*, Wiley-VCH: Weinheim (2011).

50. Picher, M., Anglaret, E., Arenal, R., Jourdain, V. *ACS Nano*, **5**, 2118–2125 (2011).

51. Mattevi, C., Wirth, C. T., Hofmann, S., Blume, R., Cantoro, M., Ducati, C., Cepek, C., Knop-Gericke, A., Milne, S., Castellarin-Cudia, C., Dolafi, S., Goldoni, A., Schloegl, R., Robertson, J. *J. Phys. Chem. C*, **112**, 12207–12213 (2008).

52. Arcos, T., Garnier, M. G., Oelhafen, P., Mathys, D., Seo, J. W., Domingo, C., Garcia-Ramos, J. V., Sánchez-Cortés, S. *Carbon*, **42**, 187–190 (2004).

53. Ghosh, R., Maruyama, T., Kondo, H., Kimoto, K., Nagai, T., Iijima, S. *Chem. Commun.*, **51**, 8974–8977 (2015).

54. Cheung, C. L., Kurtz, A., Prak, H. *J. Phys. Chem. B*, **106**, 2429 (2002).

55. Yoshida, H., Takeda, S., Uchiyama, T., Kohno, H., Homma, T. *Nano Lett.*, **8**, 2082–2086 (2008).

56. Nasibulin, A. G., Pikhitsa, P. V., Jiang, H., Kauppinen, E. *Carbon*, **43**, 2251–2257 (2005).

57. Durrer, L., Greenwald, J., Helbling, T., Muoth, M., Riek, R., Hierold, C. *Nanotechnology*, **20**, 1–7 (2009).

58. Jeong, G. H., Suzuki, S., Kobayashi, Y., Yamazaki, A., Yoshimura, H., Homma, Y. *J. Appl. Phys.*, **98**, 124311 (2005).

59. Kukovitsky, E. F., L'vov, S. G., Sainov, N. A. *Chem. Phys. Lett.*, **317**, 65–70 (2000).

60. Hoffmann, S., Sharma, R., Ducati, C., Du, G., Mattevi, C., Cepek, C., Cantoro, M., Pisana, S., Parvez, A., Cervantes-Sodi, F., Ferrari, A. C., Dunin-Borkowski, R., Lizzit, S., Petaccia, L., Goldni, A., Robertson, J. *Nano Lett.*, **7**, 602–608 (2007).

61. Gafner, S. L., Gafner, J., Zamulin, I. S., Redel, L. V. *J. Comput. Theor. Nanosci.*, **9**, 1–8 (2012).

Biography

Prof. Takahiro Maruyama graduated from the Faculty of Science at Kyoto University (1989), and he completed masters and doctoral courses in the Surface Science Laboratory under Prof. Yasuo Sakisaka and studied the electronic structure of metal and superconductor materials through angle-resolved photoemission spectroscopy. After getting his PhD at Kyoto University, Japan (1993), he was assigned to the Electronic Technical Laboratory (ETL) in Tsukuba as a postdoctoral fellow to continue to study of electronic properties of superconducting materials. Then, he moved to the University of Tsukuba (1994) and, as a lecturer, started to study of wide-bandgap semiconductors, such as ZnSe and GaN, which are materials to fabricate blue-light-emitting diodes under Prof. Katsuhiro Akimoto. Maruyama was engaged in growing crystals of these materials and characterized the electronic structures and the structures by photoemission and extended X-ray absorption fine structure (EXAFS) techniques using a synchrotron. From 2000 to 2002, he was with Ristsumeikan University and continued to study nitride semiconductors under Prof. Yasushi Nanishi. Since 2002, he has been with Meijo University and is engaged in crystal growth of semiconductors, including GaAs, GaN, and Ge, using MBE and LPE methods, collaborated with Prof. Shigeya Naritsuka, and investigated the growth mechanism of these materials in vapor and liquid phases. Since 2004, he has started the study of CNT growth, collaborating with Prof. Sumio Iijima. By developing nanotube growth under high vacuum, he has fabricated SWCNTs under vacuum and low-temperature conditions. Also, he has grown small-diameter SWCNTs using Pt catalysts and is attempting to clarify the growth mechanism of SWCNTs.

Chapter 17

Conjugated Polymers Nanostructures: Synthesis and Applications

Utpal Rana, Sanjoy Mondal, and Sudip Malik
Polymer Science Unit, Indian Association for the Cultivation of Science (IACS), Jadavpur, Kolkata 700032, India
rana.utpal@gmail.com

Conjugated polymer nanostructures with a defined size, shape, architecture, and chemical functionality have received more attention nowadays, because they are the forefront of many next-generation organic optoelectronic technologies. This chapter will discuss about the new generation of conjugated conducting polymer nanostructure, especially synthesis and application of polyaniline nanostructure in detail.

17.1 Introduction

A polymer is a macromolecule made of many repeated subunits. The term polymer first invented by a Swedish scientist Jöns Jacob Berzelius in 1983, and it is a combination of two Greek words: *poly*

Conjugated Objects: Developments, Synthesis, and Applications
Edited by Atsushi Nagai and Koji Takagi

ISBN 978-981-4774-03-1 (Hardcover), 978-1-315-22962-1 (eBook)
www.panstanford.com

(many) and *meros* (units).[1] Conventional polymers like rubber, plastic, etc., are widely used as an insulating material because of significant resistance to electrical conduction. Atoms in the polymer chain are covalently bonded by the saturated carbon compounds, there is no scope of delocalization of the valence electrons, and consequently, neither charge carriers nor path for their movements are available through the total structures; as a result they behave like insulating materials.[2]

In the 1970s, scientists accidently found that conjugated polymer (in case of polyacetylene) became a highly electrically conductive instead of insulator when a structural modification occurred, which was called doping.[3] The alternative single and double bonds in the conjugated polymer chain facilitate the delocalization of the electrons throughout the system and they act as charge carriers to make the system conductive. However, the covalently attached conjugated polymer is not conducting without any doping. Doping is the paramount importance to make the conjugated polymer chain conductive. It reduces the bandgap of the system and allows the electron to move freely throughout the conjugated system, as a result conductivity increases. Excitement about the new conjugated conducting polymer materials is generated due to their enormous potential applications in the field of electromagnetic shielding, photovoltaics, electronics, capacitors, optoelectronic devices, batteries, sensors, and actuators. For the discovery and subsequent development of conductive polymers, Alan J. Heeger, Alan MacDiarmid, and Hideki Shirakawa were awarded the Nobel Prize in Chemistry in 2000.[4]

Recently, conjugated conducting polymer–based nanostructures have received more attention over the bulk because of the significant enhancement of solubility, processability, and sustainability, which have improved performance in many application.[5] Among the different conducting polymer nanostructures, the polyaniline (PANI) nanostructure is extensively studied because of easy synthesis, simple doping/dedoping chemistry, environmental stability, and a broad area of application.[6] The PANI nanostructure is extensively used in lightweight batteries,[7] supercapacitor devices[8], electromagnetic-shielding interference[9], microwave-absorbing materials[10], sensors[11], light-emitting devices[12], antistatic and anticorrosion coatings[13], drug delivery[14], etc. The objective of this chapter is a discussion about the conjugated polymer nanostructures, especially providing insight into the recent development of PANI nanostructures.

17.2 Structures of Conjugated Conducting Polymers

The chemical structure of the conducting polymer contains alternative single or double bonds or conjugated segments coupled with heteroatoms (e.g., O, N, S) with extended π-conjugation along the polymer chains.[15] Polyacetylene is the simplest example in the conjugated polymer family. A list of the major studied conducting polymers is shown in Scheme 17.1.

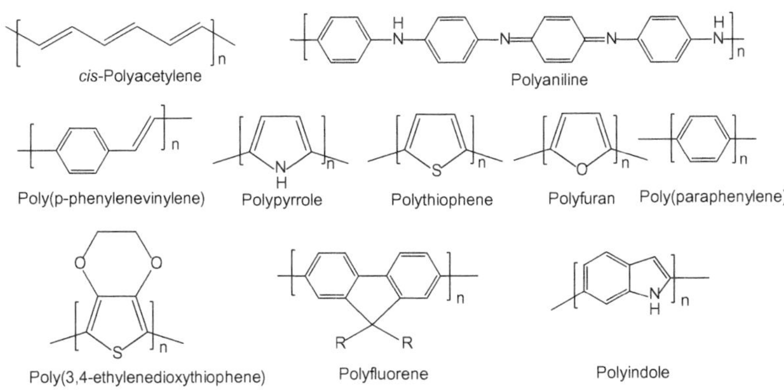

Scheme 17.1 Structure of some conjugated conducting polymers.

17.3 Conduction Mechanism in Conducting Polymers

The π-conjugation in an individual polymer chain is a basic requirement for a polymer becoming a conducting polymer. Another important thing for a polymer to become conducting is that there should be the overlap among molecular orbitals, which allows the formation of a delocalized molecular orbital. Besides this molecular orbital, the polymer must have a partially filled orbital so that the electrons can move freely throughout the lattice.

To understand the conducting behavior of conjugated polymers, it is essential to know the band structure and charge carriers. The band of the highest occupied molecular orbital (HOMO) is called the valence band, and the same of the lowest unoccupied molecular orbital (LUMO) is the conduction band. The energy difference between them is called the bandgap. The optical bandgap of a conducting polymer always controls the electronic and optical properties. The optical bandgap is inversely proportional to the conductivity of the polymer. Conducting polymers have a narrow energy bandgap and insulators have a wide energy bandgap. Electrons should have some energy to occupy in valence band and require external energy to jump from the valence band to the conduction band. To have conduction, these bands must be partially filled, as empty or full-filled bands cannot carry electricity.[16] The bandgap in a conjugated polymer is not well defined as polymers have several chain segments owing to the variation of chain length of polymers or defects related to chemical defects, morphological disorder, etc. That is why a distribution of bandgap is always seen. The observed bandgap value, which is therefore an average of all polymer chains, typically ranges from 1 to 4 eV. Conducting polymers are unusual and their electronic properties or conduction are not entirely explained by the band theory of classical semiconductors.

The conduction mechanism of a conjugated polymer can be explained by charge carriers (solitons, polarons, and bipolarons). Conducting polymers have a degenerate or nondegenerate electronic structure at the ground state. In a conjugated polymer with a degenerate ground state (e.g., polyacetylene), solitons are very important charge carriers. When a charged is surrounded by a domain or a boundary, it is called soliton and it also generates the property of a solitary wave that can migrate without deformation or dissipation. A soliton is imagined as the excitation of charge that moves from one potential well to another potential well of equivalent energy. Polyacetylene exists in two isomeric forms (*cis*- or *trans*-polyacetylene) that differ from each other by the exchange of single and double bonds. Due to the presence of two possible configurations of polyacetylene, it has a degenerate ground state in its electronic structures. In conducting polymers with aromatic rings (polyparaphenylene, PANI, polythiophene, polypyrrole, etc.) there are nondegenerate ground states and charges carry through polarons and bipolarons (confines soliton pairs). A polaron is a radical cation which is formed when an electron is removed from the top of the valence band of a conjugated polymer and is partially delocalized over a conjugated polymer segment. If an electron is removed from the polaron level (remove the unpaired electron), it creates a special dication, which is called bipolaron. Compared to a polaron, a bipolaron is doubly charged and spinless.[17] Generally, low doping (discussed in the next section) levels generate polarons, whereas higher doping levels produce bipolarons. In an electric field, both polarons and bipolarons move through the conjugated polymer chain by resonance and create narrow bipolaron bands in the bandgap (Fig. 17.1). Formation of new bandgaps between the valence band and the conduction band makes the electron conduction easy throughout the system. Therefore, attempts are always taken to diminish the bandgap in conducting polymers by introducing various dopants.

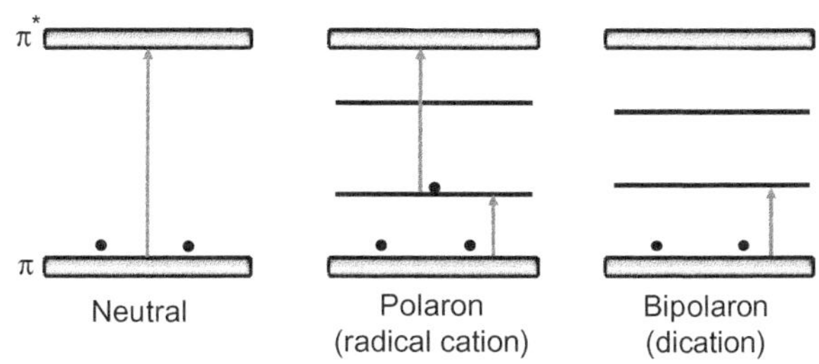

Figure 17.1 Energy level diagram of a neutral polymer, a positive polaron, and a bipolaron.

17.4 Doping in Conducting Polymers

Conjugated polymers act as insulator material due to lack of charge carriers in the conjugated polymer chains. The process of transforming a polymer to its conductive form via chemical oxidation or reduction is called doping.[18] A dopant or doping agent is a trace impurity element in a substance that can enhance the electrical properties or the optical properties of the polymer. Enhancement of the conductivity of a conjugated polymer solely depends on the reactivity of the dopant with the corresponding polymer.

The extent of enhancement of electric conductivity of a polymer primarily depends on the chemical reactivity of the dopant with the polymer. The stability of the polymer also plays an important role in it. Enhancement of conductivity directly relates to the concentration of the dopant. The reason is the enhancement of charge carrier mobility, which in turn is due to interchain interaction. On doping, positive or negative charge carriers are developed in the polymer chains. Doping results in rearrangement of polymer chains, followed by a ordered structure. The dopant concentration may be as high as 50%. A schematic presentation of doping and its effect on conductivity for polyacetylene are presented in Fig. 17.2.[19]

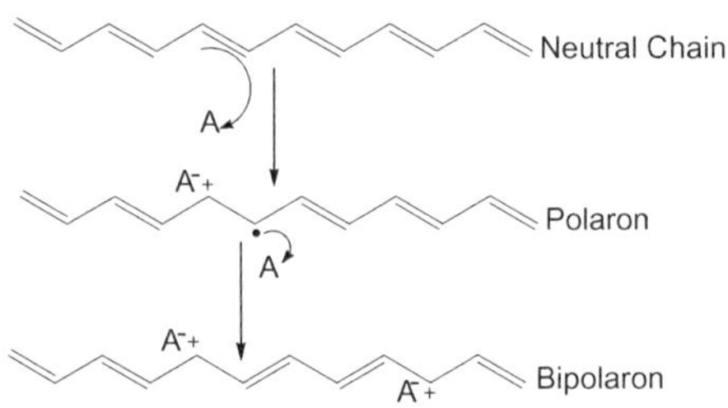

Figure 17.2 Formation of a polaron and a bipolaron in polyacetylene.

17.5 Polyaniline

PANI was discovered more than 150 years ago as *aniline black* by oxidation of aniline.[20] It captured the intense attention of the scientific community around 1970. Now, PANI is the most popular semiflexible conducting polymer in the family of conducting polymers. The fascination of PANI research arises due to the easy synthesis procedure, high electrical conductivity, good thermal stability, simple doping/dedoping chemistry, etc.[20–23] The electrical conductivity achieved is very high due to the delocalization of π-electrons throughout the polymer chain. The electrical conductivity of PANI ranges on a semiconducting level (10^0 S cm^{-1}) to metal conductivity (>10^4 S cm^{-1})[24,25] and it is higher than other common conducting polymers (<10^{-9} S cm^{-1}).

Before moving on to different properties and the micro- and nanostructures of PANI, it will be instructive to briefly review the chemical structure of PANI (Scheme 17.2). It is a polymer which consists of alternating reduced and oxidized repeat units.[26]

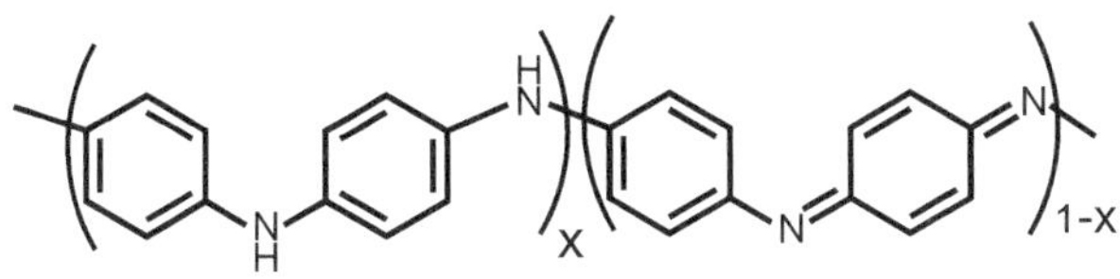

When, $x = 1$; Leucoemeraldine
$x = 0.5$; Emeraldine
$x = 0$; Pernigraniline

Scheme 17.2 Chemical structure of PANI.

Depending upon the benzenoid and the quinonoid part, PANI has three different main forms: leucoemerlidine, emerlidine, and pernigraniline (Scheme 17.2). *Leucoemeraldine* is white and clear, where $x = 1$, the fully reduced state. *Emeraldine* is blue in color, where $x = 0.5$, that is, 50% benzenoid units and 50% quinoid units. This emeraldine form basically is the emeraldine base form: when treated with acid it transfers to deep-green-colored emeraldine salt form (Scheme 17.3). *Pernigraniline* is violet/blue in color, where $x = 0$, the fully oxidized form.

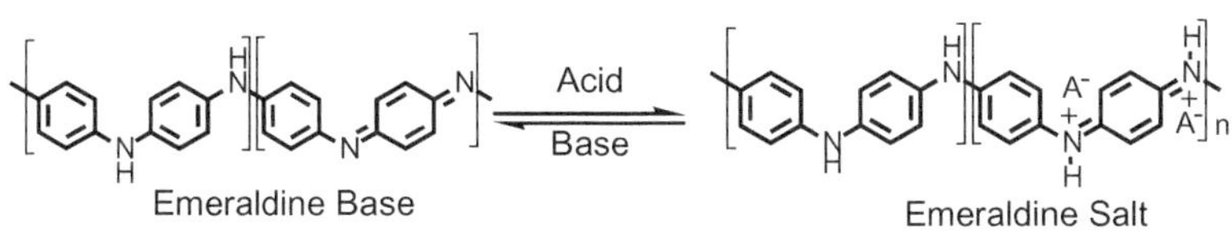

Scheme 17.3 Reversible transformations of emeraldine salt to emeraldine base by acid/base treatment.

The most common forms of PANI are one of the three states or physical mixtures of these components (Scheme 17.4). The emeraldine ($n = m = 0.5$) form of PANI, often referred to as emeraldine base (EB), is neutral; if doped (protonated by any acid) it is called emeraldine salt (ES), with the imine nitrogens protonated by an acid. The imine nitrogen atoms in the PANI chain may be protonated either in whole or in some part to generate the corresponding salts. The degree of protonation of the EB depends on its oxidation state as well as the pH of the medium. Conversion of base to salt is completely reversible by the acid and base. Complete protonation of the imine nitrogen atoms in the EB by aqueous acids results in the formation of a delocalized polysemiquinone radical cation and reduces the bandgap of the polymer; as a result higher conductivity is achieved.[20]

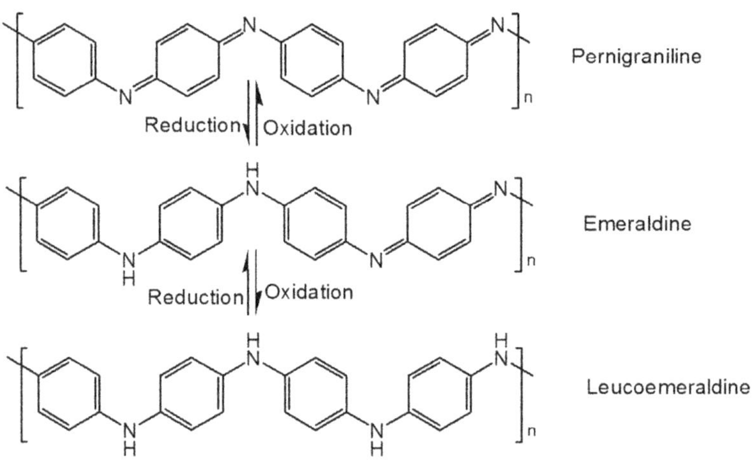

Scheme 17.4 Three different forms of PANI.

17.6 Synthesis of Polyaniline

PANI is mainly synthesized by the chemical oxidation polymerization method or electrochemical polymerization of aniline under mild conditions. Chemical oxidative polymerization of aniline is one of the traditional methods for synthesizing it in bulk quantities[20–23,27] PANI can be synthesized by several other methods like step-wise condensation, gas-phase plasma, and vapor phase deposition. Generally, chemical and electrochemical methods are used for PANI synthesis, which we discuss shortly in the next section. In this chapter, we have focused mainly on PANI nanostructure synthesis by the chemical method, which is discussed in the Morphology section (Section 17.6.4).

17.6.1 Chemical Synthesis

Chemical polymerization of aniline is a very simple one-step method by monomers, dopant acids, and oxidants. Among these reagents, the dopant is the most important reagent because it is mainly attributed to structural and electrical properties of PANI. Generally inorganic acids (HCl, HNO_3, H_2SO_4, H_3PO_4, and HF) and some organic acids are widely used as dopants for PANI. The molecular structure and acidity of dopants strongly affect the properties of the emeraldine form of PANI. In the case of a monomer, unsubstituted aniline mainly used for polymerization, sometimes *ortho*-functionalized aniline moieties are used for different derivatives of PANI. Different oxidizing agents are used for PANI synthesis; however, the most common one is ammonium peroxydisulfate, $(NH_4)_2S_2O_8$ (APS). Other oxidizing agents like gold chloride ($AuCl_3$), iron(III) chloride ($FeCl_3$), potassium permanganate ($KMnO_4$), potassium dichromate ($K_2Cr_2O_7$), etc., are also be used for polymerization. Generally a 1:1 molar ratio of aniline to oxidizing agent is responsible for polymerization. With the above 1:1 oxidant-to-aniline molar ratio, aniline gets overoxidized, which results in lower conductivity. Different factors which affect the chemical oxidation of aniline are (i) concentration of the oxidant, (ii) duration of the reaction, (iii) temperature of the medium, and (iv) pH of the medium. We have discussed these briefly in the Morphology section (Section 17.6.4).

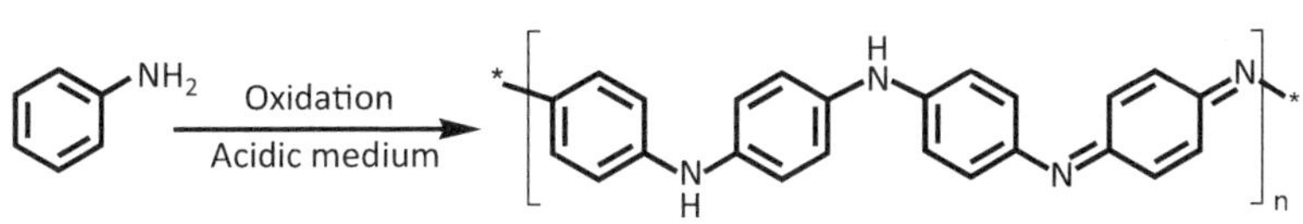

Scheme 17.5 Synthesis of PANI by the chemical oxidation method.

17.6.2 Electrochemical Synthesis

In electrochemical synthesis of PANI, an inert metal electrode is a widely used for this technique. Generally, platinum or conducting indium tin oxide glass is used as the electrode and a dilute acid solution is used as an electrolyte. The synthesized polymer is deposited on the working electrodes through the anodic oxidation of aniline in the protonated state, that is, the ES state. Polymerization can be achieved by holding the working electrode at a constant potential with respect to a standard calomel reference electrode (SCE); the oxidation potential of aniline in 2M HCl is ~0.70–0.75 V with respect to an SCE.[31] This process is known as potentiostatic growth. The synthesized PANI shows fibrillar morphology. The advantage of this method is that in situ experiments may be carried out during the synthesis.

17.6.3 Polymerization Mechanism

The aniline polymerization occurs predominantly through *para*-coupling, that is, head-to-tail coupling (Scheme 17.6). The PANI formation mechanism proceeds through a chain involving radical cation formation. There is also a chance of *ortho*-coupling, as well as chain branching, crosslinking, etc. *Ortho*-coupling usually gives to low yields and low-molecular-weight products.[32,33]

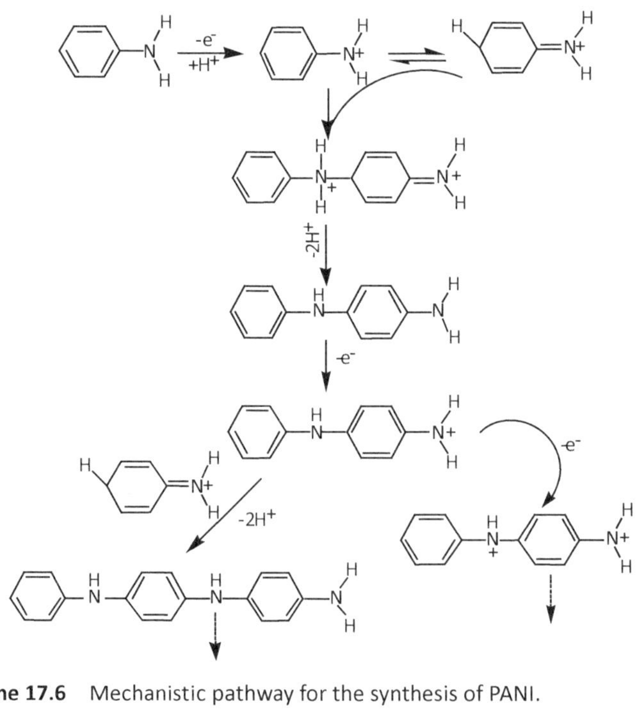

Scheme 17.6 Mechanistic pathway for the synthesis of PANI.

17.7 Properties

17.7.1 UV-Vis Spectroscopy

The UV-Vis spectroscopy study of the polymer is very important to know the relative oxidation state of PANI. PANI, when doped with acid, is called the ES state, which has three characteristic absorption peaks centered at ~350, ~440, and ~900 nm. The peak at ~350 nm can be ascribed to the π–π* transition in the benzenoid rings, while the peaks at ~440 and ~900 nm for the polaron–π* transition and the π–polaron transition, respectively (Fig. 17.3). The effect of the dopant acid concentration is reflected in the UV-Vis spectra. With the concentration of the dopant, the peak intensity at ~440 nm increases, and it indicates that the environments of the quinoid and benzenoid rings are apparently changing with dopant concentration. The peak intensity at ~900 nm (π-polaron) is also decreased with increasing concentration of the dopant acid after 1:1, implying the simultaneous shortening of the conjugation length, corresponding to the increase of the branched and crosslinked chain structures in the polymer. The effect of temperature on PANI nanostructure preparation is reflected in the UV-Vis spectra. In general, the peak of ~900 nm is weaker and narrower at higher temperature than at lower temperature, and it indicates that the polarons become much localized at higher temperature.[4]

UV-Vis spectra of the EB form of PANI show two peaks at 350 and 630 nm, which are assigned for the amine and imine segments of PANI. For a fully protonated EB form the peak at 630 nm disappeared and a new peak at ~910 nm appeared, which is assigned as the proton band is produced by proton doping, but the position of the 350 nm peak remains the same, although the intensity of the peak decreases. This indicates that the protonation process only takes place on the imine segment of the emeraldine PANI chain. The intensity of peaks at 630 nm and 910 nm strongly depends on the pH of the medium at the time of PANI preparation. The presence of these three peaks provides information about the ES state of PANI chains. The peak intensity as well as peak sharpness of π–polaron transition depends on the chain length as well as solvent polarity.[10,11]

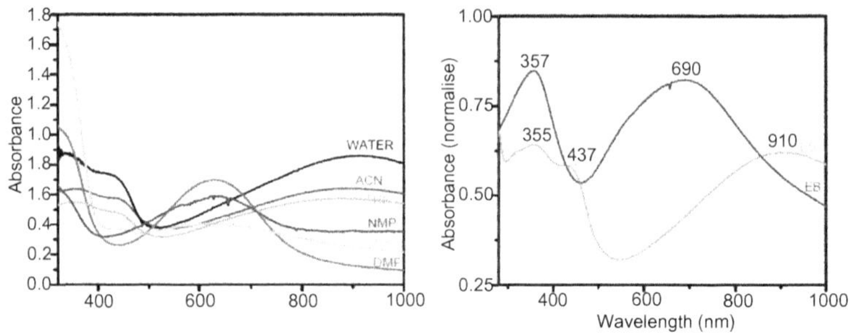

Figure 17.3 UV-Vis spectra of PANI (a) in different solvents and (b) emeraldine base and emeraldine salt form.

It is well known that the shift in the electronic transition peak is connected with the dielectric constant (ε) of the solvent. A bathochromic shift of the λ_{max} peak with an increase in the dielectric constant of the solvent is generally observed. Actually, π_B–π_Q transition is highly influenced by the dielectric constant of the solvent (solvent effect).[23]

The bandgap energy for the π–π* transition of PANI can be calculated from the UV-Vis spectra by the following equation:

$$\Delta E = \frac{hc}{\lambda_{max}}$$

where $h = 6.625 \times 10^{-34}$ J.s and $c = 3 \times 10^8$ m/s.

Here, ΔE is the bandgap energy, h is Plank's constant, c is the velocity of light, and λ_{max} is the wavelength of the absorption maxima corresponding to the first absorption band at 250–320 nm.

17.7.2 FTIR Study

Fourier transform infrared (FTIR) spectroscopy is a powerful tool to determine the structural changes that occur during doping and dedoping processes of the PANI nanostructure. The presence of typical characterization starching bands at 3432, 1580, 1496, 1300, 1140, and 800 cm^{-1} is indicating the doping of PANI (Fig. 17.4) by HCl. The characteristic IR starching bands at 1580 cm^{-1} and 1496 cm^{-1} are for γC=C of quinoid rings and benzenoid rings, respectively; 1300 cm^{-1} for γC–N of the secondary aromatic amine; and 1140 cm^{-1} and 800 cm^{-1} for aromatic in-plane and out-of-plane deformation of 1,4-disubsituted benzene.[23,30,31] The peak near 3400 cm^{-1} corresponding to the hydrogen-bonded N–H stretching vibration is strong, indicating that the polymers are highly hydrogen-bonded. The $I_{quinoid}/I_{benzenoid}$ ratio more or less provides a clue about the existence of a different form of PANI.

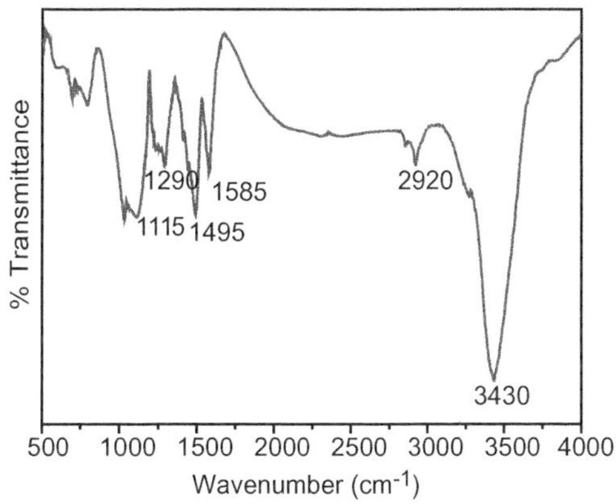

Figure 17.4 FTIR spectrum of PANI.

The ratio of the relative intensities of quinoid to benzenoid ring modes (I_B/I_Q) clearly indicates that the percentage of imine units present in the polymer chain. When I_B/I_Q = 1 that implies the emeraldine form of PANI, $I_B/I_Q \geq 1$ means leucoemeraldine, and $I_B/I_Q \leq 1$ implies the pernigraline form of PANI.

The concentration of the dopant has a big role on the structural change of PANI, which can be reflected in the IR spectra. Wan et al. have showed that with increasing concentration of the HCl dopant from 0.2 M to 2.0 M, a typical peak at 1140 cm^{-1} corresponds to the shifted peaks from 1125 (0.2 M) to 1180 cm^{-1} (2.0 M), which reveals that the electrons are more localized in samples with increased concentration of HCl. Meanwhile, the peak intensity decreases remarkably from the samples from 0.2 M to 2.0 M, which indicates the formation of the chemical crosslinks in the samples.[30] Formation of a crosslink structure of the polymer, generally formed at a higher temperature of polymerization, can be estimated from IR spectra. The samples synthesized with high concentrations of HCl exhibit increasing absorptions at around 736 and 692 cm^{-1}, which are assigned to the 1,2-substituted and 1,2,4-substituted aromatic ring, indicating the existence of branched and crosslinked structures in the samples.[31]

17.7.3 XRD Study

Crystallinity and orientation measurement of a conducting polymer by X-ray diffraction (XRD) have been of much interest because more highly ordered systems display a metallic-like conductive state. The three main diffraction peaks of crystalline PANI are at $2\theta = 6.4°$ (d = 13.8 Å), 19.2° (d = 4.6 Å), and 26.0° (d = 3.4 Å). The strong, narrow peak centered at $2\theta = 6.4°$ is related to the periodicity along the polymer chain, 19.2° is ascribed to the periodicity in the direction parallel to the polymer chain, and 26.0° is due to the periodicity in the direction perpendicular to the polymer chain.[29,34] the difference between the nanostructure of a sphere and a nanotube can be easily determined by XRD. In the case of a nanotube both peak $2\theta = 6.4°$ and 26.0° are present and it's absent in the case of a hollow sphere. The dopant highly influenced the structure of PANI. By changing the dopant nature the XRD spectral pattern changes. The XRD patterns for PANI salts doped with (a) HCl, (b) H_2SO_4, and (c) H_3PO_4 are different. The diffraction peaks of PANI-HCl are relatively sharp and strong. However, most of diffraction peaks disappear in PANI-H_2SO_4 and PANI-H_3PO_4 as the d-space corresponding to the peak at $2\theta = \sim 20°$ is in the order PANI-H_3PO_4 > PANI-HCl > PANI-H_2SO_4. The PANI chains produced under more acidic conditions have less structural defects (e.g., non-*para* linkages in PANI chains), which will cause the fractions of the crystalline phase increasing in PANI.[23]

17.7.4 Morphology

Over the past few decades, it has been become clear that structure plays a pivotal role in determining the physical properties of conducting polymers. The most striking feature is that the control of regularity and order in the polymer structure leads to remarkable enhancements in the electronic and photonic properties of these conducting polymers. An effective design strategy is aimed at controlling both the microscopic and the solid state macroscopic structure, because they collectively define the resultant band structure of a given material and thereby determine its electrical as well as optical properties. A random structure of PANI is generally seen with dilute HCl. If there is a control in polymerization of aniline by stopping the reaction at an early stage or using a template for a structure-directing agent, one may expect structured PANI with or without a template. With significant control over the nucleation step, the subsequent PANI growth produces various morphologies: (1) granules, (2) nanofibers, and (3) nanotubes. They all have very close dimensions in the range of 50–250 nm, which depend on experimental conditions, typically, and seem to be proportional to the molecular weight of PANI chains. These facts illustrate the structural similarities of all nanostructures generated by PANI. Generally there are two types of approaches for making the nanostructure, (i) the hard-template method and (ii) the soft-template or template-free method.

17.7.4.1 The hard-template method

Template polymerization or the hard-template method is usually defined as a process in which the monomer units are organized by a preorganized porous channel of a template. The most commonly used hard templates include nearly monodisperse inorganic oxide nanoparticles and polymer nanoparticles. This method requires a stepwise process that includes gaining the cores, then covering the shell, and finally removing the core material by calcination or solvent etching. A PANI nanowire prepared by a MnO_2-based hard template was reported by Alshareef et al. MnO_2 nanowires serve as both the oxidant and a template providing the scaffold from which aniline grows off (Fig. 17.5). By simply changing the morphology of the MnO_2 template, different nanosizes and shapes of PANI should be possible. Similarly CuO_2 serves as the template in order to generate PANI microstructures.[35b] Dai et al. reported the synthesis of a PANI nanosphere by a polystyrene (PS) sulphonic acid template in water (Fig. 17.6). The hard-template method is unique and it's very important in the sense of controlled PANI nanostructures. However, the removal of the template often affects the hollow structures and the postprocessing for template removal is rather tedious. A new strategy or approach is required to meet practical applications.

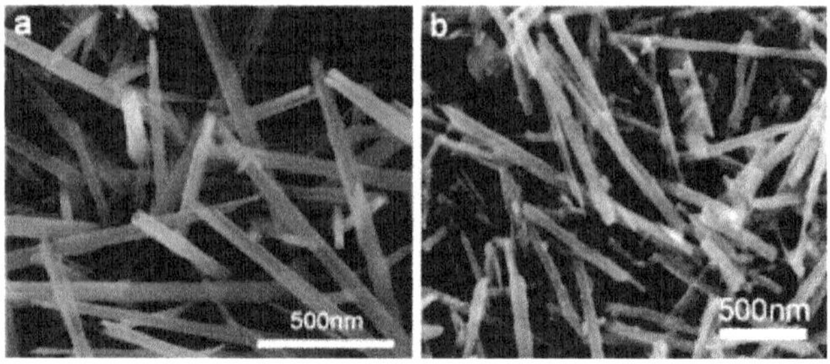

Figure 17.5 Morphology of MnO_2 hard template–based PANI morphology SEM images of (a) MnO_2 nanotubes and (b) PANI-coated MnO_2 nanotubes. Reproduced from Ref. [35] with permission from the Royal Society of Chemistry.

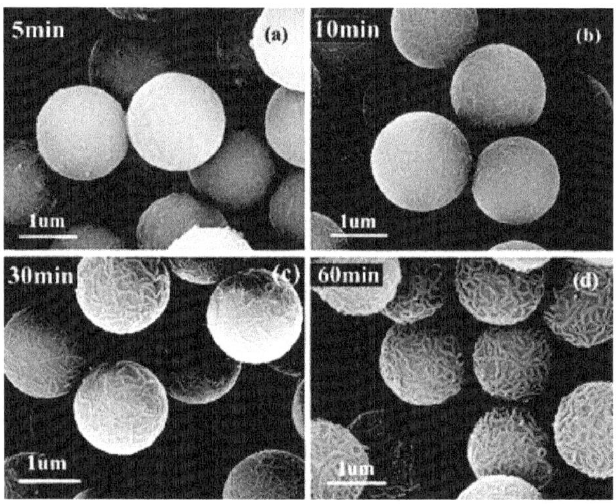

Figure 17.6 Morphology of polystyrene/PANI (PS/PANI) composite where PS is a hard template. Reproduced from Ref. [36] with permission from the Royal Society of Chemistry.

17.7.4.2 The soft-template method

The soft-template synthesis method, basically the self-assembly method, is one where nanostructure formation takes place without any template or external structure-directing materials. Structure-directing molecules like surfactants, organic acids, surfactant micelles, liquid crystals, and reverse microemulsions are generally used for nanomaterial preparation. The formation of the nanostructure of PANI depends on the reaction conditions, such as the concentration of aniline, the molar ratio of aniline to oxidant, or sometimes the oxidizing agent. Generally, the lower concentration of aniline favors the formation of nanotubes or nanofibers, while higher concentration leads to granular PANI. The soft-template method is very simple and inexpensive because of omitting a template and posttreatment for template removal. In the soft-template method the effect of the dopant acid plays a very important role in the morphology of PANI. The effect of the dopant in soft-template nanostructure PANI synthesis is discussed briefly in the next section.

17.7.4.2.1 *Chiral dopant for helical PANI*

As we discussed in the previous section, the soft-template method is a very simple one-step method for PANI nanostructure preparation, where the dopant plays a key role in the nanostructure. The effect of dopant acids on the soft-template method was nicely demonstrated by Wan et al. using chiral dopant acids. Right*- or left-handed helical nanofibers of PANI were prepared by the aniline polymerization process, where an enantiomeric D- or L-camphorsulfonic acid (D- or L-CSA) as a dopant acid (Fig. 17.7) was used. Helical PANI nanofibers were induced by chiral dopants during polymerization. The aniline-to-D-CSA or aniline-to-L-CSA molar ratio also plays a major role in helical fibrous morphology formation.

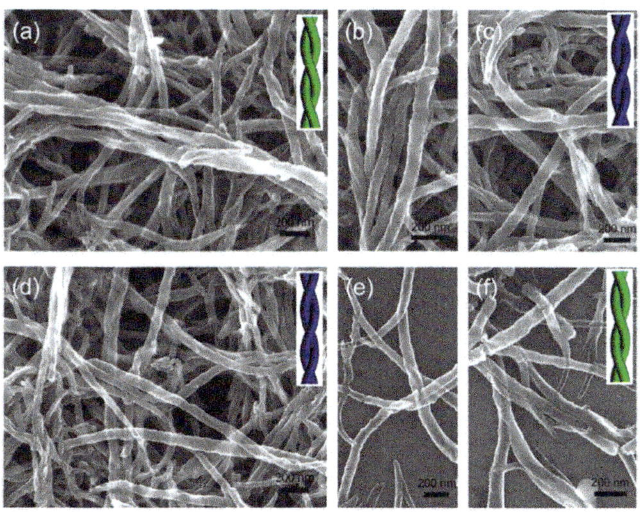

Figure 17.7 SEM images of (a–c) helical PANI nanofibers using D-CSA and (d–f) L-CSA as the dopant.[37,38] Reproduced from Ref. [37] with permission from the Royal Society of Chemistry.[37, 38]

17.7.4.2.2 *Effect of aromatic dopants on the PANI nanostructure*

Soft-template method syntheses of PANI nanostructures by aromatic dopants were extensively studied by Malik et al.[23,29,30] In the case of an aromatic acid, it acts as a dopant acid as well as a structure-directing material for nanostructure formation. Several aromatic acid dopants have been used for nanostructure preparation, which will be shown in the next section. In all cases, PANI nanofibers are observed with aromatic acids; however, for nonaromatic acids the nanostructure has granular morphology. The relation between the morphology and reaction parameters like effect of the core of the aromatic dopant, size, number of functionalities, ratio of dopant to aniline on nanofibers, etc., has been discussed briefly.[23,29,30,39]

17.7.4.2.3 *Effect of symmetry of the dopant acid*

The soft-template method synthesis of PANI nanofibers, especially PANI nanotubes, by an aromatic tetracarboxylic acid dopant was first reported by Malik et al.[30] Initially they started their journey by 1,2,4,5-benzene tetracarboxylic acid in which the core is aromatic and four acid functionality resides in the symmetrical position.[23,30,34] The reason can be explained as the formation of 1,2,4,5-benzene tetracarboxylic acid (BTCA) vesicles in water, and after addition of aniline, it interacts on both sides of the vesicle to form an aniline-filled vesicle. Partial transfer of protons from carboxylic acids to aniline generates anilinium ions that can initiate polymerization of aniline in the presence of an oxidizing agent. Several aniline-filled BTCA vesicles are aggregated as well as fused in a regular manner to form the 1D nanofiber of BTCA/PANI. For mechanistic investigation of the nanofiber formation by an aromatic dopant, a nonaromatic dopant with tetracarboxylic groups (e.g., cyclohexane tetracarboxylic acid [CTCA]) is used to know the effect of aromatic dopants on nanofiber formation. Interestingly no fibers are observed for the nonaromatic dopant (Fig. 17.8). So, the planar dopant is important for nanofiber formation.[30]

17.7.4.2.4 *Effect of the aromatic core size of a dopant acid*

Inspired from the previous result, with a different-core-size dopant, PANI nanostructures have been prepared by the soft-template method. Three cores with increasing size, from benzene to naphthalene to perylene with four –COOH groups at symmetrical positions, have been used. In all cases good fibers with tube-like morphology are observed (Fig. 17.9). The interesting result is that with the increasing core size of the dopant in soft-template polymerization, the diameter of the fibers increases. Photophysical properties of the core are also reflected in the polymer.[23,30]

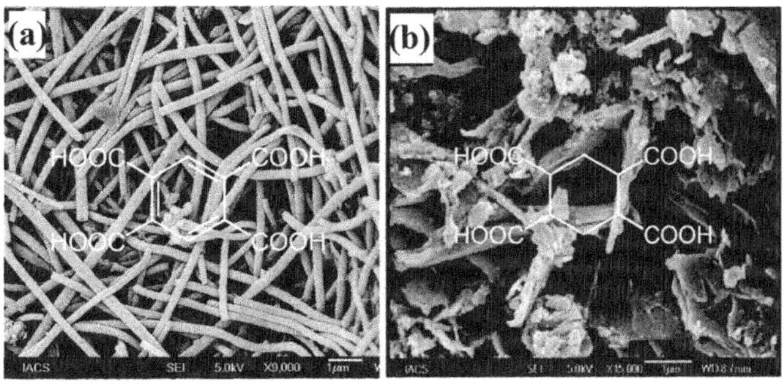

Figure 17.8 Soft-template-synthesized PANI morphology using (a) BTCA and (b) CTCA.

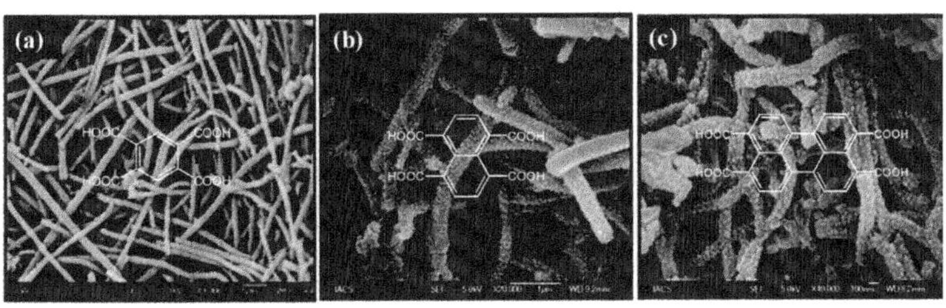

Figure 17.9 Aromatic tetracarboxylic-doped PANI: (a) benzene, (b) naphthalene, and (c) perylene.

17.7.4.2.5 *Effect of molar ratio on the morphology of PANI*

The molar ratio of monomer to dopant is very important for the morphology of the PANI nanostructure. Using a different aniline-to-dopant ratio (1.0 to 0.1), the morphology of the PANI nanostructure is changed. Except at a molar ratio of 1.00, all the other ratios produce nanofibers. However, their diameters are significantly influenced by the [BTCA]/[An] ratio; for instance, the average diameter is increased from 140 to 175 nm upon increase in the molar ratio from 0.01 to 0.5. From the morphological study, it is concluded that a 0.25 ratio is the optimum condition for making high-aspect-ratio nanotubes (Fig. 17.10). This variation of diameter is also accompanied by an enhancement in the aspect ratio of the nanostructures.[30]

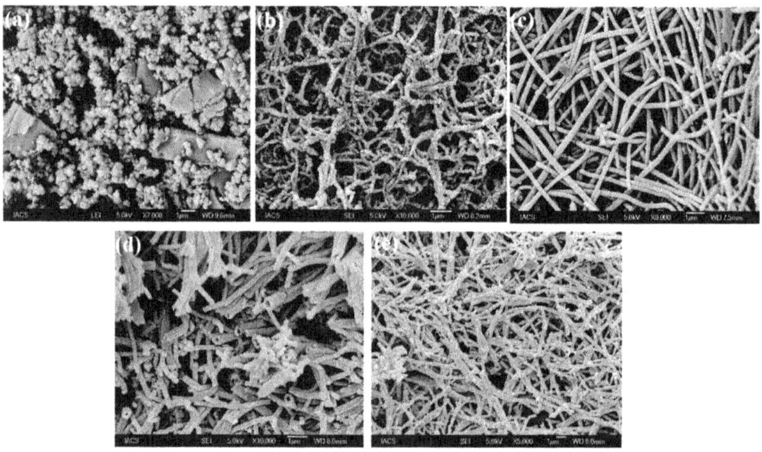

Figure 17.10 FESEM images of BTCA–PANI nanostructures having different BTCA-to-aniline molar ratios: (a) 1.00, (b) 0.50, (c) 0.25, (d) 0.10, and (e) 0.01.

17.7.4.2.6 *Effect of the number and position of functional groups on the morphology of PANI*

The mechanism of the soft-template polymerization by aromatic dopant acids has been investigated systematically by using five dopant acids, as reported by Malik et al. Firstly three acids, BTCA, 1,3,5-benzene tricarboxylic acid (BTA), and 1,4-benzene dicarboxylic acid (TA), are picked up and the number of –COOH groups is decreased from four to two in BTCA to TA.

However, all contain acid groups placed in the symmetrical position. In all cases fibers were formed and the diameter of the fibers decreased with the number of –COOH groups (Fig. 17.11a–c). To know about the symmetry effect of the –COOH group in an aromatic dopant they chose three dopants, 1,3,5-, 1,2,4-, and 1,2,3-BTAs, which have the same number of –COOH groups but different symmetry. Among the three, only 1,3,5-BTA showed the nanofiber morphology (Fig. 17.11b,d,e). The reason was explained by the formation of a vesicle-like structure of the symmetrical dopant, which acts as a soft template for 1D nanofiber formation.[40]

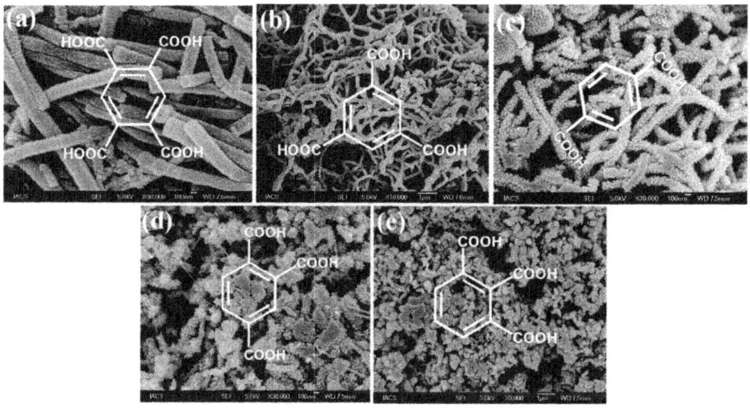

Figure 17.11 Morphology of PANI by five different dopants.

17.7.4.2.7 *Multifunctional acids contain dopant acids for PANI nanostructure*

After received the interesting results on aromatic dopants on the PANI nanostructure by the soft-template method Malik et al. extended their concepts for multifunctional –COOH-containing dopant-like graphene oxide and graphene quantum dots. In both cases very nice fibrical-like PANI with a tube-like structure (Fig. 17.12) are overwhelmingly investigated. For the first time it showed the transformation of a 2D graphene oxide sheet to a 1D PANI nanotube by the soft-template method. So, the soft-template method synthesis of the PANI nanostructure is not only limited for small organic acids, it is also applicable for large dopant-like graphene oxide.[29]

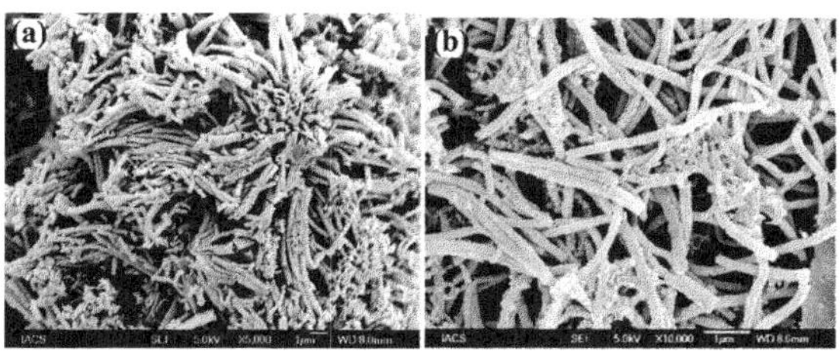

Figure 17.12 PANI nanostructure doped by (a) graphene oxide and (b) graphene quantum dot.

17.8 Application of Conducting Polymers: Polyaniline

Polymers permeate every aspect of daily life, and it is difficult to imagine society without conducting polymers, especially modern electronics. The major interest in conducting polymer research is low cost of manufacturing. The extended π-systems of conjugated polymers are highly susceptible to chemical or electrochemical oxidation or reduction. These alter the electrical and optical properties of polymers, and by controlling this oxidation and reduction, it is possible to precisely control these properties. Because of these tunable properties they are very promising materials for multifunctional applications (mentioned in the Introduction section). Conducting polymers are rapidly attracting attention in new applications with increasingly processable materials with better electrical and physical properties.[4–6] The new nanostructured forms of conducting polymers provide a big platform in this field, owing to their higher surface areas and better dispersabilities. In the next section, we discuss the application of conducting polymer nanostructures.

Conjugated polymers are in the front of many next-generation polymeric optoelectronic technologies, including polymeric photovoltaics, polymeric light-emitting diodes, and lasers. The major advantage of PANI materials is that they can be synthesized and processed on a large scale and at relatively low cost, which is a prerequisite for many applications like sensors, functional coatings, catalysts, etc.[23] Therefore, developing bulk syntheses for the PANI nanostructure would be especially important for practical reasons. It has been found that the nanostructure significantly improves the processibility of PANI and its performance in many conventional applications over the bulk counterparts. Nanostructure PANI leads to much faster and more responsive chemical sensors and ultrafast nonvolatile memory devices. Details of the application of the PANI nanostructure are discussed briefly.[6]

17.8.1 Polyaniline Nanostructure-Based Chemical Sensors

The making of high-performance chemical gas sensors is nowadays receiving immense interest because of environmental protection and homeland security. The PANI nanostructure is a very promising material for gas sensors because of very strong conductivity. It is highly conducting in the ES state and easily transforms to the less conducting EB state by doping/dedoping or reduction/oxidation. The changes of conducting to nonconducting are very sensitive in the presence of toxic acids and chemicals like HCl, HF, HNO_3, PCl_3, and BF_3. The EB form of PANI can be protonated by these acids and converted to the higher-conducting ES state; alternatively, the salt state can be deprotonated by NH_3 or organic amines and converted to the base state. In this way, by using the PANI nanostructure, one can detect toxic acids or alkaline vapors up to the parts per million (ppm) level or less. Redox-active compounds such as Cl_2, F_2, Br_2, I_2, N_2H_4, NO_x, and SO_x are also detected by PANI because they can change the oxidation state of PANI and as a result the conductivity of PANI changes. A recent report by Kaner et al. has shown that PANI, especially in nanofibers, is a highly efficient sensing material than the bulk state.[11,18,39] The major importance of the nanostructure is the unique 3D high-surface-area porous structure, which allows chemical vapors to rapidly access essentially all nanofibers upon exposure of acid or base vapors, leading to a superior sensing performance compared to conventional bulk films. Generally, nanofibers show better and faster responses due to their high porosity that allows easy diffusion of toxic gases into the film. A high aspect ratio and larger surface area of the nanofiber films show a faster response time because diffusion into small-diameter cylinders is fast. The sensitivity of the PANI nanostructure can change by film thickness; a thin film shows a better response time than thick films because more interaction between vapor molecules and PANI occur in the thin-film state.[11,41]

17.8.2 Memory Devices

A device that works exactly like the human brain is called a memory device. In computers it is used to store data and instructions required for processing. Several types of computer memory storage devices are available; among them nonvolatile memory has received more attention because in this system data is not lost when power is cut off. Developing nonvolatile memory systems comparable in speed and capacity to volatile RAM is much more important research nowadays. Use of polymer material for memory devices has several advantages than small molecule, especially a high performance rate and long-term stability. PANI nanofibers are used as a good material for nonvolatile memory devices. Recently, the R. Kaner group has discovered that gold/PANI nanofibers possess a remarkable electrically switchable bistability, which is ideal for nonvolatile flash memory devices.[42] They showed that both gold nanoparticles and PANI nanofibers are important for memory applications. Switching properties are based on an electric-field–induced charge transfer process between the PANI nanofibers and the gold nanoparticles. The gold nanoparticles can store the charge in the presence of an electric field and trap the charge in the absence of an electric field. The charge storage property and as a result memory of the composite solely depends on the nanoscale dimensions of the polymer materials and the interface between the gold nanoparticles and the PANI nanofibers.[43]

17.8.3 Corrosion Protection

Corrosion can be defined as physical interaction between a metal and its environment, which results in changes of the metal's properties. Metal corrosion results in economic damages to general infrastructures, airplanes, reservoirs, tanks, ships, etc. This interaction is often of an electrochemical (or redox) nature. The development of new materials to find a new material which has the association of different materials for corrosion protection has been an important area of research. Conducting polymers are an important and promising research area for coatings. In general, the corrosion efficacy of conducting polymers solely depends on applied conditions as well as corrosion experiments, that is, depending on the exact conditions. A conducting polymer has excellent protection capability or leads to a disastrously enhanced corrosive attack. Among the conducting polymer family, PANI has been the most widely studied for corrosion protection because of the ease of synthesis, low cost, and higher environmental stability. DeBerry (1985) was the first scientist to indicate the possibility of PANI as a corrosion inhibitor. It acts as an inhibitor due to the presence of the C=N bond on the polymer chain, which can be adsorbed on the metal surface. In most cases, PANI is also used as a protective coating. Corrosion protection properties also depend on the nature of the dopant on the PANI nanostructure. Klein et al. reported that a phosphonic acid salt of PANI is more effective for corrosion protection than sulfonic acid salts. A model has been proposed by them, which entails passivation of the metal surface through anodization of the metal by PANI, and formation of an insoluble iron dopant salt at the metal surface resulted in a strong coating.[44]

17.8.4 Rechargeable Batteries

Rechargeable lithium ion batteries with high energy density and long cycle life development are very important to meet the ever-increasing energy storage, which is very essential recently because of its application to a larger scale, particularly electric vehicles and grid-scale energy storage systems. Among different conjugated polymers, PANI is a potential cathode material for lithium ion batteries because of its easy synthesis, low cost, good electrochemical properties (oxidation/reduction), and excellent environmental stability. However, low storage capacity limits its practical application. One of the promising ways to enhance the properties of the battery application is to use 1D nanostructured PANI with appropriate dopant materials. However, the presence of electrochemically inactive anions like ClO_4^- and SO_4^{2-} significantly decreases the specific capacity as a result of limited application. So, uses of suitable doping or composite material are the most promising way to enhance the battery properties of PANI. Recently Paik et al. reported 100 nm PANI nanofibers synthesized with phosphomolybdic acid polyanion. They produced remarkably strong electrochemical performance in terms of lithium storage capacity (183.4 mAh g^{-1} at a 0.1C rate), cycling stability (80.7% capacity retention after 50 cycles), and rate capability (94.2 mAh g^{-1} at a 2C rate) compared to PANI nanofibers and a bulk PANI/polyoxometalate hybrid. The major reason for enhancement is that polyoxometalate acts as a dopant. It not only protonates aniline during the synthesis process but also improves the electrochemical properties of composites.[45] Stable Li ion battery anodes by in situ polymerization of conducting hydrogels to coat silicon nanoparticles are also reported by Wu et al., and over 90% capacity retention has been shown after 5000 cycles at a current density of 6.0 A g^{-1}.[46]

17.8.5 Supercapacitors

The development of suitable electrical storage devices is essential to meet future society and environmental energy needs. The demand for more sustainable, efficient energy storage has ignited a renewed scientific and commercial interest in advanced capacitor designs where nanotechnology is playing a critical role. Rechargeable batteries (discussed in the previous section) are indispensable energy storage devices owing to their high energy density. However, rechargeable batteries suffer from relatively low power density. From this point, supercapacitors or electrochemical capacitors are playing a pivotal role in energy storage applications due to their outstanding performance, such as high power density, long cycle life, and safe operation. Supercapacitors, also known as ultracapacitors, generally contain high-surface-area electrode materials with thin electrolytic dielectrics for very high capacitance. They are able to attain greater energy densities, while still maintaining the characteristic high power density of conventional capacitors. The first supercapacitor commercial devices were made by Nippon Electric Company (NEC), and ultracapacitor devices were made by the Pinnacle Research Institute (PRI). PANI is a famous conducting polymer for supercapacitors, owing to its high conductivity, high flexibility, easy synthesis, low cost, and multiredox state. It can offer especially high theoretical pseudocapacitances. The poor stability of PANI during the charge/discharge process restricts its real application in supercapacitors. A good dopant is the best possible way to use PANI as a supercapacitor material. As we have already discussed in the Introduction section, the major properties of PANI depend on the nature of the dopant and these will be enhanced by using a superior dopant. In this context, graphene, a single layer of sp^2-hybridized carbon atoms, is a promising material for supercapacitors due to its large surface area, great mechanical strength, and high conductivity. It can provide a long cycle life during the charge/discharge process. However, it generally shows a low capacitance value due to unavoidable aggregation of graphene nanosheets. The combination of graphene and PANI in a nanostructure could overcome the drawbacks of graphene as well as PANI and lead to good performance in supercapacitors. Graphene-

PANI in nanostructure form has shown very high capacitance than bulk material because in a nanostructure noncovalent interaction forces such as van der Waals forces, electrostatic interactions, and π–π stacking forces between PANI and graphene are very strong.[47] In a nanostructure, Wu et al. reported on a graphene/PANI nanofiber which exhibited a capacitance of 210 F/g at 0.3 A/g.[48] Recently Malik et al. reported that grapheme oxide/PANI nanotubes prepared by the soft-template method showed a capacitance of 610 F/g at 1 A/g and this depends on the aspect ratio of nanofibers. They also showed that when graphene quantum dots are used instead of graphene oxide the capacitance value reached 1044 F/g at a current density of 1 A/g with very good cyclic stability (~80% retention) after 3000 cycles.[29b] Thus, by controlling the nanostructure of graphene/PANI a specific target can be achieved.

17.8.6 Electrochromic Devices

The optical characteristics of PANI films undergo changes with corresponding changes in the structural form as well as conductivity. The optical characteristics in the oxidized states of PANI films, that is, the ES form, are color and high conductivity, while in the reduced state, that is, the EB form, they are optically transparency and low conductivity. In fact, coloration and conductivity are associated with the doping of the films. Electrochromic devices of PANI films and their applications in practical electrochromic display devices with liquid electrolytes were first reported by Kobayashi et al. The color of PANI films is reversibly changed to green by oxidation and to transparent yellow by reduction in 1 M hydrochloric acid in the potential range from −0.2 V to +0.6 V versus a saturated calomel electrode (SCE). The reversibility of the color change has been observed more than 106 times in the aforesaid potential range, with a rapid response of less than 100 msec.[49] However, the major problems are long-range stability within liquid electrolytes. So, the use of a solid state electrolyte appears to be an attractive alternative. Akhtar et al. first showed the electrochromic behavior of PANI in a solid electrolyte. The advantages claimed in their report are protection from atmospheric oxidation and easy control of solid state electrolyte composition.[50]

17.8.7 SERS-Active Substrate for Analyte Sensing

Surface-enhanced Raman scattering (SERS) is a powerful and attractive spectroscopic technique for the detection of ultrasensitive and very-low-concentration analyte (chemical and biological) molecules at the metal surface. It demonstrates single-molecule fingerprint vibration information. Generally, SERS is observed for analytes adsorbed onto a coinage metal surface (Au, Ag, etc.) with the excitation near the visible region. So, deposition of these metals on a polymer surface would be important for making a high-surface-area SERS-active substrate for very-low-concentration analyte sensing. PANI is a very important polymer among different conjugated polymers for metal-incorporated nanostructure formation, owing to the presence of many –NH groups in the polymer chain. The major advantage of PANI is its reduction potential, which is relatively higher than that of the common noble metals, as a result forming metal nanoparticles at room temperature without any stabilizing agent. Apart from these, PANI also offers good thermal and environmental stability, coupled with excellent electrical, electrochemical, and optical properties. Malik et al. recently reported sensing of trace amounts of analytes like 4-mercaptobenzoic acid and rhodamine 6G by using Ag-nanoparticle-decorated PANI nanofibers, which are prepared by a single step at room temperature. Sensitivity of the analytes depends primarily on the percentage of Ag nanoparticles on PANI fibers, as well as the aspect ratio of PANI nanofibers.[51]

17.9 Conclusions

The nanostructure of PANI (nanorods/nanowires/nanofibers) receives great attention because of its unique properties and promising applications as a nanomaterial and in nanodevices as well as plastic electronics. The properties of PANI depend on the nature of the dopant acid as well as the extent of the doping level. However, systematic studies of the relationship between conductivity, crystallinity, and morphology with dopant size are rarely seen in the literature. The effects of aromatic dopant acids on the size of the PANI nanostructure are also missing. The development of the PANI nanostructure using organic doping acids and its good application in chemical gas sensing, electrocatalytic oxidation, supercapacitor-based devices, and especially recognition of structural effects on morphology, crystallinity, and conductivity were discussed briefly in this chapter. We believe that such discussion will inspire the new generation of polymer chemists to explore new concepts of conjugated polymer nanostructures.

In this chapter, PANI is used as a model material to systematically investigate the syntheses, properties, and applications of nanofibers of conjugated polymers. We hope that graduate students will be able to use it as an introduction of the conjugated conducting polymer nanostructure.

References

1. Berzelius, J. J. *Encyclopædia Britannica Online.* Retrieved 3 August 2008.
2. Amin, M., Salman, M. *Rev. Adv. Mater. Sci.*, **13**, 93–116 (2006).
3. (a) Gorman, C. B., Ginsburg, E. J., Grubbs, R. H. *J. Am. Chem. Soc.*, **115**, 1397–1409 (1993); (b) Shirakawa, H. *Synth. Met.*, **69**, 3–8 (1995); (c) Saxon, A. M., Liepins, F., Aldissi, M. *Prog. Polym. Sci.*, **11**, 57–89 (1985).
4. (a) The Nobel Prize in Chemistry, http://www.nobelprize.org/nobel_prizes/chemistry/laureates/2000/ (2000); (b) Heeger, A. *Rev. Mod. Phy.*, **73**, 681–700 (2001); (c) Hall, N., McDiarmid, A., Heeger, A. *Chem. Commun.*, 1–4 (2003).
5. Eftekhari, A. *Nanostructured Conductive Polymers*, John Wiley & Sons (2011).
6. (a) Li, D., Huang, J., Kaner, R. B. *Acc. Chem. Res.*, **42**, 135–145 (2009); (b) Huang, J. *Pure Appl. Chem.*, **78**, 15–27 (2006).
7. Karami, H., Mousavi, M. F., Shamsipur, M. *J. Power Sources*, **117**, 255–259 (2003).
8. Zhang, K., Zhang, L. L., Zhao, X. S., Wu, J. *Chem. Mater.*, **22**, 1392–1401 (2010).
9. Na, S. I., Kim, S. S., Jo, J., Kim, D. Y. *Adv. Mater.*, **20**, 4061–4066 (2008).
10. Phang, S. W., Kuramoto, N. *Polym. Compos.*, **31**, 516–523 (2010).
11. Virji, S., Huang, J., Kaner, R. B., Weiller, B. H. *Nano Lett.*, **4**, 491–496 (2004).
12. Scott, J. C., Carter, S. A., Karg, S., Angelopoulos, M. *Synth. Met.*, **85**, 1197–1200 (1997).

13. Friend, R. H., Gymer, R. W., Holmes, A. B., Burroughes, J. H., Marks, R. N., Taliani, C., Bradley, D. D. C., Dos Santos, D. A., Brédas, J. L., Lögdlund, M., Salaneck, W. R. *Nature*, **397**, 121–125 (1999).

14. Zhou, J., Lu, Z., Zhu, X., Wang, X., Liao, Y., Ma, Z., Li, F. *Biomaterials*, **34**, 9584–9592 (2013).

15. Wallace, G. G., Spinks, G. M., Kane-Maguire, L. A. P., Teasdale, P. R. *Conductive Electroactive Polymers*, Taylor & Francis (2009).

16. (a) Jen, K. Y., Lakshmikantan, M. V., Albeck, M., Cava, M. P., Huang, W. S., MacDiarmid, A. G. *J. Polym. Sci. Polym. Lett.*, **21**, 441–448 (1983); (b) Skotheim, T. A., Elsenbaumer, R. L., Reynolds, J. R. *Handbook of Conducting Polymer*, Marcel Dekker: New York (1998).

17. Bredas, J. L., Street, G. B. *Acc. Chem. Res.*, **18**, 309–315 (1985).

18. MacDiarmid, A. G., Mammone, R. J., Kaner, R. B., Porter, S. J., Pethig, R., Heeger, A. J., Rosseinsky, D. R. *The Concept of 'Doping' of Conducting Polymers: The Role of Reduction Potentials*, Royal Society, p. 314 (1985).

19. Heeger, A. J., Schrieffer, J. R., Su, W.-P., Su, W. *Rev. Mod. Phys.*, **60**, 781 (1988).

20. Genies, E. M., Boyle, A., Lapkowski, M., Tsintavis, C. *Synth. Met.*, **36**, 139–182 (1990).

21. (a) MacDiarmid, A. G., Chiang, J. C., Richter, A. F., Epstein, A. J. *Synth. Met.*, **18**, 285–290 (1987); (b) MacDiarmid, A. J., Epstein, A. J. *Faraday Discuss. Chem. Soc.*, **88**, 317–332 (1989).

22. Chandrasekhar, P. Conducting Polymers, Fundamentals and Applications: A Practical Approach, Kluwer Academic: Boston, MA (1999).

23. Rana, U., Chakrabarti, K., Malik, S. *J. Mater. Chem.*, **22**, 15665–15671 (2012).

24. Shumaila, G. B., Lakshmi, V. S., Alam, M., Siddiqui, A. M., Zulfequar, M., Husain, M. *Curr. Appl. Phys.*, **11**, 217–222 (2010).

25. Stejskal, J., Kratochvil, P., Jenkins, A. D. *Polymer*, **37**, 367–370 (1996).

26. Chiang, J. C., Macdiarmid, A. J. *Synth. Met.*, **13**, 193–205 (1986).

27. Huang, J. X., Kaner, R. B. *Angew. Chem. Int. Ed.*, **43**, 5817–5821 (2004).

28. Razzak, J., Jafri, I., Mishra, A. K., Ramaprabhu, S. *J. Mater. Chem.*, **21**, 17601–17605 (2011).

29. (a) Rana, U., Malik, S. *Chem. Commun.*, **48**, 10862–10864 (2012); (b) Mondal, S., Rana, U., Malik, S. *Chem. Commun.*, **51**, 12365–12368 (2015).

30. (a) Rana, U., Chakrabarti, K., Malik, S. *J. Mater. Chem.*, **21**, 11098–11100 (2011); (b) Huang, K., Wan, M. X. *Chem. Mater.*, **14**, 3486–3490 (2002).

31. (a) Huang, W. S., Humphrey, B. D., MacDiarmid, A. G. *J. Chem. Soc., Faraday Trans.*, **82**, 2385 (1986); (b) Pan, L. J., Pu, L., Shi, Y., Sun, T., Zhang, R., Zheng, Y. O. *Adv. Funct. Mater.*, **16**, 1279–1288 (2006).

32. MacDiarmid, A. G., Min, Y., Oh, E. J., Scherr, E. M., Tang, X., Masters, J. G., Epstein, A. J. *Synth. Met.*, **41**, 735–739 (1991).

33. Tian, Z., Yu, H., Wang, L., Saleem, M., Ren, F., Ren, P., Chen, Y., Sun, R., Sun, Y., Huang, L. *RSC Adv.*, **4**, 28195–28208 (2014).

34. Rana, U., Paul, N. D., Mondal, S., Chakraborty, C., Malik, S. *Synth. Met.*, **192**, 43–49 (2014).

35. (a) Chen, W., Rakhi, R. B., Alshareef, H. N. *J. Mater. Chem. A*, **1**, 3315–3324 (2013); (b) Zhang, Z., Sui, J., Zhang, L., Wan, M., Wei, Y., Yu, L. *Adv. Mater.*, **23**, 2854–2857 (2005).

36. Jiang, N., Xu, Y., He, N., Chen, J., Deng, Y., Yuan, C., Han, G., Dai, L. *J. Mater. Chem.*, **20**, 10847–10855 (2010).

37. Yan, Y., Fang, J., Liang, J., Zhang, Y., Wei, Z. *Chem. Commun.*, **48**, 2843–2845 (2012).

38. Yan, Y., Yu, Z., Huang, Y., Yuan, W., Wei, Z. *Adv. Mater.*, **19**, 3353–3357 (2007).

39. Huang, J., Kaner, R. B. *J. Am. Chem. Soc*, **126**, 851–853 (2004).

40. Rana, U., Mondal, S., Sannigrahi, J., Sukul, P. K., Amin, M. A., Majumdar, S., Malik, S. *J. Mater. Chem. C*, **2**, 3382–3389 (2014).

41. (a) Anderson, M. R., Mattes, B. R., Reiss, H., Kaner, R. B. *Science*, **252**, 1412–1415 (1991).

42. (a) Baker, C. O., Shedd, B., Tseng, R. J., Martinez-Morales, A. A., Ozkan, C. S., Ozkan, M., Yang, Y., Kaner, R. B. *ACS Nano*, **5**, 3469–3474 (2011); (b) Tseng, R. J., Huang, J., Ouyang, J., Kaner, R. B., Yang, Y. *Nano Lett.*, **5**, 1077–1080 (2005).

43. Tseng, R. J., Baker, C.O., Shedd, B., Huang, J., Kaner, R. B., Ouyang, J., Yang, Y. *Appl. Phys. Lett.*, **90**, 053101 (2007).

44. Kinlen, P. J., Menon, V., Ding, Y. *J. Electrochem. Soc.*, **146**, 3690–3695 (1999).

45. Yang, H., Song, T., Liu, L., Devadoss, A., Xia, F., Han, H., Park, H., Sigmund, W., Kwon, K., Paik, U. *J. Phys. Chem. C*, **117**, 17376–17381 (2013).

46. Wu, H., Yu, G., Pan, L., Liu, N., McDowell, M. T., Bao, Z., Cui, Y. *Nat. Commun.*, **4**, 1943 (2013).

47. Wang, L., Ye, Y., Lu, X., Wen, Z., Li, Z., Hou, H., Song, Y. *Sci. Rep.*, **3**, 3568 (2013).
48. Wu, Q., Xu, Y. X., Yao, Z. Y., Liu, A. R., Shi, G. Q. *ACS Nano*, **4**, 1963–1970 (2010).
49. Kobayashi, T., Yoneyama, H., Tamura, H. *J. Electroanal. Chem.*, **161**, 419 (1984).
50. Akhtar, M., Weakliem, H. A., Paiste, R. M., Gaughan, K. *Synth. Met.*, **26**, 203–207 (1988).
51. Mondal, S., Rana, U., Malik, S. *ACS Appl. Mater. Interfaces*, **7**, 10457–10465 (2015).

Biography

Utpal Rana, PhD, is currently working as a postdoctoral researcher in the Polymer Materials Unit of the Nationl Institute for Materials Science (NIMS), Japan. His research interests are the synthesis of chiral 1D metallo-supramolecular polymers and their application as anticancer drugs as well as in DNA recognition. Dr. Rana received his MSc degree in 2008 from Calcutta University. In September 2008 he started his PhD career under the supervision of Dr. Sudip Malik in the Polymer Science Unit of the Indian Association for the Cultivation of Science under Jadavpur University at Jadavpur, Kolkata, India. The aim of his PhD research was to develop a soft-template strategy for the synthesis of the PANI nanostructure and application of the newly developed PANI nanostructure in supercapacitors, electrocatalysis, etc. For the synthesis of the PANI nanostructure various aromatic polycarboxylic acids were synthesized. Dr. Rana received the UGC-NET fellowship award in December 2007 from HRDG, India, the CSIR-SRF fellowship award in April 2012 from HRDG, India, the Young Researcher award from ICONSAT-2012, Hyderabad, India, and the best poster award from IITKGP-2011, Kharagpur, India. He has published 15 refereed articles. Coauthoring this book reflects his interest in the field of the conjugated conducting PANI nanostructure in which he completed his PhD in 2013 afrom IACS.

Sanjoy Mondal is currently pursuing PhD in the Polymer Science Unit of the IACS. His research interests mainly include the synthesis and applications of PANI and PANI/metal nanomaterials. The synthesis part contains soft-template PANI synthesis by small water-soluble organic acida, graphene oxide (GO), graphene quantum dots (GQDs), etc., as a soft-template-making agents. The applications part mainly focuses on capacitors, SERS, catalysis,

electrochemical catalysis, fuel cells, toxic material separation, etc. He obtain his BSc degree from Bankura Sammilani College under Burdwan University in 2009 and MSc degree from IIT Delhi, India, in 2011. He received the CSIR-NET fellowship award in 2012 from CSIR-HRDG, India and the best poster award in MACRO-2013 in IISc, Bangalore, India. He has published 5 articles on PANI and PANI – metal composites in international research journals.

Dr. Sudip Malik completed BSc from Ramakrishna Mission Vidyamandira (Belur Math) under the University of Calcutta, Kolkata, India, in 1996 and MSc from the same university in 1998. He received his PhD (2003) from the IACS. His thesis was on the crystallization and gelation behavior of poly(3-alkyl thiophenes), and the research was carried out under the supervision of Prof. A. K. Nandi. He moved to Institut Charles Sadron, France, in the group of Prof. J.-M. Guenet for the making of polymer materials with multiscale porosity (particularly from the syndiotactic polystyrene [sPS]/ solvent complex). Then he joined the group of Prof. S. Shinkai at Kyushu University, Japan, as JSPS researcher (2005–2007). In January 2008, he became assistant professor of the Polymer Science Unit, IACS. On leave from the IACS (2009–2011), he moved to the University of Nijmegen, the Netherlands, as a Marie Curie fellow (international category) to work with Prof. Alan E. Rowan. Currently, he is associate professor at the Polymer Science Unit, IACS. His current research interests are conducting polymers, electroluminescent polymers, small molecular hydrogels (perylene derivatives), sensors and hybrid materials for energy storage, and electronic and optoelectronic devices. He is also a recipient of the Young Associate of the Indian Academy of Science, Bengaluru, India.

Index

#0003 - 080817 - C37 - 229/152/30 [32] - CB - 9789814774031